# Methods in Enzymology

## Volume XLII
## CARBOHYDRATE METABOLISM
### Part C

# METHODS IN ENZYMOLOGY

EDITORS-IN-CHIEF

Sidney P. Colowick     Nathan O. Kaplan

*Methods in Enzymology*

*Volume XLII*

# Carbohydrate Metabolism

*Part C*

EDITED BY

## W. A. Wood

DEPARTMENT OF BIOCHEMISTRY
MICHIGAN STATE UNIVERSITY
EAST LANSING, MICHIGAN

1975

ACADEMIC PRESS   New York   San Francisco   London
A Subsidiary of Harcourt Brace Jovanovich, Publishers

ACADEMIC PRESS, INC.
111 Fifth Avenue, New York, New York 10003

*United Kingdom Edition published by*
ACADEMIC PRESS, INC. (LONDON) LTD.
24/28 Oval Road, London NW1

Library of Congress Cataloging in Publication Data
Main entry under title:

Carbohydrate metabolism.

    (Methods in enzymology, v. 9)
    Includes bibliographical references.
    1.   Carbohydrate metabolism.    2.   Enzymes.    I.   Wood,
Willis A., Date    ed.    II.   Series: Methods in
enzymology, v. 9 [etc.]    [DNLM:    1.   Carbohydrates—
Metabolism.    W1ME9615K v. 9]
QP601.C733 vol. 9 574.1'925'08s [574.1'33]    72-26891
ISBN 0-12-181942-6

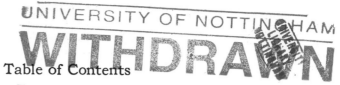
# Table of Contents

## Section I. Kinases

v

## Section II. Aldolases

## Section III. Dehydratases

## Section IV. Phosphatases

## Section V. Mutases

## Section VI. Carboxylases and Decarboxylases

## Section VII. Glycosidases

# Contributors to Volume XLII

Article numbers are in parentheses following the names of contributors. Affiliations listed are current.

HUSSEIN ABOU-ISSA (61), *Department of Biochemistry, Emory University, Atlanta, Georgia*

J. B. ALPERS (20), *Department of Biological Chemistry, Harvard Medical School, Boston, Massachusetts*

LOUISE E. ANDERSON (68), *Department of Biological Sciences, University of Illinois at Chicago Circle, Chicago, Illinois*

RICHARD L. ANDERSON (1, 6, 11, 42, 49, 74), *Department of Biochemistry, Michigan State University, East Lansing, Michigan*

T. J. ANDREWS (72), *Department of Environmental Biology, Research School of Biological Sciences, Australian National University, Canberra City, Australia*

ANN AUST (29), *Department of Biochemistry, Michigan State University, East Lansing, Michigan*

F. J. BALLARD (7), *Division of Nutritional Biochemistry, C.S.I.R.O., Adelaide, Australia*

SANTIMOY BANERJEE (9), *Central Drugs Laboratory, Calcutta, India*

T. BARANOWSKI (54), *Institute of Biochemistry and Biophysics, Medical School, Wroclaw, Poland*

ERIC A. BARNARD (2), *Department of Biochemistry, State University of New York, Buffalo, New York*

STEPHEN J. BENKOVIC (60), *Department of Chemistry, Pennsylvania State University, State College, Pennsylvania*

MOSHE BENZIMAN (32), *Department of Biological Chemistry, The Hebrew University of Jerusalem, Jerusalem, Israel*

ROBERT W. BERNLOHR (27), *Department of Microbiology, Pennsylvania State University, University Park, Pennsylvania*

ERNEST BEUTLER (8), *Division of Medicine, City of Hope Medical Center, Duarte, California, and University of Southern California School of Medicine, Los Angeles, California*

KARL-GEORG BLUME (8), *Freiburg University School of Medicine, Freiburg, Germany*

LAWRENCE BOGORAD (71), *Department of Biology, Harvard University, Cambridge, Massachusetts*

LUDWIG BRAND (16), *Department of Biology, The Johns Hopkins University, Baltimore, Maryland*

BOB B. BUCHANAN (62), *Department of Cell Physiology, University of California, Berkeley, California*

WILLIAM L. BYRNE (59), *Department of Biochemistry, University of Tennessee Center for the Health Sciences, Memphis, Tennessee*

J. CARRERAS (65, 66), *Facultad de Medicina, Instituto de Fisiologia, Barcelona, Spain*

TEH-HSING CHIU (41), *Department of Microbiology, University of Pittsburgh School of Dental Medicine, Pittsburgh, Pennsylvania*

ALBERT C. CHOU (3), *Department of Biochemistry, Michigan State University, East Lansing, Michigan*

GIUSEPPE D'ALESSIO (24), *University of Naples, Naples, Italy*

STANLEY DAGLEY (43), *Department of Biochemistry, College of Biological Sciences, University of Minnesota, St. Paul, Minnesota*

A. STEPHEN DAHMS (42, 49), *Department of Chemistry, California State University, San Diego, California*

ASIS DATTA (10), *School of Life Sciences, Jawaharlal Nehru University, New Delhi, India*

WILLIAM C. DEAL, JR. (17), *Department of Biochemistry, Michigan State University, East Lansing, Michigan*

A. GIB DeBUSK (75), *Genetics Group,*

Department of Biological Science, Florida State University, Tallahassee, Florida

K. A. DECKER (40), Biochemisches Institut der Universität, Freiburg im Breisgau, Germany

EUGENE E. DEKKER (45, 46), Department of Biological Chemistry, The University of Michigan, Ann Arbor, Michigan

J. DE LEY (48), Laboratory of Microbiology and Microbial Genetics, Faculty of Sciences, State University, Ghent, Belgium

MARGARET M. DEMAINE (60), Department of Chemistry, Pennsylvania State University, State College, Pennsylvania

C. C. DOUGHTY (21), Department of Biological Chemistry, University of Illinois at the Medical Center, Chicago, Illinois

DIANA ERSFELD (40), Department of Biochemistry, Michigan State University, East Lansing, Michigan

KAREN L. EVANS (41), Department of Microbiology, University of Pittsburgh School of Medicine, Pittsburgh, Pennsylvania

DAVID S. FEINGOLD (41), Department of Microbiology, University of Pittsburgh School of Medicine, Pittsburgh, Pennsylvania

L. E. FLANDERS (28), Searle Laboratories, Chicago, Illinois

ARTHUR M. GELLER (59), Department of Biochemistry, The University of Tennessee Center for the Health Sciences, Memphis, Tennessee

SUDHAMOY GHOSH (9), Institut de Biologie Moleculaire, Université de Paris, Paris, France

MARTIN GIBBS (36), Department of Biology, Brandeis University, Waltham, Massachusetts

JANE GIBSON (19), Section of Biochemistry, Molecular and Cell Biology, Cornell University, Ithaca, New York

AIDA GOLDSTEIN (76), Johnson & Johnson Baby Products Company, New Brunswick, New Jersey

JONATHAN GOLDTHWAITE (71), Department of Biology, Boston College, Chestnut Hill, Massachusetts

SHARON R. GRADY (45, 46), Department of Biological Chemistry, The University of Michigan, Ann Arbor, Michigan

CHARLES C. GRIFFIN (16), Department of Chemistry, Miami University, Oxford, Ohio

S. GRISOLIA (65, 66), Department of Biochemistry and Molecular Biology, University of Kansas Medical Center, Kansas City, Kansas

ARABINDA GUHA (35), The Rockefeller University, New York, New York

R. H. HAMMERSTEDT (40), Department of Biochemistry, Pennsylvania State University, University Park, Pennsylvania

THOMAS E. HANSON (11), Department of Biology, Temple University, Philadelphia, Pennsylvania

BETH A. HART (19), Department of Biochemistry, University of Vermont College of Medicine, Burlington, Vermont

M. D. HATCH (34, 55), Division of Plant Industry, C.S.I.R.O., Canberra City, Australia

J. S. HAWKER (55), Division of Horticultural Research, C.S.I.R.O., Glen Osmond, South Australia

JAMES A. HAYASHI (21), Department of Biochemistry, Rush Medical School, Chicago, Illinois

WOLFGANG HENGSTENBERG (73), Max-Planck Institut für Medizinische Forschung, Heidelberg, West Germany

PATRICIA HOFFEE (44), Department of Microbiology, University of Pittsburgh School of Medicine, Pittsburgh, Pennsylvania

B. L. HORECKER (37), Roche Institute of Molecular Biology, Nutley, New Jersey

ROGER JEFFCOAT (43), Unilever Research Laboratory, Colworth House, Sharnbrook, Bedford, England

B. CONNOR JOHNSON (51), Department of Biochemistry and Molecular Biology, University of Oklahoma Medical

Center, Oklahoma City, Oklahoma

L. JORIS (47), College of Medicine and Dentistry of New Jersey, Newark, New Jersey

JOHN JOSSE (24), Institute of Molecular Biology, Syntex Research, Palo Alto, California

VERNE F. KEMERER (16), National Institute of Arthritis, Metabolic, and Digestive Diseases, National Institutes of Health, Bethesda, Maryland

ROBERT G. KEMP (12, 13), Department of Biochemistry, Medical College of Wisconsin, Milwaukee, Wisconsin

K. KERSTERS (48), Laboratory for Microbiology and Microbial Genetics, Faculty of Sciences, State University, Ghent, Belgium

RODGER D. KOBES (45), The University of Miami School of Medicine, Miami, Florida

J. F. KOSTER (30), Department of Biochemistry, Faculty of Medicine, Erasmus University, Rotterdam, The Netherlands

NANCY KRATOWICH (61), Department of Biochemistry, University of Georgia, Athens, Georgia

GLENN D. KUEHN (69), Department of Chemistry, New Mexico State University, Las Cruces, New Mexico

SHIGERU KUROOKA (15), Dainippon Pharmaceutical Company, Ltd., Osaka, Japan

J. OLIVER LAMPEN (76), Institute of Microbiology, Rutgers University, New Brunswick, New Jersey

M. DANIEL LANE (70), Department of Physiological Chemistry, The Johns Hopkins University School of Medicine, Baltimore, Maryland

ROBERT B. LAYZER (18), Department of Neurology, University of California School of Medicine, San Francisco, California

HERBERT G. LEBHERZ (39), Laboratory for Developmental Biology, Swiss Federal Institute of Technology, Zurich, Switzerland

CAROL B. LIBBY (60), Department of

Chemistry, Pennsylvania State University, State College, Pennsylvania

G. H. LORIMER (72), Department of Environmental Biology, Research School of Biological Sciences, Australian National University, Canberra, Australia

BRUCE A. MCFADDEN (69), Department of Chemistry, Washington State University, Pullman, Washington

P. K. MAITRA (4), Tata Institute of Fundamental Research, Bombay. India

THOMAS H. MASSEY (17), Norwich Pharmaceutical Research and Development Laboratories, Norwich, New York

RUDOLPH MEDICUS (61), Department of Biochemistry, University of Georgia, Athens, Georgia

EDON MELLONI (57), Institute of Biochemistry, University of Genoa, Genoa, Italy

JOSEPH MENDICINO (61), Department of Biochemistry, University of Georgia, Athens, Georgia

GEORGE MICHAELS (33), Department of Biochemistry, Case Western Reserve University School of Medicine, Cleveland, Ohio

YORAM MILNER (33), Department of Biological Chemistry, The Hebrew University, Jerusalem, Israel

H. MÖHLER (40), Medical Research Department, Hoffman-LaRoche and Company, A. G., Basel, Switzerland

M. L. MORSE (73), Webb-Waring Lung Institute, and Department of Biophysics and Genetics, University of Colorado Medical Center, Denver, Colorado

HIROMU NISHIHARA (46), The Center for Adult Diseases, Higashinari-ku, Osaka, Japan

RICHARD E. PALMER (1, 74), Department of Biochemistry, Michigan State University, East Lansing, Michigan

EDWARD E. PENHOET (38), Department of Biochemistry, University of California, Berkeley, California

S. J. PILKIS (5), Department of Physiology, Vanderbilt University School of Medicine, Nashville, Tennessee

BURTON M. POGELL (60), Department of Microbiology, St. Louis University School of Medicine, St. Louis, Missouri

SANDRO PONTREMOLI (56, 57), Institute of Biochemistry, University of Genoa, Genoa, Italy

GIAMPIETRO RAMPONI (64), Department of Biochemistry, University of Florence Medical School, Florence, Italy

D. D. RANDALL (63), Department of Agricultural Chemistry, University of Missouri, Columbia, Missouri

RICHARD E. REEVES (31), Department of Biochemistry, Louisiana State University Medical Center, New Orleans, Louisiana

ZELDA B. ROSE (67), The Institute for Cancer Research, Fox Chase Cancer Center, Philadelphia, Pennsylvania

ORA M. ROSEN (58), Department of Medicine, Albert Einstein College of Medicine, New York, New York

WILLIAM J. RUTTER (38, 39), Department of Biochemistry and Biophysics, University of California, San Francisco, California

H. J. SALLACH (28), Department of Physiological Chemistry, University of Wisconsin Medical School, Madison, Wisconsin

VIRGINIA L. SAPICO (6, 11), Department of Biochemistry, Michigan State University, East Lansing, Michigan

L. H. SCHLOEN (28), Sloan-Kettering Institute for Cancer Research, New York, New York

Z. SCHNEIDER (51), Institute of Biochemistry, Academy of Agriculture, Wolynska, Poland

R. K. SCOPES (22, 23), Department of Biochemistry, La Trobe University, Bundoora, Victoria, Melbourne, Australia

MARVIN I. SIEGEL (70), Department of Pharmacology and Experimental Therapeutics, The Johns Hopkins University School of Medicine, Baltimore, Maryland

C. R. SLACK (34), Plant Physiology Division, Department of Scientific and Industrial Research, Palmerston North, New Zealand

DOROTHY J. SOUTH (31), Department of Biochemistry, Louisiana State University Medical Center, New Orleans, Louisiana

THOMAS G. SPRING (52), Department of Biophysical Sciences, University of Houston, Houston, Texas

G. E. J. STAAL (30), Haematological Department, State University Hospital, Utrecht, The Netherlands

EARLE STELLWAGEN (14), Department of Biochemistry, University of Iowa, Iowa City, Iowa

RICHARD STEPHENS (75), Genetics Group, Department of Biological Science, The Florida State University, Tallahassee, Florida

ALLEN C. STOOLMILLER (50), Departments of Pediatrics and Biochemistry, The University of Chicago, Chicago, Illinois

A. STROINSKI (51), Department of Plant Physiology, Academy of Agriculture, Wolynska, Poland

C. H. SUELTER (29), Department of Biochemistry, Michigan State University, East Lansing, Michigan

F. ROBERT TABITA (69), Department of Microbiology, University of Texas, Austin, Texas

JEREMY W. THORNER (26), Department of Bacteriology and Immunology, University of California, Berkeley, California

N. E. TOLBERT (63, 72), Department of Biochemistry, Michigan State University, East Lansing, Michigan

SERENA TRANIELLO (56), Istituto di Chimica Biologica dell'Universita di Ferrara, Ferrara, Italy

O. TSOLAS (47), Roche Institute of Molecular Biology, Nutley, New Jersey

F. WILLIAM TUOMINEN (27), Biochemicals Division, General Mills Chemicals, Inc., Minneapolis, Minnesota

EDGAR H. ULM (60), *Department of Drug Metabolism, Merck Institute for Therapeutic Research, West Point, Pennsylvania*

KOSAKU UYEDA (15), *Veterans Administration Hospital, and Department of Biochemistry, University of Texas Southwest Medical School, Dallas, Texas*

C. VEEGER (30), *Department of Biochemistry, Agricultural University, Wageningen, The Netherlands*

HARVEY WILGUS (14), *Department of Biochemistry, University of Iowa, Iowa City, Iowa*

JAMES M. WILLARD (36), *Department of Biochemistry, University of Vermont College of Medicine, Burlington, Vermont*

JOHN E. WILSON (3), *Department of Biochemistry, Michigan State University, East Lansing, Michigan*

FINN WOLD (52, 53), *Department of Biochemistry, University of Minnesota, St. Paul, Minnesota*

E. WOLNA (54), *Institute of Biochemistry and Biophysics, Medical School, Wroclaw, Poland*

HARLAND G. WOOD (33), *Department of Biochemistry, Case Western Reserve University School of Medicine, Cleveland, Ohio*

W. A. WOOD (40), *Department of Biochemistry, Michigan State University, East Lansing, Michigan*

AKIRA YOSHIDA (25), *Department of Biochemical Genetics, City of Hope Medical Center, Duarte, California*

SHYUN-LONG YUN (29), *Department of Biochemistry, Michigan State University, East Lansing, Michigan*

# Preface

Volumes XLI and XLII of "Methods in Enzymology" report new procedures appearing in the literature since 1965. As with Volume IX, the procedures included are for dissimilatory reactions between disaccharides and pyruvate. A few important reactions of pyruvate leading to fermentation end products are also included. The originally planned single volume became two volumes in recognition of the greatly increased number of pages needed to adequately cover this expanding field. The distribution of material between the volumes is arbitrary.

I wish to thank all of the authors for their contributions and their cooperation. It is a pleasure to recognize Ms. Patti Prokopp for her expert secretarial assistance.

W. A. Wood

# METHODS IN ENZYMOLOGY

### EDITED BY

## Sidney P. Colowick and Nathan O. Kaplan

VANDERBILT UNIVERSITY
SCHOOL OF MEDICINE
NASHVILLE, TENNESSEE

DEPARTMENT OF CHEMISTRY
UNIVERSITY OF CALIFORNIA
AT SAN DIEGO
LA JOLLA, CALIFORNIA

# METHODS IN ENZYMOLOGY

## EDITORS-IN-CHIEF

### Sidney P. Colowick      Nathan O. Kaplan

VOLUME VIII. Complex Carbohydrates
*Edited by* ELIZABETH F. NEUFELD AND VICTOR GINSBURG

VOLUME IX. Carbohydrate Metabolism
*Edited by* WILLIS A. WOOD

VOLUME X. Oxidation and Phosphorylation
*Edited by* RONALD W. ESTABROOK AND MAYNARD E. PULLMAN

VOLUME XI. Enzyme Structure
*Edited by* C. H. W. HIRS

VOLUME XII. Nucleic Acids (Parts A and B)
*Edited by* LAWRENCE GROSSMAN AND KIVIE MOLDAVE

VOLUME XIII. Citric Acid Cycle
*Edited by* J. M. LOWENSTEIN

VOLUME XIV. Lipids
*Edited by* J. M. LOWENSTEIN

VOLUME XV. Steroids and Terpenoids
*Edited by* RAYMOND B. CLAYTON

VOLUME XVI. Fast Reactions
*Edited by* KENNETH KUSTIN

VOLUME XVII. Metabolism of Amino Acids and Amines (Parts A and B)
*Edited by* HERBERT TABOR AND CELIA WHITE TABOR

VOLUME XVIII. Vitamins and Coenzymes (Parts A, B, and C)
*Edited by* DONALD B. McCORMICK AND LEMUEL D. WRIGHT

VOLUME XIX. Proteolytic Enzymes
*Edited by* GERTRUDE E. PERLMANN AND LASZLO LORAND

# Section I
# Kinases

# [1] β-Glucoside Kinase

By Richard L. Anderson and Richard E. Palmer

Cellobiose + ATP → cellobiose monophosphate + ADP
Gentiobiose + ATP → gentiobiose monophosphate + ADP

β-Glucoside kinase is an inducible enzyme that has been shown to function in the metabolism of cellobiose[1,2] and gentiobiose[3] in *Aerobacter aerogenes*. The disaccharides are phosphorylated at carbon atom 6 of the nonreducing ring.[2]

## Assay Method

*Principle.* The continuous spectrophotometric assay is based on the following sequence of reactions[2]:

$$\text{Cellobiose + ATP} \xrightarrow{\text{β-glucoside kinase}} \text{cellobiose monophosphate + ADP}$$

$$\text{Phosphoenolpyruvate + ADP} \xrightarrow{\text{pyruvate kinase}} \text{pyruvate + ATP}$$

$$\text{Pyruvate + NADH + H}^+ \xrightarrow{\text{lactate dehydrogenase}} \text{lactate + NAD}^+$$

With pyruvate kinase and lactate dehydrogenase present in excess, the rate of cellobiose phosphorylation is equivalent to the rate of NADH oxidation, which is measured by the absorbance decrease at 340 nm.

*Reagents*

Glycylglycine–NaOH buffer, 0.2 $M$, pH 7.5
$MgCl_2$, 0.1 $M$
ATP, 50 m$M$
Phosphoenolpyruvate, 0.1 $M$
NADH, 10 m$M$
Cellobiose, 0.2 $M$
Crystalline lactate dehydrogenase
Crystalline pyruvate kinase

*Procedure.* The following are added to a microcuvette with a 1.0-cm light path: 0.05 ml of buffer, 0.01 ml of $MgCl_2$, 0.01 ml of ATP, 0.01 ml of phosphoenolpyruvate, 0.025 ml of cellobiose, 0.005 ml of NADH, nonlimiting amounts of lactate dehydrogenase and pyruvate kinase, a

[1] R. E. Palmer and R. L. Anderson, *Biochem. Biophys. Res. Commun.* **45**, 125 (1971).
[2] R. E. Palmer and R. L. Anderson, *J. Biol. Chem.* **247**, 3415 (1972).
[3] R. E. Palmer and R. L. Anderson, *J. Bacteriol.* **112**, 1316 (1972).

rate-limiting amount of $\beta$-glucoside kinase, and water to a volume of 0.15 ml. The reaction is initiated by the addition of $\beta$-glucoside kinase. A control cuvette minus cellobiose measures adenosine triphosphatase and NADH oxidase activities, which must be subtracted from the total rate. A control cuvette minus ATP should also be run to check for possible cellobiose reductase activity. The rates are conveniently measured with a Gilford multiple-sample absorbance recorder. The cuvette compartment should be thermostated at 25°. Care should be taken to confirm that the rates are constant with time and proportional to the $\beta$-glucoside kinase concentration.

*Definition of Unit and Specific Activity.* One unit is defined as the amount of enzyme that catalyzes the phosphorylation of 1 $\mu$mole of cellobiose per minute. Specific activity is expressed in terms of units per milligram of protein. Protein is determined either by the ratio of absorbancies[4] at 280 and 260 nm or by the method of Lowry *et al.*[5]

*Alternative Assay Procedure.* For some studies, such as determining phosphoryl donor specificity, a different assay procedure must be used. For these purposes a phospho-$\beta$-glucosidase[6]-glucose-6-phosphate dehydrogenase-linked assay has been described.[2]


## Purification Procedure[2]

*Growth of Organism.* A uracil auxotroph of *A. aerogenes* PRL-R3 was grown at 30° in a uracil-supplemented mineral medium[7] containing 0.5% cellobiose. The 500-ml cultures were grown overnight in 3-liter Fernbach flasks on a rotary shaker.

*Preparation of Cell Extracts.* Cells were harvested by centrifugation, suspended in water, and broken by treatment for 10 min in a Raytheon 250-W, 10-kHz sonic oscillator cooled with circulating ice water. The cell extract was the supernatant obtained after centrifugation of the broken-cell suspension for 10 min at 27,000 *g*.

*General.* The following procedures were performed at 0–4°. A summary of the purification procedure is given in the table.

*Ammonium Sulfate Fractionation.* To a cell extract, a 5% (w/v) solution of protamine sulfate (pH 7.0) was added to give a final concentration of 0.37% (w/v). After 30 min, the suspension was centrifuged and

[4] O. Warburg and W. Christian, *Biochem. Z.* **310**, 384 (1941).
[5] O. H. Lowry, N. J. Rosebrough, A. L. Farr, and R. J. Randall, *J. Biol. Chem.* **193**, 265 (1951).
[6] R. E. Palmer and R. L. Anderson, *J. Biol. Chem.* **247**, 3420 (1972).
[7] V. Sapico, T. E. Hanson, R. W. Walter, and R. L. Anderson, *J. Bacteriol.* **96**, 51 (1968).

PURIFICATION OF β-GLUCOSIDE KINASE

| Fraction | Volume (ml) | Total activity (units) | Specific activity (units/mg protein) |
|---|---|---|---|
| Cell extract | 154 | 218 | 0.083 |
| Ammonium sulfate | 17 | 164 | 0.36 |
| Sephadex G-100 | 70 | 114 | 1.1 |
| $Ca_3(PO_4)_2$ gel | 20 | 50 | 2.1 |
| DEAE-cellulose | 15[a] | 45[a] | 7.6 |

[a] The actual value was multiplied by 2 to correct for the proportion of the previous step not fractionated further.

to the supernatant fraction was added crystalline ammonium sulfate to 40% saturation. The resulting precipitate was collected by centrifugation and dissolved in water.

*Chromatography on Sephadex G-100.* The above fraction was placed on a column (2 × 50 cm) of Sephadex G-100 equilibrated 0.02 M glycylglycine buffer (pH 7.5) and was eluted with the same buffer. Ten-milliliter fractions were collected, and those with most of the activity were combined.

*Calcium Phosphate Gel Adsorption and Elution.* To the above fraction were added 20 ml of calcium phosphate gel (11% solids). All the β-glucoside kinase activity was adsorbed. The gel was successively eluted with 10-ml volumes of 50 mM, 80 mM, and 0.11 M sodium phosphate buffer (pH 7.5). The first two eluates, which contained the β-glucoside kinase activity, were combined.

*DEAE-Cellulose Chromatography.* One-half of the above fraction was placed on a column (1.2 × 7.0 cm) of DEAE-cellulose equilibrated with 20 mM sodium phosphate buffer (pH 7.5). The protein was eluted by the stepwise addition of 18-ml volumes of 0.1, 0.2, 0.3, and 0.4 M NaCl in the same buffer. The active fractions, which began eluting with 0.3 M NaCl, were combined and concentrated 4-fold with a Diaflo ultrafiltration cell (Amicon Corporation). The enzyme was purified about 90-fold with about a 20% overall recovery of the activity. It was free from (<0.5%) phospho-β-glucosidase,[6] glucokinase, phosphoglucomutase, and 6-phosphogluconate dehydrogenase activities.

## Properties[2]

*Substrate Specificity.* β-Glucosides that serve as substrates are: cellobiose, gentiobiose, methyl β-D-glucoside, salicin, phenyl β-D-glucoside, arbutin, cellotriose, cellotetraose, cellobiitol, amygdalin, and sophorose.

$K_m$ values for cellobiose and gentiobiose, which are considered to be the natural substrates,[2,3] are about 1.0 m$M$ and 1.5 m$M$, respectively. Compounds that do not serve as substrates are: $p$-nitrophenyl $\beta$ D-glucoside, methyl $\alpha$-D-glucoside, sucrose, maltose, melibiose, trehalose, lactose, turanose, melezitose, raffinose, inulin, D-glucose, D-fructose, D-galactose, D-mannose, D-fucose, L-sorbose, D-ribose, L-arabinose, D-mannitol, and D-glucitol.

*pH Optimum.* Reaction velocity as a function of pH (glycylglycine buffer) is maximal at about pH 7.2.

*Estimated Molecular Weight.* The sedimentation coefficient as determined by sucrose density gradient centrifugation is about 8.0 S, suggesting a molecular weight of about 150,000.

*Stability.* Purified $\beta$-glucoside kinase retained about 65% of its activity when stored at 4° for 2 months.

# [2] Hexokinases from Yeast

*By* ERIC A. BARNARD

Hexose + ATP → hexose 6-phosphate + ADP

Classical procedures for the isolation of hexokinase from yeast[1-4] involved drying of the yeast and autolysis of the yeast cells at 37° to liberate the enzyme, followed by ammonium sulfate fractionation, in some cases adsorption on bentonite,[4] and repeated recrystallizations. Chromatography on DEAE-cellulose columns[5,6] gave a great improvement in the later stages of this procedure. The main difficulty in the preparation, however, was a general problem in isolations of enzymes from autolyzed yeast cells,[5] namely, the abundance of protease activity released from such cells, with a resulting partial degradation of the enzyme either at that stage, or after isolation in apparently pure form, due to adhering traces of protease.[7] Since the cleavages produced thus can be compatible with

[1] M. Kunitz and M. R. McDonald, *J. Gen. Physiol.* **29**, 393 (1946).
[2] L. Berger, M. W. Slein, S. P. Colowick, and C. F. Cori, *J. Gen. Physiol.* **29**, 379 (1946).
[3] K. Bailey and E. C. Webb, *Biochem. J.* **42**, 60 (1948).
[4] R. A. Darrow and S. P. Colowick, this series, Vol. 5, p. 226.
[5] N. R. Lazarus, A. H. Ramel, Y. M. Rustum, and E. A. Barnard, *Biochemistry* **5**, 4003 (1966).
[6] I. T. Schulze, J. Gazith, and R. H. Gooding, this series, Vol. 9, p. 376.
[7] A. H. Ramel, Y. M. Rustum, J. G. Jones, and E. A. Barnard, *Biochemistry* **10**, 3499 (1971).

high retention of activity, the degradation is initially easily overlooked. This difficulty has been countered by the addition of inhibitors of serine proteases, e.g., DFP[5] or phenylmethanesulfonyl fluoride (PMSF).[6] However, two problems remain: not all the proteases present are serine proteases sensitive to such inhibitors, and the addition of such inhibitors cannot be made until autolysis is complete, since if added initially the liberation of hexokinase is greatly impeded.[5,6]

In the purification scheme described here, these problems are resolved, in particular because autolysis is completely avoided, mechanical methods of disruption of the yeast cells being available in convenient form, and because other precautions are taken. However, the autolysis at 37°, if limited to 2 hr[6,8] can leave a good proportion of the hexokinase molecules untouched, since Colowick and co-workers[6,8] have found in that case that equivalent forms (which they term P-I and P-II) can be later separated, using a procedure in which PMSF is added after autolysis.[9] The final yield of intact forms of the highest activity in the nonautolytic procedure, however, is greater (see footnote 9), and the risk is much reduced of proteolytic damage due to unintentional minor fluctuations in the procedure.

Yeast (*Saccharomyces cerevisiae*) contains 3 isoenzymes of hexokinase[7], A, B, and C, all of which can be obtained by one form or other of the present preparative method. The methods employed in the present procedure to avoid proteolysis to any degree and to increase the yield are: (1) mechanical disruption of the yeast cells, and avoidance of an air-drying stage; (2) treatment with DFP at the outset, and at critical early stages; (3) use of rapid gel filtration as the method for removal of ammonium sulfate: (4) chromatography on high-capacity DEAE-cellulose for the initial stage, and then on microgranular DEAE-cellulose (DE-52) for the final separations; (5) control of pH and temperature

---

[8] F. C. Womack, M. K. Welch, J. Nielsen, and S. P. Colowick, *Arch. Biochem. Biophys.* **158**, 451 (1973).

[9] It must be borne in mind, if that procedure is used, that the initial slow drying of the yeast at room temperature[4,8] required for good results therein, can itself provide some opportunity for proteolysis by enzymes released when cellular membranes are disrupted by the drying. This stage is often, therefore, a source of variations in the yield in this alternative type of procedure for enzymic release, involving slow air-drying (over 14 days) followed by autolysis for 2 hr at 37°. Comparing the most recent form[8] of the latter procedure, the present preparation gives an average yield of activity in the crude extract about 50% greater, and final yields of the pure isoenzymes A and B that are each about 2.5 times greater. Probably these differences are due to destruction of some hexokinase (in the procedure of Womack et al.[8]) in the autolytic stages, and in the acidification at pH 4.4 and the 24-hr dialysis at pH 4.4, where the instability of the enzyme at that pH and the presence of adhering acid protease can be limiting factors.

throughout, to avoid attack by DFP-insensitive proteases; (6) final removal of adhering traces of proteases.

## Assay Method

The most convenient method is the buffered indicator method, applied by Darrow and Colowick[4] to hexokinase with cresol red as the indicator and glycylglycine as the buffer of the same $pK_a$ value (8.25 at 25°). The method described follows, with some changes, that of Darrow and Colowick.[4]

### Reagents

> 0.006% cresol red, 1.8% $MgCl_2 \cdot 6H_2O$; stir overnight to dissolve, and filter through glass wool; keeps indefinitely, in the dark.
> ATP, 0.1 $M$ disodium salt
> Glycylglycine-NaOH buffer, 0.1 $M$ pH 9.0
> NaOH, 0.1 $N$

For the stock assay solution, 2 ml of ATP, 5 ml of glycylglycine, and 10 ml of cresol red, solutions are mixed. The pH is adjusted to 8.6 with 0.1 $N$ NaOH (about 7.4 ml) and the volume to 50 ml with water. This mixture can be stored in the cold, darkened, for up to 4 days.

*Procedure.* In a cuvette of 1-cm light path, mix 2.5 ml of the stock solution and 0.4 ml of 0.2 $M$ glucose. $A_{574\ nm}$ should read between 0.7 and 0.9, at pH 8.6. The enzyme solution (5–100 $\mu$l) is added and the absorbance decrease at 574 nm is read. The change in $A_{574\ nm}$ is proportional to the amount of hexose-phosphate formed,[4] and the reaction can proceed thus up to the final pH of about 7.7. The quantity of enzyme to be added is adjusted such that the initial velocity can be determined accurately. For calibration, to the same reaction mixture (without enzyme addition) is added 100 $\mu$l of 0.1 $N$ HCl, which should give a decrease in $A_{574\ nm}$ of about 0.5. The precise value obtained thus is used in converting the readings to micromoles of acid produced.

*Units.* One unit is defined as the amount of enzyme producing from this mixture 1 $\mu$mole of acid per minute at 25°. The cuvette and the assay solutions must be thermostatted at 25.0°, since both the buffer acid dissociation and the enzymic reaction are unusually temperature-sensitive; the enzymic activity at 30° is 1.5 times that at 25°, so that units which are quoted for 30° assays must be reduced to two-thirds of their value to be on the present scale.

*Assays in Other Conditions.* The same assay can be used for other sugar substrates. Below pH 7.5, however, this assay becomes too insensitive, and for more acidic pH values and for any substrate sugar, a convenient assay is that in which the production of ADP is coupled to the pyruvate kinase and lactate dehydrogenase reactions, recording spectrophotometrically at 340 nm the oxidation of NADH. The version of Maley and Ochoa[10] is easily adapted to the hexokinase reaction in any conditions.

## Purification Procedure[11]

All operations are performed at 0–4°, except where noted. All solutions are made with doubly deionized distilled water, since heavy metal contaminants are highly deleterious.[12] The procedure is described for a batch of 1.9 kg yeast (weight of moist yeast, as packed commercially), but can easily be scaled down or up.

*Protease Inhibitor.* DFP is very satisfactory, but on account of its toxicity adequate warnings must be given to all who will handle it. DFP from Sigma Chemical Co. has been found to be satisfactory. A 1-g vial is opened cautiously in a fume hood, and its tip is instantly placed under the surface of 4 ml of propane-1,2-diol (previously dried over type 5A molecular sieves). This stock DFP solution is then about 1 $M$, and can be stored (in several well-stoppered vials, securely wrapped) at —20° or below for any period, so long as condensation of water vapor in an opened vial is carefully avoided. To avoid the risk of partial denaturation of the enzyme by a local high concentration of DFP or propanediol, the 1 $M$ DFP/propanediol solution is diluted 10-fold in buffer just before the addition; this solution can be removed from the fume hood and added from a pipette tip held below the surface of the enzyme solution, with efficient magnetic stirring. If the potency of a DFP solution is in doubt, it can be checked by measuring the rate of DFP inhibition of pure bovine α-chymotrypsin (1 mg/ml, in 40 m$M$ Tris·HCl, pH 7.8): DFP is added to give 10 μ$M$ final DFP concentration, and, after thorough mixing, samples are withdrawn at timed intervals and assayed[13] spectrophotometrically on benzoyl-L-tyrosine ethyl ester in the same buffer. A half-time for the inactivation reaction of 5 min or less (at 25°) should be observed if the nominal concentration is sufficiently correct.

[10] I. Maley and S. Ochoa, *J. Biol. Chem.* **233**, 1538 (1958).
[11] Developed from the work of colleagues in this laboratory[5,7] and recent work of A. K. Bhargava, S. Otieno, and E. A. Barnard. This work was supported by Grant GM-16726 from the Institute of General Medical Sciences, U.S. Public Health Service.
[12] Some buffer components are occasionally contaminated with heavy metals, and require to be first treated in solution with Chelex 100 chelating resin (Bio-Rad Laboratories). For example, certain samples of solid Tris, although claimed to be of analytical grade, have decreased the activity of hexokinase until so extracted. The purest grades of reagents should be used.
[13] B. C. W. Hummel, *Can. J. Biochem. Physiol.* **37**, 1393 (1959).

If it is desired to avoid the use of DFP, PMSF can be used instead.[6,8] The stock solution is 40 m$M$ in 95% ethanol, and the final concentration employed is 2 m$M$. No experience has been reported with the addition of PMSF, instead of DFP, to the extract of mechanically disrupted yeast cells, but at later stages it is effective.[6,8]

*Preparation of Crude Extract.* Two methods are available for avoiding autolysis.

METHOD A. The most efficient method involves the use of a mechanical shearing device for disrupting the yeast cells.[14] The Manton-Gaulin Submicron Disperser[15] is one form of convenient large-scale (multiliter quantity) press, which gives 80–90% cell breakage. (A conventional French press, operating at 15,000–20,000 psi, is an alternative, which can deal only with much smaller volumes.) The yeast cells (1.9 kg baker's yeast, e.g., Fleischmann's) are crumbled into 50 m$M$ Tris·HCl–1 m$M$ DFP, pH 7.0 (600–800 ml, precooled to 0°), cooled in ice, and stirred, and the suspension is passed twice through the precooled[16] press; the maximum temperature it reaches during the pressure application (about 4 min) is about 14°, after which the slurry is collected in an ice-cooled beaker.

METHOD B. The alternative method employs freeze-thawing, in toluene as a vehicle.[5] Four liters of toluene is preequilibrated with solid $CO_2$ at —75°; 1.9 kg of yeast is crumbled into this, and solid $CO_2$ is added to maintain —75° for about 4 hr. The toluene is decanted, and the thawing stage then occupies about 40 hr (for a 4-liter glass beaker container) in a cold room; for about 27 hr of this stage, the melted cell lysate is present.

In all cases, the cell debris is removed by centrifugation at 23,000 $g$, 1 hr, and the supernatant is taken as the crude extract.

*Acid and Ammonium Sulfate Fractionations.* The crude extract (in an ice-bath) is cautiously acidified with 3 $N$ acetic acid to pH 4.60. Since the enzyme is readily inactivated at pH 4.4 (at 4°), the acid should be added slowly from a burette with efficient stirring (but avoiding foam formation) to avoid high local acid concentration,[17] the process taking about 2 hr. The copious precipitate is removed at 23,000 $g$, 45 min. To

---

[14] Y. M. Rustum, A. H. Ramel, and E. A. Barnard, *Prep. Biochem.* **1**, 309 (1971).

[15] Manton-Gaulin Manufacturing Co., Everett, Massachusetts; Model 15M-8TA with Cell Rupture Assembly is currently in use in our laboratory, operated at 8000 psi.

[16] We have attached to the cell housing of the Sub-Micron Disperser two metal plates cooled by refrigeration coils; regulation of the compressor is made to a point where the cell contents, when static, are just above freezing. Under pressure, they then reach 14°.

[17] Throughout, all pH values are recorded as read (in contrast to an earlier procedure: see footnote 5) with temperature compensation on the pH meter set for the actual temperature of the sample.

the supernatant, solid ammonium sulfate is added in portions to 0.25 saturation (14.4 g/100 ml initial volume). After about 2 hr at 4°, the precipitate is removed at 23,000 $g$, 30 min or until clear. To the supernatant, further solid ammonium sulfate, 0.67 g/100 ml initial volume, is added to give 0.65 saturation. After a further 2 hr, the suspension is centrifuged and the pellets are taken up in the minimum volume (about 38 ml, here) of 5 m$M$ succinate–0.5 m$M$ EDTA, pH 5.8. The pH is adjusted to 7.0 with 2 $N$ NaOH, and DFP to 0.1 m$M$ is added. After 20 min at 4°, the pH is readjusted to 4.6 with 2 $N$ HCl, and solid ammonium sulfate added,[18] as needed, to at least 0.8 $M$. The precipitate formed during these adjustments is removed at 48,000 $g$, 10 min.

*Gel Filtration.* The solution (up to 150 ml) is desalted rapidly on a Sephadex G-25 column (70 × 6 cm) equilibrated with 5 m$M$ succinate–1 m$M$ EDTA, pH 5.8. Fractions (10–15 ml) are collected at a flow rate of 800 ml/hr. The active fractions are pooled and the pH is readjusted to 5.8 with solid Tris.[19]

*Chromatography on DEAE-Cellulose.* For the first chromatographic stage, where a relatively large amount of protein must be put on the column, a high-capacity (~0.9 meq/g) exchanger with high flow rate is required. Selectacel[20] DEAE-cellulose, type 70 or type 20, is satisfactory for this. For the subsequent chromatographic stages, higher resolution is obtained on the lower capacity microgranular DEAE-cellulose, e.g., Whatman DE-52.[21] It is essential that the ion exchangers used can give rise to a pH gradient of the form shown in Figs. 1 and 2. Such a gradient must be measured completely in the effluent fractions with each

[18] In practice, owing to the initial presence of an undefined amount of ammonium sulfate in the redissolved precipitate, it is desirable in routine preparations to add the solid salt so as to give a measured conductivity (90 millimho cm$^{-1}$ in this case) which has been shown previously to correspond to 0.8 $M$ ammonium sulfate solution in the buffer employed.

[19] At this stage the enzyme solution can be stored frozen below −20° for a few days, but in general we prefer to avoid delays until the preparation is complete, since some heterogeneity can result therefrom. Free DFP must be absent when freezing hexokinase.

[20] Schleicher and Schuell, Keene, New Hampshire. The ion exchanger is precycled with 0.5 $N$ HCl and 0.5 $N$ NaOH, and prepared and packed in the column, as described for DE-52 (see footnote 21), in the Whatman instruction leaflet IL2. With Selectacel (60 g) the column bed (2 cm in diameter) is compressed with a packing rod, so that a final height of 30 cm is achieved. This exchanger is recycled similarly before each use and can (in contrast to observations with early batches: see footnote 5) be re-used then indefinitely.

[21] H. Reeve Angel Inc., Bridewell Place, Clifton, New Jersey. Preparation, column packing and cycling are as in Whatman leaflet IL2. The column is poured in 0.2 $M$ sodium succinate–1 m$M$ EDTA, pH 5.80. DE-52 is recycled after about every third chromatogram, and can be re-used indefinitely.

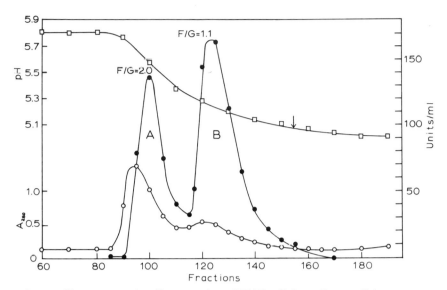

FIG. 1. Chromatography (first stage) on DEAE-cellulose. For conditions, see text. The arrow marks the point where hexokinase C would chromatograph on this column. The right-hand axis shows the enzymic activity (●——●) in units per milliliter of effluent, and the left-hand axes show (below) the protein content (○——○) and (above) the pH in the fractions. F/G (see Table II) is shown for the two peaks of activity.

batch of ion-exchanger used, and should be checked in all chromatograms in every tenth fraction collected. A much steeper decline in pH than shown, or a considerable retardation in the start of the pH decline, means (provided that the column was fully preequilibrated[22]) that the ion-exchanger batch has additional groups with pK near 6, and will not give good resolution of hexokinases in these conditions. TEAE-celluloses, in particular, have been found to show such variations between batches,[5,14] but the DEAE-celluloses mentioned have recently been fully reproducible from batch to batch.[23]

[22] Prior to each chromatography, unusual care is necessary in the exhaustive washing of the ion-exchanger columns (in the cold room) with the starting buffer, to the complete equilibration of inflow and outflow solution, judged by both pH and conductivity readings, because of the very low buffer capacity of most of the solutions used as compared to the high capacity of these columns. Similarly, it is checked each time that the conductivity and the acidity of the protein solution applied do not exceed at all those values for the starting buffer.

[23] The titration curve (see footnote 5) of the ion-exchanger should show no inflection near pH 6. A convenient criterion of the required properties in a new batch of anion exchanger is obtained by applying the gradient to the column without protein present, when the appropriate form of the pH profile (Figs. 1 and 2) should be generated.

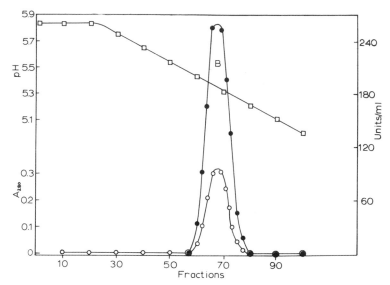

FIG. 2. Rechromatography on DE-52, of hexokinase B. Details as for Fig. 1. The pooled B peak from the first DE-52 chromatography (run under the same conditions as shown here) was applied to the column. The gradient is applied from the start, but the pH of the effluent is unchanged for about the first 100 ml.

Fibrous DEAE-cellulose (6.0 g) is suspended in 400 ml 0.2 $M$ sodium succinate, pH 5.80, and packed[20] into a $30 \times 2$ cm column. This is washed[22] with 5 m$M$ sodium succinate–1 m$M$ EDTA, pH 5.80. The desalted enzyme solution at pH 5.8, in any volume, is passed into the column overnight under a maximum hydrostatic head of 80 cm, whereupon all the activity is retained. A concave pH gradient is generated by a constant-volume mixing chamber, containing 280 ml of 5 m$M$ succinate–0.5 m$M$ EDTA, pH 5.8, and a reservoir with 1000 ml of a similar solution at pH 4.7. The flow rate is maintained at 50 ml/hr, and fractions of 5 ml are collected. Excellent results are obtained on the $30 \times 2$ cm column if the column charge does not exceed 6 g of total protein. The fractions in the peaks of hexokinases A and B (Fig. 1) are pooled, respectively, and adjusted to a total salt concentration not above 10 m$M$ and a pH of 5.8.

The DE-52 column ($17 \times 1.5$ cm) is equilibrated[22] in 10 m$M$ sodium succinate–1 m$M$ EDTA, pH 5.80. The protein solution (containing the hexokinase A or B pool) is applied as before, and the column head is then connected to a mixing chamber containing 250 ml of 10 m$M$ succinate–1 m$M$ EDTA, pH 5.80, and a reservoir containing 250 ml of that solution adjusted to pH 4.6, such that a linear pH gradient is generated (as in Fig. 2). Fractions (5 ml) are collected at a flow rate of 30 ml/hr. The maximum load on these columns has been 150 mg of protein.

The fractions having specific activity above 500 units/mg in the hexokinase B chromatogram, or above 200 units/mg in the hexokinase A chromatogram, are each pooled. Each solution is adjusted as before and adsorbed on a similar DE-52 column at pH 5.80, for a rechromatography under the same conditions (Fig. 2).

*Concentration of the Solutions.* The pooled samples of hexokinase A and of B are adjusted to pH 8.0 using solid Tris, and concentrated by retention on a very small DE-52 column (column volume 1 ml per 25 mg of protein) and displacement therefrom at high ionic strength.[24] This column is first equilibrated in 10 m$M$ Tris·HCl–1 m$M$ EDTA, pH 8.3. The enzyme is adsorbed from any volume of solution at pH 8, and after brief washing with equilibrating buffer, is displaced using 1.0 $M$ NaCl in 0.1 $M$ sodium phosphate (or any other) buffer at pH 5.5. The point of collection of the enzyme (to minimize dilution) can be determined by testing for turbidity formation with 10% trichloroacetic acid solution in the emerging drops.

*Storage, Desalting, and Crystallization.* Routinely, the pure enzyme can be stored at 4° in solution, as obtained from the concentration step, without activity loss for many weeks. It should not be frozen. The presence of 1 m$M$ EDTA is usually sufficient preservative.

If crystals are required, the protein solution (30 mg/ml) is dialyzed against ammonium sulfate (0.50 saturation) in 50 m$M$ sodium phosphate, pH 7.0, for about 6 hr. Saturated ammonium sulfate is then added to make the external solution up to 0.560 saturation. Crystallization of pure hexokinase B commences immediately. After 2 days the medium is made up to 0.565 ammonium sulfate saturation, and left at room temperature for about a week, when the crystals are harvested and stored at 4°. It is difficult to get crystals of hexokinase A from ammonium sulfate solutions in these conditions, but they can be obtained, as described[8] from Colowick's laboratory, at room temperature by gradual increase in ammonium sulfate concentration over a period of 21 days.

To remove the 1 $M$ NaCl from the concentrated solution, one can use either dialysis against any buffer, or gel filtration on Sephadex G-25, or, to avoid final dilution, ultrafiltration. The latter can be accomplished in, e.g., an Amicon stirred cell with a PM-30 membrane.[25]

*Freedom from Protease Contamination.* The hexokinases as prepared thus are free from all overtly detectable protease activity. They are

---

[24] The DE-52 in the small concentrating column is removed after use and recycled with the bulk of the DE-52, a recycled sample being taken for each occasion.
[25] Amicon Corp., Lexington, Massachusetts 02173.

TABLE I

PURIFICATION OF HEXOKINASES A AND B

| Stage | Total activity (units) | Protein (mg) | Units/mg (at 25°) | Yield (%) |
|---|---|---|---|---|
| 1. Crude extract | 350,000[a] | | | (100) |
| 2. Acid fractionation | 330,000 | | | 94 |
| 3. Ammonium sulfate | | | | |
| fractionation (i)[b] | 291,000 | | | 83 |
| (ii)[b] | 204,000 | | | 58 |
| 4. G-25 filtration | 217,000 | 13,000 | 17 | 62 |
| 5. DEAE-cellulose | | | | |
| chromatography | | | | |
| Hexokinase A | 49,900 | 624 | 80 | 14⎫ |
| Hexokinase B | 104,400 | 392 | 266 | 30⎭ 44 |
| 6. DE-52 chromatography | | | | |
| Hexokinase A | 40,700 | 244 | 170 | 12⎫ |
| Hexokinase B | 84,700 | 162 | 523 | 24⎭ 36 |
| 7. DE-52 rechromatography | | | | |
| Hexokinase A | 30,800 | 129 | 239[c] | 9⎫ |
| Hexokinase B | 79,400 | 99 | 800[d] | 23⎭ 32 |

[a] This is the average for routine preparations from 1.9 kg (wet weight) yeast. The crude extract can sometimes contain 450,000 or more units/mg, both from Fleischmann and Red Star yeasts (not stored too long). All other values here are then increased accordingly.

[b] The first fraction is the soluble phase at 0.25 $(NH_4)_2SO_4$ saturation, and the second fraction is the subsequent precipitate at 0.65 saturation, redissolved. The activity in the latter is somewhat underestimated owing to the salt present.

[c] The maximum specific activity of this species can go higher (see Table II); the values here are those of the entire pooled peak, for an average routine preparation.

[d] This is about the highest attainable; only fractions from stage 6 with a specific activity >500 units/mg were pooled. The mean specific activity of the entire B peak in stage 7 in a routine preparation is >600 units/mg.

stable in solution indefinitely (when growth of microorganisms is prevented) at 4° or 25°: the instability in solution of hexokinases described for other, early types of preparation is attributed to their distinct, active protease contamination.[5,26] This type of contamination is present in all commercial preparations of yeast hexokinase so far available.

Further, some traces of *latent* protease activity may sometimes still remain in the purified final forms, despite their chromatographic and elec-

[26] J. Gazith, I. T. Schulze, R. H. Gooding, F. C. Womack, and S. P. Colowick, *Ann. N.Y. Acad. Sci.* **151**, 307 (1968).

trophoretic homogeneity.[27] These traces, probably of a pro-proteinase, are, if present, manifested by heterogeneity of the hexokinase protein material which appears in sodium dodecyl sulfate (SDS) polyacrylamide electrophoretograms (as first reported by Pringle,[31] using a commercial hexokinase preparation); this effect is due to denaturation of the hexokinase by the SDS and activation by the SDS of the traces of latent protease.[29] The prevention of this effect by a brief prior boiling of the hexokinase-SDS solution is also diagnostic.[29,31] In any case where such latent minor proteolytic capacity is present, it does not give any cleavage of the hexokinase molecule except in strong denaturing media and does not interfere, prior to that state, with any property. If present and if its removal is required, this may be accomplished, first by inserting a stage of gel-filtration chromatography on Agarose[32] of the crude extract, as described elsewhere[7]; and second, by giving, prior to the DE-52 chromatography, a treatment (15 min) at 35°, pH 8, in the presence of 0.1 m$M$ DFP. The former treatment removes the bulk of the DFP-insensitive protease, by size difference, and the latter releases and inactivates the bound pro-proteinase.[29] All other stages are as before. The final product (hexokinase A or B) is then free of the trace of latent pro-proteinase sometimes present. Usually, however, these additional treatments are not necessary.

A typical purification is summarized in Table I.

## Properties

*Forms of Hexokinase Isolated.* The preparative procedure in the form given above yields the isoenzymes A and B (Fig. 1). A third active form, hexokinase C, is present in the crude extract of the yeast cells, in fact replacing most of B in the initial state.[7,14] Hexokinase C is eluted in the DEAE-cellulose chromatograms after, and resolved from, hexokinase B (Table II). C is convertible to B by high ionic strength at pH <5, and the routine procedure contains a step to bring this about. Hexokinases B and C appear to involve the same primary structure, whereas A has a different one.[8,27,28]

When limited proteolysis has occurred, as in some earlier methods

[27] This point has been emphasized recently by Schmidt and Colowick[28]; their evidence confirms the similar reports from this laboratory.[7,14,29,30]

[28] J. J. Schmidt and S. P. Colowick, *Arch. Biochem. Biophys.* **158**, 458 (1973).

[29] Y. M. Rustum, E. J. Massaro, and E. A. Barnard, *Biochemistry* **10**, 3509 (1971).

[30] M. Derechin, Y. M. Rustum, and E. A. Barnard, *Biochemistry* **11**, 1793 (1972).

[31] J. R. Pringle, *Biochem. Biophys. Res. Comm.* **39**, 46 (1970).

[32] Bio-Gel A-0.5 m; Bio-Rad Laboratories. Sephadex G-100 gel gives a better separation, if needed.

TABLE II

CHARACTERISTICS OF THE CHROMATOGRAPHIC PEAKS OF YEAST HEXOKINASES[a,b]

| Property | A | B | C |
|---|---|---|---|
| $F/G^c$ | $2.6^d$ | 1.1 | 1.1 |
| pH of elution[e] | 5.6 | 5.4 | 5.15 |
| Specific activity[f] | 275 | 900 | 750 |
| $K_m$ (glucose) | $0.2-0.3 \text{ m}M^{a,g,h}$ | $0.25-0.6 \text{ m}M^{a,h,i}$ | $0.6 \text{ m}M^{a,i}$ |
| $K_m$ (ATP·Mg) | $0.3 \text{ m}M^{j,k}$ | $0.29 \text{ m}M^{j,k}$ | — |
| $\dfrac{V_{max} \text{ (fructose)}}{V_{max} \text{ (glucose)}}$ | $3.0^g$ | $1.1-1.3^i$ | $1.1-1.3^i$ |
| Specific absorptivity[l] | 1.24 | 0.98 | 0.98 |

[a] A. H. Ramel, Y. M. Rustum, J. G. Jones, and E. A. Barnard, *Biochemistry* **10**, 3499 (1971).

[b] Y. M. Rustum, A. H. Ramel, and E. A. Barnard, *Prep. Biochem.* **1**, 309 (1971).

[c] Ratio of activity on glucose to activity on fructose, in the standard conditions of assay (at 27 m$M$ hexose).

[d] The F/G value of the A peak in the initial chromatographic steps can vary in the range 1.8–2.6, due to the presence of some B complexed[b] with the A. The value of 2.6 is that of pure rechromatographed A.

[e] pH values measured at 4° in the eluate of the DE-52 column. The values vary (±0.05 unit) between chromatograms, and are valid only for the pH gradient of the type shown in Fig. 2. The relative separations of the peaks are, however, always maintained.

[f] Maximum specific activity on glucose that has been attained, after repeated rechromatography (units per milligram, at 25°, pH 8.5).

[g] pH 8.5, 25°.

[h] J. Gazith, J. T. Schulze, R. H. Gooding, T. C. Womack, and S. P. Colowick, *Ann. N.Y. Acad. Sci.* **151**, 307 (1968).

[i] pH 8.5–8.0, 25°.

[j] pH 8.0, 25°.

[k] D. P. Kosow and I. A. Rose, *J. Biol. Chem.* **245**, 198 (1970).

[l] In ml mg⁻¹ cm⁻¹ at 280 nm.

of preparation and in presently available commercial specimens, enzymically active modified forms can be separated, some of which have been designated A″ and D[7,14] and others S-I and S-II.[26,28,33] S-I and S-II have nearly half the molecular weight of the native molecules,[28,33] and differ considerably from the latter in their affinity for sugar substrates.[9,34] Hence, care must be taken in studies on yeast hexokinase to determine exactly which forms are present in the preparation used.

*Properties of Hexokinases A and B.* Some of these are summarized in Table II. A comprehensive review of other properties has recently been

[33] I. T. Schulze, and S. P. Colowick, *J. Biol. Chem.* **244**, 2306 (1969).

made by Colowick.[34] Both have molecular weights of about 104,000.[30,35] The molecule consists of two identical subunits of about 52,000 molecular weight,[28,29] such that A can be termed $\alpha_2$ in structure, and B, $\beta_2$. This has recently been confirmed directly by X-ray crystallographic solutions[36,37] of the three-dimensional structure of hexokinase B, which shows, further, that the two subunits are related to each other by a quasi-2-fold axis; i.e., they are not symmetrically placed with respect to each other. Changes occur in this structure upon the binding of substrates.[37]

It should be made clear that the forms A and B when prepared as described above are each homogeneous and intact, in view of the discussion above of the effect of bound pro-proteinase, and in view also of a recent discussion[28] which states that, when samples[38] of hexokinases from this laboratory were tested, two of these had subunit molecular weights (in SDS media) of about 16,000. These two samples appear to be those made by a form of the procedure that we subsequently recognized[7,29] as having a definite contamination by bound pro-proteinase, active in SDS. While a brief abstract[39] based upon that preparation also gave a low subunit molecular weight (26,000), this was explicitly withdrawn in 1971 when the modified procedure described above was reported and when the true situation of proteolytic effects in all previously detailed preparations and the actual subunit molecular weight were documented.[7,28] That is, the preparations of A and B described here contain only the intact proteins with subunit molecular weight about 52,000. From the evidence recently reported by Colowick and co-workers,[8,28] it can be concluded that their present forms P-I and P-II are identical with our forms A and B, respectively.

The enzyme at pH values above about 7 exists appreciably in a dimer–monomer equilibrium; dissociation of the native forms is favored by low protein concentration, high ionic strength, and certain substrates or substrate-related molecules.[30,33,40]

[34] S. P. Colowick, in "The Enzymes" (P. D. Boyer, ed.), 3rd ed., Vol. IX, p. 1. Academic Press, New York, 1973.
[35] N. R. Lazarus, M. Derechin, and E. A. Barnard, *Biochemistry* **7**, 2390 (1968).
[36] T. A. Steitz, *J. Mol. Biol.* **61**, 695 (1971).
[37] T. A. Steitz, R. J. Fletterick, and K. J. Hwang, *J. Mol. Biol.* **78**, 551 (1973).
[38] These samples were, in fact, two among various types of hexokinase preparation provided to other laboratories upon request, at a period some years before the report[28] in question was made. The ones reported as giving the anomalous behavior were not subsequently identified by our laboratory as to the type of preparation (fully protected or not) originally employed.
[39] Y. M. Rustum, E. J. Massaro, and E. A. Barnard, *Fed. Proc., Fed. Amer. Soc. Exp. Biol.* **29**, 334 (1970).
[40] J. S. Easterby and M. A. Rosemeyer, *Eur. J. Biochem.* **28**, 241 (1972).

Of the epimers of D-glucose, only D-mannose is a good substrate.[41,42] Only the C-2 position of D-glucose can be modified without affecting the $K_m$ value.[42,43] D-Fructose is a good substrate, but much more so[44] with A than with B (Table II).

Yeast hexokinase possesses native nucleotide triphosphatase activity,[45] but at only $5 \times 10^{-6}$ times that of the rate of phosphorylation of glucose: this secondary activity is present in each of the native isoenzymes A and B. These species also catalyze a very slow ADP–ATP exchange in the absence of other substrates, but Solomon and Rose[46] have pointed out that the site catalyzing this exchange has quite different properties to the site of ATPase hydrolysis.

While pentoses are nonsubstrates,[47,48] D-xylose and D-lyxose can strongly increase the ATPase activity of yeast hexokinase. Other aspects of the substrate binding of yeast hexokinase are detailed elsewhere.[34,42]

Hexokinases A and B each contain eight cysteine residues, present in the reduced form, per dimer molecule.[35] Four of these react rapidly with methylmercuric iodide or $p$-chloromercuribenzoate without effect on the activity.[35] The other four (two per subunit) react under certain conditions with iodoacetate with the loss of all activity.[35,49] This latter reaction can proceed rapidly and specifically when a substrate-related alkylating agent, $N$-bromoacetyl-D-galactosamine, is employed.[50,51] It was pointed out[35] that the alkylation inactivation phenomena do not establish that the two thiols involved (per subunit) are in the active center; but the other two are clearly nonessential.

The specific absorptivity of the protein, useful for rapid determination

[41] A. Sols, G. De La Fuente, C. Villar-Palasi, and C. Assensio, *Biochim. Biophys. Acta* **30**, 92 (1958).

[42] R. K. Crane, *in* "The Enzymes" (P. D. Boyer, H. Lardy, and K. Myrbäck, eds.), 2nd ed., Vol. 6, p. 47. Academic Press, New York, 1962.

[43] E. M. Bessell, A. D. Foster, and J. H. Westwood, *Biochem. J.* **128**, 199 (1972).

[44] The ratio for fructose to glucose phosphorylation under the standard assay conditions described is, in fact, close to the ratio of $V_{max}$ values for the two substrates (3.0 for A). In the report where this was shown,[7] by a printing error (Table IV) that ratio was written as $V_{max}$ (fructose)$/C_{max}$ (glucose), instead of $V_{max}$ (fructose)$/V_{max}$ (glucose).

[45] A. Kaji and S. P. Colowick, *J. Biol. Chem.* **240**, 4454 (1965).

[46] F. Solomon and I. A. Rose, *Arch. Biochem. Biophys.* **147**, 349 (1971).

[47] G. Dela Fuente, R. Lagunas, and A. Sols. *Eur. J. Biochem.* **16**, 226 (1970).

[48] G. Dela Fuente, *Eur. J. Biochem.* **16**, 240 (1970).

[49] J. G. Jones, S. Otieno, E. A. Barnard, and A. K. Bhargava (submitted, 1974).

[50] Y. M. Rustum and E. A. Barnard, *Fed. Proc., Fed. Amer. Soc. Exp. Biol.* **30**, 1122 (Abstract) (1971).

[51] S. Otieno, E. A. Barnard, and A. K. Bhargava, *Fed. Proc., Fed. Amer. Soc. Exp. Biol* **33**, 1444 (1974).

of its concentration when relatively pure, is given in Table II for each form.[52] Very small quantities, as occur in following specific chemical modifications, are conveniently estimated by total acid hydrolysis and measurement (relative to a hexokinase hydrolyzate standard) of aspartic acid content on the amino acid analyzer.

[52] These values in Table II were redetermined, on the basis of interferometric measurements. An earlier value[5] of 0.92 mg ml$^{-1}$ cm$^{-1}$ for the specific absorptivity of B is slightly in error.

# [3] Hexokinase of Rat Brain[1]

By ALBERT C. CHOU and JOHN E. WILSON

Approximately 80% of the total hexokinase activity in rat brain homogenates is associated with the mitochondria.[2] The enzyme may be rather specifically eluted from the mitochondria with glucose-6-P or ATP,[3,4] an observation that provides the basis for this purification procedure.[5] Since mitochondrial hexokinases found in other tissues also are solubilized by similar treatment, this procedure is, in principle, applicable to purification of hexokinase from other sources.

## Assay Method

*Principle.* The hexokinase reaction is monitored by coupling to NADPH formation (followed by absorbance at 340 nm) with excess glucose-6-P dehydrogenase.

*Reagents*

Assay mix: 3.7 m$M$ glucose, 7.5 m$M$ MgCl$_2$, 11 m$M$ 1-thioglycerol, and 45 m$M$ HEPES ($N$-2-hydroxyethylpiperazine-$N'$-2-ethanesulfonic acid), pH 7.5

NADP, 25 mg/ml

Glucose-6-P dehydrogenase,[6] 100 units/ml, in 0.02% bovine serum albumin–20 m$M$ Tris Cl, pH 7.5

ATP, 0.22 $M$, pH 7.0

[1] ATP: D-hexose 6-phosphotransferase, EC 2.7.1.1.
[2] M. K. Johnson, *Biochem. J.* **77**, 610 (1960).
[3] I. A. Rose and J. V. B. Warms, *J. Biol. Chem.* **242**, 1635 (1967).
[4] J. E. Wilson, *J. Biol. Chem.* **243**, 3640 (1968).
[5] A. C. Chou and J. E. Wilson, *Arch Biochem. Biophys.* **151**, 48 (1972).

*Procedure.* In a cuvette 0.9 ml of assay mix, 0.01 ml of NADP, 0.01 ml of glucose-6-P dehydrogenase, 0.03 ml of ATP, $H_2O$, and enzyme to give a total volume of 1.0 ml are mixed. In the absence of glucose-6-P in the enzyme sample, the enzyme is routinely added last. In the case that the enzyme contains significant glucose-6-P, all components except ATP are mixed, and oxidation of endogenous glucose-6-P is allowed to go to completion (approximately 2–3 min); the hexokinase reaction is then initiated by the addition of ATP. The rate of increase in absorbance at 340 nm is recorded in a spectrophotometer with sample compartment maintained at 25°. Potential interfering enzyme activities (e.g., NADPH oxidase, 6-phosphogluconate dehydrogenase) have not presented any problems with brain extracts, but this should be verified if the assay is used with extracts from other tissues.

*Definition of Units of Activity and Specific Activity.* One unit of hexokinase activity is defined as the amount of enzyme catalyzing the formation of 1 $\mu$mole of glucose-6-P per minute at 25°. Specific activity is expressed in units per milligram of protein. Protein is routinely determined turbidimetrically by the procedure of Katzenellenbogen and Dobryszycka.[7] Crystalline bovine serum albumin standard, dissolved in the same medium as the sample, is used as a standard.

## Purification Procedure

All centrifugations, concentrations, and column chromatography operations were at 0–4°.

*Removal and Freezing of Brains.* Approximately half of the particulate hexokinase in brain homogenates is in latent form, apparently owing to entrapment of hexokinase-bearing mitochondria in the relatively impermeable nerve ending particles ("synaptosomes") formed during homogenization.[8] Various membrane-disrupting techniques are effective at exposing the latent activity. A simple procedure, convenient for the present purposes, is freezing in liquid nitrogen.

Rats are killed by decapitation, and the brains are rapidly removed, rinsed in cold 0.25 $M$ sucrose, and dropped into liquid $N_2$. Brains stored in liquid $N_2$ for up to several months have been satisfactorily used in the preparation of hexokinase.

*Preparation of Washed Brain Particles.* Frozen brains (55 g) were

[6] We have found the type V glucose-6-P dehydrogenase (from baker's yeast) sold by Sigma Chemical Co. to be very satisfactory with regard to both stability and the absence of contaminating enzymes.

[7] W. M. Katzenellenbogen and W. M. Dobryszycka, *Clin. Chim. Acta* 4, 515 (1959).

[8] J. E. Wilson, *Arch. Biochem. Biophys.* 150, 96 (1972).

thawed by immersion in 250 ml of 0.25 $M$ sucrose[9] at room temperature. When thawing was complete, the sucrose was decanted. Approximately nine volumes (495 ml) of ice-cold 0.25 $M$ sucrose was added, and the brains were homogenized (batchwise) in a Teflon–glass homogenizer (Size C, A. H. Thomas Co.). The total volume of homogenate was then adjusted to 10 volumes (550 ml) based on original brain weight. After centrifugation at 1000 $g$ for 20 min, the supernatant plus the loosely packed white layer on top of the pellet were decanted and centrifuged at 40,000 $g$ for 20 min. The supernatant was decanted taking care to retain the loosely packed material on top of the pellet. The pellet was then homogenized in cold 0.25 $M$ sucrose, the total volume was again adjusted to 550 ml, and the suspension was centrifuged at 40,000 $g$ for 20 min. This washing procedure was repeated twice more, and the final pellet resuspended by homogenizing in 1100 ml (20 volumes, based on original brain weight) of 0.25 $M$ sucrose at room temperature.

*Solubilization with Glucose 6-Phosphate.* The particle suspension was made 1 m$M$ in glucose-6-P by adding 11 ml (0.01 volume of suspension) of a 0.1 $M$ solution, which had been adjusted to pH 7.0 ± 0.2. After incubation at 25° for 1 hr with occasional gentle stirring, the suspension was centrifuged at 105,000 $g$ for 60 min. The volume of the supernatant, containing the solubilized enzyme, was noted. A solution was prepared, containing (per 100 ml of solubilized enzyme) 1.0 ml of 1 $M$ potassium phosphate, 1.0 ml of 1 $M$ glucose, 1.0 ml of 50 m$M$ EDTA, and 0.042 ml of 1-thioglycerol; the pH was adjusted to 6.9. After addition of this mixture, the solubilized enzyme was concentrated to approximately 50 ml with an Amicon ultrafiltration device (PM-10 membrane); the ultrafiltration cell was rinsed well with column buffer (see below) to ensure complete removal of the enzyme. The concentrated enzyme was centrifuged at 40,000 $g$ for 20 min to remove a precipitate that had formed during concentration.

*DEAE-Cellulose Column Chromatography.* A 2 × 25 cm column of DEAE-cellulose (Serva, Gallard-Schlessinger Mfg. Co.) was equilibrated with column buffer which contained 10 m$M$ potassium phosphate, 10 m$M$ glucose, 0.5 m$M$ EDTA, and 5 m$M$ 1-thioglycerol, pH 7.0. The concentrated enzyme was applied, then the column was washed extensively (30–36 hr) with this buffer. A linear gradient of KCl in column buffer was then started, using 300 ml of buffer in the mixing chamber, and 300 ml of buffer containing 0.3 $M$ KCl in the reservoir. The flow rate was maintained at approximately 30 ml/hr with a peristaltic pump, and 5-ml fractions were collected. The enzyme was eluted at approximately 60 m$M$ KCl.

*Final Concentration and Storage.* The peak fractions from the DEAE-

PURIFICATION OF RAT BRAIN HEXOKINASE

| Fraction | Volume (ml) | Protein (mg) | Activity (units) | Recovery (%) | Specific activity (units/ mg) |
|---|---|---|---|---|---|
| Original homogenate | 550 | —[a] | 462 | (100) | — |
| Washed brain particles | 1150 | —[a] | 369 | 80 | — |
| Glucose-6-P solubilized Enzyme, after concentration | 79 | 26.2 | 392 | 85[b] | 15 |
| DEAE-column fractions | 55.5 | 5.0 | 330 | 71 | 66 |
| Concentrated dialyzed enzyme | 4 | 5.0 | 320 | 69 | 64 |

[a] The turbidimetric protein assay used precluded protein determinations at this stage.

[b] The slight increase in total activity resulting from solubilization by glucose-6-P is consistently observed and has been previously reported [J. E. Wilson, *J. Biol. Chem.* **243**, 3640 (1968)].

column were combined and concentrated to 2–5 ml by ultrafiltration. The concentrated enzyme was dialyzed against several changes of a buffer containing 0.1 $M$ potassium phosphate, 0.5 m$M$ EDTA, 10 m$M$ glucose, and 5 m$M$ 1-thioglycerol, pH 7.0. Dialysis removed an unidentified, UV-absorbing, nonprotein contaminant that eluted from the DEAE-column in a broad peak overlapping that of hexokinase. Solid glucose was added to give a final concentration of 0.2 $M$, and the enzyme stored at −20°. The enzyme is quite stable under these conditions; no activity loss was seen after storage for several months, even with repeated freezing and thawing.

*General Comments.* The table shows the results of a typical purification. Comparable results have been obtained in more than 20 preparations, with overall recoveries ranging from 50 to 75% and final specific activities of 60–66 units/mg. The purified enzyme has been found to be homogeneous when examined by isoelectric focusing and by ultracentrifugal, electrophoretic, and immunological techniques.

## Properties of Rat Brain Hexokinase

The molecular weight determined by the sedimentation equilibrium method was 97,500 ± 500. Comparison of sedimentation velocity of hexokinase and standard proteins on sucrose density gradients gave a value of 98,000. Electrophoresis of the denatured enzyme on SDS-acrylamide gels indicated a molecular weight of 96,000–98,000; thus, the enzyme appears to consist of a single polypeptide chain. In accord with this, a single mole of N-terminal glycine was found per mole of enzyme (based on a molecular weight of 98,000). The C-terminal amino acid has been identified as serine.

The amino acid composition of the rat brain enzyme is similar to that of the bovine brain enzyme[9a] with relatively high amounts of glutamate and aspartate (amount of amidation undetermined) present while the aromatic amino acid content is rather low. The pI determined by isoelectric focusing is 6.4. The molar extinction coefficient[10] at 280 nm is $5.1 \pm 0.1 \times 10^4$ cm$^2$ mole$^{-1}$.

Under the assay conditions described above, the $K_m$ values of the purified enzyme for ATP and glucose are 0.4 m$M$ and 0.04 m$M$, respectively. Using the method of Paulus,[11] the enzyme has been found to bind 1 mole of glucose per mole of enzyme (based on molecular weight of 98,000) with a dissociation constant of 40 $\mu M$ for the enzyme–glucose complex, in excellent agreement with the kinetically determined $K_m$ value. A single binding site for glucose-6-P has also been found, with a dissociation constant of approximately 1 $\mu M$. This is somewhat lower than the value obtained by other methods (7–8 $\mu M$),[12] a discrepancy due to technical problems with the Paulus procedure.[13]

Conformational changes have been observed to result from binding of the inhibitor, glucose-6-P.[12] Glucose-6-P binding is antagonized by P$_i$, and a model has been proposed relating the conformational changes caused by binding of glucose-6-P and P$_i$ to the effects of these agents on catalytic activity and interaction with the mitochondrial membrane.

Unlike the initially solubilized, unpurified enzyme, the pure hexokinase cannot be rebound to mitochondria. The reason for this loss of binding ability during purification is presently unknown. Recent observations suggesting the involvement of enzyme-bound phospholipid in the binding

[9] The use of reagent grade sucrose throughout this procedure represents an appreciable expense. Ordinary table-grade sucrose is considerably less expensive, and has been found to be completely adequate for use during the purification procedure.

[9a] G. P. Schwartz and R. E. Basford, *Biochemistry* **6**, 1070 (1967).

[10] The enzyme appears to aggregate to some extent during prolonged frozen storage (no precipitation or loss of activity has been observed to result from aggregation), resulting in a contribution by light scattering to the $A_{280}$. The extinction coefficient given is for the freshly prepared, nonaggregated enzyme, and should be used for exact determinations only when turbidimetric contributions are negligible, i.e., when there is negligible "absorbance" at wavelengths longer than 320 nm.

[11] H. Paulus, *Anal. Biochem.* **32**, 91 (1969).

[12] J. E. Wilson, *Arch. Biochem. Biophys.* **159**, 543 (1973).

[13] A. C. Chou and J. E. Wilson, *Arch. Biochem. Biophys.*, in press.

process[14] raise the possibility that loss of enzyme-bound phospholipid during the purification may account for the inability of the purified enzyme to bind.

[14] J. E. Wilson, *Arch. Biochem. Biophys.* **154**, 332 (1973).

# [4] Glucokinase from Yeast

*By* P. K. MAITRA

D-Glucose + ATP → D-glucose 6-phosphate + ADP

## Assay Method

*Principle.* The activity of glucokinase is estimated fluorometrically by measuring the rate of formation of NADPH in a coupled assay using NADP and glucose-6-phosphate dehydrogenase.[1]

*Reagents*

Triethanolamine buffer, 50 m$M$ containing 10 m$M$ MgCl$_2$, pH 7.4
Glucose, 1 $M$
NADP, 50 m$M$
ATP, 200 m$M$, pH 7.4
Glucose-6-phosphate dehydrogenase, 140 units per milliliter of saturated (NH$_4$)$_2$SO$_4$ solution

*Procedure.* To 1 ml of triethanolamine buffer–MgCl$_2$ mixture in a cuvette the following additions are made: 10 $\mu$l of glucose, 5 $\mu$l of ATP, 2 $\mu$l of NADP, and 1 $\mu$l of the glucose-6-phosphate dehydrogenase suspension. It is convenient to prepare daily a mixture of all the above ingredients except the dehydrogenase. At the time of the assay, 1 ml of this mixture is taken in a cuvette into which 1 $\mu$l of the coupling enzyme is stirred in with a glass rod. The cuvette is monitored on the fluorometer for about a minute; the reaction is then started by adding an appropriate volume (1–25 $\mu$l) of the glucokinase preparation. The rate of increase of fluorescence is measured continuously on a strip-chart recorder (see this series, Vol. 10 [74]). In the spectrophotometric modification of the method the amounts of NADP and glucose-6-phosphate dehydrogenase are increased 5-fold. Since commercial preparations of the dehydrogenase invariably contain hexokinase, it is often necessary to take an account of the rate of NADPH produced in absence of added glucokinase.

[1] P. K. Maitra, *J. Biol. Chem.* **245**, 2423 (1970).

A variation of this assay method measures the rate of ADP formation in a coupled assay using NADH, phosphoenolpyruvate, pyruvate kinase, and lactate dehydrogenase. This is particularly useful when sugars other than glucose are used as the phosphate acceptor.

Since the assay procedure measures a nonspecific ATP-glucose-6-phosphotransferase activity, its application to crude yeast extracts is very limited owing to the presence of relatively large activity of hexokinase. However, yeast glucokinase is practically inert toward fructose, unlike yeast hexokinase. The velocity of phosphorylation of fructose with respect to that of glucose is thus a convenient, if only approximate, measure of the relative amounts of hexokinase and glucokinase present in a preparation. To measure the phosphorylation of fructose by the above procedure, glucose is replaced by a final concentration of 10 m$M$ fructose and 1 unit of purified phosphoglucose isomerase per milliliter of the assay mixture.

*Definition of Unit and Specific Activity.* A unit of glucokinase is defined as the amount of the enzyme catalyzing the conversion of 1 $\mu$mole of substrate per minute at room temperature (24°). Specific activity is expressed as enzyme units per milligram of protein determined in crude preparations by the ultraviolet absorption method of Warburg, and in the purified preparation by the Folin reagent (see this series, Vol. 3 [73]).

## Purification Procedure[2]

The enzyme solution was kept at 4° and between pH 7.4 and 7.6 unless otherwise mentioned.

*Step 1. Preparation of Crude Extract.* One kilogram of commercial dried baker's yeast (Indian Yeast Company, Calcutta) was suspended in 3 liters of 0.2 $M$ $K_2HPO_4$ prewarmed to 37° (see this series, Vol. 9 [69b]), dispensed in four 2-liter flasks and shaken gently on a rotary shaker at 37° for 1.5 hr; 0.25 ml of caprylic alcohol was added to each flask, which was then kept in the cold room for 15 min. After the bulk of the foam had subsided, the suspension was centrifuged in the cold for 15 min at 18,000 $g$ using the Sorvall GSA rotor.

*Step 2. Ammonium Sulfate Fractionation.* The supernatant, measuring 1700 ml, was adjusted to pH 7.4 with dropwise addition of 4 $N$ $NH_4OH$, and then 585 g of $(NH_4)_2SO_4$ were added slowly to a saturation of 55%. The precipitated protein was collected by centrifugation and the supernatant containing the bulk of hexokinase activity was discarded.

*Step 3. Heat Treatment and Ammonium Sulfate Fractionation.* The

[2] P. K. Maitra and Z. Lobo, unpublished observations, 1973.

bulky precipitate was dissolved in a known volume of 50 m$M$ potassium phosphate buffer, pH 7.4, containing 2 m$M$ EDTA, 2 m$M$ 2-mercapto-ethanol, and 10 m$M$ glucose. The final volume was noted (205 ml). Phenylmethanesulfonyl fluoride (PMSF), 72 mg, dissolved by shaking in 10 ml of ethyl alcohol, was added dropwise with stirring to the thick protein solution to give a final concentration of PMSF of 2 m$M$ (see this series, Vol. 9 [69b]). Stirring was continued for an additional 45 min, after which an equal volume of the above buffer was added.

The solution (430 ml) was adjusted to pH 5.0 with 8 $N$ acetic acid and divided in two portions in 500-ml Erlenmeyer flasks. One flask was heated to 45° for 5 min in a water bath with gentle shaking, chilled immediately, centrifuged in a Sorvall SS-34 rotor at 40,000 $g$ for 5 min and the yellow supernatant immediately neutralized to pH 7.4 with 4 $N$ NH$_4$OH. The second portion was treated similarly after neutralization of the first batch was completed. The supernatants were pooled and fractionated with (NH$_4$)$_2$SO$_4$ assuming the starting salt concentration to be 10% saturation: to 385 ml of the pooled neutralized solution 65 g of solid (NH$_4$)$_2$SO$_4$ were added to give 40% saturation. The collected precipitate was discarded and the supernatant (400 ml) was further treated with 61 g of (NH$_4$)$_2$SO$_4$ to give a 65% saturation. The precipitate collecting between 40 and 65% saturation of (NH$_4$)$_2$SO$_4$ was dissolved in a minimum volume of a buffer containing 10 m$M$ potassium phosphate, pH 7.4, 1 m$M$ EDTA, 1 m$M$ 2-mercaptoethanol, and 5 m$M$ glucose and dialyzed overnight against this buffer.

*Step 4. DEAE-Cellulose Chromatography.* The dialyzed protein solution was adsorbed on a DEAE-cellulose column (160 ml) previously equilibrated with the dialysis buffer. The column was washed with two column volumes of the buffer and a linear gradient of 700 ml of 0 to 0.4 $M$ KCl dissolved in the same buffer was applied thereafter at a flow rate of 20 ml per hour. Fractions containing the highest glucose-phosphorylating activity were eluted around 0.24 $M$ KCl. The active fractions were pooled, and the enzyme was precipitated with (NH$_4$)$_2$SO$_4$ to 65% saturation; the precipitate was collected by centrifugation and dissolved in 50 m$M$ Tris buffer, pH 7.5 containing 2 m$M$ 2-mercaptoethanol, 2 m$M$ EDTA, and 5 m$M$ glucose. The enzyme solution was dialyzed overnight against this buffer.

*Step 5. First DEAE-Sephadex A-50 Chromatography.* The dialyzate was adsorbed on a DEAE-Sephadex A-50 column equilibrated with the Tris dialysis buffer. The initial volume of the resin bed was 45 ml. The column was washed with 250 ml of 0.1 $M$ KCl dissolved in the equilibrating buffer. The enzyme was subsequently eluted with a linear gradient of 290 ml of 100–125 m$M$ KCl dissolved in the Tris dialysis buffer at a flow rate of 7 ml/hr. Maximal elution of enzyme was observed at 180 m$M$ KCl. Fractions with a minimum specific activity of 60 units of glucose–ATP phosphotransferase activity per $E_{280}$ were pooled, and the enzyme was precipitated with solid $(NH_4)_2SO_4$ to 65% saturation. The fluffy white precipitate of the enzyme was taken up in a minimum volume of the Tris dialysis buffer, transferred to a thin cellophane bag and dialyzed against this buffer for an hour, whereupon it dissolved.

*Step 6. Sephadex G-100 Chromatography.* The protein solution from this step was layered carefully on a column (175 ml; 55 cm length) of Sephadex G-100 resin equilibrated with the Tris dialysis buffer and eluted with the same buffer. Fractions of 3 ml were collected in 15 min. The enzyme came out of the column within its void volume. Fractions containing less than 200 units of glucokinase were discarded. The relevant fractions were pooled, the protein was precipitated with 65% saturation of $(NH_4)_2SO_4$, and the sparingly soluble gelatinous white precipitate was dialyzed overnight against Tris dialysis buffer.

*Step 7. Second DEAE-Sephadex A-50 Chromatography.* The dialyzed enzyme was finally adsorbed on a DEAE-Sephadex A-50 resin packed in a column of starting volume 15 ml. The column was washed with 60 ml of 0.1 $M$ KCl dissolved in the Tris dialysis buffer containing glucose, 2-mercaptoethanol, and EDTA. Subsequently elution was started with 150 ml of a linear gradient of 100–225 m$M$ KCl dissolved in the equilibrating buffer. Each 10-min fraction contained 2 ml. Fractions with a specific activity of less than 150 per $E_{280}$ were discarded. Maximal elution of the enzyme was observed at 193 m$M$ KCl. The enzyme activity in the pooled eluate was precipitated with $(NH_4)_2SO_4$ to 60% saturation, the precipitate collected by centrifugation and dissolved in the Tris dialysis buffer as in step 5.

The table summarizes a typical purification procedure.

*Comments on Purification Procedure*

1. The loss of enzyme activity following the heating step at low pH is variable. When the starting quantity of dried yeast is 300 g or less, it is practicable to obtain first a 0 to 40% saturation of $(NH_4)_2SO_4$ and then raise the saturation to 55%. Most of the glucokinase activity precipitates in the range of 40 to 55% saturation unlike in the example illustrated here. No loss of glucokinase in the heating step has been observed in such preparations.

2. When glucokinase is prepared in presence of PMSF from freshly harvested cells of hexokinaseless mutant of *Saccharomyces cerevisiae*[1] without taking it through the heat step, the enzyme coming from the

PURIFICATION OF YEAST GLUCOKINASE

| Step | Fractions | Volume (ml) | Enzyme[a] (units) | Protein[b] (mg) | Enzyme[a] (units/mg protein)[b] | Fructose[c]/ glucose |
|---|---|---|---|---|---|---|
| 1 | Crude extract | 1700 | 100,000 | (15,000) | (6.7), 7.1 | 1.35 |
| 2 | (NH$_4$)$_2$SO$_4$ precipitate, 0–55% saturation | 205 | 31,000 | (9,650) | (3.2) | 0.32 |
| 3 | Heat and second (NH$_4$)$_2$SO$_4$ precipitate after dialysis | 62 | 17,000 | (1,680) | (10.1) | 0.10 |
| 4 | DEAE-cellulose chromatogram | 20 | 14,300 | (550) | (26) | 0.03 |
| 5 | First DEAE-Sephadex chromatography | 6.5 | 14,000 | (137) | (102) | 0.02 |
| 6 | Sephadex G-100 chromatography | 6.7 | 9,730 | (74) | (132) | 0.002 |
| 7 | Second DEAE-Sephadex chromatography | 2.9 | 5,520 | 70 | (173), 79 | 0.002 |

[a] Enzyme refers to total glucose : ATP phosphotransferase activity.
[b] Values in parentheses have been determined by ultraviolet absorption at 260 and 280 nm. Preparations obtained after steps 1 and 7 have been assayed also by the Folin reagent (see this series, Vol. 3 [73]).
[c] The quotient fructose/glucose indicates the velocity of phosphorylation of fructose relative to that of glucose assayed as described in the procedure.

DEAE-cellulose column is often rendered insoluble in a number of buffers. Of a variety of chemicals tested, urea seems to be the most effective agent in solubilizing a part of glucokinase activity.

## Properties

Unless mentioned otherwise, the properties listed here refer to a preparation of glucokinase obtained from a hexokinaseless mutant of *S. cerevisiae*.[1]

*Substrate Specificity.* The Michaelis constants for various sugars, in mmoles per liter, follow: glucosone, 0.01; glucose, 0.03; mannose, 0.12; glucosamine, 0.74; 2-deoxyglucose, 1.46; and fructose, 31. The relative maximal velocities, respectively, are: 10, 100, 20, 9, 45, and 0.4. The $K_m$ value for ATP at 10 m$M$ glucose is 50 $\mu M$. The following sugars are inactive with ATP as the phosphate donor: galactose, 6-deoxygalactose, 6-deoxyglucose, ribose, arabinose, and a number of disaccharides. The following compounds fail to serve as phosphate donor with glucose as the acceptor: ADP, pyrophosphate, phosphoenolpyruvate, creatine phosphate and acetyl phosphate; GTP, CTP, and UTP are slightly active.

*pH Optimum.* Glucokinase displays a broad pH optimum between pH 7.5 and pH 9.6.

*Inhibitors.* N-Acetylglucosamine ($K_i$ 1.7 m$M$) and ADP ($K_i$ 0.3 m$M$) are competitive inhibitors relative to glucose and ATP, respectively.

*Inducibility.* Glucokinase is a constitutive enzyme in *S. cerevisiae*. Prior growth of the yeast in glucose or galactose leads to 50% increase in specific activity in crude extracts.

*Stability.* Yeast glucokinase suspensions containing 5 mg or more of protein per milliliter do not lose any appreciable activity over a year at 0° if kept in either phosphate or Tris buffer in presence of 5 m$M$ glucose, 1 m$M$ 2-mercaptoethanol, and 1 m$M$ EDTA at pH 7.5 as an $(NH_4)_2SO_4$ precipitate.

*Physicochemical Properties.*[2] The enzyme moves as a sharp boundary in the ultracentrifuge with indications of both lighter and heavier components. The $s_{20,w}$ value ranges between 4.2 and 19.2. The subunit molecular weight (see this series, Vol. 26 [1]) is 51,000. Glucokinase contains approximately 4 sulfhydryl groups per subunit.

*Use.* Trace quantities of glucose in fructose may be analyzed conveniently by using glucokinase.

# [5] Glucokinase of Rat Liver

*By* S. J. PILKIS

Rat hepatic glucokinase is an unusual form of hexokinase whose most striking characteristic is its low affinity for glucose.[1] Glucokinase differs from other hexokinases found in animal tissues with regard to substrate specificity,[2] $K_m$ for glucose,[2-5] product inhibition,[5] tissue distribution,[6,7] electrophoretic mobility,[6-8] immunologic attributes,[7] hormonal adaptation,[1,9-12] inactivation by dilution,[4,13] proteolytic attack,[13] and denaturation by urea.[13]

## Assay Method

*Principle.* There are a number of methods for the assay of hexokinases in animal tissues (see this series Vol. 1 [32] and [33], Vol. 5 [25] and Vol. 9 [70a]). The most reliable method, particularly in crude preparations, involves the measurement of the formation of glucose 6-phosphate in the presence of excess glucose-6-phosphate dehydrogenase to reduce $NADP^+$.

*Procedure.* A known volume of the enzyme preparation was added to a reaction mixture which had a total volume of 2.0 ml containing 0.1 $M$ Tris buffer, pH = 7.4, 0.2 m$M$ $NADP^+$, 5 m$M$ ATP; 5 m$M$ $MgCl_2$; 0.1 unit glucose-6-phosphate dehydrogenase (Boehringer), and either 0.5 m$M$ or 100 m$M$ glucose at a final temperature of 30°. Glucokinase activity was obtained by subtracting the activity obtained at 0.5 m$M$ glucose

[1] D. S. DiPietro and W. Weinhouse, *J. Biol. Chem.* **235**, 2542 (1960).

[2] D. G. Walker, *in* "Essays in Biochemistry" (P. N. Campbell and G. D. Greville, eds.), Vol. 2, p. 33. Academic Press, New York, 1966.

[3] S. J. Pilkis and M. E. Krahl, *Fed. Proc., Fed. Amer. Soc. Exp. Biol.* **25**, 523 (1966).

[4] D. G. Walker and M. J. Parry, this series, Vol. 10, p. 381.

[5] C. Gonzalez, T. Ureta, R. Sanchez, and H. Niemeyer, *Biochemistry* **6**, 460 (1967).

[6] L. Grossbard, M. Weksler, and R. Schimke, *Biochem. Biophys. Res. Commun.* **24**, 32 (1966).

[7] S. J. Pilkis, R. J. Hansen, and M. E. Krahl, *Comp. Biochem. Physiol.* **25**, 903 (1968).

[8] H. M. Katzen and R. T. Schimke, *Proc. Nat. Acad. Sci. U.S.A.* **54**, 1218 (1965).

[9] D. G. Walker, *Biochem. J.* **84**, 118P (1962).

[10] H. Niemeyer, I. Clark-Turri, N. Perez, and E. Rabajille, *Arch. Biochem. Biophys.* **109**, 634 (1965).

[11] A. Sols, A. Sillero, and T. Salas, *J. Cell. Comp. Physiol.,* Suppl. 1, **66**, 23 (1965).

[12] S. J. Pilkis, *Biochim. Biophys. Acta* **215**, 461 (1970).

[13] S. J. Pilkis, *Arch. Biochem. Biophys.* **149**, 349 (1972).

from that at 100 m$M$. The optical density changes at 340 nm obtained
in assays of the original liver extract were divided by 1.5 as approximate
correction for 6-phosphogluconate dehydrogenase. This correction can be
omitted for the purified enzyme beyond the first DEAE-Sephadex step,
as fractions at this step were found to have no detectable ability to oxi-
dize 6-phosphogluconate. Another consideration, generally ignored, is the
effect of high concentrations of glucose to inhibit one of the isozymes
of hexokinase in rat liver.[5]

*Specific Activity.* The specific activity of glucokinase was determined
by the following formula:

$$\text{Specific activity} = \frac{OD_{340} \text{ nm/min}}{0.311 \times \text{mg protein/ml}} \times 10^3$$

$$= \frac{\text{nmoles NADP}^+ \text{ reduced}}{\text{min/mg protein}} \text{ or milliunits/min/mg protein}$$

One unit of activity is defined as that amount of enzyme catalyzing
the phosphorylation of 1 $\mu$mole of glucose per minute at 30°.

## Starch Gel Electrophoresis

*Principle.* Starch gel electrophoresis has revealed most clearly that
ATP:hexose phosphotransferase activity consists of a number of enzymes
depending on the tissue and animal. Glucokinase activity is visualized
on starch gels by coupling the typical glucose-6-phosphate dehydrogenase
method described above with a color-generating oxidation-reduction dye
system.

*Procedure.* Vertical starch gel electrophoresis[14] was carried out at 0–3°
for 16 hr with a constant current of 6 V/cm across the gel. The gels
were prepared by addition of 60 g of starch in 500 ml of a pH 8.6 buffer
containing 50 m$M$ barbital and 5 m$M$ EDTA.

The gels were sliced and stained for glucose-ATP phosphotransferase
activity at 37° for 1.5 hr. The solution used for staining was made up
of the following ingredients: 100 ml of 0.1 $M$ Tris, 6 ml of 200 m$M$ $MgCl_2$,
0.5 mg of glucose-6-phosphate dehydrogenase, 330 mg of ATP, 36.8 mg
TPN, 40 mg of Nitro Blue tetrazolium, and 4 mg of phenazine methosul-
fate. The pH was adjusted to 7.6. Glucose was added to give a final concentration of either 100 m$M$ or 0.5 m$M$. Controls were run with ATP
omitted from the staining medium.

Glucokinase has the greatest mobility of the hexokinase isozymes
after starch gel electrophoresis by the above method.[7] Staining for gluco-

---

[14] O. Smithies, *Biochem. J.* **61**, 629 (1955).

kinase in polyacrylamide is more difficult, but this can be accomplished by incorporating glucose-6-phosphate dehydrogenase in the gel matrix.[15]

## Purification

The purification below is essentially a modification of that of Walker and Parry (see this series, Vol. [70a]).

*Step 1. Preparation of 105,000 g Supernatant.* Livers from 40 fed male Holtzman rats (200–300 g) were excised and subjected to the Waring Blendor at high speed for 2 min with 2 volumes of pH 7.0 buffer containing 0.15 $M$ KCl, 0.1 $M$ Tris, 5 m$M$ EDTA, 10 m$M$ $\beta$-SH EtOH, 4 m$M$ MgSO$_4$, and 50 m$M$ glucose. The resulting suspension was then homogenized in a glass–Teflon homogenizer with 6 passes. The supernatant was then spun for 120 min in the No. 30 head of the Spinco Model L at 30,000 rpm. The clear supernatant was taken for the next step. Step 2 is begun immediately since glucokinase activity is often rapidly lost from the supernatant upon standing, particularly if there is any microsomal contamination, which is difficult to avoid in large preparations.

*Step 2. $(NH_4)_2SO_4$ Fraction.* Fractionations were carried out with a solution saturated with $(NH_4)_2SO_4$ at 2–4°, containing the following additions: 10 m$M$ Tris, 5 -m$M$ MgSO$_4$, 5 m$M$ EDTA, 10 m$M$ $\beta$-SH EtOH, EDTA, 50 m$M$ glucose; the pH was adjusted to 7.0. Inclusion of glucose prevented large losses of enzyme activity during the $(NH_4)_2SO_4$ fractionation.

To each 100 ml of supernatant from step 1, 85 ml of saturated $(NH_4)_2SO_4$ solution were added with constant magnetic stirring to give 48% saturation. After 15 min, the precipitate was spun down at 9000 rpm in the GSA head in the Sorvall centrifuge. To each 100 ml of the resulting clear red supernatant was added 60 ml of saturated $(NH_4)_2SO_4$ solution to give 68% saturation. The precipitate was spun down and stored at −20°. Two or three preparations from about 100 rats were pooled at this stage. The frozen $(NH_4)_2SO_4$ precipitate could be stored up to a week with no loss of activity.

*Step 3. Dialysis.* The pooled $(NH_4)_2SO_4$ precipitate from 100 rats is dissolved in a minimum of pH 7.0 buffer containing 0.1 $M$ KCl, 0.1 $M$ Tris, 5 m$M$ EDTA, 5 m$M$ MgSO$_4$, 10 m$M$ $\beta$-EtOH, and 50 m$M$ glucose (buffer B). This volume was generally about 250 ml. This volume of solution was dialyzed with stirring for 2.5 hr against 8 liters of 0.1 $M$ KCl buffer with two changes of buffer. No loss of activity was noted on dialysis.

---

[15] R. C. Hard, *Fed. Proc. Fed. Amer. Soc. Exp. Biol.* **30**, 292 (1971).

*Step 4. Batchwise DEAE-Sephadex Treatment.* The dialyzate was further diluted to 1 liter with buffer B. To this was added 40 ml, for each 2.5 g of protein, of a slurry (supernatant decanted) of DEAE-Sephadex equilibrated with buffer B. This mixture was stirred for 30 min. This quantity of DEAE-Sephadex was sufficient to absorb at least 90% of the measurable glucokinase activity from solution. This mixture was allowed to settle for 15 min, and the bulk of the red supernatant fluid was decanted and the remaining sediment containing the absorbed glucokinase activity was poured directly into a column 5 × 50 cm. The DEAE-Sephadex was allowed to pack, buffer B was run through the column until no further protein was eluted. The column was then eluted with a 0.3 $M$ KCl buffer containing the same ingredients as buffer B. The fractions containing the bulk of the glucokinase activity were pooled and diluted 1:1 with a pH = 7 buffer containing 10 m$M$ Tris, 5 m$M$ EDTA, 5 m$M$ MgSO$_4$, 10 m$M$ $\beta$-SH EtOH, 50 m$M$ glucose.

*Step 5. DEAE-Sephadex Chromatography.* The resultant solution from step 4 was applied to a 2 × 25 cm column of DEAE-Sephadex previously equilibrated with buffer B. The sample was allowed to sink into the column slowly overnight. In the morning the elution of glucokinase was begun with a 500-ml linear gradient (pH = 7.0), 0.1 $M$ KCl to 0.6 $M$ KCl, which contained 10 m$M$ Tris, 5 m$M$ EDTA, 5 m$M$, MgSO$_4$, 10 m$M$ $\beta$-SH EtOH, and 50 m$M$ glucose. Glucokinase was normally eluted in the region of 0.25 $M$–0.35 $M$ KCl and emerged as a symmetrical peak. The flow rate was adjusted to about 20 ml/hr, and 5-ml fractions were collected, pooled, and diluted 1:2 with buffer containing no KCl.

*Step 6. DEAE-Cellulose Chromatography.* The fractions from step 5 were applied to a DEAE-cellulose column (3 × 35 cm) equilibrated with buffer B. Glucokinase activity was eluted with a linear KCl gradient identical to that used in step 5. Glucokinase was nearly always eluted as a trailing peak.

*Step 7. Concentrations by DEAE-Sephadex and Ultrafiltration.* The diluted fractions from step 6 were allowed to sink slowly into a small column (1 × 8 cm) containing DEAE-Sephadex equilibrated previously with 0.2 $M$ KCl buffer. After application of the sample the column was extensively washed with 0.2 $M$ KCl buffer. Glucokinase was eluted with 0.4 $M$ KCl buffer; a concentration of 0.3 $M$ KCl did not elute all the glucokinase activity from this column. One-milliliter fractions were collected and the active fractions were pooled. The pool was subjected to ultrafiltration against 0.4 $M$ KCl buffer until the original volume was reduced to a volume of 1–2 ml. Glucokinase from smaller preparations was often stored at this stage until material from 1000 g of liver was accumulated.

*Step 8. Gel Filtration of Glucokinase.* The Sephadex column ($1.7 \times 92$ cm) is equilibrated with 0.15 $M$ KCl, 5 m$M$ EDTA, 10 m$M$ Tris, 4 m$M$ MgSO$_4$, 10 m$M$ $\beta$-SH, EtOH, 50 m$M$ glucose pH = 7.0. The flow rate was about 8 ml/hr and 2-ml fractions were collected. $\beta$-SH EtOH is absolutely necessary at this stage. Its omission leads to complete loss of enzymic activity.

Glucokinase and the low $K_m$ hexokinases can be separated by means of Sephadex G-100 gel filtration.[13] Great advantage is afforded by the fact that glucokinase is retarded on Sephadex. On Sephadex G-100 glucokinase has an apparent molecular weight of 48,000. The retardation of glucokinase is responsible for the almost 4-fold purification achieved with this step. The activity peak does not coincide with any protein peak. At this step (8), tubes 65–70 were pooled and concentrated by ultrafiltration to about 1 ml. Such a narrow cut is taken so as to avoid as much as possible the protein peak at tube 60.

*Step 9. Starch Gel Electrophoresis.* Disc electrophoresis of the post Sephadex G-100 fraction revealed that the fraction contained three major protein bands and a number of minor components (see Fig. 1). If, however, greater amounts of protein are applied to the gel, it is seen that there are actually 13–14 bands present at this stage (Fig. 1). This finding, coupled with the noncoincidence of the glucokinase activity with any of the major protein peaks in the elution pattern, suggested that glucokinase was still a minor component.

Since it was always possible to stain for glucokinase on starch gels, it appeared that this would be the best medium to use for an electrophoretic separation. The fractions from the Sephadex step were concentrated to a small volume (200 to 500 $\mu$l) and applied to a starch gel (50 $\mu$l per slot).

The conditions for the preparation and running of the starch gel at this stage are particularly important. The gel is prepared by adding 60 g of starch to 500 ml of a pH 8.6 buffer containing 50 m$M$ barbital, 5 m$M$ EDTA, and 50 m$M$ glucose. The buffer in the electrophoresis chambers was the same as above except for the addition of 10 m$M$ $\beta$-SH EtOH. The gel was run for 18 hr at 0–4° with a constant current of 6 V/cm across the gel. Inclusion of EDTA, $\beta$-SH EtOH, and glucose are all necessary to preserve maximum enzyme activity. KCl was not included in the chambers or gel since the presence of KCl destroyed the voltage-current relations during the electrophoresis. After electrophoresis the gel was cut into a 10 cm $\times$ 10 cm square, care being taken to include all the areas where glucokinase should be located. This depended on the slots to which the sample was applied and the migration of the enzyme (usually about 9 cm under these conditions). The square slab is carefully

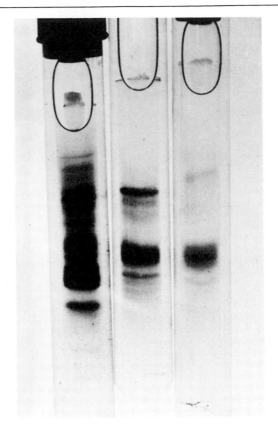

Fig. 1. Disc electrophoresis of glucokinase after passage through Sephadex G-100. The first gel on the left is from the first Sephadex step. The gel has been overloaded with protein so that all bands are visible. The middle gel represents the glucokinase fraction after recycling through the same Sephadex column. The third gel represents a second recycling through the Sephadex column. A number of minor protein bands are also seen in this fraction. Densitometer tracings of the third gel were done, and the purity of the enzyme was judged to be 80% by measuring the area under the curves.

transferred to the E-C convection-elution apparatus. The slab is rotated 90° to the original direction of migration. The protein is thus eluted from the gel in such a manner that it need move only the width of the gel. The buffer is the same as that used in the original electrophoresis chambers. A current of 250 mA was applied for 45 min. Since glucokinase has a high electrophoretic mobility (slightly greater than albumin) it is completely eluted in this time interval. Glucokinase activity was pooled and concentrated by means of ultrafiltration. The conditions of this ultrafiltration are important. If KCl (0.2 $M$ final concentration) is added to the pooled fractions and the ultrafiltration is carried out, all glucokinase activity is lost. However, as long as KCl is omitted from the pooled fractions after the elution, there is no loss of activity. This phenomenon does not occur at any stage of the purification and is at present without explanation.

The specific activity at this stage is 22 units per milligram of protein. However, the yield is only about 3% and the absolute amount of material is small.

*Recycling of Glucokinase through Sephadex G-100.* If the post Sephadex G-100 fraction before the gel elution step is recycled through G-100 once again, an excellent purification is obtained (2940 overall). Disc electrophoresis of this fraction reveals one major protein band (Fig. 1). When it is possible to obtain greater amounts of material at this stage, a recycling through Sephadex followed by elution from starch gel would probably yield a single pure protein. Recycling through G-100 a second time results in no change in the specific activity (30 units/mg) but does result in essentially only one protein band by disc electrophoresis (Fig. 1). See the table for a summary of the purification scheme.

*Notes on Purification of Glucokinase.* In order to maximize the recovery of enzyme activity during the purification it is necessary to include $K^+$, thiol, and glucose in all solutions.[4,13] $K^+$ concentration was never allowed to fall below 0.1 $M$, thiol was maintained at 10 m$M$, and glucose concentration was 50 m$M$.

PURIFICATION OF RAT HEPATIC GLUCOKINASE[a]

| Step | Protein recovery (mg) | Enzyme recovery (%) | Specific activity (units/mg protein) |
|---|---|---|---|
| 1. 105,000 $g$ supernatant | 135,000 | 100 | 0.010 |
| 2. $(NH_4)_2SO_4$ and dialysis | 37,500 | — | — |
| 3. DEAE-Sephadex, batchwise | 2,410 | 40 | 0.3 |
| 4. DEAE-Sephadex | 482 | 35 | 1.0 |
| 5. DEAE-cellulose | 165 | 30 | 2.2 |
| 6. DEAE-Sephadex-concentration | 81 | 25 | 3.4 |
| 7. Sephadex G-100 concentration (ultrafiltration) | 8.0 | 8 | 12.0 |
| 9. Starch gel electrophoresis concentration (ultrafiltration) | 1.8 | 3 | 22.0 |
| Recycling through G-100 | 1.0 | — | 29.4 |

[a] Preparation from 100 rat livers.

## Properties

Rat hepatic glucokinase has been purified 2940-fold by means of $(NH_4)_2SO_4$ fractionation, DEAE-Sephadex, DEAE-cellulose, Sephadex G-100, and elution from starch gels. The purified enzyme displays one principal band on polyacrylamide electrophoresis and is judged to be at least 80% pure at this stage. The key to the improved purification achieved here is the use of Sephadex G-100. Glucokinase is retarded on Sephadex columns and emerges beyond the main protein peak applied to the column. Further recycling results in an almost 3000-fold purification.

*Molecular Weight.* The apparent molecular weight of glucokinase is 48,000–50,000. This value is obtained by three different methods: sucrose gradients, Bio-Gel P-100, and Sephadex G-100 gel filtration.[16] The molecular weight of purified glucokinase is 53,000 by the method of SDS disc gel electrophoresis.[17] These data support the contention that the basic subunit structure of the hexokinases is a 50,000 molecular weight subunit. Interestingly, the molecular weight of the low $K_m$ hexokinases is 96,000 and, in contrast to yeast hexokinase and glucokinase, it has been impossible to demonstrate any dissociation behavior for these enzymes.[16] It is possible that glucokinase may represent an "incomplete" hexokinase molecule that is unable to assume the normal quaternary and tertiary structure typical of the low $K_m$ hexokinases.

The determination of apparent molecular weights of the hexokinases by gel filtration is critically dependent on the type of buffer system used and on ionic strength.[16] This is particularly true for hepatic glucokinase and probably accounts for the discrepancies between different laboratories in the determination of the molecular weight of the enzyme.

*Immunologic Properties.* Antibodies can be prepared against 1200-fold purified rat liver glucokinase and against commercial yeast hexokinase.[7] The antibody against rat liver glucokinase produced 60–85% inhibition of the glucokinase activity of all animals tested between turtle and man on the phylogenetic scale. There was no inhibition of glucokinases from amphibian livers. Nor was there inhibition of the low $K_m$ hexokinases from any of these livers, or of yeast hexokinase. Conversely, neither glucokinase nor the other rat liver hexokinases were inhibited by an antibody against yeast hexokinase which produced 75% inhibition of that enzyme.

*Tissue Distribution.* Glucokinase activity has been measured in livers

---

[16] S. J. Pilkis, R. J. Hansen, and M. E. Krahl, *Biochim. Biophys. Acta* **154**, 250 (1968).

[17] S. Grossman, C. G. Dorn, and V. Potter, *J. Biol. Chem.* **249**, 3055 (1974) and personal communication.

of several amphibians, reptiles, and mammals, including man.[7] It is tentatively concluded that glucokinase appeared first in the vertebrates at the level of the amphibians. Starch gel electrophoretic patterns of liver ATP:D-hexose-6-phosphotransferases revealed that some animals have more than one high $K_m$ enzyme.[7]

*General Properties.* The $K_m$ for glucose of the purified enzyme is 12 m$M$. Glucose and various sulfhydryl reagents protect the enzyme against inactivation. If the enzyme is diluted in the absence of glucose or K$^+$ ion, there is an inactivation of dilution.[4,13] Potassium prevents this activity loss. The hexokinases of liver which have low $K_m$ values for glucose do not lose activity under these conditions and do not require potassium.[13] Glucokinase, as contrasted with the low $K_m$ hexokinase, is very sensitive to proteolytic attack by trypsin and denaturation by urea.[13] The isoelectric point of the enzyme is 4.7.[17] The turnover number is 4346.[17]

# [6] D-Fructose (D-Mannose) Kinase

*By* RICHARD L. ANDERSON and VIRGINIA L. SAPICO

$$\text{D-Fructose} + \text{ATP} \rightarrow \text{D-fructose 6-phosphate} + \text{ADP}$$
$$\text{D-Mannose} + \text{ATP} \rightarrow \text{D-mannose 6-phosphate} + \text{ADP}$$

This enzyme apparently initiates the metabolism of D-fructose and D-mannose in *Leuconostoc mesenteroides.*[1] Its specificity is unique in that it phosphorylates D-fructose and D-mannose equally well, but does not phosphorylate D-glucose or many other sugars. Evidence which indicates that the activities on D-fructose and D-mannose are the result of a single enzyme has been discussed.[1]

## Assay Method

*Principle.* The continuous spectrophotometric assay is based on the following sequence of reactions[1]:

$$\begin{array}{c} \text{D-Fructose} \\ \text{or} \\ \text{D-mannose} \end{array} + \text{ATP} \xrightarrow{\text{D-fructose (D-mannose) kinase}} \begin{array}{c} \text{D-fructose 6-phosphate} \\ \text{or} \\ \text{D-mannose 6-phosphate} \end{array} + \text{ADP}$$

$$\text{Phosphoenolpyruvate} + \text{ADP} \xrightarrow{\text{pyruvate kinase}} \text{pyruvate} + \text{ATP}$$

$$\text{Pyruvate} + \text{NADH} + \text{H}^+ \xrightarrow{\text{lactate dehydrogenase}} \text{lactate} + \text{NAD}^+$$

With pyruvate kinase and lactate dehydrogenase present in excess, the rate of hexose phosphorylation is equivalent to the rate of NADH oxidation, which is measured by the absorbance decrease at 340 nm.

[1] V. Sapico and R. L. Anderson, *J. Biol. Chem.* **242**, 5086 (1967).

*Reagents*

Glycylglycine-NaOH buffer, 0.2 $M$, pH 6.9
$MgCl_2$, 0.1 $M$
ATP, 50 m$M$
Phosphoenolpyruvate, 40 m$M$
NADH, 10 m$M$
D-Mannose or D-fructose, 0.1 $M$
Crystalline lactate dehydrogenase
Crystalline pyruvate kinase

*Procedure.* The following are added to a microcuvette with a 1.0-cm light path: 0.05 ml of buffer, 0.01 ml of $MgCl_2$, 0.01 ml of ATP, 0.01 ml of phosphoenolpyruvate, 0.01 ml of hexose, 0.005 ml of NADH, non-limiting amounts of lactate dehydrogenase and pyruvate kinase, a rate-limiting amount of D-fructose (D-mannose) kinase, and water to a volume of 0.15 ml. The reaction is initiated by the addition of D-fructose (D-mannose) kinase. A control cuvette minus hexose measures adenosine triphosphatase and NADH oxidase activities, which must be subtracted from the total rate. A control cuvette minus ATP should also be run to check for possible hexose reductase activity. The reaction rates are conveniently measured with a Gilford multiple-sample absorbance recorder. The cuvette compartment should be thermostatted at 25°. Care should be taken to confirm that the rates are constant with time and proportional to the D-fructose (D-mannose) kinase concentration.

*Definition of Unit and Specific Activity.* One unit is defined as the amount of enzyme that catalyzes the phosphorylation of 1 $\mu$mole of D-fructose or D-mannose per minute at pH 6.9 and 25°. Specific activity (units per milligram of protein) is based on a spectrophotometric determination of protein.[2]

*Alternative Assay Procedure.* For some studies, such as determining nucleotide specificity, a different assay procedure must be used. For these purposes a phosphohexose isomerase–glucose-6-phosphate dehydrogenase-linked assay has been described.[1]

## Purification Procedure[1]

*Organism and Growth Conditions.* A strain of *Leuconostoc mesenteroides*, designated LM, was obtained from the culture collection of W. A. Wood, Michigan State University. It was grown at 30° without agitation in a 20-liter carboy of modified LBS medium.[3] The inoculum was

---

[2] O. Warburg and W. Christian, *Biochem. Z.* **310**, 384 (1941).
[3] R. N. Costilow, J. L. Etchells, and T. E. Anderson, *Appl. Microbiol.* **12**, 539 (1964).

1 liter of a 24-hr culture in the same medium. The cells were harvested by centrifugation about 30 hr after inoculation and were washed once with distilled water. The yield was about 2 g (wet weight) per liter.

*Preparation of Cell Extracts.* Cells were suspended in distilled water and were broken by treatment for about 15 min in a Raytheon 250-W, 10-kHz sonic oscillator cooled with circulating ice water. The supernatant obtained after centrifugation of the broken-cell suspension at 16,300 $g$ was designated the cell extract.

*General.* Unless stated otherwise, all operations were performed at 0–4°. A summary of the purification procedure is shown in the table.

*Manganous Chloride Fractionation.* $MnCl_2$ (1.0 $M$) was added with stirring to the cell extract (13 mg of protein per milliliter) to give a final concentration of 50 m$M$. The resulting precipitate was removed by centrifugation and discarded.

*First Ammonium Sulfate Fractionation.* Solid ammonium sulfate was added slowly with stirring to the above fraction without adjustment of the pH. The protein precipitating between 40 and 70% saturation was collected by centrifugation and was dissolved in water to give a protein concentration of about 3.6 mg/ml.

*Heat Treatment.* The above fraction was brought rapidly to 50° by heating in an 80° water bath, held at 50° for 3 min, and quickly cooled to 0°. The precipitated protein was removed by centrifugation and discarded.

*Second Ammonium Sulfate Fractionation.* Three volumes of saturated ammonium sulfate (pH 7.0) were added with stirring to 2 volumes of the above fraction. The resulting precipitate was collected by centrifugation and was dissolved in water to give a protein concentration of about 6.1 mg/ml.

*Fractionation with Calcium Phosphate Gel.* Calcium phosphate gel (Sigma, 11% solids) was mixed with the above fraction at a ratio of 0.04 ml of gel per milliliter of enzyme solution. The solids were removed by centrifugation and the supernatant was equilibrated with 50 m$M$ sodium phosphate buffer (pH 7.0) by passage through a column of Sephadex G-25 equilibrated with the same buffer. Fractions (5 ml) were collected, and those containing the kinase activity were pooled.

*Chromatography on DEAE-Cellulose.* The above fraction was chromatographed on a column (1.5 × 23 cm) of DEAE-cellulose (Sigma, exchange capacity = 0.9 mEq/g) equilibrated with 5 m$M$ sodium phosphate buffer (pH 7.0). Fractions (3 ml) were collected during elution of the enzyme with 225 ml of the same buffer containing ammonium sulfate in a linear gradient from 0 to 0.4 $M$. Fractions 36 through 56, which had the highest specific activity, were combined. Although a considerable

PURIFICATION OF D-FRUCTOSE (D-MANNOSE) KINASE

| Fraction | Volume (ml) | Total protein (mg) | $A_{280}:A_{260}$ | Total activity (units) | Specific activity (units/mg protein) |
|---|---|---|---|---|---|
| Cell extract | 332 | 4316 | 0.64 | 7110 | 1.65 |
| MnCl₂ | 334 | 1670 | 0.62 | 6760 | 4.04 |
| First (NH₄)₂SO₄ (40–70%) | 210 | 756 | 1.10 | 5540 | 7.32 |
| Heat | 210 | 147 | 0.61 | 2560 | 17.4 |
| Second (NH₄)₂SO₄ (pH 7.0) | 20.4 | 124 | 1.16 | 2930 | 23.6 |
| Ca₃(PO₄)₂ gel | 25.0 | 126 | 1.10 | 3900 | 31.0 |
| DEAE-cellulose, fractions 36–56 | 62.6 | 47 | 1.6 | 1140 | 24.3 |
| Sephadex G-100, fractions 46–55 | 14.3 | 4.8 | 1.7 | 365 | 76.0 |

loss of activity resulted from this step, the $A_{280}:A_{260}$ ratio increased from 1.1 to 1.6.

*Chromatography on Sephadex.* The above fraction was concentrated by precipitating the protein with ammonium sulfate and dissolving the precipitate in 3.5 ml of water. This solution was then chromatographed on a column (2.7 × 38 cm) of Sephadex G-100 equilibrated with 0.2 *M* ammonium sulfate (pH 7.0). Fractions (1.5 ml) were collected during elution with neutral 0.2 *M* ammonium sulfate. (Equilibration and elution with either water or 50 m*M* glycylglycine buffer, pH 7.0, resulted in a considerable loss in kinase activity.) Fractions 46–55 contained about 5% of the activity of the cell extract; the kinase in these fractions was about 50-fold purified. Concentration of the pooled Sephadex G-100 fractions with ammonium sulfate followed by chromatography on Sephadex G-75 did not increase the specific activity. Electrophoresis on polyacrylamide gel removed a minor protein impurity and resulted in an apparently homogeneous enzyme.[1]

### Properties[1]

*Specificity.* D-Mannose ($K_m$ = 0.4 m*M*) and D-fructose ($K_m$ = 0.4 m*M*) are phosphorylated at equivalent rates at pH 6.9. Phosphorylation of the following compounds could not be detected at 70 m*M* concentrations: D- and L-glucose, 2-deoxy-D-mannose, L-mannose, L-fructose, D-allose, D-altrose, D- and L-galactose, L-sorbose, D-lyxose, D- and L-xylose,

D- and L-arabinose, D-ribose, D- and L-ribulose, D- and L-xylulose, L-rhamnose, D-gluconate, D-glucuronate, D-galacturonate, D-glucitol, D-mannitol, D- and L-arabitol, ribitol, and xylitol.

*pH Optimum.* Phosphorylation of D-mannose is optimal at about pH 6.9. Phosphorylation of D-fructose is maximal over a broader pH range, extending from about pH 7.0 to pH 8.5.

*Estimated Molecular Weight.* The sedimentation coefficient as determined by sucrose density gradient centrifugation is about 4.1 S, suggesting a molecular weight of about 47,000.

*Stability.* The purified enzyme is stable for several months when stored at 0° in the presence of 0.2 $M$ $(NH_4)_2SO_4$, pH 7.0.

## [7] Galactokinase from Pig Liver

### By F. J. BALLARD

$$\text{D-Galactose} + \text{ATP} \rightleftharpoons \text{D-galactose 1-phosphate} + \text{ADP}$$

Galactokinase (EC 2.7.1.6) is present at high activities in livers of fetal or suckling mammals and decreases at weaning so that the activity in adults is between 10 and 20% of the maximum attained during development.[1,2] For this reason livers from 2-week-old piglets are used as a source of enzyme.

### Assay Method

*Principle.* ADP formed in the reaction from ATP and galactose is measured spectrophotometrically at 340 nm as the rate of NADH oxidation in the presence of phosphoenolpyruvate, pyruvate kinase, lactate dehydrogenase, and NADH. L-Cysteine is included since it is an activator of the enzyme.

*Reagents*

Triethanolamine-HCl, 0.3 $M$, pH 7.8
KCl, 1 $M$
ATP, 50 m$M$, pH 7.8
$Mg_2Cl$, 60 m$M$
K-phosphoenolpyruvate, 20 m$M$
D-Galactose, 20 m$M$

[1] F. J. Ballard, *Biochem. J.* **98**, 347 (1966).
[2] R. Cuatrecasas and S. Segal, *J. Biol. Chem.* **240**, 2382 (1965).

NADH, 10 m$M$

L-Cysteine, 150 m$M$, pH 7.8

Pyruvate kinase, 10 units/ml

Lactate dehydrogenase, 10 units/ml

NaF, 0.5 $M$

*Procedure.*[1] One milliliter of triethanolamine buffer and 0.1 ml of each of the other solutions are added to a cuvette of 1-cm light path together with sufficient water to bring the final volume to 2.95 ml. After preincubation for 5 min at 37°, 50 $\mu$l of galactokinase solution is added and the reduction in absorbance at 340 nm is followed with the cuvette maintained at 37°. One unit of galactokinase activity is the amount of enzyme that will catalyze the oxidation of 1 $\mu$mole of NADH per minute after appropriate blanks have been subtracted.

*Application to Measurements in Crude Preparations.* Any reaction that leads to ADP production will give an apparently greater activity. For this reason, NaF is included in all assays where adenosine triphosphatase is present and blanks with galactose omitted must be performed. Interference from ADP-generating reactions can be eliminated by using a radioactive assay in which the [14]C-labeled galactose 1-phosphate formed from [[14]C]galactose is separated on DEAE-cellulose paper strips.[3,4]

## Purification Procedure

Since liver galactokinase loses activity unless stored in an ammonium sulfate suspension, the following procedure[1] is completed in 30 hr. The enzyme stability is enhanced by the inclusion of dithiothreitol (DTT) and phenylmethyl sulfonyl fluoride (PMSF) in the solutions. All steps are carried out at 0–4°.

*Preparation of Liver Cytosol Fraction.* Liver (100 g) from 2-week-old piglets is chilled in ice and homogenized in 4 volumes of a solution containing 0.25 $M$ sucrose, 50 m$M$ Tris pH 7.5, 0.25 m$M$ PMSF, and 1 m$M$ DTT using a Potter-Elvehjem type homogenizer. The homogenate is centrifuged at 100,000 $g$ for 30 min to obtain the clear cytosol fraction.

*Ammonium Sulfate Fractionation.* To each 100 ml of cytosol is added 17.6 g of $(NH_4)_2SO_4$ with constant stirring and with the pH maintained at 7.5 by the dropwise addition of unneutralized 1 $M$ Tris. After standing for 30 min the suspension is centrifuged at 10,000 $g$ for 20 min, and the precipitate is discarded. The percent saturation of $(NH_4)_2SO_4$ is

[3] J. R. Sherman and J. Adler, *J. Biol. Chem.* **238**, 873 (1963).

[4] F. J. Ballard, *Biochem. J.* **101**, 70 (1966).

increased from 30% to 50% by the further addition of 12.7 g of $(NH_4)_2SO_4$ to each 100 ml of supernatant. The precipitate is collected by centrifugation as above and dissolved in 50 ml of 50 m$M$ Tris pH 7.5 containing 0.25 m$M$ PMSF and 1 m$M$ DDT (solution A).

*Sephadex G-100 Chromatography.* The Sephadex is suspended in solution A and poured to give a column of dimensions 16 cm$^2$ × 45 cm. After at least 1 liter of solution A has passed through the column, the 30% to 50% saturation $(NH_4)_2SO_4$ fraction is applied to the column and eluted with solution A at a flow rate of 2 ml/min.

*DEAE-Cellulose Chromatography.* Those fractions from the Sephadex column which contain approximately 70% of the recovered activity are combined and chromatographed on a 4.9 cm$^2$ × 25 cm column of DEAE-cellulose (Whatman DE 52) that has previously been washed with 100 ml of 10 m$M$ EDTA pH 7.5 followed by 200 ml of solution A. The enzyme fraction is washed into the column bed with a small volume of solution A and eluted by 500 ml of a linear gradient of solution A containing from 0 to 1 $M$ KCl. Galactokinase is eluted at approx. 0.2 $M$ KCl.

*Second Ammonium Sulfate Fractionation.* Sufficient $(NH_4)_2SO_4$ is added to the pooled galactokinase fractions from the DEAE-cellulose column to bring the solution to 40% saturation (24.3 g/100 ml). After standing for 30 min, the suspension is centrifuged at 10,000 $g$ for 20 min and the precipitate is discarded. Further $(NH_4)_2SO_4$ is added to the supernatant to establish 60% saturation (13.2 g/100 ml); the precipitate formed is collected by centrifugation. This precipitate is suspended in 5 ml of saturated $(NH_4)_2SO_4$ containing 1 m$M$ EDTA and 1 m$M$ DDT at pH 7 and stored at 0°. A summary of this procedure is given in the table.

### Properties

*Stability.* The galactokinase suspension in saturated $(NH_4)_2SO_4$ retains at least 80% of its activity during storage for 1 week at 0°.

*Specificity of the Reaction.*[2,4] Galactose may be replaced in the reaction by either 2-deoxygalactose or galactosamine, but none of the following sugar or sugar derivatives are phosphorylated: mannose, fructose, glucose, 2-deoxyglucose, 6-deoxy-D-galactose, 6-deoxy-L-galactose, lactose, or $N$-acetylgalactosamine. 2-Deoxy-ATP can act as phosphate donor in addition to ATP, but no reaction is obtained with ITP, GTP, UTP, CTP, TTP, deoxy-CTP, or deoxy-GTP. The divalent metal ion reactivity decreases in the order $Mn^{2+}$, $Mg^{2+}$, $Ca^{2+}$, and $Fe^{2+}$, but $Zn^{2+}$, $Co^{2+}$, $Cd^{2+}$, and $Ni^{2+}$ are not reactive.

PURIFICATION OF LIVER GALACTOKINASE

| Fraction | Volume (ml) | Protein[a] (mg) | Galacto-kinase (units) | Purification (fold) | Units/mg | Recovery (%) |
|---|---|---|---|---|---|---|
| Cytosol | 345 | 7160 | 265 | 1 | 0.037 | 100 |
| Ammonium sulfate, 30–50% precipitate | 50 | 2530 | 202 | 2.2 | 0.080 | 76 |
| Sephadex G-100 | 62 | 312 | 125 | 11 | 0.40 | 47 |
| DEAE-cellulose | 45 | 13 | 80 | 168 | 6.2 | 30 |
| Ammonium sulfate, 40–60% precipitate | 5 | 5.0 | 52 | 281 | 10.4 | 20 |

[a] Protein was measured by absorbance at 280 and 260 nm [O. Warburg and W. Christian, *Biochem. Z.* **310**, 384 (1941)].

*Kinetic Constants.* Measurements of initial velocity in the presence or in the absence of reaction products have been used to propose an ordered reaction mechanism for galactokinase[4] in which galactose combines with an initial enzyme–MgATP$^{2-}$ complex followed by the release of galactose 1-phosphate and then MgADP$^-$. Michaelis constants at pH 7.8 are galactose, 0.58 m$M$, and MgATP$^{2-}$, 0.17 m$M$. Substrate inhibition occurs with galactose concentrations above 1 m$M$ while free Mg$^{2+}$ ion is a competitive inhibitor of MgATP$^{2-}$ with a $K_i$ of 25 m$M$. Of the reaction products, ADP$^{2-}$ (or MgADP$^-$) is a competitive inhibitor of MgATP$^{2-}$ with a $K_i$ of 0.2 m$M$, and a noncompetitive inhibitor of galactose, while galactose 1-phosphate is a noncompetitive inhibitor with respect to either galactose or MgATP$^{2-}$.

*Other Properties.* Galactokinase has a pH optimum between 7.5 and 8.0 in Tris or triethanolamine buffers.[4] Enzyme activity is stimulated 2- to 4-fold by 5 m$M$ cysteine, glutathione, $\beta$-mercaptoethanol, or DTT and is inhibited by $p$-chloromercuribenzoate.

## [8] Galactokinase from Human Erythrocytes

*By* KARL-GEORG BLUME and ERNEST BEUTLER

$$\alpha\text{-D-Galactose} + \text{ATP} \rightarrow \alpha\text{-D-galactose 1-phosphate} + \text{ADP}$$

As in other tissues from other mammals and in microorganisms, human red cell galactokinase (ATP:D-galactose-1-phosphotransferase, EC 2.7.1.6) catalyzes the first step in the conversion of galactose to glucose: the phosphorylation by ATP of $\alpha$-D-galactose to $\alpha$-D-galactose 1-phosphate. In human tissues the galactokinase reaction is unidirectional toward galactose phosphorylation.

### Clinical Significance

The activity of the enzyme of normal human red cells is high at birth and decreases so that by the age of approximately 4 years it has reached the adult value, approximately one-third of that observed in the newborn.[1] Kinetic alteration of human galactokinase during maturation have been described,[2,3] and a fetal enzyme has been separated chromatographically from adult enzyme.[3] Induction of the enzyme in cultured human

---

[1] W. G. Ng, G. N. Donnell, and W. R. Bergren, *J. Lab. Clin. Med.* **66**, 115 (1965).
[2] C. K. Mathai and E. Beutler, *Enzymologia* **33**, 224 (1967).
[3] S. K. Srivastava, K. G. Blume, C. van Loon, and E. Beutler, *Arch. Biochem. Biophys.* **150**, 191 (1972).

amniotic cells[4] and fibroblasts[5] has been achieved by the addition of galactose to the growth medium. However, galactose intake does not influence the level of enzyme in human red cells.[6] Blood samples containing a high percentage of reticulocytes have somewhat higher levels of enzyme than the blood of normal subjects.[6]

Hereditary deficiency of galactokinase was first discovered in 1965.[7] So far, almost 20 patients with this inborn error of metabolism have since been identified. The disease is transmitted as an autosomal recessive disorder: homozygotes or doubly mutant individuals have a marked galactokinase deficiency, whereas heterozygotes are partially deficient. The gene frequency for galactokinase deficiency is not accurately known, but a survey suggested that as many as 1% of the population studied might be heterozygous.[8] The disorder is characterized by galactosemia, galactosuria, and cataract formation in early childhood. It has been suggested that heterozygotes for galactokinase deficiency may also be at risk for cataract formation, but clear-cut evidence for this has not been obtained.[9] A 30% decrease in the average enzyme activity has been found in Negro populations.[10] This difference in galactokinase activity implies the existence of racially determined enzyme polymorphism.

The diagnosis of galactokinase deficiency can be established by measuring the enzyme activity in hemolysates[7] or lysates from cultured human amniotic cells[4] or fibroblasts.[5] Since normal human galactokinase exhibits rather low specific activity spectrophotometric methods are not satisfactory. However, the use of radioactively labeled galactose permits accurate measurement of galactokinase in human tissue extracts as well as in hemolysates.[6]

## Assay Method

*Principle.* The procedure is a modification[6] of a method which has been described for the determination of galactokinase from *Escherichia coli.*[11] It depends upon the incubation of [$^{14}$C]galactose with ATP and the enzyme. The reaction is stopped by the addition of a large excess of nonradioactive galactose. The product, [$^{14}$C]galactose 1-phosphate, is

[4] P. F. Benson, S. Blunt, and S. P. Brown, *Lancet* 1, 106 (1973).
[5] F. Zacchello, P. F. Benson, S. Brown, P. Croll, F. Gianelli, and T. P. Mann, *Nature (London) New Biol.* 239, 95 (1972).
[6] E. Beutler, N. V. Paniker, and F. Trinidad, *Biochem. Med.* 5, 325 (1971).
[7] R. Gitzelmann, *Lancet* 2, 670 (1965).
[8] J. S. Mayes and R. Guthrie, *Biochem. Genet.* 2, 219 (1968).
[9] E. Beutler, F. Matsumoto, W. Kuhl, A. Krill, N. Levy, R. Sparkes, and M. Degnan, *N. Engl. J. Med.* 288, 1203 (1973).
[10] T. A. Tedesco, R. Bonow, K. Miller, and W. J. Mellman, *Science* 178, 176 (1972).
[11] J. R. Sherman and J. Adler, *J. Biol. Chem.* 238, 873 (1963).

measured as the radioactivity remaining on ion exchange paper after elution of nonphosphorylated galactose from the paper with water.

*Preparation of Substrate.* Since many commercially available preparations of [$^{14}$C]galactose are contaminated with labeled hexoses other than galactose that would cause errors, it is necessary to purify the substrate for galactokinase reaction. The contaminants are phosphorylated by hexokinase and ATP and are then removed on DEAE-cellulose. A small column is prepared by placing a small quantity of glass wool into the outlet of a 1-ml syringe. Microgranular DEAE (Whatman DE 52) slurry to give a volume of 0.2–0.3 ml is placed into the barrel. To decrease nonspecific absorption of galactose the column is washed with 0.5 ml of 3.8 m$M$ galactose solution followed by 5 ml of water. [1-$^{14}$C]Galactose or [U-$^{14}$C]galactose is diluted to contain approximately 10 $\mu$Ci/ml. Two microliters of 1 $M$ Tris · HCl buffer pH 8.0, 5 $\mu$l of 0.1 $M$ MgCl, 100 $\mu$l of hexokinase diluted in water to contain 2 units/ml, and 10 $\mu$l of 60 m$M$ neutralized ATP are added to each milliliter of the [1-$^{14}$C]galactose solution. After 60 min at 37°, the reaction mixture is passed through the column, which is then washed with sufficient additional water to give five times the original volume. The radioactive galactose solution which is eluted from the column contains approximately 2 $\mu$Ci/ml and is stable when frozen.

*Preparation of Reaction Mixture.* A partial reaction mixture is prepared by mixing the following reagents in the proportions shown: 1 $M$ Tris · HCl, pH 7.4, 400 $\mu$l; 100 m$M$ NaF, 100 $\mu$l; 100 m$M$ MgCl$_2$, 100 $\mu$l; 7.6 m$M$ galactose, 100 $\mu$l; 2 $\mu$Ci/ml [1-$^{14}$C]galactose, 200 $\mu$l; 120 m$M$ ATP, 100 $\mu$l.

The mixture is incubated at 37° for 4 hr and then stored in the frozen state. It is particularly important that freshly prepared galactose solutions not be used in the assay. An equilibrium between $\alpha$- and $\beta$-form of galactose which is the result of mutarotation requires incubation for 4 hr at 37° or boiling for 10 min.

*Preparation of Hemolysates.* Blood samples drawn into ACD,[12] heparin or EDTA are washed two times in 0.9% saline solution and then hemolysed by the addition of 4 volumes of stabilizing solution containing 7 m$M$ $\beta$-mercaptoethanol and 2.7 m$M$ neutralized EDTA, to 1 volume of packed red cells. Stroma is removed by centrifugation at 5000 $g$ and 4° for 15 min. The hemoglobin content of the supernatant is measured, and galactokinase activity is determined. Uncentrifuged hemolysates or even whole blood with a modified reaction mixture[13] may be used with

---

[12] Acid-citrate-dextrose solution containing 7.3 g of citric acid, 22 g of sodium citrate · 2H$_2$O and 24.5 g of glucose per liter.

[13] E. Beutler and F. Matsumoto, *J. Lab. Clin. Med.* **82**, 818 (1973).

the same results.[6] Leukocyte contamination of red cell preparations does not significantly influence the result.[1]

*Assay Procedure.* Hemolysate, 100 μl, is mixed with 100 μl of the partial reaction mixture described above. Two 50-μl aliquots of the reaction mixture are removed immediately, and each is mixed on a spot plate or in a small test tube with 20 μl of 1 *M* galactose. Then 50 μl of each mixture are spotted on a circle of DEAE paper (Whatman DE-81) 15–20 mm in diameter. One of the spotted papers is dropped immediately into a small cup of water (0 min sample), while the other is permitted to dry (100% sample). The remainder of the reaction mixture is now incubated for 1 hr at 37°, and another 50-μl aliquot (60-min sample) is treated in a fashion identical with the 0-min sample. The 0-min sample and the 60-min sample are placed, while wet, on a sintered-glass funnel

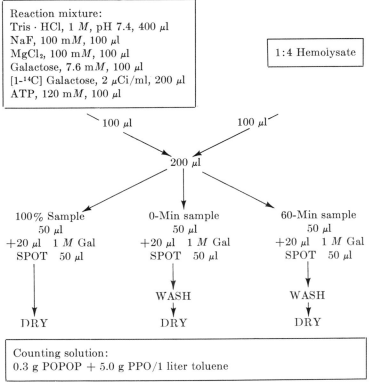

COUNT AND CALCULATE

Fig. 1. The determination of erythrocyte galactokinase using [¹⁴C]galactose: Schematic summary [E. Beutler, N. V. Paniker, and F. Trinidad, *Biochem. Med.* **5,** 325 (1971)].

and washed with 600–800 ml of distilled water. The samples are permitted to dry and are ready for counting; this is accomplished in a scintillation solution consisting of toluene containing 0.03% $p$-bis-2-(5-phenyloxazolyl)benzene (POPOP) and 0.5% 2,5-diphenyloxazole (PPO). The procedure is summarized in Fig. 1.

*Linearity.* When various dilutions of normal hemolysates are made the percentage of galactose phosphorylated increases in an approximately linear fashion.[6] When fairly concentrated hemolysates are used, a considerable degree of quenching is observed: there is approximately 55% quenching with a 50% hemolysate, 20% quenching with a 25% hemolysate, and little or no quenching with a 12% hemolysate. Drying samples at too high temperature produces an increase in the amount of quenching because of the change in color of hemoglobin under these conditions. The phosphorylation of galactose is linear with time over a 2.5-hr period.[6]

*Definition of Unit and Specific Activity.* One unit of galactokinase is defined as the amount of enzyme which phosphorylates 1 $\mu$mole of galactose per minute under the conditions of assay. The red cells of normal human adults have been found to contain $0.0297 \pm 0.0049$ unit of galactokinase per gram of hemoglobin (mean ± standard deviation). Some differences in apparent activity are observed with different batches of [$^{14}$C]galactose, probably because of varying degrees of contamination with the keto sugar (tagatose) or the 2-epimer (talose).

## Purification Procedure[14]

All procedures are carried out at 4°. All centrifugations are performed for 20 min at 37,500 $g$. The following buffers are used during the purification procedure. Buffer A: 10 m$M$ $KH_2PO_4$–$K_2HPO_4$, 7 m$M$ $\beta$-mercaptoethanol, 0.5 m$M$ $Na_2$-EDTA, final pH 7.2; buffer B: 100 m$M$ $KH_2PO_4$–$K_2HPO_4$, 7 m$M$ $\beta$-mercaptoethanol, 0.5 m$M$ Na–EDTA, final pH 7.2; buffer C: 10 m$M$ $KH_2HPO$/$K_2HPO_4$, 7 m$M$ $\beta$-mercaptoethanol, 0.5 m$M$ $Na_2$-EDTA, final pH 6.7.

*Preparation of Hemolysate.* One-hundred milliliters of washed red cells are mixed with 400 ml of buffer A and frozen and thawed twice. After centrifugation the sediment is discarded.

*DEAE-Sephadex Chromatography.* The hemolyzate is applied to a DEAE-Sephadex A-50 column (2.5 × 35 cm) equilibrated with buffer A. Hemoglobin is eluted with 3.5 liters of buffer A. Galactokinase is then eluted by descending chromatography using a 500-ml linear gradient between buffer A and buffer B at a flow rate of 30 ml/hr. Fractions containing galactokinase are pooled.

[14] K. G. Blume and E. Beutler, *J. Biol. Chem.* **246**, 6507 (1971).

PURIFICATION PROCEDURE FOR HUMAN RED CELL GALACTOKINASE

| Fraction | Volume (ml) | Protein (mg/ml) | Specific activity (mU/mg) | Purification (fold) |
|---|---|---|---|---|
| Hemolysate | 500 | 47.5 | 0.021 | 1 |
| DEAE-Sephadex A 50 eluate | 66 | 1.42 | 4.30 | 204 |
| Ammonium sulfate, 35–60% precipitate | 1.8 | 19.2 | 9.29 | 440 |
| Sephadex G-75 filtrate | 16 | 0.41 | 28.2 | 1336 |
| CM-Sephadex C-50 eluate | 18 | 0.06 | 81.2 | 3850 |

*Ammonium Sulfate.* The DEAE-Sephadex A-50 eluate is fractionated with solid $(NH_4)_2SO_4$. The precipitate which forms at 35% saturation (20.9 g/100 ml) is discarded. Galactokinase is precipitated between 35 and 60% (16.4 g/100 ml) ammonium sulfate saturation and is redissolved in a small amount of buffer A.

*Sephadex G-75 Filtration.* The ammonium sulfate 35 to 60% precipitate is then chromatographed through a Sephadex G-75 column (1.5 × 90 cm), equilibrated with buffer A using ascending chromatography with a flow rate of 8.0 ml/hour. Fractions containing galactokinase are combined.

*CM-Sephadex Chromatography.* The Sephadex G-75 filtrate is passed through a CM-Sephadex C-50 column (1.5 × 30 cm), equilibrated with buffer C, with descending chromatography at a flow rate of 4.2 ml/hr. Whereas contaminating proteins were absorbed, galactokinase is not bound to the cation exchanger. Fractions containing galactokinase are pooled and concentrated by ultrafiltration or by ammonium sulfate precipitation at 60% saturation.

The cumulative purification ranges between 3300 and 3900-fold, the yield between 14 and 19%. The table summarizes a typical enzyme preparation including all steps described. Purification of fetal galactokinase from human placenta can be achieved by means of a similar method.[3]

## Properties

*Purity.* By analytical disc electrophoresis in polyacrylamide (20 and 50 $\mu$g protein per gel column) a single protein band is found with the purified enzyme when stained with Coomassie blue. Of the enzymes of red cell glycolysis, hexose monophosphate shunt, glutathione and galactose metabolism only traces of triosephosphate isomerase activity (representing less than 0.1% of the specific activity of triosephosphate isomerase in crude hemolysates) are present in the final preparation.

*Stability.* Intact red cells lose less than 5% of galactokinase activity when stored in acid-citrate-dextrose at 4° for 3 weeks.[6] When glucose to give a final added concentration of about 20 mM is added to blood in EDTA solution, the enzyme is stable for at least 20 days at 4°, 6 days at 25°, and 4 days at 30°. In hemolysates the enzyme is unstable, the activity decreases about 30% in 24 hr. After purification the enzyme preparations lose about 10% of activity within 6 days when stored at 4° in 3.0 M ammonium sulfate.

*Other Properties.* The $K_m$ of galactokinase preparations from red cells for galactose is $126 + 11$ $\mu M$ (mean $\pm$ standard deviation), for ATP $342 + 63$ $\mu M$. Identical results have been obtained for galactokinase from crude hemoylsates from adult red cells. $K_m$ values of galactokinase from cord blood red cell hemolysates and after partial purification are significantly higher for galactose: $225 \pm 8$ $\mu M$.[2] Similar results have been obtained with extracts and purified preparations of galactokinase from human placenta,[3] indicating different kinetic properties of human fetal and adult galactokinase. Both forms of the enzyme can be separated by ion exchange chromatography.[3] The pH optimum for the enzymic phosphorylation of galactose is pH 7.6 in hemolysates and pH 7.8 after purification. The molecular weight of galactokinase determined by Sephadex G-75 and Sephadex G-200 chromatography is 55,000 in crude extracts and after purification. By sodium dodecyl sulfate–polyacrylamide electrophoresis of 3600- to 3850-fold purified preparations and staining with Coomassie blue the molecular weight of the assumed monomer of red cell galactokinase is found to be 26,000. The isoelectric point of the enzyme determined by isoelectrofocusing in a pH 5–8 stabilized sucrose gradient is pH 5.7. The ultraviolet spectrum of purified galactokinase exhibits a minimum at 251 nm and a maximum at 280 nm.

# [9] N-Acetylmannosamine Kinase

By Santimoy Banerjee and Sudhamoy Ghosh

$$N\text{-Acetylmannosamine} + ATP \rightarrow N\text{-acetylmannosamine-6-P} + ADP$$

N-Acetylmannosamine kinase has been previously purified from rat liver.[1,2] The enzyme does not appear to be synthesized in bacteria if they

[1] W. Kundig, S. Ghosh, and S. Roseman, *J. Biol. Chem.* **241**, 5619 (1966).
[2] W. Kundig and S. Roseman, this series, Vol. 8 [28].

are grown in the presence of glucose, but a number of bacteria are capable of synthesizing the enzyme when glycerol replaces glucose in their growth medium.[3] A simple procedure for the purification of N-acetylmannosamine kinase from glycerol-grown *Salmonella typhimurium* is described below.

### Assay Method

*Principle.* ADP produced by the action of N-acetylmannosamine kinase is determined by the procedure of Kornberg and Pricer[4] which utilizes a coupled reaction sequence catalyzed by pyruvate kinase and lactate dehydrogenase as shown below:

$$ADP + \text{P-enolpyruvate} \rightarrow ATP + \text{pyruvate}$$
$$\text{Pyruvate} + NADH + H^+ \rightarrow \text{lactate} + NAD^+$$

Oxidation of NADH to NAD$^+$ is measured spectrophotometrically by the absorbance decrease at 340 nm. The assay is carried out in two steps. In the first step, the kinase catalyzes the formation of ADP for a certain period of time, after which the reaction is stopped by heating. In the second step, an aliquot of the first reaction mixture is assayed for ADP by the coupled pyruvate kinase–lactate dehydrogenase NADH–oxidation system. This assay is not suitable for determining the activity of N-acetylmannosamine kinase in crude fractions which may contain an interfering amount of ATPase.

An alternate method for the assay of kinase in crude extracts is based on the principle of removing the phosphorylated sugar, after the reaction is over, by the addition of $ZnSO_4$ and $Ba(OH)_2$ solutions and measuring the free sugar left in the supernatant fluid. N-Acetylmannosamine can be estimated by the Morgan-Elson color reaction with some modifications.[5]

*Reagents*

N-Acetylmannosamine, 40 m$M$
ATP, 50 m$M$
MgCl$_2$, 0.1 $M$
Tris·HCl buffer, pH 7.6, 0.5 $M$
Potassium phosphate buffer, pH 7.6, 0.2 $M$
P-enolpyruvate, 25 m$M$
NADH, 10 m$M$

---

[3] S. Banerjee and S. Ghosh, *Eur. J. Biochem.* **8**, 200 (1969).
[4] A. Kornberg and W. E. Pricer, Jr., *J. Biol. Chem.* **193**, 481 (1951).
[5] J. L. Reissig, J. L. Strominger, and L. F. Leboir, *J. Biol. Chem.* **217**, 959 (1955).

Lactic dehydrogenase, crystalline (commercial), 0.1 mg/ml. It contains normally enough pyruvate kinase for satisfactory assay.

$ZnSO_4$, 0.38 $M$

$Ba(OH)_2$, saturated solution

Sodium tetraborate, saturated solution. pH adjusted to 8.8.

*p*-Dimethylaminobenzaldehyde. A stock solution is prepared by dissolving 10 g of *p*-dimethylaminobenzaldehyde in a mixture of 90 ml of glacial acetic acid and 10 ml of 10 $N$ hydrochloric acid. This is stored in a refrigerator. Before use, the stock solution is diluted 10-fold with glacial acetic acid. Fresh stock solution should be made every month.

*Procedure.* The reaction mixture (first step) contains 0.025 ml of *N*-acetylmannosamine, 0.02 ml of ATP, 0.01 ml of $MgCl_2$, 0.025 ml of Tris·HCl buffer and the enzyme fraction in a total volume of 0.2 ml. This is incubated at 37° for 30 min. The reaction is stopped either by heating at 100° for 1 min, or (for the alternate assay) by adding 1 ml of $ZnSO_4$ solution and an equivalent amount of saturated $Ba(OH)_2$ solution (approximately 1 ml required); exact amount to be added should be found out by a prior titration using phenolphthalein as indicator.

For the spectrophotometric assay of ADP (second step), the heat-killed reaction mixture is diluted to 1 ml with $H_2O$ and an aliquot of 0.1 ml from this is added to a mixture containing potassium phosphate buffer (0.1 ml), $MgCl_2$ (0.02 ml), P-enolpyruvate (0.01 ml), NADH (0.01 ml), and water to make a final volume of 1 ml in a cuvette (1-cm light path). Initial absorbance at 340 nm is noted, and 0.01 ml of lactate dehydrogenase solution is added. The absorbancy decreases and attains a constant value, usually within a minute. Reaction mixtures (first step) containing heated enzyme or no *N*-acetylmannosamine are used as control for the second-step ADP assay. In the alternate method, the precipitate formed owing to the addition of $ZnSO_4$ and $Ba(OH)_2$ solutions is centrifuged off and the amount of residual sugar in the supernatant is estimated as follows. To an aliquot of the supernatant solution (0.4 ml) containing *N*-acetylmannosamine (0.02–0.1 $\mu$mole), 0.2 ml of sodium tetraborate solution is added and heated for 10 min at 100°. After cooling, 2.5 ml of diluted *p*-dimethylaminobenzaldehyde reagent is added, incubated at 37° for 10 min, and cooled immediately in an ice-bath. Absorbance at 585 nm is determined and compared to a standard curve prepared with pure *N*-acetylmannosamine. Difference in concentration between the control (without ATP or enzyme) and the experimental reaction tube is taken as the amount of *N*-acetylmannosamine phosphorylated.

*Definition of Unit and Specific Activity.* One unit of enzyme is defined as that amount which phosphorylates 1 $\mu$mole of $N$-acetylmannosamine per minute. Specific activity is units per milligram of protein.

## Purification Procedure

The procedure described has been carried out in our laboratory several times with reproducible results. Unless otherwise stated, all operations are carried out between 0° and 4°. Phosphate buffers used in the fractionation of the enzyme contain 1m$M$ EDTA and 0.1% 2-mercaptoethanol.

*Growth of Cells.* Salmonella typhimurium LT-2 is grown in a medium containing the following (per liter): $K_2HPO_4$, 10.5 g; $KH_2PO_4$, 4.5 g; sodium citrate, 0.5 g; $MgSO_4 \cdot 7H_2O$, 0.1 g; $(NH_4)_2SO_4$, 1.0 g; and glycerol, 4.0 g. pH of the medium is adjusted to 7.2. Cells are grown at 37° with aeration up to the stationary phase, harvested, washed with 1% KCl solution, and stored frozen until used. Yield: 4–5 g of cells (wet weight) per liter.

*Preparation of Crude Extract.* Frozen cells of *S. typhimurium* (20 g) are ground with fine glass powder (30 g) in a mortar and pestle until it becomes highly viscous; 60 ml of 20 m$M$ potassium phosphate buffer, pH 7.6, are added slowly to the viscous mass with thorough mixing. After centrifuging the mixture at 14,000 $g$ for 30 min, the supernatant fluid is retained. The residue is again extracted with 30 ml of the same buffer and centrifuged at 14,000 $g$ for 30 min. The two supernatant fractions are combined (crude extract, 80 ml).

*Protamine Sulfate Treatment.* Ten milliliters of 2% protamine sulfate (w/v) are slowly added to 80 ml of the crude extract with constant stirring. After 10 min the mixture is centrifuged. The precipitated impurities are discarded and the supernatant fraction (protamine supernatant, 85 ml) is retained for further treatments.

*Heat Treatment.* The protamine sulfate fraction is brought to 25°, and its pH is adjusted to 5.5 with 1 $M$ acetic acid with the help of a pH meter. The solution is taken in a thin-walled 50-ml conical flask about 20 ml at a time and heated for a total period of 2.5 min in a 70° water bath with gentle swirling. It takes about 1.75 min to reach 68° by this method. The conical flask is immediately cooled to 0° in ice after the heat treatment. The heat-treated enzyme solution is centrifuged at 14,000 $g$ for 10 min, and the supernatant fraction is retained (heated enzyme).

*Ammonium Sulfate Treatment.* Solid $(NH_4)_2SO_4$ (20.9 g) is added to the heat-treated enzyme fraction (81 ml) with constant stirring (45% saturation). After 10 min in ice, inactive protein is removed by centrifugation. The supernatant fraction is made 65% saturated with the addition

PURIFICATION OF N-ACETYLMANNOSAMINE KINASE

| Fraction | Total protein (mg) | Total activity (units) | Specific activity (units/ mg × 10) | Purification (fold) |
|---|---|---|---|---|
| Crude extract[a] | 460 | 3.50 | 0.077 | 1.0 |
| Protamine Supernatant | 289 | 3.55 | 0.123 | 1.6 |
| Heated enzyme | 106 | 3.55 | 0.333 | 4.4 |
| $(NH_4)_2SO_4$ | 6.25 | 2.10 | 3.33 | 44 |
| DEAE-cellulose Eluate | 0.30 | 0.58 | 19.3 | 258 |

[a] Enzyme activity is determined by the alternate $ZnSO_4$–$Ba(OH)_2$ assay procedure.

of solid $(NH_4)_2SO_4$ (11.4 g). After 10 min in ice, the precipitate is collected by centrifugation and dissolved in 12 ml of 5 m$M$ phosphate buffer, pH 7.6. This is dialyzed for 16 hr against 1 liter of the same buffer (ammonium sulfate fraction, 12.5 ml).

*DEAE-Cellulose Chromatography.* The dialyzed ammonium sulfate fraction (10 ml) is passed through a DEAE-cellulose column (1.2 × 6.4 cm), previously equilibrated with 20 m$M$ KCl–10 m$M$ potassium phosphate buffer, pH 7.6. The active enzyme is then fractionated by subjecting the column to a linear gradient between 125 ml of 50 m$M$ KCl–10 m$M$ potassium phosphate buffer, pH 7.6 and 125 ml of 0.2 $M$ KCl–10 m$M$ potassium phosphate buffer, pH 7.6. The eluate is collected in 5-ml fractions. The active fractions (between 55 ml and 130 ml) are pooled, dialyzed against 5 m$M$ potassium phosphate buffer, pH 7.6, for 5 hr and again adsorbed on a small DEAE-cellulose column (5 ml packed volume). The enzyme is eluted with 8.5 ml of 0.25 $M$ KCl–20 m$M$ potassium phosphate buffer, pH 7.6. This concentrated enzyme fraction is designated as DEAE-cellulose eluate.

A summary of a typical purification is shown in the table.

## Properties

*Specificity.* The DEAE-cellulose eluate fraction is highly specific for N-acetylmannosamine. The following sugars are found to be inactive as substrate: glucose, mannose, fructose, glucosamine, N-acetylglucosamine, and N-acetylgalactosamine. Also, ATP could not be replaced by CTP, UTP, or ITP as phosphoryl donor.

*Kinetic Constants.* The $K_m$ values for N-acetylmannosamine and ATP are 0.75 m$M$ and 1.15 m$M$, respectively. The enzyme has a broad pH optimum between 6.2 and 8.0.

*Stability.* The DEAE-cellulose eluate fraction loses about 30% of its activity in 7 days during storage at 4°. The enzyme is stable to dialysis.

*Metal Ion Requirement.* A divalent metal ion is essential for the enzyme activity. The relative efficiency of the following cations are: $Mg^{2+}$, 100; $Mn^{2+}$, 50; $Co^{2+}$, 30; and $Ca^{2+}$, 0.0.

*Distribution of the Enzyme.* The enzyme is detected in *Escherichia coli, Staphylococcus aureus,* and *S. typhimurium* when the bacteria are grown with glycerol or peptone and beef extract as carbon source. No enzyme is found in the extracts when these bacteria are allowed to grow in the presence of glucose.

# [10] *N*-Acetylglucosamine Kinase from Hog Spleen

## By Asis Datta

ATP + *N*-acetylglucosamine → ADP + *N*-acetylglucosamine 6-phosphate + H$^+$

A kinase from hog spleen which catalyzes ATP-dependent phosphorylation of *N*-acetylglucosamine has been purified 3500-fold.[1] Kinetic studies indicate that *N*-acetylglucosamine kinase is a regulatory enzyme, and *N*-acetylglucosamine 6-phosphate, the product of the reaction and UDP-*N*-acetylglucosamine, the end product of *N*-acetylglucosamine metabolism both act on the enzyme as feedback inhibitor.[2]

### Assay Method

*Principle.* *N*-Acetylglucosamine kinase can be assayed by two methods. Method A involves the specific removal of phosphorylating sugar by the addition of $ZnSO_4$ and $Ba(OH)_2$ solutions after enzyme incubation and the disappearance of free sugar from the assay mixture during enzyme reaction.[3] This procedure is used throughout the enzyme purification. Method B involves the estimation of the amount of ADP formed in the kinase reaction by coupling with a system of phosphoenolpyruvate, pyruvate kinase, lactic dehydrogenase, and NADH and following the decrease in optical density at 340 nm. Method B cannot be used to assay crude extracts because of the presence of interfering enzymes.

---

[1] A. Datta, *Biochim. Biophys. Acta* **220**, 51 (1970).
[2] A. Datta, *Arch. Biochem. Biophys.* **142**, 645 (1971).
[3] J. L. Reissig, J. L. Strominger, and L. F. Leloir, *J. Biol. Chem.* **217**, 959 (1955).

### Reagents

METHOD A: Glycine–NaOH buffer, 0.5 $M$, pH 9.0

ATP, 0.1 $M$; the solution is adjusted to pH 7.6 with $NaHCO_3$

$MgCl_2$, 0.1 $M$

$N$-Acetyl-D-glucosamine (NAGA), 0.1 $M$

$ZnSO_4$, 5%

$Ba(OH)_2$, 0.15 $N$

Reagents for the determination of $N$-acetylhexosamines[3]

METHOD B: In addition to substrates and buffer as described above, the following reagents are required for the estimation of ADP:

NADH, 10 m$M$

Phosphoenol pyruvate, 25 m$M$

Potassium phosphate buffer, 50 m$M$, pH 7.6

$MgCl_2$, 0.1 $M$

Potassium fluoride, 0.1 $M$

Lactic dehydrogenase, containing pyruvate kinase

### Procedure

ASSAY A. Mixtures of the following in final volumes of 0.2 ml are used for assay: NAGA, 10 μl; ATP, 10 μl; $MgCl_2$, 25 μl; glycine-NaOH buffer, pH 9.0, 25 μl; enzyme fraction; and distilled water. The reaction mixtures are incubated for 10 min at 37°. After incubation, the reaction is stopped by adding 1.0 ml of $ZnSO_4$, followed by the addition of an equivalent amount of 0.15 $N$ $Ba(OH)_2$ solution. After mixing thoroughly, it is centrifuged at low velocity for few minutes and 0.2 ml of the supernatant is taken for the estimation of residual NAGA. Standards, boiled enzyme controls, and complete but unincubated mixtures are also assayed in the same way. For color reading, any conventional spectrophotometer or photocolorimeter, at 585 nm, can be used.

ASSAY B. This procedure is used with the purified enzyme and involves the estimation of ADP formed in the kinase reaction. The incubation mixture contains the components described in enzyme assay A. After incubation for 10 min at 37°, the reaction is terminated by heating at 100° for 2 min. Aliquots (0.1 ml) of these reaction mixtures are added to a mixture containing 0.5 ml potassium phosphate buffer, pH 7.6, 0.2 ml potassium fluoride, 0.02 ml $MgCl_2$, 0.01 ml PEP, and 0.01 ml of NADH in a final volume of 1 ml in a Beckman cuvette. After the initial determination of absorbancy at 340 nm, 0.01 ml of lactic dehydrogenase containing pyruvate kinase is added; the absorbancy at 340 nm is followed until it attains a constant value (within 2 min). Control tubes contain heat-denatured enzyme, or lack $N$-acetylglucosamine.

*Definition of Unit and Specific Activity.* One unit of enzyme is defined as the quantity that catalyzes the phosphorylation of 1 μmole of NAGA per minute, under the assay conditions described above. Specific activity (units per milligram of protein) is based on a spectrophotometric determination of protein.[4]

## Purification Procedure

Unless otherwise indicated, all operations are conducted between 0° and 4°, and all buffers contain 1 μmole of EDTA and 10 μmoles of 2-mercaptoethanol per milliliter.

*Step 1. Crude Extract.* Hog spleen, collected at the abattoir is immediately chilled in ice. The organs can be stored in ice for as long as 5 days or in the deepfreeze at −18° for several months, before kinase extraction, without much loss of activity. The spleen (50 g) is homogenized in a Waring Blendor with 100 ml of 30 m$M$ potassium phosphate buffer (pH 7.6) for two periods of 30 sec each. The mixture is centrifuged at 18,000 $g$ for 30 min, and the supernatant is collected.

*Step 2. Protamine Sulfate.* The kinase is precipitated from 100 ml of the collected supernatant by the addition of 15 ml of a 2% protamine sulfate solution (dissolved at 37°) over 10 min with continuous but gentle stirring. The protamine precipitate is collected by centrifugation at 18,000 $g$ for 10 min and washed once with 100 ml and again with 50 ml of 0.1 $M$ Tris. The enzyme is then extracted from the residue by stirring with increasing concentrations of potassium phosphate buffer (pH 7.6). The first few extractions, each consisting of 50 ml of 20, 40, and 50 m$M$ potassium phosphate buffer (pH 7.6) are rejected. Active enzyme is obtained by four further extractions with 50 ml each of 0.075, 0.075, 0.075, and 0.1 $M$ potassium phosphate buffer (pH 7.6).

*Step 3. Ammonium Sulfate.* Ammonium sulfate is recrystallized from water containing 0.1 $M$ EDTA; a saturated solution of the crystals in water is adjusted with NaOH until a 1:4 dilution in $H_2O$ is at pH 7.6 when measured with a glass electrode. To the protamine extract (200 ml), 28.8 g of solid ammonium sulfate are added; the preparation is allowed to stand for 10 min, then insoluble proteins are removed by centrifugation for 10 min at 16,000 $g$. The ammonium sulfate concentration of the supernatant is increased to 50% saturation by the addition of solid ammonium sulfate (15.8 g of ammonium sulfate is required per 100 ml). The precipitate, collected by centrifugation at 16,000 $g$ for 10 min, is then suspended in 5 ml of 20 m$M$ potassium phosphate buffer (pH 7.0) ;

---

[4] O. H. Lowry, N. J. Rosebrough, A. L. Farr, and R. J. Randall, *J. Biol. Chem.* **193**, 265 (1951).

to the suspension, saturated ammonium sulfate solution is added slowly until a final saturation of 25% is reached. After stirring for 10 min, insoluble proteins are removed by centrifugation for 10 min at 16,000 $g$. Enough ammonium sulfate is then added with constant stirring to the resulting supernatant to give 50% saturation (1.58 g is required per 10 ml). After 10 min of standing, the resulting precipitate is collected by centrifugation and dissolved in 30 ml of 20 m$M$ potassium phosphate buffer (pH 7.0).

*Step 4. Charcoal Treatment.* The solution obtained above (30 ml) is treated with 65 mg of acid-washed charcoal (protein to charcoal ratio 1:3). After 10 min, the supernatant is collected by centrifugation and dialyzed against 5 m$M$ potassium phosphate buffer (pH 7.6) for 24 hr.

*Step 5. Calcium Phosphate Gel.* Calcium phosphate gel (9 ml; dry weight 16.2 mg/ml) is sedimented by centrifugation. The sediment is resuspended in 30 ml of dialyzed charcoal fraction (protein to gel ratio 1:8) by gentle stirring with a glass rod. After 10 min, the residue is collected by centrifugation. After preliminary washing with water (30 ml), the active enzyme is eluted from the residue first with 30 ml of 50 m$M$, then with 15 ml of 50 m$M$ and finally with 15 ml of 75 m$M$ potassium phosphate buffer (pH 7.6). These extracts are then combined, and 18.78 g of ammonium sulfate is added to bring a final saturation of 50%. The precipitate is collected by centrifugation, dissolved, and dialyzed against 50 m$M$ potassium phosphate buffer (pH 7.6).

*Step 6. DEAE-Cellulose Chromatography.* The enzyme solution obtained above (2 ml) is applied to a DEAE-cellulose column (1.5 × 7.5 cm) which has been equilibrated with 0.02 $M$ KCl–0.01 $M$ potassium phosphate buffer (pH 7.6). After a preliminary washing of the column with 25 ml of 0.1 $M$ potassium phosphate buffer (pH 7.6), the active enzyme is eluted with 15 ml of 0.2 $M$ potassium phosphate buffer (pH 7.6). The enzyme is then concentrated by ammonium sulfate as in the

PURIFICATION OF N-ACETYL-D-GLUCOSAMINE KINASE FROM HOG SPLEEN

| Fraction | Total units | Activity yield (%) | Specific activity | Purification factor |
|---|---|---|---|---|
| Crude extract | 63.8 | 100 | 0.025 | 1 |
| Protamine sulfate | 57.2 | 90 | 0.67 | 27 |
| Ammonium sulfate | 48.8 | 76 | 2.29 | 91 |
| Charcoal | 48.0 | 75 | 2.73 | 109 |
| Calcium phosphate gel | 45.2 | 71 | 15.60 | 624 |
| DEAE-cellulose | 13.4 | 21 | 90 | 3600 |

previous step and dialyzed against 50 m$M$ potassium phosphate buffer (pH 7.6).

The purification procedure is summarized in the table.

## Properties

*Stability*. Enzyme stored at 4° in 20 m$M$ potassium phosphate buffer (pH 7.6)–1 m$M$ EDTA–10 m$M$ 2-mercaptoethanol, is stable for at least 1 week. The enzyme is unstable to freezing and thawing and labile to heat and low pH. Approximately 50% of the total activity is lost on heating at 70° for 2 min.

*Specificity*. The enzyme displays a high degree of specificity for $N$-acetylglucosamine ($K_m$, 1.1 m$M$), and it does not act on a variety of other sugars, including $N$-acetylmannosamine, $N$-acetylgalactosamine, glucosamine, glucose, mannose, galactose, and fructose. The phosphoryl donor, ATP ($K_m$, 1.8 m$M$) cannot be replaced by UTP, CTP, ADP, or phosphoenol pyruvate. Only GTP has some donating capacity (30% of ATP).

A metal ion is required for activity. At 10 m$M$ concentrations of metal salts (twice the ATP concentration), the relative rates of $N$-acetylglucosamine phosphorylation are as follows: $Mg^{2+}$, 100; $Co^{2+}$, 58; $Mn^{2+}$, 47; $Ca^{2+}$, 0; and $Ba^{2+}$, 0.

*Optimum pH*. The enzyme exhibits a broad pH optimum between 8.6 and 9.4.

*Inhibitors*. The enzyme is inhibited by reagents such as $p$-chloromercuribenzoate, and this inhibition is reversed by sulfhydryl reagents. The other inhibitors are $PP_i$, $P_i$, ADP, UTP, CTP, UDP-$N$-acetylglucosamine and $N$-acetylglucosamine-6-P. Inhibition due to ADP is competitive with ATP, $K_i$ being 0.9 m$M$. $N$-Acetylglucosamine-6-P, the product of the reaction and UDP-$N$-acetylglucosamine, the end product of the pathway of $N$-acetylglucosamine metabolism, are noncompetitive inhibitors. Maximum inhibitory effect of these metabolites occurs around pH 7.5, whereas the kinase reaction is optimal within a pH range of 8.6–9.4. This may have physiological implications.

# [11] D-Fructose-1-phosphate Kinase

*By* Richard L. Anderson, Thomas E. Hanson,
and Virginia L. Sapico

D-Fructose 1-phosphate + ATP → D-fructose 1,6-diphosphate + ADP

This enzyme was discovered independently by Hanson and Anderson[1] in *Aerobacter aerogenes* and by Reeves *et al.*[2] in *Bacteroides symbiosis.* Its role in D-fructose metabolism in *A. aerogenes* has been firmly established by biochemical and genetic techniques[1,3-5]. This article describes the purification and properties of this inducible enzyme from *A. aerogenes.*[6,7]

## Assay Method

*Principle.* The continuous spectrophotometric assay is based on the following sequence of reactions[6,7]:

$$\text{D-Fructose 1-phosphate} + \text{ATP} \xrightarrow{\underset{\text{kinase}}{\text{D-fructose-1-phosphate}}} \text{D-fructose 1,6-diphosphate} + \text{ADP}$$

$$\text{D-Fructose 1,6-diphosphate} \xrightarrow{\text{aldolase}} \text{D-glyceraldehyde 3-phosphate} + \text{dihydroxyacetone phosphate}$$

$$\text{D-Glyceraldehyde 3-phosphate} \xrightarrow{\underset{\text{isomerase}}{\text{triosephosphate}}} \text{dihydroxyacetone phosphate}$$

$$\text{Dihydroxyacetone phosphate} + \text{NADH} + \text{H}^+ \xrightarrow{\underset{\text{dehydrogenase}}{\alpha\text{-glycerophosphate}}} \alpha\text{-glycerol phosphate} + \text{NAD}^+$$

With fructose -1,6-diphosphate aldolase, triosephosphate isomerase, and α-glycerophosphate dehydrogenase in excess, the rate of D-fructose 1-phosphate phosphorylation is equivalent to one-half the rate of NADH oxidation, which is measured by the absorbance decrease at 340 nm.

[1] T. E. Hanson and R. L. Anderson, *J. Biol. Chem.* 241, 1644 (1966).
[2] R. E. Reeves, L. G. Warren, and D. S. Hsu, *J. Biol. Chem.* 241, 1257 (1966).
[3] T. E. Hanson and R. L. Anderson, *Proc. Nat. Acad. Sci. U.S.* 61, 269 (1968).
[4] V. Sapico, T. E. Hanson, R. W. Walter, and R. L. Anderson, *J. Bacteriol.* 96, 51 (1968).
[5] N. E. Kelker, T. E. Hanson, and R. L. Anderson, *J. Biol. Chem.* 245, 2060 (1970).
[6] V. Sapico and R. L. Anderson, *J. Biol. Chem.* 244, 6280 (1969).
[7] V. Sapico and R. L. Anderson, *J. Biol. Chem.* 245, 3252 (1970).

*Reagents*

Glycylglycine–NaOH buffer, 0.4 $M$, pH 7.5
KCl, 1.0 $M$
MgCl$_2$, 0.2 $M$
ATP, 0.1 $M$
NADH, 0.01 $M$
D-Fructose 1-phosphate, potassium salt, 0.2 $M$
Crystalline D-fructose-1,6-diphosphate aldolase
Crystalline    $\alpha$-glycerol-phosphate    dehydrogenase–triosephosphate
    isomerase

*Procedure.* The following are added to a microcuvette with a 1.0-cm light path: 0.025 ml of buffer, 0.01 ml of KCl, 0.005 ml of MgCl$_2$, 0.005 ml of ATP, 0.005 ml of NADH, 0.005 ml of D-fructose 1-phosphate, non-limiting amounts of aldolase and $\alpha$-glycerol-phosphate dehydrogenase-triosephosphate isomerase, a rate-limiting amount of D-fructose-1-phosphate kinase, and water to a volume of 0.15 ml. The reaction is initiated by the addition of D-fructose-1-phosphate kinase. A control cuvette minus ATP measures NADH oxidase and possible D-fructose-1-phosphate reductase activities, which must be subtracted from the total rate. The rates are conveniently measured with a Gilford multiple-sample absorbance-recording spectrophotometer. The cuvette compartment should be thermostatted at 25°. Care should be taken to confirm that the rates are constant with time and proportional to the D-fructose-1-phosphate kinase concentration.

*Definition of Unit and Specific Activity.* One unit is defined as the amount of enzyme that catalyzes the phosphorylation of 1 $\mu$mole of D-fructose 1-phosphate per minute. Specific activity (units per milligram of protein) is based on a spectrophotometric determination of protein.[8]

*Alternative Assay Procedure.* For some studies, such as determining phosphoryl acceptor specificity, a different assay procedure must be used. For these purposes, a pyruvate kinase-lactate dehydrogenase-linked assay has been described,[6] but this should be modified to include KCl.[7]

## Purification Procedure[6]

*Organism and Growth Conditions.* In the preparation described here, D-fructose-1-phosphate kinase was obtained from a D-fructose-6-phosphate kinase-less mutant of a uracil auxotroph of *A. aerogenes* PRL-R3. However, the wild-type strain could be used as well. The cells were grown

---

[8] O. Warburg and W. Christian, *Biochem. Z.* **310,** 384 (1941).

PURIFICATION OF D-FRUCTOSE-1-PHOSPHATE KINASE

| Fraction | Volume (ml) | Total protein (mg) | Total activity[a] (units) | Yield (%) | Specific activity (units/mg protein) |
|---|---|---|---|---|---|
| Cell extract | 900 | 7200 | 105 | (100) | 0.15 |
| Protamine sulfate supernatant | 1050 | 8400 | 86 | 82 | 0.10 |
| $(NH_4)_2SO_4$ precipitate (30–60%) | 48 | 3464 | 103 | 99 | 0.30 |
| Sephadex G-200 | 595[b] | 680 | 79 | 76 | 1.12 |
| $Ca_3(PO_4)_2$ gel | 336 | 54 | 66 | 63 | 12.2 |
| $(NH_4)_2SO_4$ precipitate (0–70%) | 4 | 20 | 48 | 46 | 17.0 |
| pH 4.6 supernatant | 4 | 9.2 | 43.5 | 41 | 47.0 |

[a] These values were obtained before the activating effect of $K^+$ was known [V. Sapico and R. L. Anderson, *J. Biol. Chem.* **244**, 6280 (1969)]. With KCl added to the assay mix, the values would be higher [V. Sapico and R. L. Anderson, *J. Biol. Chem.* **245**, 3252 (1970)].

[b] Corrected for the portion of the previous step not subjected to further fractionation.

at 30° with aeration in a mineral medium[4] supplemented with 0.005% uracil and 0.5% D-fructose. The cells were harvested by centrifugation.

*Preparation of Cell Extracts.* Washed cells were suspended in distilled water and broken by treatment for 10 min in a Raytheon 250-W, 10-kHz sonic oscillator cooled with circulating ice water. The cell extract was the supernatant obtained after centrifugation of the broken-cell suspension for 10 min at 27,000 *g*.

*General.* The following procedures were performed at 0–4°. A summary of the purification procedure is shown in the table.

*Protamine Sulfate Precipitation.* Crystalline ammonium sulfate was added to the cell extract (8 mg of protein per milliliter) to a concentration of 0.2 *M*. Nucleic acids were then precipitated by the slow addition of 20% by volume of 2% (w/v) protamine sulfate solution (pH 7.0) and were discarded after centrifugation.

*Ammonium Sulfate Fractionation.* Crystalline ammonium sulfate was added slowly with stirring to the protamine sulfate supernatant, and the precipitate which formed at 30% saturation was removed by centrifugation and discarded. Additional ammonium sulfate was added, and the protein which precipitated between 30 and 60% saturation was collected by centrifugation and dissolved in water.

*Sephadex G-200 Chromatography.* A 16-ml sample of the above fraction was layered on a column (45 × 4.5 cm) of Sephadex G-200 equilibrated with 20 m$M$ sodium phosphate buffer, pH 7.5. The same buffer solution was used to elute 400-drop fractions. Fractions 43–52 were combined.

*Calcium Phosphate Gel Fractionation.* The combined Sephadex G-200 fraction (1.2 mg of protein per milliliter) was treated with 10% by volume of calcium phosphate gel containing 62 mg of solids per milliliter. The suspension was stirred for 5 min and centrifuged. The collected solids were washed with 20 m$M$ sodium phosphate buffer (pH 7.5) and eluted twice (10 min each time) with 75 m$M$ sodium phosphate buffer (pH 7.5). The two eluates were discarded. D-Fructose-1-phosphate kinase was then eluted from the gel with 0.15 $M$ sodium phosphate buffer (pH 7.5) for 15 min. The elution was repeated resulting in a recovery of 80% of the adsorbed D-fructose-1-phosphate kinase. The two eluates were combined.

*Ammonium Sulfate Precipitation.* The above fraction was concentrated by precipitating the protein with ammonium sulfate (70% saturation) and dissolving the collected precipitate in water.

*pH Fractionation.* The pH of the above fraction (5 mg of protein per ml) was carefully adjusted, with constant stirring, to pH 4.6 with 7.5% acetic acid. The supernatant obtained upon centrifugation was adjusted to pH 7.5 with 0.1 $M$ NaOH. D-Fructose-1-phosphate kinase activity in this preparation was about 315-fold purified over that in the cell extract and was free from adenylate kinase, mannitol-1-phosphate dehydrogenase, D-fructose-1,6-diphosphatase, and D-fructose-6-phosphate kinase.

## Properties[6,7]

*Substrate Specificity.* The enzyme appears to be specific for D-fructose 1-phosphate ($K_m = 0.3$ m$M$). Phosphorylation of the following compounds could not be detected: D-fructose, D-fructose 6-phosphate, L-fructose 1-phosphate, D-glucose 6-phosphate, D-glucose 1-phosphate, D-mannose 6-phosphate, and L-sorbose 1-phosphate.

*pH Optimum.* Activity as a function of pH is maximal at about pH 7.5.

*Metal Requirement.* Mg$^{2+}$ at a concentration at least twice that of ATP is required for maximal activity. K$^+$ at 40 m$M$ increases the $V_{max}$ about 3-fold and decreases the $K_m$ for D-fructose 1-phosphate and ATP about 3-fold.

*Estimated Molecular Weight.* The molecular weight, as determined by Sephadex G-100 chromatography, was estimated to be about 75,000.

*Stability.* Purified D-fructose-1-phosphate kinase in 0.1 $M$ (NH$_4$)$_2$SO$_4$ (pH 7.5) is stable for several months at −20°.

## [12] Phosphofructokinase from Rabbit Liver

*By* Robert G. Kemp

Fructose 6-phosphate + ATP → fructose 1,6-diphosphate + ADP + H⁺

### Assay Method

The assay for liver phosphofructokinase is based upon the coupling of fructose-1,6-diP formation to the oxidation of NADH through the use of aldolase, triosephosphate isomerase, and α-glycerophosphate dehydrogenase. Each mole of fructose diphosphate leads to the oxidation of 2 moles of NADH.

The reagents, enzyme dilution, and assay procedure exactly follows the method described in this volume for the assay of muscle phosphofructokinase.[1] When assaying crude extracts of liver, the endogenous oxidation of NADH must be determined in the absence of added fructose-6-P. The value for this rate is then used to correct the values obtained in the presence of fructose-6-P. The contaminant responsible for the endogenous oxidation of NADH is removed in the early steps of the enzyme purification procedure. One unit of phosphofructokinase is defined as that amount of enzyme that catalyzes the conversion of 1 μmole of fructose-6-P to fructose diphosphate per minute at pH 8.2 and 26° under the assay conditions.

### Purification Procedure

The following procedure is described for the purification of frozen rabbit liver and the data are typical for a preparation of 2 pounds of liver. This procedure has been repeated successfully more than 50 times. Several preparations employing this procedure have been carried out with fresh rabbit liver with essentially the same results.

*Step 1. Extraction.* In a refrigerated room, 2 pounds (910 g) of frozen livers from young (8–12 weeks) New Zealand white rabbits of mixed sex (Pel-Freez Biologicals Inc., Rogers, Arkansas) are broken into small pieces with a meat cleaver. The pieces, which must be less than 1 cm thick, are placed in a 1-gallon Waring Blendor containing 2 volumes (1820 ml) of ice-cold 30 m$M$ KF, 10 m$M$ EDTA, and 0.1 m$M$ dithiothreitol, all adjusted to pH 7.5 with NaOH. The mixture is homogenized on a medium speed setting for 45 sec. An additional 1 volume (910 ml) of

---

[1] R. G. Kemp, this volume [13].

the fluoride extraction medium is added, and the mixture is again homogenized for 45 sec. The homogenate is centrifuged for 30 min at 10,000 $g$ and 3°. The supernatant is carefully decanted through glass wool, avoiding contamination of the supernatant solution with the soft sediment. The sediment is discarded and the pH of the supernatant, which is usually between 6.6 and 6.8, is adjusted with stirring to pH 7.5 by the dropwise addition of 1 $M$ Tris base.

Step 2. Heat Treatment. The supernatant from the preceding step (2700 ml) is divided into two equal portions in 2-liter flasks. The flasks are placed in a water bath at 75°, and the contents are stirred efficiently. The temperature of the extract is allowed to rise to 59° and the temperature is maintained between 57° and 59° for 3 min by removing the flasks from the bath to allow cooling to 57° and then returning them to the bath until the temperature of the contents once again reaches 59°. An alternative procedure is to employ two baths, one of which is maintained at 58°. The extract is heated to 58° in a 75° bath and then the flasks are transferred to the 58° bath for 3 min. At the end of the heat treatment period, the flasks are placed in an ice bath and the contents are stirred rapidly until the temperature drops to 10° or lower. The suspensions are combined and centrifuged for 30 min at 10,000 $g$ and 3°. The supernatant is decanted and the pellets are resuspended in 600 ml of the homogenizing buffer. This suspension is centrifuged as above, the sediment is discarded, and this supernatant is combined with the first (total volume of 2600 ml).

Step 3. Ammonium Sulfate Precipitation. All subsequent steps are performed at 4°. To the previous fraction, solid ammonium sulfate is slowly added with stirring to a final saturation to 50% (29.1 g/100 ml initial solution). The suspension is allowed to stir overnight, and the sediment is subsequently collected by centrifugation for 30 min at 10,000 $g$. The sediment is dissolved in 300 ml of a solution at pH 8.0 containing 50 m$M$ Tris-phosphate, 2 m$M$ EDTA, 100 m$M$ ammonium sulfate, 0.1 m$M$ ATP, and 0.1 m$M$ dithiothreitol. The solution is dialyzed against this buffer for 2 hr and any insoluble material is then removed by centrifugation at 17,000 $g$ for 20 min. Solid ammonium sulfate (12.7 g/100 ml initial solution) is slowly added to the supernatant fraction to a final saturation of 26%. After all the ammonium sulfate is added, the suspension is stirred for 1 hr and subsequently centrifuged at 17,000 $g$ for 20 min. The pellet is dissolved in 120 ml of the Tris-phosphate buffer described above and then dialyzed for 18 hr against this buffer with one change of the dialysis medium. Any material that precipitates during the dialysis is removed by centrifugation.

Step 4. DEAE-Sephadex Chromatography. A column with a 2-cm

diameter is packed to a height of 37 cm with DEAE-Sephadex A-50 previously equilibrated with the Tris-phosphate buffer described in step 3. The fraction from step 3 is applied to the column at a rate of 20–30 ml per hour. After the sample has entered the gel bed, the column is washed with the Tris-phosphate buffer until the absorbance at 280 nm of the effluent drops below 0.1. The enzyme is eluted with linear a gradient of 0.1 to 1 $M$ ammonium sulfate in a pH 8.0 buffer containing 50 m$M$ Tris-phosphate, 2 m$M$ EDTA, 0.1 m$M$ ATP, 0.1 m$M$ dithiothreitol. To prevent excessive tailing of the enzyme during elution and at that same time to achieve optimum separation of the enzyme from a protein with a red pigment that elutes immediately before the phosphofructokinase, a concave elution gradient is employed. The gradient is generated by the use of 400 ml of the high salt buffer flowing by gravity from a reservoir (9.5 cm) into an open mixing chamber of a larger diameter (11 cm) that contains 600 ml of the low-salt buffer. The enzyme is eluted from the column at a flow rate of 15–25 ml/hr. The red protein elutes at about 0.14 $M$ ammonium sulfate and provides a visual marker for the phosphofructokinase that appears immediately after the red peak and is located between 0.24 and 0.52 $M$ ammonium sulfate. Ammonium sulfate concentrations may be estimated from conductivity measurements in which the column fractions may be compared with buffer containing known concentrations of ammonium sulfate. Fractions containing more than 0.4 units/ml are pooled and the enzyme is precipitated by slowly adding solid ammonium with constant stirring to a final saturation of 38%. The suspension is stirred for 1 hr and the precipitate is subsequently collected by centrifugation for 20 min at 17,000 $g$. The precipitate is dissolved in 6 ml of the Tris-phosphate buffer described in step 3.

*Step 5. Gel Exclusion Chromatography.* It is usually convenient to pool several preparations at this stage for the final purification step. That which follows describes the procedure for a single preparation and, if preparations are pooled, a proportionately larger column is employed. Solid ammonium sulfate is slowly added with constant stirring to the solution from step 4 to achieve a final saturation of 26%. After 30 min with stirring at 4°, the precipitate is collected by centrifugation and dissolved in 1.0 ml of 50 m$M$ Tris-phosphate buffer (pH 8.0) containing 2 m$M$ EDTA, 100 m$M$ ammonium sulfate, 0.1 m$M$ ATP, and 0.5 m$M$ dithiothreitol. Any insoluble material is removed by centrifugation and the supernatant is applied to a column of Bio-Gel A-1.5 $M$ (1.5 × 80 cm) that was previously equilibrated with the same Tris-phosphate buffer. The flow rate is adjusted to 15 ml/hr and 2-ml fractions are collected. Fractions containing more than 0.5 unit/ml are pooled and are concentrated by precipitation with ammonium sulfate at 38% final satu-

PURIFICATION OF PHOSPHOFRUCTOKINASE FROM RABBIT LIVER[a]

| Procedure step | Volume (ml) | Total units | Protein (mg/ml) | Specific activity (units/mg) | Yield (%) |
|---|---|---|---|---|---|
| Extract | 2700 | 1200 | 1.9 | 0.023 | — |
| Heated supernatant | 2600 | 850 | 4.5 | 0.073 | 71 |
| Ammonium sulfate | 130 | 820 | 3.5 | 1.8 | 68 |
| DEAE-Sephadex chromatography | 6 | 620 | 4.3 | 24 | 52 |
| Agarose chromatography | 3 | 570 | 2.1 | 90 | 48 |

[a] Typical data obtained with 910 g of frozen liver.

ration. The precipitate is dissolved in 1 or 2 ml of 50 m$M$ Tris-phosphate (pH 8.0) buffer containing 2 m$M$ EDTA, 100 m$M$ ammonium sulfate, 0.1 m$M$ ATP, and 0.5 m$M$ dithiothreitol.

The data obtained from a typical procedure are shown in the table. Some fractions from the agarose column have specific activities slightly over 100 units/mg, but the pooled enzyme usually has a specific activity of 85–95 with an overall yield for the procedure of about 50%.

## Properties

*Stability.*[2] Concentrated solutions ($>2$ mg/ml) of liver phosphofructokinase may be stored in the same buffer used for dissolving the precipitated fraction following the Bio-Gel chromatography (step 5). The enzyme will gradually lose activity over a period of several weeks, but most of the activity can be recovered following dialysis overnight against a fresh solution of the same buffer. Preparations have been stored for several months with little irreversible activity loss. Ammonium sulfate, dithiothreitol, and ATP all contribute to the stability of dilute solutions of liver phosphofructokinase. If diluted to a concentration of 0.7 unit/ml in a pH 7.0 buffer consisting of 25 m$M$ glycyglycine, 25 m$M$ $\beta$-glycero-P, 1 m$M$ EDTA, 90% of the initial activity is lost in 3 days. No activity is lost over this period if dithiothreitol (0.1 m$M$), ATP (0.1 m$M$), and ammonium sulfate (5 m$M$) are included in the buffer. Each of the three agents alone provides some protection against loss of activity on storage.

*Comparison with Muscle Phosphofructokinase.*[2,3] The purified liver phosphofructokinase migrates toward the anode on zone electrophoresis more rapidly than does phosphofructokinase from muscle. The liver en-

[2] R. G. Kemp, *J. Biol. Chem.* **246**, 245 (1971).
[3] M. Y. Tsai and R. G. Kemp, *J. Biol. Chem.* **248**, 785 (1973).

zyme is similar to the muscle enzyme in pH optimum, activation by $K^+$ and $NH_4^+$, and substrate affinity at pH 8.2. Striking differences, however, are observed in the regulatory properties at pH 7.0. The liver enzyme is more inhibited by ATP, less sensitive to the deinhibiting action of AMP, ADP, and cyclic 3',5'-AMP, less inhibited by citrate, phosphoenolpyruvate, phosphocreatine, and 3-phosphoglycerate, and more inhibited by 2,3-diphosphoglyceric acid than muscle phosphofructokinase.

Antiserum prepared by injecting guinea pigs with purified muscle phosphofructokinase showed a partial reaction with the purified liver enzyme, whereas little cross-reaction was observed with antiserum to liver phosphofructokinase toward muscle enzyme. It has been proposed[3] that muscle phosphofructokinase be designated phosphofructokinase A and that the liver enzyme be designated phosphofructokinase B.

*Other Preparations.* Massey and Deal[4] have reported a procedure for the rapid preparation of phosphofructokinase of high specific activity from pig liver. Tarui *et al.*[5] have reported the purification of rabbit erythrocyte phosphofructokinase to a specific activity of 139 units/mg. According to studies of the distribution of phosphofructokinase isozymes,[3] the enzyme present in erythrocytes is the B isozyme and thus should be identical to liver phosphofructokinase.

[4] T. H. Massey and W. C. Deal, Jr., *J. Biol. Chem.* **248,** 56 (1973). Also see this volume [17].
[5] S. Tarui, N. Kono, and K. Uyeda, *J. Biol. Chem.* **247,** 1138 (1972).

# [13] Phosphofructokinase from Rabbit Skeletal Muscle

*By* ROBERT G. KEMP

Fructose 6-phosphate + ATP → fructose 1,6-diphosphate + ADP + $H^+$

## Assay Method

*Principle.* The assay for phosphofructokinase is derived from the method originally described by Racker[1] and is based upon the coupling of fructose-1,6-diP formation to the oxidation of NADH with the use of aldolase, triosephosphate isomerase, and α-glycerophosphate dehydrogenase. Each mole of fructose diphosphate formed leads to the oxidation of 2 moles of NADH.

[1] E. Racker, *J. Biol. Chem.* **167,** 843 (1947).

*Reagents*

Magnesium buffer mixture consisting of 75 m$M$ glycylglycine, 75
m$M$ disodium $\beta$-glycerophosphate, 3 m$M$ disodium EDTA, 18 m$M$
MgCl$_2$, and 9 m$M$ ammonium sulfate, all adjusted to pH 8.2

Fructose-6-P, 30 m$M$, adjusted to pH 8.2

ATP, 30 m$M$, adjusted to pH 8.2

NADH, 10 mg/ml

Dithiothreitol, 3 m$M$

Auxiliary enzyme solution consisting of aldolase (36 units/ml),
triosephosphate isomerase (18 units/ml), and $\alpha$-glycerophosphate
dehydrogenase (18 units/ml), in 25 m$M$ glycylglycine, 25 m$M$
$\beta$-glycerophosphate (pH 8.0)

*Enzyme Dilution.* The phosphofructokinase is diluted in 25· m$M$
$\beta$-glycerophosphate, 25 m$M$ glycylglycine, 1 m$M$ EDTA, 0.1 m$M$ dithiothreitol, all at pH 8.0.

*Procedure.* A reaction mixture of 3.0 ml is made up as follows: 1.0
ml of magnesium buffer mixture, 1.5 ml of H$_2$O, 0.1 ml of fructose-6-P,
0.1 ml of ATP, 0.05 ml of NADH, 0.1 ml of dithiothreitol, 0.05 ml of
auxiliary enzymes. After several minutes for temperature equilibration
in the 10 mm light path cuvette in a thermostatted spectrophotometer
at 26°, 0.1 ml of the diluted phosphofructokinase is added. After a brief
lag, the optical density change at 340 nm is recorded. The units can be
calculated from the rate of optical density change per minute based on
the fact that 1 unit of enzyme in the reaction mixture will produce a
rate of optical density change of 0.24 per minute. One unit of phosphofructokinase is defined as that amount of enzyme that catalyzes the conversion of 1 $\mu$mole of fructose-6-P to fructose diphosphate per minute
at pH 8.2 and 26° under the above-described reaction conditions.

## Purification from Fresh Muscle[2]

The procedure for the preparation of phosphofructokinase combines
and modifies several steps in the methods originally published by Parmeggiani *et al.*[3] and by Ling *et al.*[4] The use of fluoride in the extraction
medium and the isopropanol precipitation step were taken from the
method of Ling *et al.*,[4] and the heat treatment step and the crystallization

[2] R. G. Kemp and P. B. Forest, *Biochemistry* **7**, 2596 (1968).

[3] A. Parmeggiani, J. Luft, D. S. Love, and E. G. Krebs, *J. Biol. Chem.* **247**, 4625
(1966).

[4] K.-H. Ling, V. Paetkau, F. Marcus, and H. A. Lardy, this series, Vol. 9, p. 425.

procedure were first employed by Parmeggiani *et al.*[3] The quantities given are typical for this procedure.

*Step 1. Extraction.* A large albino rabbit is killed by an injection of sodium pentobarbital (250 mg) into the external marginal vein of an ear. The animal is immediately bled by cutting the blood vessels of the neck, and the muscles of the hind legs and back are quickly removed and chilled in ice. The muscle mass is passed through a meat grinder in a cold room, weighed (600 g), and then homogenized for 30 sec in a Waring Blendor in the presence of 2 volumes (1200 ml) of 30 m$M$ KF, 4 m$M$ EDTA, and 1 m$M$ dithiothreitol, all at pH 7.5. An additional 1 volume (600 ml) of this extraction medium is added, and the mixture is homogenized a second time for a 30-sec period. The mixture is centrifuged at 4° for 30 min at 10,000 $g$. The supernatant fraction is passed through glass wool to remove lipid particles and the pH (usually about 6.6) is adjusted to 6.8 by the dropwise addition of 1 $M$ Tris. The volume of extract is about 1500 ml.

*Step 2. Precipitation with Isopropanol.* The extract is divided into two equal portions and placed in a dry ice–ethanol bath at about —6° to —8°. When the temperature of the extract drops to 0°, the dropwise addition of two-tenths volume of cold isopropyl alcohol is begun while maintaining efficient stirring. As the volume of alcohol added increases, the temperature of the extract is allowed to decrease to —5°. Freezing of the solution should be avoided; and if ice starts to form on the wall of the flask or as crystals in solution, the flask should be removed from the ethanol bath. After all the isopropanol is added, the mixture is stirred at —5° for an additional 20 min and the precipitate is collected by centrifugation at —5° for 30 min at 10,000 $g$. The white precipitate is dissolved in 100 ml (about $\frac{1}{15}$ of the volume of fraction 1) of 0.1 m$M$ Tris-phosphate, pH 8.0, 0.2 m$M$ EDTA, and 0.2 m$M$ fructose-1,6-diP. This suspension is dialyzed for 20 hr against the same buffer to further reduce the amount of isopropanol present in the fraction.

*Step 3. Heat Treatment.* The dialyzed suspension is transferred to a flask and is placed in a 70° water bath. The suspension is rapidly stirred and the temperature of the preparation is brought to 57° and held at 57–59° for 3 min. During this period it is usually necessary to occasionally remove the flask from the bath when the temperature of the contents reaches 59° and to return it to the bath when the temperature drops to 57°. Following the 3-min heat treatment, the flask is placed in an ice bath and the suspension is rapidly stirred until the temperature drops to 5°. The suspension is then centrifuged for 20 min at 4° and at 15,000 $g$ and the supernatant is retained. The pellet is resuspended in 15 ml of the previously mentioned Tris-phosphate buffer and the suspension is centrifuged as before. The supernatant is combined with the supernatant fraction from the original heat treatment. The combined supernatants (110 ml) were clear and yellow.

*Step 4. Ammonium Sulfate Fractionation.* The flask containing the solution from step 3 is placed in an ice bath on a magnetic stirrer. Solid ammonium sulfate is slowly added to 38% saturation (21.3 g/100 ml of initial solution), and the solution is allowed to stir for 30 min before the precipitate is removed by centrifugation for 20 min at 10,000 g. The supernatant is removed, and the enzyme is precipitated by the further addition of 10.1 g solid ammonium sulfate per 100 ml (55% saturation). The sediment is collected by centrifugation for 20 min at 10,000 g and is dissolved in a buffer consisting of 50 m$M$ glycerophosphate (pH 7.2) containing 2 m$M$ EDTA and 2 m$M$ ATP to a final volume of 10 ml.

*Step 5. Crystallization.* Crystallization was carried out by a procedure very similar to that described by Parmeggiani et al.[3] The solution from step 4 is dialyzed in the cold against the glycerophosphate–EDTA–ATP buffer at pH 7.2 to which has been added 21.3 g of ammonium sulfate per 100 ml (38% saturation). Any precipitate that forms within 4 hr is removed by centrifugation. Crystallization starts after about 1 week. Several days after the beginning of the crystallization the dialysis bag containing enzyme is transferred to the pH 7.2 buffer to which has been added 22.6 g of ammonium sulfate per 100 ml (40% saturation). The entire crystallization period can be shortened to several days by seeding, and seed crystals should be set aside to facilitate subsequent crystallizations. When no further material crystallizes, the suspension is centrifuged at low speed, the supernatant is discarded, and the crystals are dissolved in the glycerophosphate buffer (8 ml). After any insoluble material is removed by centrifugation, the enzyme is recrystallized by dialysis

TABLE I

PURIFICATION OF PHOSPHOFRUCTOKINASE FROM FRESH RABBIT MUSCLE[a]

| Procedure step | Volume (ml) | Total units | Protein (mg/ml) | Units/mg | Yield (%) |
|---|---|---|---|---|---|
| Extract | 1500 | 60,000 | 14 | 2.9 | — |
| Isopropanol precipitate | 110 | 49,000 | 33 | 13.5 | 82 |
| Heat treatment | 110 | 48,000 | 5.5 | 79 | 80 |
| Ammonium sulfate precipitate | 11 | 43,000 | 37 | 105 | 72 |
| 1st crystallization | 9 | 34,000 | 24 | 160 | 57 |
| 2nd crystallization | 9 | 30,000 | 21 | 160 | 50 |

[a] Typical data obtained with 600 g of fresh muscle.

against ammonium sulfate at 38% saturation as described above. For the recrystallization procedure it is usually unnecessary to employ ammonium sulfate above 38% saturation to achieve a nearly quantitative recovery.

The data obtained from a typical procedure are shown in Table I.

### Purification from Frozen Muscle

If frozen rabbit skeletal muscle is homogenized as described for fresh tissue, very little activity is recovered in the extract. For unknown reasons, the enzyme is recovered in the insoluble fraction that is removed in the initial centrifugation step. Mansour[5] has described the purification of phosphofructokinase from frozen sheep hearts, in which procedure the activity was also located in the particulate fraction. The procedure described below employs the observation of Mansour[5] that heart phosphofructokinase can be solubilized upon incubation with MgATP and a thiol compound. The location of the phosphofructokinase in the particulate fraction provides some advantages; principally, ease of preparation, and the fact that the first EDTA extract can be used for the preparation of other enzymes. In our laboratory, we have prepared at various times aldolase, α-glycerol-P dehydrogenase, fructose-1,6-diphosphatase, and glyceraldehyde-3-P dehydrogenase from the initial extract.

*Step 1. Extraction.* One pound of frozen muscle from young rabbits (Pel-Freez Biologicals, Rogers, Arkansas) is chopped into small pieces with a meat cleaver and passed through a meat grinder in a cold room. The muscle is homogenized for two 30-sec periods in a Waring Blendor in the presence of 3 volumes (1350 ml) of a solution of 1 m$M$ EDTA that has been adjusted to pH 7.5. The mixture is centrifuged at 4° and 10,000 $g$ for 30 min. The supernatant is discarded or set aside for other enzyme isolation procedures and the sediment is resuspended in two volumes (900 ml) of 1 m$M$ EDTA. The mixture is again centrifuged at 4° and 10,000 $g$ for 30 min. The supernatant is discarded and the sediment is suspended in 2 volumes (900 ml) of a pH 8.0 buffer consisting of 50 m$M$ Tris·HCl containing 50 m$M$ MgSO$_4$, 0.1 m$M$ EDTA, 0.5 m$M$ ATP, and 1 m$M$ dithiothreitol. The suspension is divided into two portions in 1-liter flasks and the two portions are carried through the heat-treatment separately. The flask containing the suspension is transferred to a 70° water bath and the suspension is rapidly stirred until the temperature of the thick slurry reaches 57°. The temperature of the flask contents is maintained between 57° and 59° for 3 min. The flasks are removed and placed in a slurry of ice and the contents are stirred until the temperature

[5] T. E. Mansour, this series, Vol. 9, p. 430.

drops to 10°. The mixture is centrifuged at 4° and at 10,000 $g$ for 20 min. The supernatant solution is retained, and the sediment is resuspended in two volumes (900 ml) of the Tris–Mg–EDTA–ATP–dithiothreitol buffer and the mixture is centrifuged for 20 min at 10,000 $g$. The supernatant solution is combined with that of the first sedimentation (1900 ml).

*Step 2. First Ammonium Sulfate Fractionation.* Solid ammonium sulfate (21.3 g/100 ml) is added slowly to the ice-cold extract while efficient stirring is provided. After all the ammonium sulfate is dissolved, the suspension is stirred for an additional 30 min before the sediment is removed by centrifugation for 20 min at 10,000 $g$. The enzyme is then precipitated from the supernatant solution by the further addition of 10.1 g of solid ammonium sulfate per 100 ml of solution. The sediment is dissolved in 30 ml of 50 m$M$ glycerophosphate (pH 7.2) containing 2 m$M$ EDTA and 2 m$M$ ATP. The preparation is dialyzed overnight in the cold against the glycerophosphate–EDTA–ATP buffer.

*Step 3. Second Ammonium Sulfate Fractionation.* The dialyzed solution from step 2 is placed in an ice bath and 21.3 g ammonium sulfate per 100 ml solution is slowly added with efficient stirring. The solution is stirred for an additional 30 min before the sediment is removed by centrifugation. The supernatant is placed in an ice bath and 5.2 g ammonium sulfate per 100 ml is slowly added (47% saturation). After 30 min the enzyme is collected by centrifugation at 10,000 $g$ and is dissolved in 50 m$M$ glycerophosphate (pH 7.2) containing 2 m$M$ EDTA and 2 m$M$ ATP to a final volume of 10 ml.

*Step 4. Crystallization.* The preparation is less pure than that obtained at a similar stage in the purification from fresh muscle (after step 4 of that procedure) and crystallization without the use of seed crystals

TABLE II

PURIFICATION OF PHOSPHOFRUCTOKINASE FROM FROZEN RABBIT MUSCLE[a]

| Procedure step | Volume (ml) | Total units | Protein (mg/ml) | Units/mg | Yield (%) |
|---|---|---|---|---|---|
| Heated extract | 1900 | 42,000 | 1.8 | 12 | — |
| First ammonium sulfate precipitate | 36 | 35,000 | 42 | 23 | 83 |
| Second ammonium sulfate precipitate | 10 | 32,000 | 58 | 55 | 76 |
| 1st crystallization | 9 | 24,000 | 21 | 127 | 57 |
| 2nd crystallization | 9 | 20,000 | 15 | 150 | 48 |

[a] Typical data obtained with 450 g of frozen muscle.

has never been attempted. Otherwise, the crystallization procedure is identical to that described for the enzyme from fresh muscle. After removing the material that precipitates after 4 hr of dialysis against the glycerophosphate–EDTA–ATP buffer to which has been added 21.3 g of ammonium sulfate per 100 ml (38% saturation), seed crystals are added and crystallization begins within 24 hr. Recrystallization is carried out as described in the fresh muscle procedure.

The data obtained from a typical procedure are shown in Table II.

## Properties

By all criteria that have been examined, the properties of phosphofructokinase from fresh and frozen muscle procedures are identical. The procedure for fresh muscle has been repeated successfully more than 50 times and the more recently developed procedure for frozen muscle has been employed with good yield for 20 preparations.

*Stability.* Crystalline phosphofructokinase is very stable when stored at 4°. One preparation was found to have retained almost all its initial activity after 6 years of storage. At high protein concentrations (10–20 mg/ml), phosphofructokinase solutions are very stable in Tris-phosphate buffer (pH 8.0) or glycylglycine-glycerol phosphate buffer (pH 8.0). Stability is greatly diminished at lower pH. The enzyme is also much less stable at lower protein concentrations. At a concentration of 5 $\mu$g/ml, more than half of the initial activity is lost after 3 days at pH 8.2 in 25 m$M$ glycylglycine, 25 m$M$ glycerophosphate, and 1 m$M$ EDTA. The addition of ATP (0.1 m$M$) and dithiothreitol (0.1 m$M$) provides complete stability over this period of time. Fructose-6-P, fructose-1,6-diP, and AMP also help retain activity in dilute phosphofructokinase solutions.

*Specific Activity and Yield.* The specific activity of the enzyme after the second crystallization varies from 155 to 170, although some preparations from frozen muscle are as low as 150. The yield of first crystals of phosphofructokinase is about 50 units per gram of muscle (wet weight) of fresh muscle and 45 units per gram of frozen muscle. Ling et al.[4] reported a yield of 44 units per gram of fresh muscle with their procedure. The method described here provides a yield similar to that described previously while providing a rapid and easily reproduced procedure.

## [14] Phosphofructokinase[1] of Yeast

*By* Earle Stellwagen and Harvey Wilgus

$$\text{Fructose-6-P} + \text{ATP} \rightarrow \text{fructose-1,6-P}_2 + \text{ADP} + \text{H}^+$$

## Assay

*Principle.* During purification of yeast phosphofructokinase, enzymic activity is most conveniently measured spectrophotometrically by coupling the production of fructose-1,6-P$_2$ with oxidation of NADH using an excess of aldolase, triosephosphate isomerase, and $\alpha$-glycerol phosphate dehydrogenase.[2] Since the substrate ATP is also a negative effector for yeast phosphofructokinase, it is preferable to use a nucleoside triphosphate which is a good substrate but relatively poor negative effector, e.g., ITP, to measure maximal catalytic rates of aliquots of fractions obtained during execution of the purification procedure. Two alternative procedures for measurement of yeast phosphofructokinase activity are useful for special purposes. The production of ADP catalyzed by phosphofructokinase can be coupled with oxidation of NADH measured spectrophotometrically using excess pyruvate kinase and lactate dehydrogenase,[3] and the production of protons can be measured by maintaining a constant pH in an unbuffered assay solution.[4]

*Reagents*

Substrate solution: 56 m$M$ Tris, 3.3 m$M$ fructose-6-P, 1.67 m$M$ ITP, 5 m$M$ MgSO$_4$, 10 m$M$ ammonium sulfate, and 5.6 m$M$ 2-mercapoethanol adjusted to pH 7.6 and stored at 4°. The exact concentration of fructose-6-P can be measured spectrophotometrically.[5]

NADH, 4.4 m$M$, freshly prepared and adjusted to pH 6.0

Aldolase from rabbit muscle, as a crystalline suspension in 2.0 $M$ ammonium sulfate, 10 mg of protein per milliliter, approximately pH 6 (Boehringer Mannheim).

Glycerolphosphate dehydrogenase/triosephosphate isomerase from rabbit muscle, approximately 1:6 in catalytic activities, as a crystalline suspension in 2.8 $M$ ammonium sulfate, 10 mg protein/ml, approximately pH 6 (Boehringer Mannheim).

[1] ATP:D-fructose-6-phosphate 1-phosphotransferase, EC 2.7.1.11.
[2] E. Racker, *J. Biol. Chem.* 167, 843 (1947).
[3] R. Jauch, C. Riepertinger, and F. Lynen, *Hoppe-Seyler's Z. Physiol. Chem.* 351, 74 (1970).
[4] J. E. Dyson and E. A. Noltmann, *Anal. Biochem.* 11, 362 (1965).
[5] H. J. Hohorst *in* "Methods of Enzymatic Analysis" (H. U. Bergmeyer, ed.), p. 134. Academic Press, New York, 1963.

*Procedure.* A solution of the analytical enzymes is prepared by diluting 0.2 ml of the aldolase suspension and 0.04 ml of the glycerol-phosphate dehydrogenase/triosephosphate isomerase suspension to 1.0 ml with 50 m$M$ Tris·HCl buffer, pH 7.6, containing 5 mg/ml of bovine serum albumin. The phosphofructokinase assay solution is formulated in a cuvette by mixing 0.9 ml of the substrate solution with 0.05 ml of 4.4 m$M$ NADH and 0.05 ml of the analytical enzyme solution. The cuvette containing the assay solution is then placed in a thermostatable sample compartment of a recording spectrophotometer and adjusted to and maintained at 25°. Phosphorylation of fructose-6-P is initiated by addition of a small aliquot of yeast phosphofructokinase solution sufficient to give an initial decrease of 0.05 to 0.50 absorbance units at 340 nm per minute. The yeast phosphofructokinase may be diluted in 100 m$M$ potassium phosphate containing 2 m$M$ fructose-6-P, 3 m$M$ 2-mercaptoethanol, and 0.5 m$M$ EDTA, adjusted to pH 7.1, prior to assay for enzymic activity. Contaminating NADH oxidase activity is measured in a similar fashion by addition of substrate solution in place of the analytical enzyme solution.

*Definition of Unit and Specific Activity.* One unit of enzyme is defined as that amount of enzyme which catalyzes the conversion of 1 $\mu$mole of fructose-6-P to fructose-1,6-P$_2$ per minute at 25°, while specific activity is expressed as units per minute per milligram of protein. The protein concentrations of aliquots from steps 1–5 in the purification procedure are determined by the biuret method while the Lowry method is used in steps 6–8. Bovine serum albumin is used as a standard for both methods. The protein concentration of the purified enzyme solutions used for physical chemical measurements can be calculated from their absorbance at 280 nm using an extinction of 1.09 $\pm$ 0.11 ml mg$^{-1}$ cm$^{-1}$.

## Purification Procedure

Fresh baker's yeast can be obtained from almost any commercial bakery as a pressed wet cake. The purification procedure described here represents an amalgam of four published procedures.[6-9] This procedure is efficient and reproducible and produces a yeast phosphofructokinase preparation with minimal proteolytic contamination. An alternative purification procedure has been reported by Jauch *et al.*[3] All operations in the purification procedure described here are done at 4°, unless noted

[6] T. J. Lindell and E. Stellwagen, *J. Biol. Chem.* **243**, 907 (1968).
[7] W. Atzpodien and H. Bode, *Eur. J. Biochem.* **12**, 126 (1970).
[8] H. Wilgus, J. R. Pringle, and E. Stellwagen, *Biochem. Biophys. Res. Commun.* **44**, 89 (1971).
[9] K. Nissler, S. Friedrich, and E. Hofmann, *Acta Biol. Med. Ger.* **28**, 739 (1972).

otherwise, and all centrifugation is done at 27,300 $g$. Buffer A contains 50 m$M$ potassium phosphate and 0.1 m$M$ EDTA adjusted to pH 7 5; buffer B contains 10 m$M$ potassium phosphate and 2 m$M$ MgSO$_4$ adjusted to pH 6.8; and buffer C contains 25 m$M$ potassium phosphate, 10 m$M$ MgSO$_4$ and 0.2 m$M$ ATP adjusted to pH 6.8. All three buffers also contain 1 m$M$ 2-mercaptoethanol and 1 m$M$ phenylmethanesulfonyl fluoride (PMSF). The PMSF is added as an aliquot from a 0.33 $M$ solution in 95% ethanol. Analytical details for a typical purification are shown in Table I.

*Step 1. Extraction.* A suspension of 1 kg of yeast per 1.5 liters of buffer A is ruptured by circulating it nine times through a Manton-Gaulin type 15A homogenizer (Manton-Gaulin Mfg. Co., Everett, Massachusetts) set at 8000 psi. The homogenizer is cooled by a circulating refrigerant so that the temperature of the yeast homogenate does not exceed 20°. Alternatively, a colloid mill[6] or a sonic oscillator[9] can be used to rupture the yeast. The homogenized solution is adjusted to pH 6.4 by addition of 1 $N$ KOH and centrifuged for 30 min.

*Step 2. Protamine Sulfate Precipitation.* A freshly prepared solution of protamine sulfate (Krishell Laboratories, Inc., Portland, Oregon), 1.5% (w/v), containing 1 m$M$ PMSF is adjusted to pH 6.4 with 1 $N$ KOH. This solution is added slowly with stirring to the supernatant from step 1 until a concentration of 1 g of protamine sulfate per 15 g of protein is achieved. The resulting precipitate is removed by centrifugation and discarded.

*Step 3. First Ammonium Sulfate Fractionation.* Solid ammonium sulfate is added (224 g/liter) slowly with gentle stirring to the supernatant from step 2. After standing for 20 min, the precipitate is removed by centrifugation for 40 min. More solid ammonium sulfate is then added (60 g/liter) to the supernatant. The suspension is again allowed to stand for 20 min and then centrifuged for 40 min. The precipitate is dissolved in buffer B and dialyzed for 2.5 hr against two 60-volume portions of buffer B. The dialyzed fraction is then diluted to a protein concentration of about 35 mg/ml.

*Step 4. Acetone Fractionation.* The solution obtained from step 3 is cooled to 0° and chilled (−15°) acetone is added slowly to a concentration of 27% (v/v) while the temperature is lowered to −6°. The resulting precipitate is removed by centrifugation at −6° for 30 min. Chilled acetone is again added to the supernatant to increase its concentration to 35% (v/v) while the temperature is lowered to −8°. The resulting suspension is centrifuged at −8° for 30 min and the precipitate is dissolved in buffer C to a concentration of 10 mg/ml.

*Step 5. Second Ammonium Sulfate Fractionation.* Solid ammonium sulfate is added to a concentration of 256 g/liter. The resulting suspension is allowed to stand for 20 min, then centrifuged for 30 min; the precipitate is discarded. More solid ammonium sulfate is added (55 g/liter) to the supernatant, the resulting suspension is equilibrated for 20 min and then centrifuged for 30 min. The precipitate is dissolved in a minimum volume of buffer C (about 5–10 ml).

*Step 6. Exclusion Chromatography.* The enzyme solution is applied to either a 4 × 82 cm column of Bio-Gel 1.5 *M* (Bio-Rad Laboratories, Richmond, California) or a 5 × 70 cm column of Sepharose 4B or 6B (Pharmacia Fine Chemicals, Piscataway, New Jersey) equilibrated with buffer C. The column is eluted with buffer C using a flow rate of about 1.3 ml/min. Fractions containing 5.4 ml are collected and fractions 88–112 which contain phosphofructokinase activity are pooled.

*Step 7. DEAE-Cellulose Chromatography.* The pooled fractions are applied to a 2.2 × 41 cm column of DEAE-cellulose (Whatman DE-52, Reeve Angel, Clifton, New Jersey) equilibrated with buffer C. The column is eluted with 400 ml of buffer C containing a linear gradient of KCl between 20 and 200 m*M* using a flow rate of about 1 ml/min. Fractions containing 4.2 ml are collected. Phosphofructokinase activity elutes as a single component located in fractions 50 to 73, and these fractions are pooled.

*Step 8. Hydroxyapatite Chromatography.* The pooled fractions are applied to a 2.2 × 13 cm column of hydroxyapatite (Bio-Gel HT, Bio-Rad Laboratories) equilibrated with buffer C. The column is developed batchwise with successive 125-ml portions of buffer C containing 100 m*M*, 125 m*M*, and 225 m*M* potassium phosphate. Fractions containing 4.2 ml are collected. Phosphofructokinase activity emerges as a single component spanning fractions 68 to 75. These fractions are pooled and dialyzed against buffer C devoid of PMSF. The protein concentration of the dialyzed solution can be conveniently increased by ultrafiltration.

## Properties

*Purity.* The purified enzyme migrates as a single component when subjected to polyacrylamide electrophoresis at pH values of 6.1, 7.5, and 9.1.[6-9] The protein sediments as a single symmetrical boundary when examined by velocity sedimentation[3,6-9] and produces a protein gradient after equilibrium sedimentation characteristic of a homogeneous preparation.[6,9] Unfortunately, the purified enzyme contains a proteolytic contaminant which is not inhibited by phenylmethanesulfonyl fluoride or by diisopropyl fluorophosphate, which appears to be tightly bound to phosphofructokinase and is active in concentrated detergent solutions.[8,9]

TABLE I

PURIFICATION OF PHOSPHOFRUCTOKINASE FROM YEAST

| Step | Volume (ml) | Protein (mg) | Total units of activity | Specific activity (units/mg) | Recovery (%) |
|---|---|---|---|---|---|
| 1. Extraction | 710 | 89,800 | 28,400 | 0.32 | 100 |
| 2. Protamine sulfate | 900 | 45,900 | 25,900 | 0.57 | 91 |
| 3. First ammonium sulfate | 38 | 6,610 | 25,800 | 3.91 | 91 |
| 4. Acetone | 48 | 2,110 | 22,000 | 10.0 | 78 |
| 5. Second ammonium sulfate | 5 | 575 | 18,000 | 31.3 | 63 |
| 6. Bio-Gel 1.5 m | 139 | 154 | 14,600 | 94.4 | 51 |
| 7. DEAE-cellulose | 45 | 76 | 9,720 | 128 | 34 |
| 8. Hydroxyapatite | 23 | 30 | 4,510 | 148 | 16 |

Some success in removing the proteolytic contaminant using a denatured protein affinity column[10] or a Cibacron blue affinity column[10a] has been reported.

*Stability.* The enzyme preparation can be stored in buffer C at $-15°$ for at least 4 weeks without loss of activity.[7] The presence of ligands, particularly fructose-6-P, ATP, or $NH_4^+$, is critical to the retention of activity during dialysis or storage at pH values ranging from 6.0 to 9.0 at $4°$.[11] Complexation of the enzyme with these ligands presumably makes it less susceptible to proteolysis. However, even in the presence of one or more of these ligands, the sedimentation coefficient of the enzyme decreases from about 20 S to 17 S during enzyme purification after storage for 10 days.[12]

*Catalytic Properties.* Yeast phosphofructokinase is a K-type allosteric enzyme typified by a sigmoidal dependence of the catalytic rate on the concentration of fructose-6-P under most assay conditions. Fructose-6-P,[13] 5'-AMP,[14] and $NH_4^{+15}$ are the positive effectors, and ATP[13] and citrate[16] are the principle negative effectors. Most other common nucleoside phosphates and hexose phosphates are poor effectors, if they are effectors at all. The enzyme can be desensitized to ATP inhibition either by incubation with NaF together with a separate protein fraction obtained from yeast[17,18] or by limited tryptic hydrolysis.[19] Low concentrations of $NH_4^+$ and to a lesser extent $K^+$ but not $Rb^+$ or $Na^+$ stimulate yeast phosphofructokinase activity without altering cooperatively.[7,15] The apparent $K_D$ values for $NH_4^+$ and $K^+$ stimulation are 3 m$M$ and 10 m$M$, respectively.[15] Higher concentrations of $NH_4^+$ but not $K^+$ alter

[10] W. Diezel, K. Nissler, W. Heilmann, G. Kopperschläger, and E. Hofmann, *FEBS Lett.* **27**, 195 (1972).

[10a] W. Diezel, H. J. Böhme, K. Nissler, R. Freyer, W. Heilmann, G. Kopperschläger, and E. Hofman, *Eur. J. Biochem.* **38**, 479 (1973).

[11] St. Liebe, W. Diezel, G. Kopperschläger, and E. Hofmann, *Acta Biol. Med. Ger.* **28**, 39 (1972).

[12] G. Kopperschläger, W. Diezel, M. Prausche, and E. Hofmann, *FEBS Lett.* **22**, 133 (1972).

[13] E. Viñuela, M. L. Salas, and A. Sols, *Biochem. Biophys. Res. Commun.* **12**, 140 (1963).

[14] A. Ramaiah, J. A. Hathaway, and D. E. Atkinson, *J. Biol. Chem.* **239**, 3619 (1964).

[15] R. D. Mavis and E. Stellwagen, *J. Biol. Chem.* **245**, 674 (1970).

[16] M. L. Salas, E. Viñuela, M. Salas, and A. Sols, *Biochem. Biophys. Res. Commun.* **19**, 371 (1965).

[17] E. Viñuela, M. L. Salas, M. Salas, and A. Sols, *Biochem. Biophys. Res. Commun.* **15**, 243 (1964).

[18] W. Atzpodien, J. M. Gancedo, V. Hagmaier, and H. Holtzer, *Eur. J. Biochem.* **12**, 6 (1970).

[19] M. L. Salas, J. Salas, and A. Sols, *Biochem. Biophys. Res. Commun.* **31**, 461 (1968).

TABLE II
CONSTANTS FOR THE CONCERTED ALLOSTERIC MODEL[a]

| Ligand | $K_R$ (mM) | $K_T$ (mM) |
|---|---|---|
| Fructose-6-P | 0.060 | 10.2 |
| ATP (substrate) | 0.13 | 0.13 |
| ATP (effector) | 1.0 | 0.05 |
| ITP (effector) | 2.0 | 0.02 |

[a] $n = 4$; $L_0 = 100$; $\Delta G' = 2.7$ kcal/mole.

cooperativity. The nucleoside triphosphate substrates and effectors appear to bind to the enzyme as $Mg^{2+}$ complexes.[15]

·The pH optimum for catalysis is dependent on the concentrations of both substrates, decreasing with increasing concentrations of fructose-6-P[20] but increasing with increasing concentrations of ATP.[7] As the pH of the assay solution is lowered, both the cooperativity as reflected in the Hill coefficient for fructose-6-P and the concentration of fructose-6-P necessary to half-saturate the enzyme, $S_{1/2}$, are decreased.[20] At pH 6.4, yeast phosphofructokinase exhibits linear double reciprocal plots of $1/v$ vs $1/[\text{fructose-6-P}]$ whose variations with [ATP][20] suggests an ordered sequential mechanism.

A systematic study of the influence of various effectors on the Hill coefficient and $S_{1/2}$ value for fructose-6-P measured at pH 7.6 has been reported by Mavis.[21] The data were analyzed using the saturation function of the allosteric model of Monod et al.[22] as extended by Rubin and Changeux.[23] The good agreement between the values calculated in the same fashion from kinetic measurements of *Escherichia coli* phosphofructokinase catalysis[24] and the association constants subsequently measured by equilibrium dialysis[25] establishes the potential reliability of this approach. The constants calculated for yeast phosphofructokinase using the concerted allosteric model are shown in Table II. The model predicts that the enzyme contains four protomers per oligomer, $n$, and that the T conformation predominates by about a factor of about 100 in the absence of all effectors, $L_0$. The substrate fructose-6-P has about a 150-fold greater affinity for the R conformation while ATP and ITP as negative

[20] R. Freyer, M. Kubel, and E. Hofmann, *Eur. J. Biochem.* **17**, 378 (1970).
[21] R. D. Mavis, Ph.D. Thesis, University of Iowa, 1970.
[22] J. Monod, J. Wyman, and J.-P. Changeux, *J. Mol. Biol.* **12**, 88 (1965).
[23] M. M. Rubin and J.-P. Changeux, *J. Mol. Biol.* **21**, 265 (1966).
[24] D. Blangy, H. Buc, and J. Monod, *J. Mol. Biol.* **31**, 13 (1968).
[25] D. Blangy, *Biochimie* **53**, 135 (1971).

effectors have about a 10- to 20-fold greater affinity for the T conformation. The activator 5'-AMP binds equally well to the ATP regulatory site on both conformations, thereby functioning as an apparent positive affector by competing with the negative effector ATP for the regulatory site. Theoretical curves generated from these values are in reasonable agreement with the experimental values. This kinetic analysis suggests that within the range in the physiological concentrations of fructose-6-P, ATP, and 5'-AMP reported by Betz and Moore,[26] yeast phosphofructokinase is under constant inhibition by ATP with variation in fructose-6-$P$ being the major means of control.

*Physical Chemical Properties.* An extensive study[11] of the sedimentation properties of yeast phosphofructokinase indicates that the molecular size of the purified enzyme is reversibly dependent on protein concentration, on ligand concentrations, and on the pH of the solvent even in the "physiological" pH range. While the physical parameters of several polymeric forms of the enzyme have been reported by several laboratories,[3,6,11,27] the documented[10,12] occurrence of proteolysis during enzyme purification, storage, and analysis makes the significance of these parameters relative to those of the native enzyme very questionable. Since the dissociated protein is even more susceptible to proteolysis,[6] molecular weight measurements of the component polypeptide chains of the enzyme are even more suspect. An additional complication is the apparent polymerization of the dissociated polypeptide chains in detergent solutions such as sodium dodecyl sulfate.[12] Given these complications, it is not surprising that wide ranges of values have been reported for the molecular weights of the individual polypeptide chains[6,12,28] and of various polymeric states. Definition of the tertiary and quaternary structures of the yeast phosphofructokinase and their relationship with the allosteric properties thus awaits a procedure for inhibition or removal of all proteolytic activities during enzyme purification, storage, and analysis. Similarly, although the amino acid composition of the enzyme has been determined in two laboratories[27,28] and the results are in good agreement, this composition probably represents that of a proteolyzed derivative of the native enzyme, not that of the native enzyme.

[26] A. Betz and C. Moore, *Arch. Biochem. Biophys.* **120**, 268 (1967).
[27] G. Kopperschläger, I. Lorenz, W. Diezel, J. Marguard, and E. Hofmann, *Acta Biol. Med. Ger.* **29**, 561 (1972).
[28] T. J. Lindell, Ph.D. Thesis, University of Iowa, 1969.

# [15] Phosphofructokinase from *Clostridium pasteurianum*

By Kosaku Uyeda and Shigeru Kurooka

$$\text{Fructose 6-phosphate} + \text{ATP} \xrightarrow{\text{Mg}^{2+}} \text{fructose 1,6-bisphosphate} + \text{ADP} \qquad (1)$$

## Assay[1]

*Principle.* Phosphofructokinase activity, Eq. (1), is determined by coupling with aldolase, Eq. (2), triosephosphate isomerase, Eq. (3), and α-glycerophosphate dehydrogenase, Eq. (4).

$$\text{Fructose 1,6-bisphosphate} \rightleftharpoons \text{dihydroxyacetone phosphate}$$
$$+ \text{glyceraldehyde 3-phosphate} \qquad (2)$$
$$\text{Dihydroxyacetone phosphate} \rightleftharpoons \text{glyceraldehyde 3-phosphate} \qquad (3)$$
$$\text{Dihydroxyacetone phosphate} + \text{NADH} + \text{H}^+ \rightarrow \alpha\text{-glycerophosphate} + \text{NAD}^+ \qquad (4)$$

Thus, each mole of fructose bisphosphate formed by phosphofructokinase leads to the oxidation of 2 moles of NADH. Since crude extracts of *Clostridium pasteurianum* contain mannitol-1-phosphate dehydrogenase [Eq. (5)] which interferes with the phosphofructokinase assay, the assay rate of NADH formation is determined therefore in the presence and in the absence of ATP. The difference between these rates gives phosphofructokinase activity.

$$\text{Fructose 6-phosphate} + \text{NADH} + \text{H}^+ \rightleftharpoons \text{mannitol 1-phosphate} + \text{NAD}^+ \qquad (5)$$

Alternatively, the enzyme activity may be determined by coupling with pyruvate kinase [Eq. (6)] and lactate dehydrogenase [Eq. (7)].

$$\text{ADP} + \text{phosphoenolpyruvate} \rightarrow \text{ATP} + \text{pyruvate} \qquad (6)$$
$$\text{Pyruvate} + \text{NADH} + \text{H}^+ \rightleftharpoons \text{NAD}^+ + \text{lactate} \qquad (7)$$

Thus each mole of ADP formed by phosphofructokinase leads to the oxidation of 1 mole of NADH.

*Reagents*[2]

Imidazole·HCl, 0.2 $M$, pH 7.0 (50 m$M$)
Bovine serum albumin, 1 mg/ml (100 $\mu$g)
Dithiothreitol, 0.1 $M$, a fresh solution (10 m$M$)
Fructose 6-phosphate, 0.1 $M$ (2.5 m$M$)
ATP, 50 m$M$ (0.5 m$M$)
NADH 2.7 m$M$ (0.16 m$M$)
MgCl$_2$ 0.1 $M$ (6 m$M$)

---

[1] E. Racker, *J. Biol. Chem.* **167**, 843 (1947).
[2] Final concentrations are enclosed in parentheses.

Auxiliary coupling enzyme solution is made up as follows: 0.4 ml of aldolase (10 mg/ml), 0.1 ml of α-glycerophosphate dehydrogenase (10 mg/ml), and 0.04 ml of triosephosphate isomerase (10 mg/ml) are mixed. Since these enzymes are in ammonium sulfate the requirements for ammonium is satisfied.

*Dilution.* The enzyme to be assayed is diluted in 50 m$M$ potassium phosphate, pH 8.0, 0.2 m$M$ EDTA and 10 m$M$ dithiothreitol.

*Procedure.* A reaction mixture of 1 ml is made up as follows: 0.25 ml of imidazole buffer, 0.06 ml of $MgCl_2$, 0.025 ml of fructose 6-phosphate, 0.01 ml of ATP, 0.06 ml of NADH, 0.1 ml of dithiothreitol, 0.1 ml of bovine serum albumin, 0.005 ml of the auxiliary enzyme solution. The above reaction mixture is pipetted into a cuvette with 10-mm light path, and the volume is made up to 1 ml with $H_2O$. The assay is performed at 28° in a spectrophotometer with a recorder. The reaction is initiated with addition of phosphofructokinase. The reaction rate is recorded as a change in absorbance at 340 nm. After a short lag period the rate is constant, and this part of the rate curve is employed for determination of the enzymic activity. The units of phosphofructokinase are obtained by multiplying the absorbance change per minute by 0.081.

When a crude extract of *C. pasteurianum* which contains mannitol-1-phosphate dehydrogenase is assayed, an additional reaction mixture containing the same assay mixture as above minus ATP is pipetted in a cuvette. The difference between the rate of reaction obtained in this assay mixture and that of complete reaction mixture gives the phosphofructokinase activity.

*Units.* One unit of enzyme activity is defined as the amount of the enzyme that catalyzes the formation of 1 μmole of fructose 1,6-bisphosphate per minute. Protein is determined by a modification[3] of the Lowry phenol reagent method with crystalline serum albumin as a standard.

## Purification[4]

*Clostridium pasteurianum* was grown in a medium containing sucrose-ammonium sulfate as described by Lovenberg *et al.*[5] and stored frozen at −90°. All operations were carried out at 2–4° unless otherwise stated.

*Step 1. Crude Extract.* Frozen cells of *C. pasteurianum* (230 g) were suspended in 500 ml of a mixture of 50 m$M$ potassium phosphate (pH 9) and 1 m$M$ EDTA. The suspension was divided into three equal

[3] J. C. Rabinowitz and W. E. Pricer, Jr., *J. Biol. Chem.* **237**, 2898 (1962).
[4] K. Uyeda and S. Kurooka, *J. Biol. Chem.* **245**, 3315 (1970).
[5] W. Lovenberg, B. B. Buchanan, and J. C. Rabinowitz, *J. Biol. Chem.* **238**, 3899 (1963).

batches, and the cells were disrupted by sonic oscillation with a Bronwill sonic oscillator for 7 min each. The sonically treated extract was then centrifuged at 21,500 *g* for 1 hr in a Sorvall centrifuge. The pH of the supernatant solution was adjusted to 7.0 by addition of 10 *N* KOH.

*Step 2. Heat Treatment.* The crude extract (520 ml) was heated for 10 min in a 58° bath, chilled to 5° immediately at the end of the period, and centrifuged for 30 min at 21,500 *g*.

*Step 3. Isopropanol Fractionation.* The supernatant solution from the heat treatment was then transferred to an ice-salt bath at −5°, and 420 ml of isopropanol (at −5°) were added dropwise with continuous stirring. A special care was taken so that the solution would not freeze. After 10 min the solution was centrifuged at 16,300 *g* for 10 min and the precipitate was discarded. To the supernatant solution were added 570 ml of isopropanol (at −5°) as above, and the mixture was stirred for 10 min at −5°. The precipitate was removed by centrifugation at 16,300 *g* for 10 min and dissolved in 100 ml of a mixture of 50 m*M* potassium phosphate (pH 8) and 1 m*M* EDTA. The protein concentration of the enzyme solution was diluted to 10 mg/ml with the addition of the above buffer. The final volume was 240 ml.

*Step 4. First Ammonium Sulfate.* To the above enzyme solution were added gradually 240 ml of saturated (25°) ammonium sulfate solution, and the solution was stirred gently for 15 min. The precipitate was removed by centrifugation at 21,500 *g* for 10 min and dissolved in 10 ml of a mixture of 50 m*M* potassium phosphate (pH 8), 1 m*M* EDTA, and 14 m*M* 2-mercaptoethanol.

*Step 5. Second Ammonium Sulfate.* The above enzyme solution was diluted with addition of 15 ml of a mixture of 50 m*M* potassium phosphate (pH 8), 1 m*M* EDTA, and 14 m*M* 2-mercaptoethanol to adjust protein concentration to 7 mg/ml. The final volume of the solution was 25 ml. To the enzyme solution, 20.5 ml of saturated ammonium sulfate solution were added with stirring and the suspension was allowed to stand for 10 min. The precipitate was removed by centrifugation at 21,500 *g* for 10 min and dissolved in 5 ml of a mixture of 50 m*M* potassium phosphate (pH 8) and 1 m*M* EDTA.

*Step 6. DEAE-Cellulose Chromatography.* In order to remove excess salt, the enzyme solution was applied to a Sephadex G-25 column (1.5 × 27 cm) which had been equilibrated with 50 m*M* potassium phosphate (pH 8), EDTA (1 m*M*) and 14 m*M* 2-mercaptoethanol and was eluted from the column with the same buffer mixture. The enzyme fractions were combined (20 ml) and adsorbed on a DEAE-cellulose (microgranular DE-52) column (1.7 × 10 cm) which had been equilibrated with the same buffer mixture. The column was washed with 7 ml of the buffer mixture containing 50 m$M$ NaCl until the absorbance of the solution at 280 nm was less than 0.05. The enzyme was then eluted from the column with the buffer mixture containing 0.15 $M$ NaCl, and 3.5-ml fractions were collected on an automatic fraction collector. The enzyme usually appeared in the fractions between 6 and 20, and these fractions were combined. The enzyme was then concentrated to about 5 ml with a Diaflo concentrator using XM-50 or PM-30 membrane (Amicon, Boston, Massachusetts). The enzyme was precipitated with addition of 6 ml of saturated ammonium sulfate solution and centrifuged, and the precipitate was dissolved in 1.5 ml of the buffer mixture.

*Step 7. First Crystallization.* To the above enzyme solution were added about 160 mg of powdered ammonium sulfate with continuous stirring. When slight turbidity appeared, 2 $\mu$moles of ATP were added and crystals began to form in a few minutes. The enzyme solution was then allowed to stand overnight at 2°.

*Step 8. Second and Third Crystallization.* The crystals of the enzyme were then removed by centrifugation and dissolved in 1.5 ml of the same buffer mixture. The enzyme was recrystallized with addition of ammonium sulfate according to the above procedure (step 7). The recrystallization was repeated once more under the same conditions. The enzyme could also be crystallized without the addition of ATP in step 7, but the crystals formed much more slowly. The enzyme was then stored as a crystalline suspension at 2°.

## Properties

*Stability.* A crystalline suspension of clostridial phosphofructokinase in 45% saturation of ammonium sulfate is stable for at least 2 months. After 2 weeks of storage, however, the enzyme showed a change in the activity such that a considerable lag period was observed during the enzyme assay. This lag period is eliminated if the enzyme is preincubated with 10 m$M$ dithiothreitol.

*Specific Activity.* The specific activity of crystalline phosphofructokinase is 160.

*Cation Requirement.* Clostridial phosphofructokinase requires both $Mg^{2+}$ and $NH_4^+$ or $K^+$. The $K_m$ values for $NH_4^+$ and $K^+$ are $1.8 \times 10^{-4}$ $M$ and $2 \times 10^{-2}$ $M$, respectively.

*Molecular Properties.* The crystalline phosphofructokinase behaves as a homogeneous protein both electrophoretically and ultracentrifugally. The sedimentation constant, $s_{20w}$ is 7.8 S. The molecular weight by high speed sedimentation equilibrium method is 144,000 ± 2000. The protein dissociates to subunits in 8 $M$ urea or 7 $M$ guanidine. A treatment of

Purification of Phosphofructokinase from *Clostridium pasteurianum*

| Fractionation step | Total volume (ml) | Total activity (units) | Protein (mg/ml) | Specific activity (units/mg) | Recovery (%) |
|---|---|---|---|---|---|
| 1. Crude extract | 520 | 4200 | 22.2 | 0.36 | 100 |
| 2. Heat treatment | 472 | 4060 | 10.7 | 0.84 | 97 |
| 3. Isopropanol | 100 | 3600 | 25.0 | 1.44 | 86 |
| 4. First ammonium sulfate | 10 | 2760 | 22.5 | 12.2 | 66 |
| 5. Second ammonium sulfate | 5 | 2400 | 11.7 | 42 | 57 |
| 6. DEAE chromatography and concentration | 5.8 | 1200 | 1.3 | 160 | 29 |
| 7. First crystallization | 1.5 | 860 | 3.6 | 160 | 21 |
| 8. Second crystallization | 1.5 | 748 | 3.2 | 156 | 19 |
| 9. Third crystallization | 1.0 | 724 | 4.5 | 160 | 18 |

the enzyme with succinic anhydride also results in dissociation into sub-units. The molecular weight of the subunit has been determined to be about 35,000 by the sedimentation equilibrium method and acrylamide gel electrophoresis. Based on tryptic peptide mapping and acrylamide gel electrophoresis, these subunits appear to be identical.

*Ultraviolet Light Absorption Spectra.* The enzyme absorbs maximally at 275 nm at pH 7 and 290 nm in 0.1 $N$ NaOH.

*pH Optimum.* The enzyme has optimum activity at a broad range of 7.0–8.2.

*Kinetic Properties.* The $K_m$ value for ATP at infinite fructose 6-phosphate concentration is $5.5 \times 10^{-5}$ $M$. Unlike many phosphofructokinases, clostridial enzyme is not inhibited by ATP. The enzyme shows sigmoidal kinetics with respect to fructose 6-phosphate. ADP is a positive effector and alters this sigmoidal kinetics to Michaelis-Menten kinetics. GTP, UTP, ITP, and CTP also serve as substrates.

# [16] Phosphofructokinase from *Escherichia coli*[1]

By Verne F. Kemerer, Charles C. Griffin, and Ludwig Brand

ATP + fructose 6-phosphate → ADP + fructose 1,6-diphosphate + H$^+$

## Assay Methods

### Coupled Assay[2]

*Principle.* The routine assay employed to monitor the purification of phosphofructokinase is based on conversion of the reaction product, fructose 1,6-diphosphate (FDP) to $\alpha$-glycerophosphate in the presence of aldolase (EC 4.1.2.13), triosephosphate isomerase (EC 5.3.1.1), $\alpha$-glycerophosphate dehydrogenase (EC 1.1.1.8), and $\beta$-NADH. The rate of disappearance of NADH is followed spectrophotometrically at 340 nm. Two micromoles of NADH are oxidized per micromole of FDP formed.

*Reagents.* The composition of the assay mixture is shown in Table I.

*Procedure.* Components of the assay (Table I) are mixed in a cuvette (1-cm pathlength), and the oxidation of NADH is followed spectrophotometrically at 340 nm. With impure enzyme preparations, a background

---

[1] ATP:D-fructose-6-phosphate 1-phosphotransferase, EC 2.7.1.11.

[2] C. C. Griffin, B. N. Houck, and L. Brand, *Biochem. Biophys. Res. Commun.* **27**, 287 (1967).

## TABLE I
### Standard Assay Mixture

| Component | Volume (ml) | Final concentration (m$M$) |
|---|---|---|
| Tris Cl, 0.3 $M$, pH 8.2 | 0.5 | 80[a] |
| MgCl$_2 \cdot$6H$_2$O, 10 m$M$ | 0.3 | 1 |
| H$_2$O | 1.2 | — |
| F6P, 0.1 $M$ | 0.2 | 6.7 |
| Auxiliary enzyme mixture[b] | 0.3 | — |
| NADH, 1.5 m$M$ (in 0.3 $M$ Tris Cl pH 8.2) | 0.3 | 0.15 |
| ATP, 30 m$M$ (pH7) | 0.1 | 1 |
| Phosphofructokinase[c] | 0.1 | — |

[a] Includes amount added with NADH.

[b] Aldolase and a mixture of $\alpha$-glycerophosphate dehydrogenase and triosephosphate isomerase [crystalline suspensions in (NH$_4$)$_2$SO$_4$, Boehringer Mannhein or Sigma] are mixed with 10 m$M$ Tris Cl–2 m$M$ EDTA (pH 8) buffer to yield a solution containing 1.5 mg of aldolase per milliliter and 0.5 mg of $\alpha$-glycerophosphate dehydrogenase–triosephosphate isomerase per milliliter. This solution is freed from ammonium sulfate by dialysis against the same buffer. The auxiliary enzyme mixture occasionally contains a contaminating phosphoglucose isomerase (EC 5.3.1.9) activity that may significantly reduce the concentration of F6P in any lengthy preincubation.

[c] Phosphofructokinase is diluted in 0.1 $M$ sodium phosphate–0.1 $M$ ammonium chloride (pH 7.4).

rate of NADH oxidation is obtained for 2–3 min without ATP in the reaction mixture. Adenosine triphosphate is then added to initiate the phosphofructokinase-catalyzed reaction. The background rate which is due to contaminating mannitol-1-phosphate dehydrogenase and other oxidoreductases is subtracted from the rate obtained after addition of ATP.

*Units and Specific Activity.* One unit of activity is defined as that amount of enzyme which catalyzes the formation of 1 $\mu$mole of FDP per minute under the conditions of the standard assay. For calculating specific activity (units/mg) protein concentration is estimated by the microbiuret method.[3]

### pH-Stat Assay[4,5]

*Principle.* At pH 8.5 and above phosphofructokinase catalyzes the stoichiometric release of a proton in the reaction. The pH-stat assay,

[3] S. Zamenhof, this series, Vol. 3 [103].

[4] J. E. Dyson and E. A. Noltmann, *Anal. Biochem.* 11, 362 (1965).

[5] A. Ku and C. C. Griffin, *Arch. Biochem. Biophys.* 149, 361 (1968).

which avoids the use of auxiliary enzymes, is based on the volume of standard base required to maintain constant pH in an essentially un-buffered reaction mixture. This assay is more usually employed for detailed kinetic studies than for monitoring the purification of phosphofructokinase.

*Reagents.* A typical reaction mixture contains the following components in a final volume of 4–8 ml.

KCl, 80 m$M$
MgCl$_2$, 10 × [ATP]
ATP, variable
F6P, variable

*Procedure.* Components are mixed in a thermostated titration vessel under a continuously renewing atmosphere of nitrogen. The vessel is fitted with calomel and glass electrodes and a titrant delivery tip for addition of standard base. After adjustment of the pH of the reaction mixture to 8.5, the reaction is initiated by addition of phosphofructokinase. The volume of standard base required to maintain pH 8.5 is recorded as a function of time.

## Purification

All operations are carried out at 4° unless otherwise noted. Centrifugations are performed at 13,000 $g$ for 15 min.

*Step 1. Homogenization.* One pound of well-thawed *E. coli* B (three-quarter log phase, minimal medium, Grain Processing, Muscatine, Iowa) is homogenized at low speed in a 4-liter, jacketed Waring Blendor for 45 min in the presence of 1500 ml of 50 m$M$ glycylglycine–1 m$M$ EDTA (pH 7.0) and 1 kg of acid-washed glass beads (Potters Industries, Carlstadt, New Jersey, No. P-010). After settling for 10–15 min, the supernatant is decanted and the residue is washed by homogenization for 5 min with 500 ml of the above buffer. The supernatant and wash liquor are combined and centrifuged.

*Step 2. Autolysis.* Magnesium chloride is added to a final concentration of 4 m$M$. The mixture is heated to 30° with stirring and 6 mg of deoxyribonuclease I (EC 3.1.4.5) and 1 mg of ribonuclease I (EC 3.1.4.22) (Worthington, No. 2007 codes D, and No.. 5650 code RAF, respectively) are added. The mixture is maintained at 30° for 3 hr with gentle stirring. Aliquots are removed before and after autolysis to measure the absorbance at 260 nm after precipitation of protein and nucleic acid with 3.5% perchloric acid. Usually a 3-fold increase in soluble OD$_{260}$ is observed during hydrolysis.

*Step 3. Heat.* The autolyzed preparation is heated as rapidly as possible to 50° and rapidly cooled to 6° in a jacketed, stainless steel vessel. After cooling the material is centrifuged. Phosphofructokinase is stable for at least 10 min at 50°, somewhat unstable at 55° and immediately loses all activity at 60°.

*Step 4. Precipitation by Ammonium Sulfate.* Ammonium sulfate (475 g/liter) is added with stirring over a period of 40 min. After a 20-min equilibration, the suspension is centrifuged. The precipitate is dissolved in 0.1 $M$ Tris $SO_4$–0.1 m$M$ ATP–10 $\mu M$ FDP–1.5 m$M$ 2-mercaptoethanol (pH 7.5) and dialyzed for 12 hr against two 4-liter changes of the same buffer.

*Step 5. Acetone Fractionation.* The dialyzate is cooled to 0° and one-fourth of its volume of acetone (−15°) is added over a 5-min period with rapid stirring. After an additional 15 min in a bath at −5° the preparation is centrifuged at −5°, and a $5/11$ volume (based on the volume of the supernatant) of acetone is added. The suspension is cooled to −10°, stirred for 15 min and centrifuged at −10°. The precipitate is suspended in 300 ml of 50 m$M$ sodium pyrophosphate (pH 9.0) and gently stirred for 1 hr and centrifuged to remove undissolved material.

*Step 6. QAE-Sephadex Chromatography I.* The clear supernatant from step 5 is applied to a 4.0 × 40 cm column of QAE-Sephadex A25 equilibrated with 50 m$M$ sodium pyrophosphate–50 m$M$ ammonium sulfate–2 m$M$ 2-mercaptoethanol (pH 9.0) at a flow rate of 100 ml/hr. The column is washed for 3 hr with the starting buffer, and the enzyme is eluted with a linear gradient from 50 m$M$ to 0.60 $M$ ammonium sulfate in a total volume of 3600 ml. The mean elution molarity is 0.2.

*Step 7. Acid Precipitation.* Pooled phosphofructokinase activity from the column is concentrated by lowering the pH to 4.4 with 2 $M$ acetic acid. After centrifugation, the precipitate is suspended in 30 ml of 0.1 $M$ Tris $SO_4$–1 m$M$ EDTA–0.1 m$M$ ATP–10 $\mu M$ FDP–2 m$M$ 2-mercapto-ethanol (pH 8.0), dialyzed against 20 volumes of the same buffer for 3 hr and centrifuged.

*Step 8. QAE-Sephadex Chromatography II.* The phosphofructokinase preparation is applied to a 2.5 × 30 cm column of QAE-Sephadex A25 equilibrated with 0.1 $M$ Tris $SO_4$–1 m$M$ EDTA–2 m$M$ 2-mercaptoethanol (pH 8.0) at a flow rate of 30 ml/hr. The column is rinsed for 2 hr with equilibrating buffer and the flow rate increased to 45 ml/hr. A 1000-ml linear potassium sulfate gradient from 0.0 to 0.4 $M$ is used for elution. The mean elution molarity is 0.1. The pooled enzyme is concentrated to 10 ml in an Amicon ultrafiltration cell (Model UF52 with PM 30 membrane).

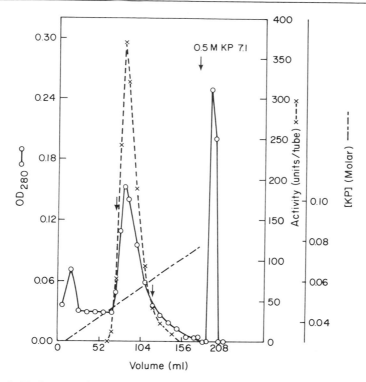

Fig. 1. Hydroxyapatite chromatography of *Escherichia coli* phosphofructokinase. Activity (x—x) eluted from a 1.1 × 25 cm column with a 250-ml linear potassium phosphate (pH 7.1) gradient to 0.1 *M*. Activity collected from 73–120 ml was pooled. After 177 ml were eluted from column the buffer concentration was raised to 0.5 *M* potassium phosphate.

*Step 9. Hydroxyapatite Chromatography.* The material from step 8 is dialyzed against two 250-ml changes of 30 m*M* potassium phosphate–2 m*M* 2-mercaptoethanol (pH 7.1) for a total of 6 hr. It is applied to a 1.1 × 25 cm column of hydroxyapatite[6] equilibrated with the above buffer at a flow rate of 10 ml/hr. After a 2-hr rinse with the equilibrating buffer, the enzyme is eluted with a 250-ml linear gradient from 0.03 to 0.1 *M* in potassium phosphate (pH 7.1). A typical elution profile for this column is shown in Fig. 1. The pure enzyme is concentrated in the Amicon ultrafiltration cell and dialyzed against 0.1 *M* potassium phosphate–1 m*M* EDTA–0.32 m*M* dithiothreitol–0.1 m*M* ATP–10 μ*M* FDP (pH 7.1). Phosphofructokinase is stable for a minimum of 40 days at 4° in this buffer; it is stable for at least 4 months when frozen at −15°

[6] G. Bernardi, this series, Vol. 22 [29].

TABLE II
PURIFICATION OF *Escherichia coli* PHOSPHOFRUCTOKINASE

| Fraction | Volume (ml) | Activity (units) | Protein (mg) | Specific activity (units/mg) | Yield (%) | Purification (fold) |
|---|---|---|---|---|---|---|
| Crude extract | 1980 | 7600 | 21,000 | 0.362 | 100 | 1 |
| Autolysis/heat | 1860 | 5850 | 14,400 | 0.406 | 77 | 1.1 |
| $(NH_4)_2SO_4$ | 328 | 5560 | 14,400 | 0.386 | 73 | 1.1 |
| Acetone | 266 | 4880 | 6,960 | 0.700 | 64 | 1.9 |
| QAE I | 506 | 3480 | 152 | 22.8 | 46 | 63 |
| Acid | 28.6 | 2780 | 93 | 29.9 | 37 | 82 |
| QAE II | 9.25 | 2300 | 12.7 | 181 | 30 | 500 |
| Hydroxyapatite | 3.75 | 1840 | 7.0 | 263 | 24 | 728 |

in this buffer with 20% sucrose added. A summary of the purification procedure is outlined in Table II.

## Properties [2,7-11]

*Gel Electrophoresis.* Purified phosphofructokinase migrates as a single, sharp band in 7.5% polyacrylamide gels. The mobility relative to bromophenol blue is 0.43 in the Ornstein-Davis[12] system (resolving gel pH 9.5) and 0.65 in the Robard-Chrambach[13] system (pH 7.8). Preparations with specific activities of 200–220 show several faint satellite bands.

*Cation Requirement.* *Escherichia coli* phosphofructokinase requires $Mg^{2+}$ as do other ATP-dependent phosphotransferases. Presumably, the Mg-ATP complex is the true substrate for the enzyme although the enzyme may require free $Mg^{2+}$ as well. Inhibition by ATP is not observed if the $Mg^{2+}$ concentration is high enough to complex all the ATP. In addition, phosphofructokinase exhibits a requirement for a monovalent cation. This requirement is satisfied by $NH_4^+$ or, at higher concentrations, $K^+$.

*Substrate Specificity.* The enzyme can utilize a variety of nucleoside triphosphates as phosphoryl donors. Deoxy-ATP is as effective as ATP.

[7] D. E. Atkinson and G. Walton, *J. Biol. Chem.* **240,** 757 (1965).
[8] D. Blangy, H. Buc, and J. Monod, *J. Mol. Biol.* **31,** 13 (1968).
[9] D. Blangy, *FEBS Lett.* **2,** 109 (1968).
[10] C. C. Griffin, unpublished observation.
[11] V. Kemerer and L. Brand, unpublished observations.
[12] B. Davis, *Ann. N.Y. Acad. Sci.* **121,** 404 (1964).
[13] D. Rodbard and A. Chrambach, *Anal. Biochem.* **40,** 95 (1971).

$K_m$'s for the other nucleotides are approximately an order of magnitude higher. Only F6P has been shown to be an acceptor in the reaction. The borohydride reduction products (hexitol 6-phosphates) are not substrates. Other analogs have not been examined.

*Metabolic Effectors.* The catalytic activity of phosphofructokinase responds sigmoidally to increasing concentrations of F6P. At low concentrations nucleoside diphosphates like ADP and GDP convert this sigmoidal response to a hyperbolic one. At higher concentrations ADP acts as a product inhibitor, competitive with respect to ATP and noncompetitive with F6P as the variable substrate. Neither 5'-AMP nor cyclic-3',5'-AMP are positive effectors for the *E. coli* enzyme. Phosphoenolpyruvate is a feedback inhibitor, but citrate has no effect. Fructose 1,6-diphosphate is not an activator of *E. coli* phosphofructokinase, and product inhibition by this ester cannot be observed at concentrations as high as 5–10 m$M$ in the pH-stat assay. Pyruvate, phosphogluconate, and ribulose 5-phosphate have no effect provided that the concentration of magnesium is kept high.

*Molecular Weight.* Phosphofructokinase from *E. coli* B has a molecular weight of 141,000–142,000[9,14] and appears to be composed of 4 subunits of identical molecular weight. Unlike the mammalian enzyme, the enzyme for *E. coli* does not appear to undergo any freely reversible aggregations and dissociations. The sedimentation constant remains essentially unchanged with enzyme concentrations from 0.5 to 5 mg/ml.

*Fluorescent Derivatives.* Dansyl chloride (DNS, 1-dimethylaminonaphthalene 5-sulfonyl chloride) may be covalently attached to phosphofructokinase (ca. 1 mg/ml) in 50 m$M$ sodium bicarbonate–1 m$M$ EDTA–0.1 m$M$ ITP–1 m$M$ F6P (pH 9.2) at 4°. The reaction is terminated after 90 min by addition of a 100-fold excess of 2-mercaptoethanol. After centrifugation, the conjugated enzyme is separated from reactants by desalting on a column of Sephadex G-25. With 10-, 20-, and 50-fold molar excesses of dansyl chloride, dansyl phosphofructokinase conjugates with molar ratios (DNS/phosphofructokinase) of 1.5, 2.7, and 8 have been prepared. Molar ratios are estimated from an extinction coefficient if $3.4 \times 10^3$ for DNS at 350 nm.[15] The specific activity after dansylation is approximately 60% of the native enzyme. All three conjugates show similar excitation and emission spectra with excitation maxima at 350 nm and emission maxima at 515 nm.

The dansylated enzymes exhibit kinetic properties similar to the native enzyme. As shown in Fig. 2, the derivatives are allosteric with respect to F6P and are inhibited by high concentrations of ATP. Phosphoenol-

[14] C. K. Marschke and R. W. Bernlohr, *Arch. Biochem. Biophys.* **156**, 1 (1973).
[15] R. Chen, *Anal. Biochem.* **25**, 412 (1968).

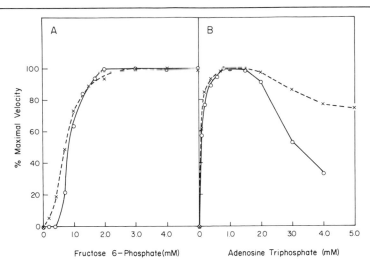

Fig. 2. (A) Effect of F6P concentration on the activity of native (O——O) and octadansyl (x--x) phosphofructokinases from *Escherichia coli*. (B) Effect of ATP on the activity of native (O——O) and octadansyl (x--x) phosphofructokinases. The coupled enzyme assay was used. Reaction mixtures contained 240 μmoles of Tris Cl pH 8.2, 3 μmoles of MgCl₂, 0.44 μmole of β-NADH, 450 μg of aldolase, and 150 μg of α-glycerophosphate dehydrogenase–triosephosphate isomerase in 2.9 ml at 29°. In (A) the ATP concentration was 4 m$M$ and in (B) the F6P concentration was 1 m$M$.

pyruvate is much less potent an inhibitor for the dansylated enzymes. In all cases ADP is able to relieve ATP inhibition completely at concentrations below 1 m$M$ in the presence of 4 m$M$ ATP and 1 m$M$ F6P.

## Acknowledgment

This work was supported by Grant No. GM 11632 from the National Institute of General Medical Sciences and Grants No. P-610 from The American Cancer Society and AM 13883 from The National Institute of Arthritis and Metabolic Diseases.

# [17] Phosphofructokinases from Porcine Liver and Kidney and from Other Mammalian Tissues[1]

*By* Thomas H. Massey and William C. Deal, Jr.

Fructose 6-phosphate + ATP → fructose 1,6-diphosphate + ADP + H⁺

## I. Porcine Liver and Kidney

### Assay

*Principle.* This enzyme may be assayed using one of the following: (a) a pH-stat,[2] (b) pyruvate kinase and LDH, to measure ADP produced, (c) aldolase, triosephosphate isomerase, and glyceraldehyde-3-phosphate dehydrogenase, to measure FDP produced, and (d) aldolase, triosephosphate isomerase, and α-glycerophosphate dehydrogenase[3] to measure FDP produced. The first method[2] has the advantage that it is the only continuously recording assay which does not require auxiliary enzymes. We prefer the last method[3] for routine use, and it is described below.

### Reagents[4]

Tris·HCl, 0.5 $M$, pH 8.0 (50 m$M$) or
Imidazole, 0.5 $M$, pH 7.0 (50 m$M$)
MgCl₂, 50 m$M$ (5 m$M$)
KCl, 0.5 $M$ (50 m$M$)
2-Mercaptoethanol, pure (50 m$M$)
Aldolase, ammonium sulfate suspension, 10 mg/ml
α-Glycerolphosphate dehydrogenase and triosephosphate isomerase
    mixture, crystalline ammonium sulfate suspension, 10 mg/ml
NADH, 20 m$M$ (0.2 m$M$)
ATP (potassium, 0.1 $M$, pH 8.0 (or 7.0), (2 m$M$) or (0.1 m$M$ vs
    1.0 m$M$) or (0.002 m$M$ to 10 m$M$)

[1] T. H. Massey and W. C. Deal, Jr., *J. Biol. Chem.* **248**, 56 (1973). This research was supported in part by NIH postdoctoral fellowship AM-49051, NIH Postdoctoral Traineeship GM-1091, by NIH grant GM-11170, and by the Michigan Agricultural Experiment Station (Hatch 932, Publication No. 6940).
[2] J. E. Dyson and E. A. Noltmann, *Anal. Biochem.* **11**, 362 (1965).
[3] E. Racker, *J. Biol. Chem.* **167**, 843 (1947).
[4] Final concentrations in the assay are given in parentheses at the right; where two choices are given, the first is for the standard assay (pH 8.0) or *maximum activity,* and the second is for assay of *allosteric activity.*

Fructose 6-phosphate (potassium), 0.1 $M$, pH 8.0 (or 7.0), (4 m$M$)
or (0.1 m$M$) or (0.002 m$M$ to 6 m$M$)
Phosphofructokinase solution, 0.1 mg/ml (0.2 $\mu$g/ml to 1 $\mu$g/ml)

Two types of assays may be performed. The first type is carried out
at pH 8.0 and measures *maximum catalytic activity;* the second type
(allosteric) is performed at pH 7.0 and measures the activity under conditions where the enzyme shows *allosteric* properties, namely, inhibition
by excess ATP or citrate and activation by fructose 6-phosphate, cAMP
and other activators.[5,6] In both cases, the decrease in absorbance at 340
nm is measured.

*Standard Assay Procedure.* A stock assay reaction mixture is made
up fresh daily as follows: 1 ml of Tris·HCl, 1 ml of MgCl$_2$, 1 ml of
KCl, 35 $\mu$l of 2-mercaptoethanol, 0.1 ml of aldolase, 0.05 ml of dehydrogenase–isomerase mixture, 0.1 ml of NADH, 0.2 ml of ATP, and 6.5 ml
of H$_2$O. For assays, 0.5 ml of this reaction mixture is added to the cuvette,
followed by 1–5 $\mu$l of pure PFK solution, or 0.02 to 0.1 unit. After mixing,
the background reaction is monitored for 1 minute, or until it is clearly
negligible. Then, the reaction is initiated with 20 $\mu$l of fructose 6-phos-
phate. Ammonium sulfate is carried into the assay (about 30 m$M$) with
the coupling enzyme solutions; this is sufficient to ensure maximum activation of the enzyme.[7]

*Allosteric Assay Procedure.* The stock assay reaction mixture is the
same as for the standard assay except that imidazole buffer, pH 7.0, is
used, ATP is omitted initially, and ammonium ion is removed[7] from the
solutions of coupling enzymes. For qualitative survey work both suspensions of coupling enzymes are centrifuged and, after careful complete
decantation of the supernatant, the centrifuged pellets are dissolved in
the stock assay mixture; this generally reduces the ammonium sulfate
concentration to negligible levels. However, for extremely careful work,
such as determination of kinetic constants, the coupling enzymes are each
dissolved in 1 ml of 10 m$M$ buffer, combined, and dialyzed for 1 hr at
0°. The dialyzed enzymes are added to the other components of the stock
assay reaction mixture, and then 4.65 ml of water is added. The final
reaction mixture contains low fructose 6-phosphate (0.1 m$M$). To analyze
for ATP inhibition, several different assays are run, using various concentrations of ATP (0.002 m$M$ to 10 m$M$). Generally the liver PFK activity

---

[5] D. P. Bloxham and H. A. Lardy, *in* "The Enzymes" (P. Boyer, ed.), 3rd ed.,
Vol. 8, p. 239. Academic Press, New York, 1973.

[6] T. E. Mansour, *Curr. Top. Cell Regul.* **5**, 1 (1972).

[7] High concentrations of ammonium ions can wipe out the allosteric inhibition of
PFK by ATP.

in the presence of 1 m$M$ ATP is only 10% of that in the presence of 0.1 m$M$ ATP; a comparison of PFK activity at these two ATP concentrations provides a quick qualitative test for allosteric inhibition by ATP. To test for the ability of cAMP or AMP to overcome the ATP inhibition, assay at the various ATP concentrations in the presence of 1 m$M$ cAMP or 1 m$M$ AMP. To obtain *sigmoidal* kinetics, the ATP concentration is fixed at 1 or 2 m$M$ and various concentrations of fructose-6-P are used (0.002 m$M$ to 6 m$M$).

### Protein Determination[8]

The procedure is that of Mejbaum-Ketzenellenbogen and Dobryszycka,[8] with very minor modifications.[9] It is as sensitive as the Lowry method, and it has the special advantage that it is not influenced by mercaptoethanol. It is influenced by NAD and perhaps other nucleotides, but these are no problem in this preparation.

### *Reagents*

> Tannin reagent. Dissolve 2 ml of phenol in 98 ml of 1 $N$ HCl. Add 10 g of tannic acid and keep mixture warm until dissolved.
> Gum arabic solution. Dissolve 0.1 g of gum arabic in 100 ml of warm water.
> Standard protein solution (100 $\mu$g/ml). Dissolve 5 mg of crystalline bovine serum ablumin in 50 ml of 0.9% NaCl.

Store all reagents at 0°. Precipitates form in the tannin reagent, and sometimes in the gum arabic reagent, upon cooling or standing; filter (Whatman No. 1) before use. All reagents are available from Sigma.

*Procedure.*[8] Prepare protein standards covering the range from 20 to 100 $\mu$g/ml. For each test sample, standard or unknown: (a) pipette 1.0 ml of protein solution into a 13 × 100 test tube at room temperature; (b) add 1 ml of tannin reagent and let stand 20 min[9] at room temperature[9]; (c) add 2 ml of gum arabic solution to stabilize the turbidity (stable for 1 day); (d) read turbidity at 650 nm, after zeroing against a reagent blank containing 1.0 ml of water instead of protein solution; (e) read protein concentration off standard curve (run standard curve daily).

---

[8] Adapted, with permission, from W. Mejbaum-Katzenellenbogen and W. M. Dobryszycka, *Clin. Chim. Acta* **4**, 515 (1959).

[9] The original procedure[8] calls for incubation at 30° for 10 min.

## Purification

The purification procedure described below can be carried out completely in 8 hr with 1 pig liver or 6–8 pig kidneys. For preparations on a large scale (3–10 livers), it is convenient to extend the preparation over 2 days, ending the first day after step 2. Unless otherwise indicated, centrifugation is carried out at 4° in the Sorvall GSA rotor at 9000 rpm (13,000 $g$) for 15 min or in the Sorvall SS-34 rotor at 18,000 rpm (39,000 $g$) for 5 min.

*Tissues.* Preliminary tests indicate that seasonal effects or aging, or both, are important factors affecting subsequent yield and ease of purification. Tissues obtained from winter animals, and frozen within 20 min after sacrifice of the animal give the best precipitation characteristics, the best yields, and the easiest purification. Single excised livers or 6–8 kidneys are immediately slipped into prelabeled polyethylene freezer bags and packed in dry ice in an ice chest for transport to the laboratory, where they are stored at −20°. Tissues stored for up to 6 months show no appreciable loss of activity or other adverse changes.

*Buffers. Homogenization buffer* contains 50 m$M$ Tris-Cl, 50 m$M$ 2-mercaptoethanol and 5 m$M$ EDTA at pH 8.0. Buffer A contains 50 m$M$ Tris-Cl, 50 m$M$ 2-mercaptoethanol, 5 m$M$ $MgCl_2$, 0.1 m$M$ ATP, and 0.1 m$M$ FDP at pH 8.0. Buffer B contains 110 m$M$ ammonium sulfate, in addition to the components in buffer A.

*Step 1. Homogenization.* Before use in a preparation, the frozen tissue is partially thawed by standing overnight at 4°. One partially frozen pig liver (about 1200 g) is sliced up with a knife and the pieces are homogenized in homogenization buffer (2 liters/kg tissue), for 30 sec (three 10-sec bursts) in a Waring Blendor. The homogenate is centrifuged at 13,000 $g$ for 15 min, the supernatant fluid is removed by aspiration, and the pellet is discarded. The supernatant fluid has a pH of 7.1–7.3 and contains some of the fluffy layer from the pellet.

*Step 2. Heat and Alcohol Treatment.* First, 2-mercaptoethanol (1 ml/100 ml of homogenate supernatant) and then 95% ethanol (20 ml/100 ml of homogenate supernatant) are added and the mixture is brought to 41–42° and incubated for 45 min. After centrifugation at 13,000 $g$ for 15 min, the clear, red supernatant is filtered (Whatman No. 4) using a Büchner filter funnel.

*Step 3. Magnesium-Alcohol Precipitation.* The filtered supernatant from step 2 is cooled to less than 3° and 1 $M$ $MgCl_2$ (5 ml/100 ml is added slowly, with stirring. Incubation for 30 min at 0° precipitates PFK completely. The suspension is centrifuged at 13,000 $g$ for 15 min, the supernatant is removed by aspiration, and the brown pellets are resuspended in buffer A (2 ml/100 g of original wet tissue) at 4° and combined.

*Step 4. Washing the Insoluble Enzyme.* The crude PFK suspension from step 3 is further dispersed at 4° in a 50-ml centrifuge tube with a Teflon pestle and then centrifuged at 39,000 $g$ for 5 min. The supernatant fluid is gently swirled to resuspend the upper fluffy layer of the pellet, transferred to a separate 50-ml centrifuge tube, dispersed with the Teflon pestle and centrifuged again. The second supernatant fluid is again gently swirled to resuspend the upper fluffy layer of the second pellet and discarded. The first and second hard brown pellets are combined and resuspended in buffer A (2 ml/100 g of original wet tissue) at 4°. The suspension is then dispersed with a Teflon pestle and again centrifuged at 39,000 $g$ for 5 min at 4°. The supernatant fluid is discarded. In this fashion the material insoluble in buffer A is washed 4 times or until the supernatants after centrifugation are colorless and clear. A hard, brown pellet is obtained at this stage.

*Step 5. Extraction and Reprecipitation by Dilution.* In the original method,[1] the final pellet from step 4 is suspended in buffer B (0.2 ml/100 g of original wet tissue) at 25°, gently mixed for 10 min taking care to prevent foaming, and centrifuged at 12,000 $g$ for 5 min at room temperature. The resulting pellet is reextracted by the same procedure. The two supernatant fluids are combined, diluted with 4 volumes of buffer A and incubated at 0° for 30 min to allow precipitation of phosphofructokinase. A recent modification[10] of the original method[1] greatly increases the overall yield: the pellet from step 4 is extracted two times with a minimal volume of buffer B (0.2 ml/1000 g of original wet tissue) at 25° and then two more times with a larger volume of buffer B (0.7 ml/1000 g of original wet tissue); the extracts are combined and *dialyzed* 2 hr at 0° against buffer A (instead of being diluted). Then enzyme precipitates as very fine needles. Upon swirling, the precipitated enzyme suspension exhibits a sheen like that seen in ammonium sulfate suspensions of crystalline enzymes.

The enzyme suspension is centrifuged in a 12-ml glass centrifuge tube at 39,000 $g$ for 5 min at 4°. The extraction-precipitation process is repeated with this pellet and succeeding pellets until a white pellet is obtained upon centrifugation of the precipitated enzyme. The crystalline enzyme is stored in buffer A at a concentration of about 4 mg/ml. It is dissolved and reprecipitated weekly with freshly prepared solutions. It is more stable kept under nitrogen or in dithiothreitol.[10]

Results from typical preparations of liver and kidney PFK are shown in Table I.[1]

[10] J. L. Trujillo and W. C. Deal, Jr., *Fed. Proc.* 33, 1384 (1974).

TABLE I

PREPARATION OF CRYSTALLINE PHOSPHOFRUCTOKINASES FROM PORCINE LIVER AND KIDNEY[a]

| Fractions | Pig liver (1375 g) | | | | Pig kidney (1072 g) | | | |
|---|---|---|---|---|---|---|---|---|
| | Total volume (ml) | Total protein[b] | Total units[c] | Specific activity[d] (units/mg) | Total volume (ml) | Total protein[b] | Total units[c] | Specific activity[d] (units/mg) |
| Homogenate supernatant | 3150 | 189 g | 1190 | 0.0063 | 2260 | 87 g | 1230 | 0.014 |
| Heat-alcohol filtrate | 2750 | 62 g | 468 | 0.0074 | 2000 | 16 g | 905 | 0.057 |
| Magnesium-alcohol suspension | 63 | 3.5 g | 428 | 0.123 | 30 | 0.64 g | 503 | 0.79 |
| Washed enzyme | 29 | 14 mg | 197 | 14.1 | 22 | 83 mg | 404 | 4.9 |
| Final suspension (crystalline) | 10 | 2.1 mg | 210 | 100 | 8 | 1.8 mg | 230 | 128 |

[a] Adapted from T. H. Massey and W. C. Deal, Jr., *J. Biol. Chem.* **248**, 56 (1973).

[b] Protein is measured using a modification of the tannic acid-turbidity method [W. Mejbaum-Katzenellenbogen and W. M. Dobryszycka, *Clin. Chem. Acta* **4**, 515 (1959)].

[c] One unit of enzyme activity is defined as the quantity of enzyme which phosphorylates 1 μmole of fructose 6-phosphate per minute; in the assay described, this is equivalent to the quantity of enzyme which results in the oxidation of 2 μmoles of NADH per minute.

[d] Specific activity is defined as units per milligram of protein.

## Properties

*Structural Properties.* In the protein concentration range of 3–8 mg/ml, the liver PFK exhibits a two-peak ($s_{20,w}^{0.5\%} = 55$ S and 64 S) schlieren pattern in sedimentation velocity studies at 23° in the Model E ultracentrifuge; it is an associating-dissociating system in rapid equilibrium.[1] The molecular weight of the purified enzyme is above 3.5 million and is concentration dependent.[1,10] The enzyme becomes progressively more dissociated as the temperature is increased over the range from 4° to 35°[10]; however, the dissociation is only partial, because the lowest sedimentation coefficient observed is 45 S.[10] This enzyme is one of the largest phosphofructokinases known; it also differs considerably from the widely studied rabbit muscle PFK, which exhibits a 3-peak system (13, 18, 30 S).[11]

*Purity and Electrophoretic Properties.*[1] The crystalline liver PFK is pure, since it exhibits only a single band on SDS gel electrophoresis. The isoelectric point of pig liver PFK is pH 5.0. Interestingly, in imidazole buffers, maximum precipitation of pig liver PFK by $Mg^{2+}$ occurs in the pH range from 7.2 to 7.5; hence this is not an isoelectric precipitation. The native enzyme electrophoreses into conventional polyacrylamide gels only in the presence of ATP or ADP (or, to a lesser extent, fructose 6-phosphate); even then, it migrates as a broad patch, demonstrating again its tendency to association–dissociation reactions.[1]

*Catalytic Properties.*[12] In the presence of 5 m$M$ MgCl$_2$, the $K_m$ for ATP is 10 $\mu M$ at pH 7.0 in 50 m$M$ imidazole Cl, and 25 $\mu M$ at pH 8.0 in 50 m$M$ Tris Cl, in the presence and in the absence of 1.0 m$M$ AMP. The $K_m$ for fructose 6-phosphate is 23 $\mu M$ at both pH values, but the activity curve is sigmoidal (Hill coefficient $n = 2.8$) in the absence of AMP and hyperbolic ($n = 0.9$) in the presence of 1 m$M$ AMP. In the absence of AMP, using the pH 7.0 system, the $K_i$ for ATP is approximately 0.25 m$M$. In the pH 7.0 assay system, an ATP concentration of 1 m$M$ gives a 90% inhibition of the activity observed with 0.1 m$M$ ATP.

## II. Other Mammalian Tissues[13]

The previously described procedure is useful for purification of phosphofructokinases from a large number of different tissues, but modifications are necessary in some cases. This section describes the types of modifications which have been successful with a large number of mammalian tissues and explains the principles for application of the modifica-

[11] K.-H. Ling, F. Marcus, and H. A. Lardy, *J. Biol. Chem.* **240**, 1893 (1965).
[12] T. H. Massey and W. C. Deal, Jr., unpublished results.
[13] T. H. Massey and W. C. Deal, Jr., unpublished results.

tions to other tissues. Results from a survey of purification properties of phosphofructokinases from a wide variety of tissues are presented in Table II.[13] In all except the first two cases, only limited studies were carried out to demonstrate the feasibility of the approach; it should be possible to easily purify the partially purified enzymes to homogeneity, with a minimum of additional effort. In addition, other workers[14-16] have demonstrated the widespread applicability of this purification procedure approach: phosphofructokinases have been purified to homogeneity from goat mammary gland by Ramaiah[14] and from erythrocytes by Rose[15]; also PFK from pig spleen has been purified to a specific activity of 4-8 international units/mg.[16]

### Key Purification Principles

*Precipitation of Unwanted Protein.* As Bloxham and Lardy[5] have pointed out, many PFK purification procedures utilize a heat step to precipitate unwanted protein; frequently PFK is then itself precipitated with ethanol at temperatures below 0°. Our basic procedure[1] differs from these in that 16% ethanol is present in the heat step (41–42°) for precipitation of unwanted proteins. Under proper conditions, almost all mammalian PFK's survive such a step well, generally with less than 50% loss; on the other hand, PFK's from lower forms[17] of life generally do not. *This step is so universally successful that modifications in it seem undesirable.*

*Selective Precipitation of PFK. This is the key step where differences in PFK properties from different tissues are exhibited and hence, where modifications are useful to maximize selectivity in precipitation.* Actually it is convenient to consider that there are two key PFK precipitation steps, namely (a) the initial precipitation of PFK and (b) subsequent precipitation of PFK. The reason is that more stringent precipitating conditions are necessary in the initial precipitation, probably because of the higher ionic strength of the initial solution and because of the presence of solubilizing metabolites (ATP, fructose 6-phosphate, and other compounds).[1]

*Very rarely will modifications in the initial precipitation step be useful, because we are already using the next-to-strongest precipitating con-*

---

[14] A. Ramaiah, private communication, 1973.
[15] I. A. Rose, private communication, 1973.
[16] P. E. Hickman and M. J. Weidmann, *FEBS Lett.* **38**, 1 (1973).
[17] Extracts of brewer's yeast and earthworm (*Lumbricus terrestris*) lose all PFK activity during the heat treatment. T. Massey and W. C. Deal, unpublished results.

TABLE II
SURVEY OF PURIFICATION PROPERTIES OF PHOSPHOFRUCTOKINASES
FROM VARIOUS SOURCES[a]

| Tissue | $Mg^{2+}$-precipitable[b] from the heat-alcohol supernatant? | Insoluble in buffer A?[c] | Final purity (units/mg) |
|---|---|---|---|
| Pig liver | Yes | Yes | 100[f] |
| Pig kidney | Yes | Yes | 128[f] |
| Pig heart | Yes | No[d-1] | 43 |
| Pig small intestine | Yes | No | — |
| Pig muscle | No[b] | No | — |
| Pig adipose tissue | No[b] | No | — |
| Pig brain | Yes (partially)[b] | No | — |
| Sheep liver | Yes | Yes | 28 |
| Sheep heart | Yes | No | — |
| Rabbit liver | Yes | No[d-2] | 39[e] |
| Bovine liver | Yes | No[d-3] | Good |
| Bovine kidney | Yes | Yes | 41 |
| Bovine heart | Yes | No[d-4] | Good[f] |
| Human liver[g] | Yes | — | — |
| Rat liver[h] | Yes | — | >70 |

[a] T. H. Massey and W. C. Deal, Jr., unpublished results.

[b] In all cases where some or all of the PFK is *not* precipitable with $MgCl_2$, the addition of 10–25 m$M$ $MnCl_2$ *does* produce precipitation. However, contaminating coprecipitated protein usually hampers subsequent further purification.

[c] Standard buffer A contains 50 m$M$ Tris·HCl, pH 8.0 (at 20°), 5 m$M$ $MgCl_2$, 50 m$M$ β-mercaptoethanol, 0.1 m$M$ ATP, and 0.1 m$M$ FDP.

[d] Not insoluble in buffer A, but insoluble in *modified* buffer A, containing the following additions or substitutions:

[d-1] 5% ethanol plus 5 m$M$ imidazole, pH 7.0 in place of the Tris buffer at pH 8.0.

[d-2] 9% ethanol and 20 m$M$ imidazole, pH 7.0 in place of the Tris buffer. PFK is extracted in 30 m$M$ ammonium sulfate and diluted to precipitate the enzyme.

[d-3] 5 m$M$ $MnCl_2$, in addition to the 5m$M$ $MgCl_2$. PFK was extracted in 0.1$M$ ammonium sulfate and diluted 10-fold to precipitate the enzyme.

[d-4] 10 m$M$ imidazole at pH 7.0 in place of the Tris buffer. The pellet from the first Mg precipitation is extracted with 25% saturated ammonium sulfate (144 g plus 1 liter water) and then precipitated in 50% saturated ammonium sulfate (313 g plus 1 liter of water). After precipitation the pellet is dissolved in modified buffer A and the enzyme is precipitated by dialysis against the same buffer.

[e] SDS gel electrophoresis shows two bands (about 50–50 mixture).

[f] The preparation is white and crystalline.

[g] J. L. Trujillo and W. C. Deal, Jr., unpublished results.

[h] W. W. Farrar and W. C. Deal, Jr., unpublished results.

*ditions possible (see below).* The two major possibilities for modification are: (a) addition of 25 m$M$ MnCl$_2$ and (b) dialysis, against 10 m$M$ buffer containing the divalent cations and alcohol. As discussed below, the Mn$^{2+}$ precipitation is much less specific than Mg$^{2+}$ precipitation; therefore it should be used only as a last resort. The volume of solution in large-scale preparations generally makes it impractical to dialyze in the initial precipitation step.

*However, modifications in subsequent precipitation will be both useful and necessary.* The reason is that we want to use the mildest possible precipitating conditions (those which precipitate PFK and yet precipitate the least amount of non-PFK protein); the conditions which meet this criterion vary from tissue to tissue (Table II). A knowledge of the key precipitation characteristics of phosphofructokinases is necessary to serve as a guide to development of logical modifications of the subsequent precipitation steps. Phosphofructokinases from various tissues tested (Table II) have the following characteristics at 0° in the precipitation systems used[13]:

System 1. All are precipitated by 25 m$M$ MnCl$_2$ in the presence of 16% ethanol; however, precipitation with Mn$^{2+}$ is not very selective (contaminating proteins are also precipitated), and it may be difficult to purify PFK to homogeneity.

System 2. All, except pig muscle and pig adipose tissue, are precipitated by 50 m$M$ MgCl$_2$ in the presence of 16% ethanol; the precipitation is fairly selective and precipitated PFK is generally easy to purify further.

System 3. Many are precipitated with buffer A, which contains no ethanol and only 5 m$M$ MgCl$_2$. This precipitation is the most selective, and weakest, precipitation system of all (by weakest, we mean it precipitates the least amount of total protein). If precipitation does not occur with it as is, the following modifications (recommended for trial in the order given) progressively increase the strength of the precipitating system:

Variable a. Dialyze the enzyme solution, instead of diluting it, to reduce the ionic strength; this keeps the protein concentration high—always desirable where feasible.

Variable b. Reduce the buffer concentration to 10 m$M$.

Variable c. Adjust the pH to that giving maximum precipitation, to be determined empirically; initially pH 7.0–7.5 is a good choice.

Variable d. Add 1 $M$ MgCl, in increments, to a higher final concentration up to 50 m$M$.

Variable e. Add 95% ethanol, in increments, to a final concentration up to 16% (v/v); this essentially yields System 2.

Variable f. Add ATP to a final concentration of 0.5 to 2 m$M$, keeping the $Mg^{2+}$ at least 5 m$M$ greater than the ATP. This may help remove the endogenous solubilizer, fructose 6-phosphate.

Variable g. Add MnCl, in increments, up to a final concentration of 25 m$M$; this essentially yields System 1.

In these purification procedures, EDTA must be avoided, since free $Mg^{2+}$ is necessary for precipitation of PFK. Precipitation with $Mn^{2+}$ may be useful only for precipitating pure PFK which has been so diluted that other methods fail to precipitate it.

*Solubilization of Unwanted Protein.* Basically this solubilization involves washing the PFK pellet from the preceding step, under the minimum conditions necessary to keep the PFK precipitated while unwanted proteins are solubilized. The conditions generally will be the same as in the preceding step or, as the preparation become homogeneous, milder precipitation conditions may be used, thereby solubilizing more unwanted protein.

*Selective Metabolite Solubilization of PFK and Selective Precipitation by Removal of Metabolites. This is the second step where selective modification is potentially very useful.* The basic original procedure[1] utilizes the selective solubilization by 110 m$M$ ammonium sulfate of PFK insoluble in ammonium sulfate concentrations below about 10 m$M$; the only obvious modification in this ionic strength-dependent solubilization is to vary the concentration of ammonium sulfate used for solubilization and empirically determine, and use, the minimum concentration giving complete solubilization. Increasing the ammonium sulfate concentration above 110 m$M$ is generally unfavorable, since it increases the solubility of contaminating proteins.

However, all PFKs tested thus far also show metabolite-dependent solubilization properties.[1,13] For example, with no added ammonium sulfate, pig liver PFK is completely insoluble in 0.1 m$M$ FDP but it is completely soluble in 0.1 m$M$ fructose 6-phosphate or 1 m$M$ ATP.[1]

With this type of modification, pig liver PFK has been purified[1] to a specific activity of 53. Although not homogeneous, like enzyme obtained by our standard procedure, the PFK is highly purified. Furthermore, there may be cases where this modification will provide the best purification, so the method[1] used with pig liver PFK is provided as an example or guide for designing such an experiment. The "washed" enzyme from a single liver (1.7 kg) is resuspended in 17 ml of buffer A and made 2 m$M$ (or lower) in fructose 6-phosphate to selectively dissolve the enzyme; the solution is then centrifuged to remove insoluble protein impurities. Then ATP is added to a final concentration of 2.1 m$M$. This initiates conversion of the solubilizer, fructose 6-phosphate, to the nonsolubilizer, fructose 1,6-diphosphate. Within 10–15 sec, the solution becomes cloudy as PFK begins to precipitate. After standing 30 min at 0°, the heavy precipitate is collected by centrifugation. Then the steps of solubilization by fructose 6-phosphate and subsequent precipitation by addition of ATP are repeated. It seems likely that, with some tissues, this will be a very useful modification, since it potentially possesses a very high degree of specificity.

*Other Preparations.* Phosphofructokinase has been partially purified from rabbit liver[18,19] and sheep liver.[20] The procedure described here is the first to yield either homogeneous or crystalline PFK from mammalian liver,[1] mammalian kidney,[1] or from goat mammary gland[14]; it has also provided the first substantially purified preparation of PFK from pig[13] and bovine[13] heart, bovine liver[13] and kidney,[13] pig spleen,[16] as well as the highest specific activity of sheep liver[13] PFK. Homogeneous crystalline PFK has been obtained from chicken liver.[21] Preparations for PFK from other sources are referenced in two recent reviews[5,6] and in other articles in this series.[22–24]

[18] A. Ramaiah and G. A. Tejwani, *Biochem. Biophys. Res. Commun.* **39**, 1149 (1970).
[19] Robert G. Kemp, *J. Biol. Chem.* **246**, 245 (1971).
[20] D. J. H. Brock, *Biochem. J.* **113**, 235 (1969).
[21] Norio Kono, K. Uyeda, and R. M. Oliver, *J. Biol. Chem.* **248**, 8592 (1973).
[22] K.-H. Ling, V. Paetkau, F. Marcus, and H. A. Lardy, this series, Vol. 9 [77a].
[23] T. E. Mansour, this series, Vol. 9 [77b].
[24] A. Sols and M. L. Salas, this series, Vol. 9 [77c].

# [18] Phosphofructokinase from Human Erythrocytes

*By* ROBERT B. LAYZER

Fructose 6-phosphate + ATP → fructose 1,6-diphosphate + ADP

## Assay

*Principle.*[1] Phosphofructokinase (PFK) activity is assayed spectrophotometrically by coupling the reaction with aldolase, triosephosphate isomerase, and $\alpha$-glycerophosphate dehydrogenase, so that 2 moles of NADH are oxidized for each mole of fructose diphosphate formed.

*Reagents.* The composition of the reaction mixture is shown in Table I. An ancillary enzyme solution is prepared by mixing 0.3 ml of rabbit muscle aldolase (10 mg/ml), 0.1 ml of a mixture of $\alpha$-glycerophosphate dehydrogenase and triosephosphate isomerase (9.5 mg/ml), and 0.5 ml

[1] E. Racker, *J. Biol. Chem.* **167**, 843 (1947).

TABLE I
REAGENTS FOR ASSAY OF PHOSPHOFRUCTOKINASE ACTIVITY

| Reagent | Amounts used for 10 ml of reaction mixture | Concentration in reaction mixture (mM) | Concentration in cuvette (mM) |
|---|---|---|---|
| Tris·HCl, 40 mM, pH 8.0 | 2.0 ml | 80 | 40 |
| MgSO₄, 40 mM | 1.0 ml | 4 | 2 |
| ATP, 0.10 M, pH 6.8 | 0.2 ml | 2 | 1 |
| H₂O | 6.8 ml | — | — |
| Dithiothreitol | 3.1 mg | 2 | 1 |
| NADH | 3.6 mg | 0.5 | 0.25 |
| Na₂-fructose 6-phosphate, 10 mM | — | — | 1 |

of 50 mM Tris·HCl, pH 8.0. The reagents are obtained from Sigma Chemical Co. except for dithiothreitol, which was obtained from Calbiochem Corporation.

*Procedure.* The reaction mixture (without NADH) is prepared in advance as detailed in Table I, and stored frozen until needed. After thawing and addition of NADH, the mixture can be used for one day. The ancillary enzyme mixture is usually stable for 2 weeks. For assay of PFK activity, 0.5 ml of reaction mixture, 0.02 ml of ancillary enzyme mixture, the sample, and water to make 0.9 ml are mixed in a 1-ml cuvette. The reaction is started by addition of 0.1 ml of fructose 6-phosphate (or water in the control cuvette.) The rate of oxidation of NADH is followed at 25° in a spectrophotometer equipped with a recorder and temperature control device, by measuring the rate of decline of optical density at 340 nm. After a brief lag, the rate is linear and is proportional to PFK concentration if the rate of optical density change is less than 0.20 per minute.

*Units.* One unit of PFK catalyzes the conversion of 1 μmole of fructose 6-phosphate to fructose 1,6-diphosphate per minute at pH 8.0 and 25°. To calculate the number of units of PFK activity in the cuvette, multiply the optical density change per minute by 0.0804 (that is, $1/6.22 \times 1/2$).

## Purification[2]

All procedures are carried out at 0–5° unless otherwise stated.

*Step 1. Preparation of Hemolysate.* Three to four units of whole blood

[2] An earlier version of this method is found in R. B. Layzer, L. P. Rowland, and W. J. Bank, *J. Biol. Chem.* **244**, 3823 (1969).

(obtained as fresh as possible from the blood bank, but usually outdated) are centrifuged for 10 min at 3500 $g$. The plasma and buffy coat are removed by suction. The red cells are washed twice in 0.15 $M$ NaCl buffered with 5 m$M$ K phosphate, pH 7.0. The washed, packed red cells are then mixed with an equal volume of distilled water containing 0.2% saponin and 1 m$M$ EDTA and are allowed to stand for 30 min.

*Step 2. Batch DEAE-Cellulose.* This step removes most of the hemoglobin. The hemolysate is mixed with an equal volume of 5 m$M$ K phosphate, pH 7.0, containing 0.1 m$M$ EDTA and DEAE-cellulose (Whatman DE-52) 75 g/liter. After 1 hr of intermittent gentle mixing, the DEAE-cellulose is centrifuged for 10 min at 5000 $g$ and the supernatant discarded. The DEAE-cellulose is washed in a large Büchner funnel with about 6 liters of 5 m$M$ K phosphate, pH 7.0, containing 0.1 m$M$ EDTA, followed by about 2 liters of 50 m$M$ K phosphate, pH 8.0,[3] with 0.1 m$M$ EDTA, until only a faint red color remains. To elute the PFK, the DEAE-cellulose is mixed with 0.25 $M$ K phosphate, pH 8.0,[3] with 0.2 m$M$ EDTA, in a volume equal to half the original hemolysate volume. After 30 min the mixture is centrifuged at 5000 $g$ for 10 min and the supernatant is filtered through glass wool and reserved. The elution is repeated with another portion of eluting buffer (one-fourth the original hemolysate volume) and the supernatants are combined.

*Step 3. Ammonium Sulfate Precipitation.* Solid ammonium sulfate is added gradually to 40% saturation (24.3 g/100 ml), as the pH is kept at 8.0 by addition of NaOH. The suspension is equilibrated for 45 min and then centrifuged at 10,000 $g$ for 20 min. The pellet is dissolved in 0.1 $M$ K phosphate, pH 8.0, containing 0.2 m$M$ EDTA, 0.2 m$M$ AMP, 0.5 m$M$ fructose 1,6-diphosphate, and 0.7 m$M$ dithiothreitol, in a volume approximately one-eighth the original hemolysate volume.

*Step 4. Heat Treatment.* The solution is heated with constant stirring at 65° for 20 min, cooled in an ice bath to 0–5°, and centrifuged at 10,000 $g$ for 30 min. The pellet is discarded. To the supernatant, solid ammonium sulfate is added to 35% saturation (20.9 g/100 ml), the suspension is equilibrated for 45 min, and precipitated PFK is centrifuged down at 10,000 $g$ for 20 min. The pellet is redissolved in a small volume of 50 m$M$ Tris-phosphate, pH 8.0, containing 0.2 m$M$ EDTA, 0.2 m$M$ AMP, 2.0 m$M$ fructose 6-phosphate, and 0.7 m$M$ dithiothreitol.

*Step 5. DEAE-Cellulose Chromatography.* The sample is dialyzed in a hollow-fiber mini-beaker (Bio-Rad) by allowing buffer to flow through the fibers under gravity for 30–45 min and applied to a 2.5 × 45 cm column of DEAE-cellulose (Whatman DE-52) equilibrated with the same

---

[3] K. W. Wenzel, J. Gauer, G. Zimmermann, and E. Hofmann, *FEBS Lett.* **19**, 281 (1972).

TABLE II

PURIFICATION OF HUMAN ERYTHROCYTE PHOSPHOFRUCTOKINASE

| Step | Volume (ml) | Total activity (units) | Protein (mg) | Specific activity (units/mg) | Yield (%) |
|------|------|------|------|------|------|
| 1. Hemolyzate | 1200 | 612 | 153,000 | 0.004 | 100 |
| 2. Batch DEAE-cellulose | 850 | 447 | 2,240 | 0.20 | 73 |
| 3. Ammonium sulfate | 150 | 422 | 584 | 0.73 | 69 |
| 4. Heat treatment | 8.0 | 367 | 111 | 3.3 | 60 |
| 5. DEAE-cellulose | 8.0 | 268 | 28 | 9.6 | 44 |
| 6. Blue Dextran | 1.5 | 206 | 2.3 | 90 | 34 |

buffer. After washing with 200 ml of the starting buffer, PFK is eluted overnight with a linear gradient formed from 250 ml of 50 m$M$ Tris-phosphate, pH 8.0, and 250 ml of 600 m$M$ Tris-phosphate, pH 8.0, with the same additives as in the starting buffer. Ten-milliliter fractions are collected and the tubes with peak activity are pooled. Solid ammonium sulfate is then added to 60% saturation (39.0 g/100 ml.) The suspension is equilibrated for 90 min, and centrifuged at 10,000 $g$ for 20 min. The pellet is redissolved in 50 m$M$ K phosphate, pH 7.1, containing 0.5 m$M$ EDTA, 2 m$M$ fructose 6-phosphate, and 0.7 m$M$ dithiothreitol.[4]

*Step 6. Affinity Chromatography with Blue Dextran 2000®.* This step was adapted from the procedure described by Kopperschläger et al.[5] Blue Dextran 2000 (Pharmacia) is incorporated into polyacrylamide gel as follows: 10 g of Cyanogum 41 is dissolved in 80 ml of 50 m$M$ K phosphate, pH 7.1, next, 1.3 g of Blue Dextran 2000 is dissolved, and the volume is made up to 100 ml with the buffer. $N,N,N',N'$-Tetramethyl-ethylenediamine, 0.03 ml, and ammonium persulfate, 70 mg, are added, and the gel is allowed to polymerize at room temperature in a beaker. The solid gel is passed through a 50-mesh wire screen, and small particles are removed by straining through a 100-mesh screen as well as by repeated settling in buffer in a cylinder. A 1.5 × 30 cm column of gel is washed overnight with 50 m$M$ K phosphate, pH 7.1, containing 0.5 m$M$ EDTA, 2 m$M$ fructose 6-phosphate, and 0.7 m$M$ dithiothreitol. The PFK sample, after dialysis against the same buffer in a hollow-fiber mini-beaker, is applied to the column and washed in with 100 ml of the same buffer. PFK is eluted with 200 ml of the same buffer containing, in addi-

[4] AMP is omitted from the buffer because it prevents PFK from adhering to the Blue Dextran gel in the next step.
[5] G. Kopperschläger, W. Diezel, R. Freyer, S. Liebe, and E. Hofmann, *Eur. J. Biochem.* **22**, 40 (1971).

tion, 5 m$M$ ATP. Tubes with peak activity are pooled and concentrated by ultrafiltration (Diaflo, Amicon). For storage, the final buffer is adjusted to pH 8.0 and made to contain 1 m$M$ EDTA, 1 m$M$ AMP, 2 m$M$ fructose 6-phosphate, and 1 m$M$ dithiothreitol.

A summary of the purification appears in Table II.

## Properties

*Stability.* At 5°, human erythrocyte PFK stored in the final buffer loses about 30% of its activity per month. As is the case with purified PFK from other sources, loss of activity is hastened by dilution, low pH, and absence of stabilizing substances such as AMP, fructose 6-phosphate, fructose 1,6-diphosphate, EDTA, and dithiothreitol.

*Purity.* The purified enzyme migrates as a single band under electrophoresis in polyacrylamide gels containing sodium dodecyl sulfate. Double immunodiffusion in agarose gels, using antiserum against partly purified human erythrocyte PFK,[2] gives a single precipitation line with purified PFK, and multiple lines with partly purified PFK.

*Physical Properties.* From SDS-acrylamide gel electrophoresis, the subunit molecular weight is 76,000 in our laboratory, but it was reported by Wenzel *et al.* to be 104,000.[6] Assuming that the smallest active form of human PFK is a tetramer (like rabbit PFK),[7] its molecular weight is therefore probably between $3.0 \times 10^5$ and $4.2 \times 10^5$. In sucrose density gradient centrifugation studies, the molecular weight of human muscle PFK was estimated to be $3.8 \times 10^5$.[2] Erythrocyte PFK could not be evaluated by this method because of its marked tendency to aggregate, even at a concentration of 0.1 unit/ml.

Erythrocyte PFK migrates as a single band during electrophoresis on cellulose acetate membranes, in vertical starch gels, and in agarose gels.[2] The enzyme penetrates polyacrylamide gels very poorly, probably because of the tendency to aggregation.

*Kinetics.*[2] In the absence of ammonium sulfate, Lineweaver-Burk plots for one substrate yield a series of parallel lines when activity is measured at different fixed concentrations of the second substrate. The $K_m$ and $V_{max}$ for each substrate are increased by raising the concentration of the other substrate. A limiting $K_m$ ($K_m'$) for each substrate can be obtained from a secondary plot of $1/K_m$ against $1/[\text{second substrate}]$. The $K_m'$ for ATP is 0.12 m$M$, and the $K_m'$ for fructose 6-phosphate is 45 $\mu M$. In the standard assay, activity is maximal at pH 8.5 and markedly diminished at pH 7.0. The enzyme is inhibited by ATP in physiological

[6] K. W. Wenzel, G. Zimmermann, J. Gauer, W. Diezel, S. Liebe, and E. Hofmann, *FEBS Lett.* **19**, 285 (1972).

[7] M. Y. Tsai and R. G. Kemp, *J. Biol. Chem.* **248**, 785 (1973).

concentrations, and this inhibition is counteracted by increasing the concentration of fructose 6-phosphate. Erythrocyte PFK is relatively insensitive to inhibition by citrate.

*Other Preparations.* Human muscle PFK has been prepared[2] in nearly pure form by the procedure of Ling *et al.*[8] with minor modifications. Wenzel *et al.*[3] reported the preparation of completely pure human erythrocyte PFK with SA 136 units/mg in a yield of 51%. Their method differs from that described above primarily in the last two steps, which consist of gel chromatography on Sephadex G-200 followed by Sepharose 4B.

[8] K.-H. Ling, F. Marcus, and H. A. Lardy, *J. Biol. Chem.* **240**, 1893 (1965).

## [19] Ribulose-5-phosphate Kinase from *Chromatium*[1]

*By* BETH A. HART and JANE GIBSON

D-Ribulose 5-phosphate + ATP → D-ribulose 1,5-diphosphate + ADP

### Assay Method

*Principle.* The assay depends on measurement of ADP formed during the reaction. This can be followed continuously by enzymic coupling to the oxidation of NADH measured spectrophotometrically at 340 nm by phosphoenol pyruvate, pyruvate kinase, and lactic dehydrogenase added to the ribulose-5-phosphate kinase reaction mixture.[2] Discontinuous measurements can be made by adding the same mixture to samples of heat-inactivated ribulose-5-phosphate kinase reaction mixture and measuring $\Delta OD_{340\ nm}$ after completion of the coupled reaction. An assay based on increase in alkali-labile phosphate[3] must be used if ADP is present in the reaction mixture.

*Reagents*

Buffer:salts containing triethanolamine:HCl, pH 7.8, 0.5 $M$; $MgCl_2$, 0.2 $M$ and KCl, 0.4 $M$
Ribose 5-phosphate, Na salt, 50 m$M$
*or* Ribulose 5-phosphate, Na salt, prepared from D-gluconate-6-phosphate,[4] 16 m$M$
ATP, diNa salt, pH 7.8, 50 m$M$
Reduced glutathione, pH 7.8 (GSH), 20 m$M$

[1] B. A. Hart and J. Gibson, *Arch. Biochem. Biophys.* **144**, 308 (1971).
[2] E. Racker, *Arch. Biochem. Biophys.* **69**, 300 (1957).
[3] J. Hurwitz, A. Weissbach, B. L. Horecker, and P. Z. Smyrniotis, *J. Biol. Chem.* **218**, 769 (1956).
[4] S. Pontremoli and G. Mangiarotti, *J. Biol. Chem.* **237**, 643 (1962).

Phosphoenol pyruvate, tricyclohexylamine salt (PEP), 25 m$M$

NADH, 1.28 m$M$

Solution containing 50 $\mu$g/ml crystalline lactic dehydrogenase (LDH) and 25 $\mu$g/ml crystalline pyruvate kinase (PK)

Excess spinach ribose-5-phosphate isomerase[3]

Ribulose 5-phosphate may also be generated as an equilibrium mixture by incubating the ribose 5-phosphate solution with excess spinach ribose-5-phosphate isomerase at 38° for 5 min. Isomerase is inactivated by rapid heating to 100° for 90 sec; the equilibrium mixture should be generated fresh daily.

*Procedure A: Continuous Assay.* The reaction mixture (0.1 ml of buffer:salts, 0.1 ml of ribulose 5-phosphate, 0.1 ml of ATP, 0.1 ml of GSH, 0.1 ml of PEP, 0.1 ml of NADH, 0.1 ml of LDH:PK mixture, and 0.2 ml of water) is prewarmed to 37° in the thermostated compartment of a recording spectrophotometer, and the reaction started by addition of 0.1 ml of a suitable dilution or ribulose-5-phosphate kinase. The change in absorbance at 340 nm is followed for 5–10 min; a linear rate is usually established within 0.5 min after adding the enzyme. This assay is suitable for following activity during purification, and has the advantage of maintaining a constant ATP concentration.

*Procedure B: Discontinuous Assay.* The primary reaction mixture (0.15 ml of buffer:salts, 0.15 ml of ribulose 5-phosphate, 0.15 ml of ATP, 0.15 ml of GSH, and 0.75 ml of water) is prewarmed to 37°, and the reaction started by addition of 0.15 ml of diluted ribulose-5-phosphate kinase. At appropriate intervals, usually 15 or 30 sec, 0.2-ml samples are withdrawn with an automatic pipette, transferred rapidly to small tubes held in a boiling water bath, and kept at 100° for exactly 90 sec. After cooling to 37°, 0.8 ml of the secondary reaction mixture (0.1 ml of buffer:salts, 0.1 ml of PEP, 0.1 ml of NADH, 0.1 ml of LDH:PK solution, 0.4 ml of water for each sample tube) is added, and the absorbance at 340 nm is read after 10 min. The quantity of ribulose-5-phosphate kinase in the primary mixture is adjusted so that at least four samples are taken during the linear part of the reaction. This assay is preferable for most kinetic experiments.

*Definition of Unit.* One unit is defined as the quantity of enzyme which forms 1 $\mu$mole of ADP in 1 min at 37° in the reaction mixtures given above.

## Purification

*Growth of Cultures. Chromatium* (ATCC 17899) or similar strain D is grown in completely filled 10-liter bottles, closed with rubber stoppers, or in a 14-liter New Brunswick fermenter vessel preflushed with nitrogen gas end gently stirred at about 50 rpm, in medium of the following composition: 1 g of $NH_4Cl$, 1 g of $KH_2PO_4$, 0.5 g of $MgCl_2 \cdot 6H_2O$, 0.01 g of $CaCl_2 \cdot 2H_2O$, 4 g of $NaHCO_3$, 3 g of $Na_2S_2O_3 \cdot 5H_2O$, 0.5 g of $Na_2S \cdot 9H_2O$, and 10 ml of trace element solution (500 mg of EDTA, 200 mg of $FeSO_4 \cdot 7H_2O$, 10 mg of $ZnSO_4 \cdot 7H_2O$, 3 mg of $MnCl_2 \cdot 4H_2O$, 30 mg of $H_3BO_3$, 20 mg of $CoCl_2 \cdot 6H_2O$, 1 mg of $CuCl_2 \cdot 2H_2O$, 2 mg of $NiCl_2 \cdot 6H_2O$, 3 mg of $Na_2MoO_4 \cdot 2H_2O$ in 1 liter of distilled water) for each liter of distilled water; the final pH is adjusted to 7.8 if necessary. The $NaHCO_3$, $Na_2S_2O_3$, and $Na_2S$ are made up as concentrated solutions, autoclaved separately, and added to the bulk of the medium after sterilizing. Medium for large-scale cultures can be used without sterilizing since it supports the growth of few other organisms. A 10–20% inoculum in the same medium is used, and the cultures are illuminated with 60-W incandescent bulbs or the New Brunswick neon lamp manifold and maintained at 30–34°. The cultures are harvested as soon as sulfur granules have disappeared from the cells, usually about 4 days, washed once in 0.1 $M$ triethanolamine:HCl, pH 7.8 and resuspended by adding 4 ml of 0.1 $M$ triethanolamine, 10 m$M$ $MgCl_2$, pH 7.8 for each gram wet weight of cells. The suspension can be kept frozen at −20° for at least 6 months without significant loss of activity. The following steps are carried out at 0–4° unless otherwise stated.

*Cell Breakage and Heat Treatment.* The thawed cell suspension (usually 500–800 ml) is sonicated for a total of about 2 min, at not more than 10°, until microscopic checks indicate very few whole cells. Large debris is removed by centrifugation at 32,000 $g$ and the strongly colored supernatant solution recentrifuged for 3 hr at 90,000 $g$ to sediment membrane fragments. The supernatant is treated with 5 $\mu g$/ml each of crystalline ribonuclease and deoxyribonuclease at 37° for 15 min. Solid $(NH_4)_2SO_4$ (13.4 g/100 ml) is dissolved with stirring, and the solution is then heated rapidly to 50° for 2 min. After chilling in ice, the precipitate is removed by centrifugation at 32,000 $g$. Further $(NH_4)_2SO_4$ (21.1 g/100 ml) is added to the supernatant to precipitate the enzyme, and the pellet obtained after centrifuging dissolved in 0.1 $M$ triethanolamine:HCl, 0.65 m$M$ dithiothreitol (TEA:DTT), using approximately one-tenth the volume of the original cell extract.

*Acetone Precipitation.* The above solution is desalted and freed from ribose-5-phosphate isomerase by passage through a Sephadex G-100 column (2.6 × 44 cm) equilibrated with TEA:DTT; ribulose-5-phosphate kinase emerges from the column with the void volume. Solid $(NH_4)_2SO_4$ (5 g/100 ml) is added to the pooled active fractions. A total of 0.75 volume acetone, kept at −40° in an acetone:dry-ice bath, is then added dropwise to the stirred enzyme solution, the temperature of which is gradually lowered from 0° to −15°. Stirring is continued for 30 min after addition is complete, and the precipitate is collected by centrifugation at 37,000 g at −20°. The yellow supernatant is discarded, and residual acetone is removed from the walls of the centrifuge tube by wiping and evaporation in a stream of nitrogen; severe losses are encountered if acetone is not removed completely. The precipitate is extracted successively with two 8-ml portions of 0.75, 0.5, and of 0.3 saturated $(NH_4)_2SO_4$ made up in TEA:DTT; a large part of the precipitate remains undissolved. The most active extracts, usually those in 0.3 saturated $(NH_4)_2SO_4$, are pooled, and ribulose-5-phosphate kinase is precipitated by the addition of 2.5 g $(NH_4)_2SO_4$/10 ml. The centrifuged pellet is dissolved in a minimal volume of 50 mM Tris·HCl, pH 7.4, containing 0.65 mM dithiothreitol (Tris:DTT) and desalted by passage through a small Sephadex G-200 column equilibrated with the same buffer.

*DEAE-Cellulose Chromatography.* The light yellow effluent from the G-200 column is applied to a DEAE-cellulose column (2 × 28 cm) equilibrated with Tris:DTT, and eluted by stepwise washing with 3 column volumes each of Tris:DTT, 50 mM KCl in Tris:DTT, and 0.15 M KCl in Tris:DTT. Fractions with the highest specific activities are pooled, and ribulose-5-phosphate kinase concentrated and precipitated by suspending an open dialysis sac containing $(NH_4)_2SO_4$ crystals (4.7 g for each 10 ml) in the enzyme solution. The precipitate which forms in the approximately 4-fold concentrated enzyme solution is collected, dissolved in a minimal volume Tris:DTT, and desalted by passage through a small

PURIFICATION OF RIBULOSE-5-PHOSPHATE KINASE FROM *Chromatium*

| Fraction | Volume (ml) | Activity[b] | | Purification (fold) | Recovery (%) |
| | | Total (units) | Specific (units/mg protein) | | |
| --- | --- | --- | --- | --- | --- |
| Extract | 660 | 7286 | 0.85 | — | 100 |
| High-speed centrifugation and heat treatment | 40 | 4480 | 3.1 | 3.0 | 61 |
| Acetone | 9 | 2032 | 49.7 | 58.5 | 28 |
| DEAE-cellulose | 6 | 752 | 137.6 | 162.0 | 10 |

[a] Reproduced with permission from B. A. Hart and J. Gibson, *Arch. Biochem. Biophys.* **144,** 308 (1971).

[b] Activity was determined by the discontinuous spectrophotometric assay using triethanolamine buffer at pH 7.2, and is expressed as micromoles of ADP formed × $min^{-1}$ × $ml^{-1}$ at 37° in the assay mixture described in the text.

Sephadex G-25 column. The purified colorless solution is stored in ice under nitrogen.

A summary of the procedure is given in the table.

## Properties

*Stability.* The purified enzyme is stable for at least 3 months in either TEA:DTT or Tris:DTT when stored as above. Freezing and thawing cause more rapid loss in activity. The best preparations have a specific activity of 140 units/mg protein, measured by the Lowry procedure,[5] are at least 90% pure as judged by acrylamide gel electrophoresis in 4.5 $M$ urea, and are free of ribose-5-phosphate isomerase. Molecular weight estimated from density gradient centrifugation is about 240,000.

*Cofactors and Specificity.* Complete dependence on divalent cations, $Mg^{2+}$ and $Mn^{2+}$ being most effective, has been demonstrated. Since the enzyme rapidly inactivated in the absence of DTT, it is probable that its activity depends on SH groups. The enzyme is specific for ATP, and only 10% of the reaction rate with ATP is given in GTP. A pH optimum in the discontinuous assay is observed at 7.2.

*Kinetics and Effectors.* Sigmoidal V vs. [ATP] or [Ru-5-P] plots are obtained, indicating cooperativity between binding sites. Apparent $K_m$ is 0.22 m$M$ for ribulose 5-phosphate and 0.7 m$M$ for ATP. Of ten nucleotides tested, only 5'-AMP, and to a lesser extent 5'-ADP, is inhibitory. The inhibition is pH dependent, with $I_{0.5}$ for 5'-AMP decreasing from 0.3 m$M$ at pH 7.6 to 0.9 m$M$ at pH 7.2. Glyceraldehyde 3-phosphate is also a potent inhibitor with $I_{0.5}$ of $3.6 \times 10^{-5}$ at pH 7.2. Phosphoenol pyruvate (2.5 m$M$) is not inhibitory at pH 7.6 but at a pH of 7.2 the $I_{0.5}$ for PEP is 0.85 m$M$. At both pH values PEP potentiates the inhibitory effect of 5'-AMP about 5-fold. Direct binding studies show that PEP enhances binding of 5'-AMP, and that 5'-ATP binding is reduced by the presence of 5'-AMP.

[5] See this series, Vol. 3 [73].

# [20] Phosphoribokinase from *Pseudomonas saccharophila*

By J. B. ALPERS

Ribose 5-phosphate + ATP → ribose 1,5-diphosphate + ADP + H$^+$

## Assay Method

*Principle.* The assay is performed in two stages. Ribose 1,5-diphosphate production is measured by the ability of a boiled reaction product from the first stage to stimulate the phosphoglucomutase reaction (second stage). The identity of the product of the first stage may be confirmed[1] by chromatography, by comparing its mobility with that of authentic ribose diphosphate on paper electrophoresis, and by its acid lability.

*Reagents for first stage*

Tris buffer, 1 $M$, pH 9.1
MgCl$_2$, 3 m$M$
Ribose 5-phosphate, 5 m$M$
ATP, 5 m$M$

*Reagents for second stage*

Tris buffer, 1 $M$ pH 7.4
MgCl$_2$, 0.2 $M$
Glucose 1-phosphate, 0.2 $M$
EDTA, 0.1 $M$
Gelatin
Phosphoglucomutase from rabbit muscle

*Procedure.* In a final volume of 0.25 ml, introduce Tris buffer, pH 9.1, 0.02 ml; MgCl$_2$, 3 m$M$, 0.05 ml; ribose 5-phosphate, 0.05 ml; ATP, 0.05 ml; and an appropriate quantity of enzyme to produce at least four times the activity present in the zero time control mixture (see below). After incubation for 60 min at 37°, the reaction is terminated by boiling for 2 min in a water bath. Control mixtures are boiled immediately upon addition of enzyme. The usual 60-min incubation procedure produces between 8 and 16 nmoles of ribose 1,5-diphosphate.

The cofactor activity of ribose 1,5-diphosphate is assayed colorimetrically in stage two. A convenient phosphoglucomutase reaction system contains, in a final volume of 0.6 ml: Tris buffer, pH 7.4, 0.03 ml; MgCl$_2$,

[1] L. R. DeChatelet and J. B. Alpers, *J. Biol. Chem.* **245**, 3161 (1970).

0.2 $M$, 0.02 ml; EDTA, 0.025 ml; glucose 1-phosphate, 0.02 ml; gelatin, 0.03 mg; and 0.02 ml of boiled incubation product from stage one. The reaction is started by the addition of 0.1 IU of phosphoglucomutase (as assayed with saturating glucose 1,6-diphosphate), and incubated for 15 min at 37°. The reaction is terminated after 15 min by boiling, and 1.5 ml of Somogyi's reagent[2] is added to each incubation mixture. Mixtures are placed in a boiling water bath for 15 min, then in an ice bath for 5 min before 1.5 ml of arsenomolybdate reagent[3] is added, followed by 3.0 ml water. Samples are read in a colorimeter at 540 nm.

The quantity of ribose diphosphate produced is based on a comparison with standards of *glucose* diphosphate run with each experiment. (One micromole of glucose diphosphate is approximately equivalent to 6 $\mu$moles of ribose diphosphate in terms of cofactor activity.) The limit of sensitivity for ribose diphosphate is about 0.21 nmole. Standards containing 0, 42, and 83 $\mu$moles of glucose diphosphate are assayed with each set of experimental samples.

*Behavior of Crude Extracts.* A crude sonic extract incubated with ATP produced significant amounts of diphosphate from ribose 5-phosphate as assayed in this fashion. Ribose 1-phosphate is less effectively utilized; glucose 6-phosphate and glucose 1-phosphate are poor diphosphate precursors. Assays of sonic extract for phosphoribomutase and phosphoglucomutase demonstrate the presence of both enzymes. Hence, diphosphate production from ribose 1-phosphate may involve its prior conversion to ribose 5-phosphate. This is supported by experiments on the specificity of more purified fractions. All crude extracts of *P. saccharophila* examined showed a marked superiority (as high as 200-fold) with ribose-5-P over glucose-6-P as substrate for diphosphate production. (The significance of ribose 1,5-diphosphate production in this organism is uncertain.)

*Definition of Unit.* One unit of enzyme is defined as that amount which produces 1 $\mu$mole of ribose diphosphate per minute at pH 9.1.

## Purification Procedure

All steps (see the table) are performed at 0–4°. Centrifugation, unless otherwise specified, is at 35,000 $g$ for 15 min.

*Preparation of Bacterial Extracts. Pseudomonas saccharophila* ATCC 9114 is grown on the sucrose medium of Doudoroff[4] for 30 hr at 30° with aeration. Cultures of 1 liter are grown with shaking in 2-liter Erlenmeyer

---

[2] M. Somogyi, *J. Biol. Chem.* **160**, 61 (1945).
[3] N. Nelson, *J. Biol. Chem.* **153**, 375 (1944).
[4] M. Doudoroff, this series, Vol. 1, p. 226.

PURIFICATION OF PHOSPHORIBOKINASE FROM *Pseudomonas saccharophila*

| Procedure | Protein (mg/ml) | Specific activity $\times 10^3$ (units/mg protein) | Total units |
|---|---|---|---|
| Initial extract | 26.3 | 0.250 | 3.89 |
| Streptomycin, supernatant solution | 21.8 | 0.375 | 5.19 |
| pH 5 precipitate | 18.3 | 1.18 | 2.53 |
| Ammonium sulfate, 0–40% | 9.2 | 2.22 | 2.43 |
| Ammonium sulfate, 20–30% | 1.7 | 9.86 | 1.96 |

flasks; larger preparations (e.g., 14 liters), are grown in a New Brunswick Microferm fermentor. The cells are collected by centrifugation at 4°, washed twice in Tris buffer, 1 mM, pH 7.4, and resuspended in 2.5 volumes of sodium phosphate buffer, 33 mM, pH 6.9. The suspended cells are disrupted by sonic oscillation at 1.7A in a MSE Model 60 sonic oscillator at 0°. Larger quantities of cells (up to 100 ml) may be disrupted at 6 A in a Branson sonic oscillator. After disruption, the material is centrifuged at 35,000 $g$ for 15 min to remove unbroken cells and cell debris. Centrifugation is repeated to obtain a clear supernatant solution, which is stored at —56° until use.

*Precipitation with Streptomycin Sulfate.* To 50 ml of crude sonic extract (26 mg of protein per milliliter), 5.0 ml of a freshly prepared solution (10%) of streptomycin sulfate are added. After stirring for 15 min, the mixture is centrifuged and the precipitate discarded. This step consistently results in a significant increase in total activity, which could indicate the removal of an inhibitor.

*Precipitation at pH 5.* The supernatant solution from the previous step is adjusted to pH 5.0 by the careful addition of 0.2 M acetic acid, with constant stirring. After 15 min, the material is centrifuged and the supernatant solution discarded. The precipitate is suspended in 10 ml of phosphate buffer, 33 mM, pH 6.7, and stirred for several hours. The material may be stored overnight at this stage, if necessary. A small amount of undissolved precipitate is removed by centrifugation. The activity lost in this step (30–40%) is accounted for in the pH 5 supernatant. However, using a lower pH results in the precipitation of other proteins.

*First Ammonium Sulfate Fractionation.* The dissolved pH 5 precipitate is brought to 40% saturation by the addition of solid ammonium sulfate, and stirred for 15 min. After centrifugation, the precipitate, which contains all the enzyme activity, is dissolved in 10 ml of the same phosphate buffer.

*Second Ammonium Sulfate Fractionation.* Solid ammonium sulfate is added to 20% saturation,[5] and the solution is stirred for 15 min. The slight precipitate observed after centrifugation is discarded. The solution is then brought to 30% saturation with solid ammonium sulfate, stirred again for 15 min and centrifuged. The precipitate, dissolved in 10 ml of phosphate buffer, has a specific activity about 40-fold higher than that of the crude extract. Attempts at purification by column chromatography (carboxymethylcellulose, diethylaminoethylcellulose, and Sephadex G-100) were not successful.

## Properties

*Specificity.* The enzyme is specific for ribose 5-phosphate. Ribose 1-phosphate, glucose 6-phosphate, and glucose 1-phosphate are inactive. Deoxyribose 5-phosphate does not generate a product that can stimulate the phosphoglucomutase reaction (although no evidence is available that deoxyribose 1,5-diphosphate, if produced, would in fact do so). As ATP substitutes, ITP, UTP, ADP, and 3',5'-cyclic AMP were ineffective. Adenosine tetraphosphate was one fourth as effective as ATP.

*Effect of $Mg^{2+}$ Cations.* EDTA causes a total inhibition, which is completely reversed by the addition of excess $MgCl_2$.

*pH Optimum.* Maximal activity is observed at pH 9.1. The enzyme is about twice as active in Tris·HCl as in glycine at a given pH.

*Stability.* Activity declines about 25% during 6 months' storage at −56° in 33 m$M$ sodium phosphate buffer, pH 7.0.

*Kinetic Properties.* The apparent $K_m$ for both ATP and for ribose 5-phosphate at pH 9.1 is about 1 m$M$. Concentrations of either substrate higher than 2 m$M$ are inhibitory. The same results with ATP are obtained whether it is varied alone or both ATP and $Mg^{2+}$ are varied together to maintain a constant ratio.

[5] Based on tables in B. M. Pogell and R. W. McGilvery, *J. Biol. Chem.* **208**, 149 (1954) which are for 0°.

## [21] D-Glycerate 3-Kinase from *Escherichia coli*

By C. C. DOUGHTY and JAMES A. HAYASHI

D-Glycerate + ATP $\rightleftharpoons$ 3-phosphoglycerate + ADP

### Assay Method

*Principle.* The rate of 3-phosphoglycerate production is determined by titration of the hydrogen ion, which is produced during phosphorylation and is stoichiometrically equivalent to the amount of 3-phosphoglycerate formed.

*Reagent Solutions*

Calcium DL-glycerate, 0.5 $M$
KF, 1.0 $M$
MgSO$_4$, 0.25 $M$
ATP, 0.1 $M$ (solution titrated to pH 7.4 before use)
KOH, 0.010 $N$

*Procedure.* The reaction mixture, in a total volume of 5.0 ml, is made up in the titration vessel. It contains: calcium DL-glycerate, 0.1 ml (50 $\mu$mole); KF, 0.1 ml (100 $\mu$mole); MgSO$_4$, 0.1 ml (25 $\mu$mole); ATP, 0.1 ml (10 $\mu$mole); enzyme, 0.02–1.0 unit; water, to 5 ml final volume. The reaction mixture minus ATP, is titrated to pH 7.40. The reaction is started by adding ATP and is followed titrimetrically by adding 0.010 $N$ KOH to maintain the pH at 7.40. A pH-stat such as the Radiometer Autotitrator, Model TTT1 (Radiometer, Copenhagen) may be used. Under these conditions, the assay is linear.

*Definition of Enzyme Unit.* A unit of enzyme is that amount of enzyme producing 1.0 $\mu$mole of product per minute in the above assay system. The specific activity is expressed as units per milligram of protein. Protein is measured spectrophotometrically[1] or by the biuret method.[2]

### Purification Procedure

*Growth Medium.* The medium contains, in each liter: 1.0 g of NH$_4$Cl, 6.0 g of Na$_2$HPO$_4$, 2.0 g of KH$_2$PO$_4$, 2 g of NaCl, 0.2 g of MgSO$_4$·H$_2$O, and 9.5 g of calcium DL-glycerate (approximately 67 m$M$ in glycerate); pH 7.0.

[1] C. C. Doughty, J. A. Hayashi, and H. L. Guenther, *J. Biol. Chem.* **241**, 568 (1966).
[2] O. Warburg and W. Christian, *Biochem. Z.* **310**, 384 (1941).

*Culture of Bacteria. Escherichia coli* (Crook's strain) ATCC No. 8739 is maintained on agar slants (2% agar in medium) at 4°. The organism is inoculated into 50 ml of medium in an Erlenmeyer flask and grown on a rotary shaker at 37° for 24 hr. The culture is then poured into 400 ml of medium in a 2-liter flask and grown as before. After 24 hr, the 400-ml inoculum is transferred to a carboy, in a 37° bath, containing 8 liters of liquid medium. The culture is aerated during growth by bubbling filtered compressed air through the medium. The cells are harvested after 18 hr growth using a continuous flow centrifuge kept at 4°. The usual yield was about 3.8 g wet weight of cells per liter.

*Preparation of Cell-Free Extracts.* A suspension of bacterial cells is made in 50 m$M$ phosphate buffer, pH 7.0 (30 g wet weight of cells in 100 ml of buffer) and placed in the 400 ml homogenizer chamber of a Sorvall Omnimixer (Ivan Sorvall, Inc., Norwalk, Connecticut). Sixty milliliters of glass beads (0.1 mm diameter) are added and the chamber is immersed in an ice bath. Alternate cycles of grinding (10 sec) and cooling (20 sec) at an autotransformer setting of 70 V were used. The process may be unattended by using an "on-off" timer such as the Flexopulse, Eagle Signal Corp., (Moline, Illinois). Cell disruption is usually complete after 3 hr of operation. The beads are separated by sedimentation (500 $g$ for 10 min, at 4°) and washed twice with one-half the original volume of buffer. The combined extract and washings is clarified by centrifugation (29,000 $g$, 20 min). The cell-free extract contains 18–20 mg of protein per milliliter, and its enzyme activity is stable almost indefinitely when stored at −40°. This is a convenient point in the procedure to stockpile appreciable amounts of the enzyme. Unless stated otherwise, all subsequent operations are conducted at 0° or 4°.

*Step 1. Protamine Sulfate Precipitation.* A 1% protamine sulfate solution, pH 7.0, is added slowly with stirring to the ice-cooled extract until 1 mg of protamine sulfate has been added for each 20 mg of protein in the extract. The mixture is stirred for an additional 20 min, and the precipitate is sedimented by centrifugation (10,000 $g$, 10 min) and then discarded.

*Step 2. Ammonium Sulfate Fractionation.* Solid $(NH_4)_2SO_4$ (22.6 g/ 100 ml) is added slowly to the extract until 40% saturation is reached. After an additional 20 min stirring, the precipitate is collected by centrifugation (10,000 $g$, 10 min) and redissolved in a minimal volume of 5 m$M$ potassium phosphate buffer, pH 7.0, and dialyzed against the same buffer.

*Step 3. DEAE-Sephadex Column Chromatography.* After dialysis, the sample (3–4 g protein) is applied to a DEAE-Sephadex downward flow column (45 × 450 mm) which is equilibrated against 10 m$M$ potassium phosphate, pH 7.0. The same buffer was used for elution with the ionic strength increased in a stepwise fashion by adding NaCl. The amounts of buffer and their NaCl molarities were 60 ml, 0.0 $M$; 30 ml, 0.1 $M$; 30 ml, 0.3 $M$; 50 ml, 0.5 $M$; 40 ml, 0.7 $M$; 40 ml, 1.0 $M$. The fractions collected were 0.8 ml each.

The bulk of the activity emerges after 80 ml, and with virtually all the activity after 120 ml. The greater part of the protein ($>80\%$) emerges after 120 ml, but it is all inactive material.

*Step 4. Ammonium Sulfate Reprecipitation.* The enzyme in the pooled, active fractions from step 3 is precipitated between 20 and 35% saturation by the slow addition of solid $(NH_4)_2SO_4$ (10.6 g/100 ml and 8.3 g/100 ml). After centrifugation (12,000 $g$, 15 min) the sediment is dissolved in 5 ml of 50 m$M$ potassium phosphate buffer, pH 7.0.

*Step 5. Ethanol Precipitation.* After dialysis of the enzyme fraction against water, the enzyme is twice fractionally precipitated with ethanol at −10°. In the first fractionation, the material precipitating between 15 and 50% (w/v) ethanol is collected, redissolved in 3 ml of water and reprecipitated with ethanol between 19 and 45% (w/v) ethanol. Upon standing at 4° or at −20°, the enzyme crystallizes.

A summary of the purification is given in the table.

PURIFICATION OF D-GLYCERATE 3-KINASE FROM *Escherichia coli*

| Fraction | Total protein (mg) | Specific activity $\mu$moles/ min/mg protein) | Total activity (units) | Yield (%) | Purification (fold) |
|---|---|---|---|---|---|
| Cell extract | 22,310 | 0.050 | 1117 | 100 | 1.0 |
| Protamine sulfate precipitate | 18,800 | 0.056 | 1050 | 94 | 0.9 |
| Ammonium sulfate precipitate | | | | | |
| 0 to 20% saturation | 172 | 0.350 | 61 | 5.4 | 7 |
| 20 to 30% saturation | 1,365 | 0.460 | 632 | 56.6 | 9 |
| 30 to 40% saturation | 1,664 | 0.095 | 158 | 14.2 | 2 |
| 40 to 60% saturation | 4,120 | 0.004 | 16 | | |
| 60 to 80% saturation | 6,750 | 0.001 | 7 | | |
| DEAE-Sephadex column[a] | 12.0 | 63.8 | 765 | 68.5 | 1275 |
| Ammonium sulfate precipitate, 20 to 35% saturation | 10.0 | 66.5 | 665 | 59.3 | 1330 |
| Ethanol | | | | | |
| First precipitate, 15 to 50% | 4.0 | 115 | 460 | 41.2 | 2300 |
| Second precipitate, 19 to 45% | 2.0 | 210 | 420 | 37.2 | 4200 |

[a] The enzyme-containing material was fractionated on the column in three batches. The values indicated are the sum of the results obtained for the three batches.

## Properties[1]

The product of the reaction, 3-phosphoglycerate, has been identified and distinguished from the 2-phosphoglycerate by paper chromatography[4] and enzymically.[1] The Michaelis constant is 0.24 m$M$, and optimal activity is shown in the range of pH 7.0–7.5. The enzyme has a divalent cation requirement which is satisfied by 0.1 m$M$ $Mg^{2+}$ and $Co^{2+}$, with lesser activity with $Mn^{2+}$, $Fe^{2+}$, or $Ca^{2+}$. EDTA, 10 $\mu M$ completely inhibits activity, as did $p$-hydroxymercuribenzoate and iodoacetate at 10 m$M$.

The enzyme is stable in solution at pH 7.0 and temperatures below 10°. The crude enzyme has been stored for up to 2 years at −40° without loss of activity. The crystalline enzyme has been kept up to 4 weeks in 45% ethanol with only slight loss of activity.

[3] A. G. Gornall, C. J. Bardawill, and M. M. David, *J. Biol. Chem.* **177**, 751 (1949).
[4] R. W. Hansen and J. A. Hayashi, *J. Bacteriol.* **83**, 679 (1962).

# [22] 3-Phosphoglycerate Kinase of Skeletal Muscle

## *By* R. K. SCOPES

MgATP + 3-phosphoglycerate $\rightleftharpoons$ MgADP + 1,3-diphosphoglycerate

## Assay Method

*Principle.* The reaction is coupled to the previous enzyme in the glycolytic pathway, glyceraldehyde-3-phosphate dehydrogenase, resulting in the oxidation of 1 molecule of NADH for every molecule of 3-phosphoglycerate phosphorylated. In the conditions of assay the equilibrium of the two coupled reactions greatly favors NADH oxidation, provided that a phosphate buffer is not employed. Magnesium ions are used in a slight excess over the ATP concentration to form the MgATP complex which is the substrate,[1] but a large excess of magnesium is avoided since free $Mg^{2+}$ acts as an uncompetitive inhibitor of the (yeast) enzyme.[2] In addition, monovalent metal ions are present ($K^+$) as these stimulate activity to a small extent.[3] The same reaction rate is obtained over the pH range 6 to 9[4]; Tris, triethanolamine, and imidazole are equally satisfactory as buffers.

[1] M. Larsson-Razinkiewicz, *Biochim. Biophys. Acta* **85**, 60 (1964).
[2] M. Larsson-Raznikiewicz, *Eur. J. Biochem.* **17**, 183 (1970).
[3] M. Larsson-Raznikiewicz and J. R. Jansson, *FEBS Lett.* **29**, 345 (1973).
[4] W. K. G. Krietsch and T. Bücher, *Eur. J. Biochem.* **17**, 568 (1970).

*Reagents*

1. The stock buffer consists of:
   Triethanolamine, 30 m$M$
   3-phosphoglycerate, 10 m$M$ (K salt)
   KCl, 50 m$M$ $\Big\}$ adjusted to pH 7.5 with HCl
   MgSO$_4$, 5 m$M$
   Na$_2$EDTA, 0.2 m$M$
   To this are added:
   Bovine serum albumin 0.2 mg/ml
   20–40 $\mu$g/ml Glyceraldehyde phosphate dehydrogenase of specific activity in the range 50–100 IU/mg.
   NADH, ca. 0.1 m$M$
   This reaction mixture is stable for 2 to 3 hours at room temperature.
2. ATP, 50 m$M$, adjusted to pH 7.5 with KOH

*Procedure.* To 2.3 ml of reaction mixture, add 0.2 ml of ATP, and place in spectrophotometer at 340 nm to check for any blank rate due to contamination with enzyme.

Add 0.5–1 $\mu$l of suitably diluted enzyme[5] to give a $\Delta A_{340}$ of 0.1–0.3 per minute. The temperature in the cuvette is measured immediately afterward, so that the rate can be calculated for a standard temperature.

*Units.* One unit of activity is that amount of enzyme causing the phosphorylation of 1 $\mu$mole of 3-phosphoglycerate at 30°[6]. A temperature correction is applied if necessary, at the compound rate of 6% per degree.

In the conditions described here, the rate should theoretically be about 80% of $V_{max}$, assuming a rapid-random equilibrium mechanism of reaction.

*Protein Measurement.* Small measured volumes of samples are dialyzed against water for a few hours, then transferred quantitatively to a volumetric flask. The samples are diluted with 5 m$M$ potassium phosphate buffer, pH 7.0, containing 0.1 $M$ K$_2$SO$_4$ to a concentration of about 20 $\mu$g/ml, and the value of absorbance read at 205 nm.[7] Extinction coefficients for 1 mg/ml of 31 are used in the early stages, and 29 for the purified fractions. Prior dialysis removes possible interfering compounds, although in practice this interference is slight except for the initial ex-

---

[5] It is essential that the enzyme be diluted with a solution containing serum albumin as a stabilizer. Assay stock buffer containing 0.2 mg/ml bovine serum albumin is most suitable.

[6] In accordance with the revised recommendations of the Enzyme Commission, 30° has been chosen rather than 25°.

[7] R. K. Scopes, *Anal. Biochem.* **59**, 277 (1974).

tract, and can be omitted for routine purposes. Specific activity is given in units per milligram protein.

## Purification

*Step 1. Extraction.* This procedure was originally developed for the isolation of phosphoglycerate kinase from either rabbit or pig skeletal muscle.[8] It has been successfully applied with only minor modifications in isolating the enzyme from ox, sheep, horse, and the Australian marsupials, brushtail possum (*Trichosurus vulpecula*) and wombat (*Vombatus*). The source of material was not always a freshly killed animal, and quite satisfactory preparations have been made from frozen-stored muscle and from meat purchased from butchers. The only allowance to be made is that "meat" as opposed to fresh "muscle" contains a substantial amount of lactate due to postmortem glycolysis resulting in a low pH, which requires extra base to be included in the extraction medium. The optimum extracting conditions are for the homogenate to be stirred, at a temperature of about 20°, at pH 6.7 ± 0.2 for 30 min before centrifugation. Fresh muscle, freed as far as possible from connective tissue and fat, is homogenized in a Waring-type blender with 2½ volumes of 2% v/v glycerol,[9] containing 2 mM EDTA and 10 mM Tris. Meat extractant contains 30 mM Tris, and after homogenization the pH is adjusted upwards if necessary with 1 M Tris to pH 6.7. Downward adjustment of an over high pH is not normally necessary as glycolytic metabolism during stirring will cause further acidification.

The homogenate is transferred to large centrifuge buckets for centrifugation (5000 g for 15 min is adequate); alternatively, when dealing with large quantities it may be preferable to obtain the extract by squeezing through cheesecloth, followed by a centrifugation to separate out gross particulate matter and fat. The extract should be filtered through glass wool to remove fat particles; its volume should be close to the volume of extractant used. The extract can be frozen-stored at −20° for several months if required.

*Step 2. Ammonium Sulfate Fractionation.* The fraction required is that precipitating between 2.35 and 3.0 M ammonium sulfate (59–75% saturation) at pH 5.4, temperature 15–20°. Although this ammonium sulfate cut does not achieve as much purification as in some other procedures

---

[8] R. K. Scopes, *Biochem. J.* 113, 551 (1969).
[9] Glycerol, 2% w/v, is approximately isoosmotic, and so should minimize lysis of erythrocytes trapped in the tissue; a more valuable function of the glycerol is the protection it affords if the extract is subsequently frozen for a long period before working up.

(e.g., see footnote 4), the overall recovery is high. The fractionation may be carried out in several steps, retaining the precipitates at each step for purification of other enzymes,[8] or with only slightly poorer recovery in two steps as follows. The extract pH is adjusted to 5.4 with 5 $M$ acetic acid, and 385 g ammonium sulfate added per liter. After stirring for at least 30 min, the suspension is centrifuged (5000–10,000 $g$ for 15 min), and the precipitate is discarded. A further 110 g ammonium sulfate is dissolved per liter of supernatant, and after stirring the second precipitate is collected by centrifugation as before. The supernatant can be used for preparing glyceraldehyde phosphate dehydrogenase used in the assay of phosphoglycerate kinase.

Step 3. Heat Treatment. The second precipitate from the ammonium sulfate fractionation is dissolved in a buffer that contains 3-phosphoglycerate and ammonium sulfate, both of which stabilize the enzyme to heat denaturation. The buffer consists of 50 m$M$ sodium acetate + 1 m$M$ EDTA + 1.0 $M$ ammonium sulfate + 5 m$M$ 3-phosphoglycerate, pH 5.4, and the solution is diluted with this buffer until the volume is approximately 20% of the volume of the original extract. The flask containing the solution is then placed in water at about 65% and agitated continuously until the temperature of the solution reaches 55°.[10] The flask is then transferred to a 55° water bath and kept at that temperature with continuous agitation for 10 min. It is then cooled rapidly, and the heavy residue of denatured protein removed by centrifugation. To maximize recovery at this step, the residue can be extracted with a small volume of 1.2 $M$ ammonium sulfate and the extract combined with the previous supernatant. The combined solutions are filtered through glass wool to remove flecks of denatured protein. Then 40 g of ammonium sulfate per 100 ml of solution are dissolved in, and after stirring 30 min the precipitate collected by centrifugation.

Step 4. CM-Cellulose Chromatography. Before chromatography, excess ammonium sulfate is removed by extensive dialysis against 0.2 m$M$ EDTA, pH 8.0. Alternatively, if the volume can be handled, desalting may be carried out using a Sephadex G-25 column. The buffer used for the chromatography was chosen for maximum buffering power with the least ionic strength; in addition it can be used to advantage in cases where prior treatment with DEAE-cellulose is desirable.[11] Whatman

[10] The heat treatment at 55° is suitable for rabbit and pig enzyme, but it may result in an unacceptable loss in certain species. Thus it is advisable to make preliminary small-scale heating trials at temperatures from 50° upward to find the highest temperature which does not result in a loss of more than 20% of the activity.

[11] The muscles of certain species (e.g., the horse) contain substantial quantities of myoglobin, which is partly oxidized by the heat treatment and may cause overloading and general interference with the CM-cellulose chromatography. The fol-

microgranular CM52 is equilibrated with the buffer, which consists of 10 m$M$ Tris $+$ 0.2 m$M$ EDTA, adjusted to pH 6.5 with cacodylic acid. The column size should be 15–20 cm high and about 1 cm$^2$ cross section for every 100 ml of original extract. Thus the fraction originating from 1 kg of muscle could be accommodated adequately on a column 15 cm high and 5 cm in diameter. One-tenth volume of 0.1 $M$ Tris-cacodylate pH 6.5 is added to the dialyzed fraction, which is further diluted with the column buffer to about 25% of the original extract; the pH is adjusted to 6.5 if necessary, and the fraction is briefly centrifuged or filtered to remove any insoluble matter. The fraction is then run onto the column at a flow rate of about 12 ml/hr per square centimeter cross section, and washed in with about the same volume of buffer before commencing a gradient. A linear gradient in KCl is set up, the incremental rate of increase in KCl concentration being calculated from the formula: (1.5 m$M$/ml)/(column cross section in cm$^2$). For example, for a 10 cm$^2$ column, 500 ml of 0.15 $M$ KCl would mix into 500 ml of column buffer, giving a rate of increase of 0.15 m$M$ KCl per milliliter of elution gradient.

Phosphoglycerate kinase of rabbit is eluted at about 50 m$M$ KCl. The pig enzyme comes off at about 30 m$M$, and the horse enzyme at about 60 m$M$, reflecting their relative isoelectric points, for which horse $>$ rabbit $>$ pig. Contaminating enzymes, in particular pyruvate kinase, overlap with phosphoglycerate kinase peak, and are removed in the next step. The amount of contaminant depends very much on the animal species involved; the pig enzyme was about 80% pure after CM-cellulose, whereas the horse enzyme made up less than half of the protein at this stage.

The most important separation achieved is from triosephosphate isomerase, which is more acid and not adsorbed on the CM-cellulose. Triosephosphate isomerase has a similar molecular weight, 53,000,[12] to phosphoglycerate kinase, 48,000, and so is scarcely separated by gel filtration. It is possible that some mammalian triosephosphate isomerase may be sufficiently basic to be adsorbed on the CM-cellulose and eluted partially with the phosphoglycerate kinase, in which case the final preparation

---

lowing treatment removes much of the pigment and substantially improves the chromatography. After dialysis the solution is made 10 m$M$ in Tris base, and the pH is adjusted to 8.0 with cacodylic acid. DEAE-cellulose, preequilibrated with the same buffer, is added batchwise to the solution until the bulk of the pigment has been adsorbed. The nonadsorbed protein is recovered by filtration and washing, then adjusted to pH 6.5 with cacodylic acid for application to the CM-cellulose column.

[12] P. H. Corran and S. G. Waley, *FEBS Lett.* **30**, 97 (1973).

would remain contaminated. However, this does not occur with the species so far investigated.

*Step 5. Gel Filtration.* The enzyme from the CM cellulose column is collected and concentrated before applying to a gel filtration column. Ideally this is done by ultrafiltration; alternatively the enzyme is salted out with ammonium sulfate (60 g/100 ml), collecting the precipitate and dissolving in a small volume of Tris buffer (see below). In the latter case care must be taken when applying the dense solution to the column to avoid gravitational boundary instabilities. A column of dimensions 8 cm$^2$ × 90 cm is suitable for enzyme originating from up to 1 kg of muscle. The volume applied should be no more than 3–4% of the column volume (i.e., about 25 ml). Preequilibration and elution are carried out with a Tris buffer (20 mM Tris + 0.15 M KCl + 0.2 mM EDTA, pH 8.0), at a flow rate of about 6 ml/hr per square centimeter cross section. The gel used can be Sephadex G-150 or G-100; even G-75 has been used successfully; alternatively the corresponding Bio-Gel grades can be used. Since the only protein in muscle in substantial quantity of similar molecular weight to phosphoglycerate kinase is triosephosphate isomerase (see above), the enzyme is substantially pure after the Sephadex column.

*Step 6. Crystallization.* The enzyme peak from the Sephadex column is salted out with ammonium sulfate (60 g/100 ml), and the precipitate collected by centrifugation. This is taken up in as small a volume as possible of the assay stock buffer (lacking phosphoglycerate) and placed in ice. Ice-cold saturated ammonium sulfate is added until a permanent turbidity develops; a drop or two of water is added to the solution, which is then briefly centrifuged in the cold. The solution is then transferred to a flask containing a few drops of saturated ammonium sulfate, and this is left in the cold. Crystals normally form overnight as a silky sheen of fine needles[13]; eventually the needles may grow to be several mm in length. In particular the horse enzyme will grow crystals suitable for X-ray diffraction studies,[14] especially in the presence of a small amount of 1,4-dioxan (up to 3%).

Table I summarizes the purification procedure.

As indicated in Table I, the crystals after dissolution lose some activity; however, crystals stored for over a year lost little more than is lost initially, and can have specific activities still of the order of 800 units/mg.

Table II lists some comparative data on the preparation procedure.

[13] If crystallization does not occur overnight, warming at room temperature for a few hours may initiate crystallization.

[14] C. C. F. Blake, P. R. Evans, and R. K. Scopes, *Nature (London) New Biol.* **235**, 195 (1972).

## TABLE I
PHOSPHOGLYCERATE KINASE FROM RABBIT SKELETAL MUSCLE

| Fraction (from 1 kg muscle) | Total protein (mg) | Total activity (units × 10³) | Specific activity (units/mg) | Yield (%) |
|---|---|---|---|---|
| 1. Extract | 58,000 | 1140 | 20 | 100 |
| 2. Ammonium sulfate | 14,200 | 840 | 59 | 74 |
| 3. After heat | 3,900 | 604 | 152 | 53 |
| 4. After CM-cellulose | 810 | 564 | 695 | 50 |
| 5. Sephadex peak | 530 | 516 | 975 | 45 |
| 6. Crystals | 410 | 310 | 760 | 27 |

## TABLE II
COMPARATIVE DATA ON PREPARATION

| Fraction (from 1 kg of muscle) | Total activity (units × 10³) | | | |
|---|---|---|---|---|
| | Rabbit | Pig | Horse | Wombat |
| 1. Extract | 1140 | 930 | 670 | 550 |
| 5. Sephadex peak | 516 | 365 | 214 | 215 |

Each Sephadex fraction had a specific activity greater than 900 units/mg.

### Properties

Phosphoglycerate kinase is the only enzyme of the glycolytic pathway that is monomeric,[15] and X-ray diffraction analysis demonstrates that although it appears to be structurally formed from two globular units, these have different structures and are both part of the one polypeptide chain.[14,16] The molecular weight of the muscle enzyme is close to 48,000; this is not significantly different from the value for the yeast,[4] erythrocyte,[17] or silver beet[18] enzymes. Electrophoretically the enzyme appears quite conservative; relatively little difference in electrophoretic mobility on starch gel is found in a wide range of vertebrates, although mamma-

---

[15] This statement may be modified if the enzyme glucose-6-phosphate 1-epimerase becomes an accepted member of the glycolytic pathway and, as seems likely, proves to be monomeric.

[16] P. L. Wendell, T. N. Bryant, and H. C. Watson, *Nature (London) New Biol.* **240**, 130 (1972).

[17] A. Yoshida and S. Watanabe, *J. Biol. Chem.* **247**, 440 (1972).

[18] S. Cavell and R. K. Scopes, unpublished observations.

lian enzymes are rather less acidic than those of fish. The yeast enzyme is electrophoretically only slightly acid, but the chloroplast enzyme from silver beet is much more acid.[18] A second phosphoglycerate kinase allele, expressed only in animal testes, has been described, and this appears to be less conservative electrophoretically.[19]

The purified enzyme has a specific activity in the conditions described close to 1000 units/mg—this applies to all species tested. The yeast, silver beet, and erythrocyte enzymes have similar specific activities. In the other direction, i.e., in the direction of glycolysis, the activity is some 2–2.5 times greater,[4] but relatively little work has been carried out on the kinetics in this direction because of the lability of the substrate, 1,3-diphosphoglycerate. Extensive kinetic studies in the gluconeogenic direction have been carried out only with the yeast enzyme.

In view of the marked similarity of the reported Michaelis constants and other properties for the muscle and yeast enzymes,[4] it is most likely that the mechanisms involved are the same in each case.

[19] J. L. Vandeberg, D. W. Cooper, and P. J. Close, *Nature (London) New Biol.* **243**, 48 (1973).

# [23] 3-Phosphoglycerate Kinase of Baker's Yeast

## By R. K. Scopes

$$MgATP + 3\text{-phosphoglycerate} \rightleftharpoons MgADP + 1,3\text{-diphosphoglycerate}$$

## Assay

The assay principle and procedure is as given in the preceding article.

## Purification

Preparation of a yeast extract can be carried out by any of the standard methods, e.g., toluene autolyzate, or autolyzate of air-dried cells, but probably the easiest and least messy is to make the extract by cytolysis with ammonia.[1,2] Although fresh yeast can be used, it is most convenient to use vacuum- or nitrogen-packed cans of "active dried" yeast available commercially from bakeries.

[1] E. de la Morena, I. Santos, and S. Grisolia, *Biochim. Biophys. Acta* **151**, 526 (1968).
[2] R. K. Scopes, *Biochem. J.* **122**, 89 (1971).

All stages of this isolation procedure are carried out at room temperature.

*Step 1. Extraction.* One liter of 0.5 $M$ NH$_4$OH containing 1 g of EDTA (disodium salt) is stirred vigorously while 450 g of granulated dried yeast is added slowly over a period of 30–60 min. Any lumps forming are dispersed, and stirring is continued for several hours; then the preparation is allowed to stand overnight. The optimum temperature for this procedure is around 20°; in the cold room cytolysis is not always complete. The next day, stirring is recommenced, and 800 ml of 0.5 $M$ lactic acid are mixed in. The pH is then measured and adjusted to 7.0 ± 0.5 using 5 $M$ lactic acid that has been titrated to pH 3.5 with NH$_4$OH. The mixture is then centrifuged at at least 6000 $g$ for 30 min (preferably about 20,000 $g$ for 15 min), and the orange-brown supernatant is decanted from the residue. The extract should be about 1250 ml, and contain on an activity basis over 2 g of phosphoglycerate kinase.

*Step 2. Ammonium Sulfate Fractionation.* The original procedure[2] made use of two successive ammonium sulfate fractionations, which will be described here. However, equally good recovery of enzyme has been made by a simpler procedure, which in addition also allows another enzyme, glucose-6-phosphate dehydrogenase, to be separated into a different fraction.[3]

Ammonium sulfate, 100 g/liter, is dissolved into the extract, then the pH is further lowered, to 4.3 (checking with a 10× diluted sample) using cold 5 $M$ lactate, pH 3.5. After 10–15 min, the mixture is centrifuged (as above) and the precipitate discarded. A further 250 g ammonium sulfate is then dissolved per liter of supernatant; after stirring for 30 min, the precipitate is collected by centrifugation. This precipitate is dissolved in 800 ml of 40 m$M$ Tris base + 1 m$M$ EDTA, to raise the pH to 7.5 ± 0.2 (further 1 $M$ Tris is added if necessary) ; 340 g of ammonium sulfate per liter of solution is dissolved into the preparation, and after 30 min of stirring the precipitate is removed by centrifugation. To the supernatant, a further 150 g of ammonium sulfate per liter is added, and the precipitate is collected as before. These two fractionations, at pH 4.3 and 7.5, represent ammonium sulfate cuts of 18–56% and 60–82% saturation, respectively, indicating completely different solubility char-

---

[3] This simpler procedure has not been tested frequently. The pH of the extract is adjusted to 5.5 with 5 $M$ lactate pH 3.5, and an ammonium sulfate fraction precipitating between 50% and 63% saturation is removed for preparation of glucose-6-phosphate dehydrogenase.[4] The 63% supernatant is then adjusted to pH 4.3 using the 5 $M$ lactate, and checking the pH on a diluted sample. The precipitate forming is collected and dissolved in a little 1 m$M$ EDTA; the pH is brought to 7.0 with 1 $M$ Tris. This preparation is then ready for dialysis and step 3.

[4] R. K. Scopes, *Biochem. J.* **134**, 197 (1973).

acteristics of the enzyme at the two pH's. The final precipitate is dissolved in a little 1 mM EDTA, pH 7.0, and dialyzed extensively against 2 or 3 changes of 5 liters of 1 mM EDTA pH 7.0

Step 3 Removal of Nucleic Acids. If the dialyzed fraction from step 2 is adjusted to pH 6.0 prior to ion-exchange chromatography, turbidity develops due to interactions of nucleic acids. This can be avoided by treating with protamine sulfate. After dialysis, a solution of 1 g of protamine sulfate in 50 ml of water, adjusted to pH 7.0 with Tris, is added; the resulting precipitate is removed by centrifugation.[5] Then 0.01 volume of 1 M Tris is added to the solution, and the pH lowered to 6.0 with cacodylic acid.[6] The solution is then diluted to about 800 ml with the chromatography buffer, which consists of 10 mM Tris plus 1 mM EDTA, adjusted to pH 6.0 with cacodylic acid.[6]

Step 4. CM-Cellulose Chromatography. A column packed with CM-cellulose equilibrated in the Tris–EDTA–cacodylate buffer is prepared. The dimensions should be about 5 cm in diameter and 20–30 cm long to accommodate the protein from up to 600 g of dried yeast; preparations on a smaller or larger scale must be scaled appropriately. The fraction is run on to the column at a flow rate of approximately 12 ml per hour per square centimeter of cross section, and washed in with a little buffer. A linear gradient in KCl is commenced, with incremental KCl concentration of 0.1 mM/ml; the flow rate is maintained at between 8 and 10 ml per hour per square centimeter of cross section. The enzyme is eluted between 50 and 70 mM KCl, preceded by a substantial quantity of protein which merges with the enzyme peak. Tubes containing more than 300 units of enzyme per milliliter are combined, and the protein is salted out by adding 60 g of ammonium sulfate per 100 ml. (If available, ultrafiltration equipment can achieve a more satisfactory concentration of the protein after the CM-cellulose chromatography.) The precipitated protein is collected by centrifugation and dissolved in a small volume of 1 mM EDTA, pH 7.0. The volume at this stage should be between 40 and 50 ml.

Step 5. Gel Filtration. To accommodate the whole of the fraction from the preceding step, a column with a volume of at least 1.5 liters is necessary. If not available, the fraction must be divided for successive application to a smaller column. Sephadex G-100 or G-150 is most suitable for the gel filtration, and a buffer consisting of 0.15 M KCl containing 1 mM EDTA and 10 mM imidazole pH 7.0 can be used; the composition of this buffer is not critical, although the ionic strength should be at least

[5] Treatment with protamine is not an essential step, but it avoids clogging of the column in step 4 with precipitated material.

[6] Succinic acid can be used as the buffering anion.

0.1. Two peaks of protein are obtained, the enzyme being in the most retarded fraction. At the peak and subsequent tubes the enzyme is pure, with specific activity close to 1000, but some larger molecular weight contaminants are introduced by including the whole enzyme peak. For optimum yield and purity, the earlier tubes containing activity are retained separately, and rerun on the column.

*Step 6. Crystallization.* The tubes containing the enzyme after gel filtration are pooled, and salted out with 60 g of ammonium sulfate per 100 ml. The precipitate is collected by centrifugation and dissolved in about 15 ml of 20 m$M$ phosphate buffer, pH 6.5. Saturated ammonium sulfate is added until the solution becomes milky; a few drops of phosphate buffer are then added until this milkiness begins to clear. The solution is then centrifuged briefly, and the supernatant is placed in a 100-ml conical flask for crystallization at room temperature. If after a few hours the solution is still completely clear, a few drops of saturated ammonium sulfate may be added. Crystallization may take several days, and a thick suspension of fine needles is obtained. After collection of these crystals, any impurities are left in the supernatant; however, some loss of activity usually occurs, as with the muscle enzyme,[7] after crystallization. Larger polyhedric crystals which have been used for X-ray crystallography[2,8] can be grown at a pH just over 7.0, in the presence of about 1% 1,4-dioxane, and an ammonium sulfate concentration of close to 68% saturation.

A summary of the isolation procedure is shown in the table.

ISOLATION OF 3-PHOSPHOGLYCERATE KINASE FROM 450 G OF
GRANULATED BAKER'S YEAST

| Fraction | Total protein (mg) | Total activity (units × 10³) | Specific activity (units/mg) | Yield (%) |
|---|---|---|---|---|
| 1. Ammoniacal extract | 49,000 | 2300 | 47 | 100 |
| 2. (a) Ammonium sulfate pH, 4.3 | 26,500 | 1800 | 68 | 78 |
| (b) Ammonium sulfate pH, 7.5 | 17,500 | 1620 | 92 | 70 |
| 3. Protamine treatment | 17,200 | 1620 | 94 | 70 |
| 4. CM-cellulose | 2,400 | 1220 | 510 | 53 |
| 5. Gel filtration | 1,420 | 1160 | 820 | 50 |
| 6. Crystals | 1,180 | 1000 | 850 | 43 |

[7] R. K. Scopes, *Biochem. J.* **113**, 551 (1969).
[8] P. L. Wendell, T. N. Bryant, and H. C. Watson, *Nature (London) New Biol.* **240**, 134 (1972).

## Properties

The properties of yeast phosphoglycerate kinase are very similar to those of the muscle enzyme in most respects.[9,10] X-Ray diffraction studies on the yeast enzyme have confirmed the bilobal structure found for the horse muscle protein, and demonstrate that the long divergence of evolution has not led to any major alteration of the tertiary structure. The molecular weight is not significantly different from that of mammalian enzymes (45,000–50,000). The amino acid compositions, although remarkably similar, differ in one major respect: the sulfur-containing amino acids are poorly represented in the yeast enzyme. Thus, there are only 3 methionines compared with 13 in the mammalian enzymes (rabbit muscle and human erythrocyte), and, more significantly, only 1 cysteine compared with 8–10.[10] This cysteine residue, although close to the active site, is not required for enzyme activity, and treatment with PCMB, DTNB, or $Hg^{2+}$ does not result in loss of any activity. This contrasts with the rapid loss of activity by the muscle enzyme when treated with any thiol-reacting compound. Tryptophan and tyrosine residue contents are 0.5 and 2 times the mammalian enzymes, respectively, otherwise differences in amino acid composition are hardly significant.

Extensive kinetic studies have been carried out on the back reaction catalyzed by the yeast enzyme. The results can be summarized as follows: The true substrates for the reaction are $MgATP^{2-}$ and 3-phosphoglycerate$^{3-}$, with weak inhibitions by $ATP^{4-}$, Mg-phosphoglycerate, and $Mg^{2+}$.[11,12] Other cations activating the enzyme (by forming the substrate $MeATP^{2-}$) include $Mn^{2+}$, $Ca^{2+}$, $Zn^{2+}$, $Co^{2+}$, $Cd^{2+}$, and $Ni^{2+}$, but free $Zn^{2+}$ ions were particularly potent at inhibiting the reaction.[12] The evidence from these kinetic studies favors a rapid random equilibrium type of reaction mechanism; further support for this mechanism comes from inhibitor studies on the erythrocyte enzyme.[13]

[9] W. K. G. Krietsch and T. Bücher, *Eur. J. Biochem.* **17**, 568 (1970).
[10] R. K. Scopes, *in* "The Enzymes" (P. J. Boyer, ed.), 3rd ed., Vol. 8, p. 335. Academic Press, New York, 1973.
[11] M. Larsson-Raznikiewicz, *Biochim. Biophys. Acta* **85**, 60 (1964).
[12] M. Larsson-Raznikiewicz, *Eur. J. Biochem.* **17**, 183 (1970).
[13] C. S. Lee and W. J. O'Sullivan, *Proc. Aust. Biochem. Soc.* **6**, 18 (1973).

## [24] Phosphoglycerate Kinase and Phosphoglyceromutase from *Escherichia coli*

By GIUSEPPE D'ALESSIO and JOHN JOSSE

Glycerate-1,3-$P_2$ + ADP $\rightleftarrows$ glycerate-3-P + ATP
Glycerate-3-P + glycerate-2*,3-$P_2$ $\rightleftarrows$ glycerate-2,3-$P_2$ + glycerate-2*-P

The two enzymes ATP:3-phospho-D-glycerate 1-phosphotransferase, EC 2.7.2.3, and 2,3-diphospho-D-glycerate:2-phospho-D-glycerate phosphotransferase, EC 2.7.5.3, which catalyze succeeding steps in the glycolytic pathway, are discussed in a single chapter because purified fractions of each are obtained in good yields from the same isolation procedure.[1] In fact, the procedure outlined below yields an additional early fraction from which pure glyceraldehyde phosphate dehydrogenase may be obtained also.[1]

### Assay Methods

*Principles*. Activities of both enzymes are measured by use of coupled reactions linking the relevant reaction in exergonic sequence ultimately to one consuming reduced pyridine nucleotide in stoichiometric amounts, a process easily monitored by changes in light absorption at 340 nm. Phosphoglycerate kinase is assayed in the reverse direction of glycolysis, starting with glycerate-3-P and ATP and coupling to the glyceraldehyde-phosphate dehydrogenase-catalyzed reduction of glycerate-1,3-$P_2$ with consequent consumption of NADH and trapping of glyceraldehyde-3-P as the hydrazone.[2] Phosphoglyceromutase is assayed in the forward direction, starting with glycerate-3-P and glycerate-2,3-$P_2$ and coupling, with use of enolase, ADP, and pyruvate kinase, to the lactic dehydrogenase-catalyzed reduction of pyruvate with NADH.[3]

*Reagents*

Tris-chloride buffer, 1 *M*, pH 7.5
$MgSO_4$, 0.15 *M*
KCl, 2 *M*
EDTA, 0.3 *M*, neutral

[1] G. D'Alessio and J. Josse, *J. Biol. Chem.* **246**, 4319 (1971).
[2] H. Adam, *in* "Methods of Enzymatic Analysis" (H. U. Bergmeyer, ed.), p. 539. Academic Press, New York, 1965.
[3] R. Czok and L. Eckert, *in* "Methods of Enzymatic Analysis" (H. U. Bergmeyer, ed.), p. 229. Academic Press, New York, 1965.

Hydrazine, 60 m$M$ (freshly dissolved and neutralized with NaOH)
Glycerate-3-P, 50 m$M$, neutral
Glycerate-2,3-P$_2$, 10 m$M$, neutral (see Vol. 3 [37]; Vol. 6 [71])
ATP, 10 m$M$
ADP, 10 m$M$
NADH, 10 m$M$, in 0.01 $N$ NaOH
0.5% Glyceraldehyde phosphate dehydrogenase. Centrifuge 0.1 ml
  of crystalline suspension of commercial rabbit skeletal muscle en-
  zyme (10 mg of protein per milliliter) and dissolve sediment in
  0.2 ml of 20 m$M$ Tris chloride, pH 8, containing 2 m$M$ EDTA
  and 2 m$M$ 2-mercaptoethanol; prepare fresh daily.
Lactic dehydrogenase, 0.75%. Crystalline suspensions of commercial
  rabbit skeletal muscle lactic dehydrogenase, pyruvate kinase, and
  enolase may be used directly.
Pyruvate kinase, 1.0%
Enolase, 1.0%

*Procedures.* Phosphoglycerate kinase is assayed by addition of the
following solutions to a 1.5-ml quartz cuvette with a 1-cm light path:
0.02 ml of Tris buffer, 0.02 ml of MgSO$_4$, 0.02 ml of EDTA, 0.02 ml
of hydrazine, 0.02 ml of glycerate-3-P, 0.01 ml of ATP, 0.01 ml of
NADH, 0.01 ml of glyceraldehyde-phosphate dehydrogenase, and water
at 23° to make a final volume of 1 ml.[2]

Phosphoglyceromutase is assayed in a similar cuvette by addition of
0.02 ml of Tris buffer, 0.05 ml of MgSO$_4$, 0.03 ml of KCl, 0.02 ml of
glycerate-3-P, 0.01 ml of glycerate-2,3-P$_2$, 0.02 ml of ADP, 0.02 ml of
NADH, 0.001 ml of lactic dehydrogenase, 0.001 ml of pyruvate kinase,
0.01 ml of enolase, and water at 23° to make a final volume of 1 ml.[3]

With both enzyme assays the solution is incubated at room temperature (23°) for a minute or two, and the reaction is initiated by the addition of enzyme (0.005–0.02 ml) diluted in 20 m$M$ Tris chloride, pH 8,
containing 2 m$M$ EDTA. Linearly decreasing optical absorption of
NADH at 340 nm is followed in a Cary recording spectrophotometer or
comparable instrument for 1–2 min. Both assays can be used with crude
*E. coli* extracts, and there are no difficulties with interfering side
reactions.

*Definition of Unit and Specific Activity.* One unit of enzyme trans-
forms 1 $\mu$mole of substrate per minute under the stated assay conditions
($\epsilon = 6.22 \times 10^3$ $M^{-1}$ cm$^{-1}$ at 340 nm for NADH[4]). Specific activity is ex-

[4] A. Kornberg and W. E. Pricer, Jr., *in* "Biochemical Preparations, (E. E. Snell,
ed.), Vol. 3, p. 20. Wiley, New York, 1953.

pressed as units per milligram of protein determined either according to Lowry et al.[5] or by a microbiuret method (see Vol. 3 [103]).

## Purification Procedure

All operations are conducted at 0° to 5° unless otherwise specified; centrifugations are performed at 15,000 $g$ for 30 min. Tris-EDTA buffer refers to 20 m$M$ Tris chloride, pH 8, containing 2 m$M$ EDTA. Procedures are summarized in the table. A large-scale procedure is described, starting with 90 liters of culture medium and yielding 50–100 mg of each enzyme; however, all operations can be reproducibly scaled down proportionately.

*Step 1. Growth of Bacteria and Preparation of Extract. E. coli* (Hfr Hayes, strain K12-3000,[6] from the collection of J. Monod, Pasteur Institute) are grown to stationary phase at 37° with vigorous aeration in a medium containing, per liter: $Na_2HPO_4$, 5 g; $KH_2PO_4$, 12.5 g; ammonium sulfate, 2 g; glucose, 10 g; and yeast extract, 10 g. After 12–15 hr of growth cells are harvested with a continuous flow centrifuge with a yield of ca. 5 g per liter of culture. The cell paste can be stored at −10° for over 3 years without loss of enzymic activities. Cells (450 g, wet weight) are disrupted by high speed stirring with glass beads[7] (Super-

PURIFICATION OF ENZYMES

| Step (and fraction) | Protein (mg/ml) | PGK[a] | | | PGM[a] | |
|---|---|---|---|---|---|---|
| | | Specific activity (units/mg protein) | Total activity (units) | | Specific activity (units/mg protein) | Total activity (units) |
| 1. Extract (I) | 15 | 0.7 | 25,000 | | 1.7 | 60,000 |
| 2. Streptomycin (II) | 7 | 1.1 | 24,000 | | 2.4 | 50,000 |
| 3. Ammonium sulfate (III) | 60 | 3.3 | 24,000 | | 5.6 | 40,000 |
| 4. Sephadex G-150 (IV) | 1.8 | 7.9 | 13,000 | | 12 | 20,000 |
| 5. DEAE-cellulose[b] (V-K) | 11 | 98 | 7,000 | | | |
| (V-M) | 25 | | | | 124 | 13,000 |

[a] PGK, phosphoglycerate kinase; PGM, phosphoglyceromutase.
[b] See Fig. 1.

[5] O. H. Lowry, N. J. Rosebrough, A. L. Farr, and R. J. Randall, *J. Biol. Chem.* **193**, 265 (1951).
[6] B. J. Bachman, *Bacteriol. Rev.* **36**, 525 (1972).
[7] Washed twice in 1 m$M$ EDTA, then in distilled water 5 times. The beads were recovered by filtration and dried at 200°.

brite, average diameter 200 $\mu$m, Minnesota Mining and Manufacturing Company; 1.3 kg) in 450 ml of Tris-EDTA buffer. The procedure is carried out in a 1-gallon stainless steel blending container with an outer jacket through which polyethylene glycol at —5° is circulated, the temperature of the extract never exceeding 10°. The mixture is stirred at slow speed to form a thick suspension and then blended for 20 min at 75% of maximum speed. Tris-EDTA buffer (1.35 liters) is added, and the mixture is stirred at slow speed for an additional 10 min. After the beads have settled for 10 min, the liquid is decanted, and the beads are washed by slow speed stirring with 1 liter of Tris-EDTA buffer. The combined extract and wash are centrifuged, and the supernatant liquid is decanted (fraction I, 2.4 liters).

*Step 2. Streptomycin Precipitation of Inactive Materials.* Streptomycin sulfate (Charles Pfizer and Company; 31 g in 625 ml of Tris-EDTA buffer) is added with stirring to fraction I. The suspension is slowly stirred an additional 15 min and then centrifuged; the supernatant fluid is collected (fraction II, 3 liters).

*Step 3. Ammonium Sulfate Fractionation.* Solid ammonium sulfate (Nutritional Biochemicals Corporation, enzyme grade; 1.2 kg) is stirred into the 3 liters of fraction II. The precipitate is removed by centrifugation, and 0.7 kg of ammonium sulfate is added with stirring to the supernatant. After this suspension has stood overnight (15–18 hr), it is centrifuged, and the precipitate is collected and dissolved in 120 ml of Tris-EDTA buffer. The solution of concentrated protein is clarified by a final centrifugation (fraction III).

*Step 4. Sephadex G-150 Gel Filtration.* Fraction III (120 ml) is gently applied to the top of a column (70 cm$^2$ × 120 cm) of Sephadex G-150 gel which has been equilibrated with Tris-EDTA buffer. The same buffer is then pumped through the column at a rate of 60 ml/hr, and fractions of 35–60 ml are collected. After the void volume has passed through the column, a broad peak of protein is eluted, the latter half of which contains both enzymes. Fractions containing enzymic activity are combined (fraction IV, ca. 925 ml). (The first half of the protein peak contains glyceraldehyde-phosphate dehydrogenase.[1])

*Step 5. DEAE-Cellulose Chromatography.* A column (3.8 cm$^2$ × 25 cm) of DEAE-cellulose (Whatman DE-52, microgranular) is equilibrated with Tris-EDTA buffer, and fraction IV is applied to the top of the bed. After the column has been washed with 5 bed volumes of Tris-EDTA buffer, a linear salt gradient (0 to 0.15 $M$ NaCl in a total volume of 1 liter of Tris-EDTA buffer) is applied. The average flow rate is 30 ml/hr, and fractions of 5.3 ml are collected. The phosphoglyceromutase and phosphoglycerate kinse activities elute separately at concentrations

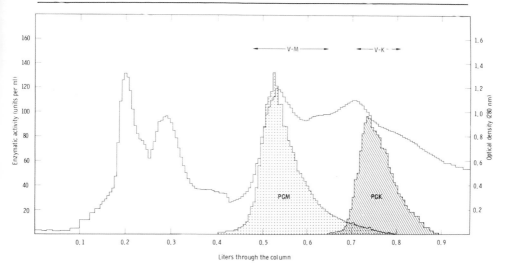

Fɪɢ. 1. Chromatography on DEAE-cellulose of fraction IV. Fractions V-K and V-M are obtained from eluates pooled as shown by the arrows. Details are given in the text. PGM, phosphoglyceromutase; PGK, phosphoglycerate kinase.

of 80 m$M$ and 0.11 $M$ NaCl, respectively, although there is a consistent and unexplained trailing of the phosphoglyceromutase peak into the phosphoglycerate kinase zone (Fig. 1). Fractions containing the respective activities are combined as shown in Fig. 1 and concentrated by pervaporation in a Schleicher and Schuell collodion bag filtration apparatus. The concentrated protein solutions (fraction V-K containing phosphoglycerate kinase, 6.5 ml, and fraction V-M containing phosphoglyceromutase, 4.2 ml) are clarified by centrifugation if turbidity is present.

## Properties

*Purity and Stability of Fractions.* The final fraction of phosphoglyceromutase appears homogeneous when examined by polyacrylamide gel electrophoresis, but phosphoglycerate kinase fraction V-K is impure,[1] including the obvious contamination with phosphoglyceromutase (Fig. 1). Attempts to purify phosphoglycerate kinase further have yielded unstable fractions with poor recoveries of activity. The final fractions of both enzymes are stable for 6 months or more at 0°, but intermediate fractions of phosphoglycerate kinase during the purification are of variable stability, especially in column eluates of low protein concentration. It is recommended that the entire procedure be carried out in sequence; this requires 2 weeks or less if the columns have been prepared in advance.

*Specificity.* The reverse phase of the phosphoglycerate kinase-catalyzed reaction is totally dependent upon the presence of ATP, divalent cation (e.g., $Mg^{2+}$), and glycerate-3-P. Isotopic studies with $^{32}P$ have shown that the terminal or $\gamma$-phosphate of ATP is transferred to glycerate-3-P to form glycerate-1,3-$P_2$.[1] The phosphoglyceromutase-catalyzed reaction requires added glycerate-2,3-$P_2$ when tested with highly purified preparations of glycerate-3-P (see Vol. 6 [71]).

*Molecular Weights and Sedimentation Coefficients.* Molecular weights of phosphoglycerate kinase and phosphoglyceromutase, determined during high speed sedimentation equilibrium,[8] are $43,700 \pm 1100$ (SD) and $56,300 \pm 1200$ (SD), respectively; $s_{20,w}^{\circ}$ values are 3.5 S and 4.8 S, respectively. The contaminants present in purified phosphoglycerate kinase (fraction V-K) do not significantly interfere with these determinations.

[8] D. A. Yphantis, *Biochemistry* **3**, 297 (1964).

# [25] Human Phosphoglycerate Kinase[1]

*By* AKIRA YOSHIDA

1,3-Diphospho-D-glycerate + ADP $\rightleftharpoons$ 3-phospho-D-glycerate + ATP

## Assay Method

*Principle.* The formation of 1,3-diphospho-D-glycerate from 3-phospho-D-glycerate may be indirectly measured by the oxidation of NADH in the presence of excess glyceraldehyde-3-phosphate dehydrogenase (EC 1.2.1.12; 1,3-disphospho-D-glycerate + $NADH \rightleftharpoons$ D-glyceraldehyde-3-phosphate + $P_i$ + NAD). The decrease in optical density arising from the oxidation of NADH may be recorded.

*Reagents*

3-Phospho-D-glycerate, sodium salt, 0.1 $M$
ATP, sodium salt, 0.1 $M$
NADH, 10 m$M$
Tris-chloride buffer, 0.1 $M$, pH 7.5, containing $MgCl_2$, 10 m$M$
Glyceraldehyde-3-phosphate dehydrogenase (from rabbit muscle), 1 mg (40–80 units) per milliliter

All reagents, except for buffer solution and glyceraldehyde-3-phosphate dehydrogenase, should be stored frozen. Glyceraldehyde-3-phosphate dehydrogenase (crystalline suspension in ammonium sulfate com-

[1] EC 2.7.2.3.

mercially available) is diluted in water to obtain the above concentration and stored at 0–4°.

*Enzyme.* The enzyme is dissolved in 10 m$M$ Tris chloride, pH 7.5, or in 10 m$M$ Na$_2$HPO$_4$–KH$_2$PO$_4$, pH 7.0, to obtain 0.1–10 units of enzyme per milliliter (see definition below).

*Procedure.* Tris-chloride buffer, 0.1 $M$, pH 7.5, containing 10 m$M$ MgCl$_2$, 0.8 ml; 0.1 ml of 0.1 $M$ 3-phospho-D-glycerate; 0.05 ml of 0.1 $M$ ATP; 0.02 ml of 10 m$M$ NADH; 0.01 ml of glyceraldehyde-3-phosphate dehydrogenase; 0.002–0.02 ml of enzyme solution; and water to a final volume of 1.0 ml are placed in a cell having a 1-cm light path.

The decrease of absorbancy at 340 nm is recorded on a logarithmic recorder connected to the output of a spectrophotometer.

*Units.* One unit of activity caused 1,3-diphosphoglycerate to be produced (or NADH to be oxidized) at a rate of 1 $\mu$mole per minute at 25°. The units of enzyme in the reaction mixture are obtained from the linear stationary rate of diminution of absorbancy at 340 nm using the following equation:

$$\text{Units of enzyme in assay mixture} = (\text{OD change in 1 minute})/6.22$$

Specific activity is expressed as the number of enzyme units per milligram of protein. The forward reaction, i.e., rate of phosphorylation of ADP, may also be used in the assay.[1] However, this method involves several problems as discussed in reference 2, and it is sensitive to interference from contamination with other enzymes. The assay using the backward reaction was used in the course of the enzyme purification.

## Purification Procedure[3]

All the procedures, unless otherwise specified, were carried out at 0–4°. All buffer solutions and distilled water used for the preparation contained 1 m$M$ 2-mercaptoethanol and 1 m$M$ EDTA (neutralized to pH 7.0).

*Hemolysis of Red Blood Cells.* Fresh or outdated human blood with acid-citrate-dextrose anticoagulant can be used for the preparation. The blood was centrifuged at 15,000 $g$ for 20 min, and supernatant plasma was removed with a pipette attached to an aspirator bottlt. The red blood cells were washed wice with 2 volumes of 0.9% NaCl solution and were mixed with an equal volume of citrate-glycerate solution (19.4 g of potassium citrate; 3.1 g of NaH$_2$PO$_4$; 2.8 g of Na$_2$HPO$_4$; and 400 ml of glycerol in 1000 ml). The mixture was dialyzed against 10 volumes of water. The outside water was changed after 3 hr, and dialysis was continued

[2] T. Bücher, this series, Vol. 1, p. 415.
[3] A. Yoshida and S. Watanabe, *J. Biol. Chem.* **247,** 440 (1972).

overnight. The enzyme activity of the hemolysate thus obtained was 135–210 units per gram of hemoglobin.

Hemolysates prepared by simply mixing the packed red cells with 2 volumes of water had about 100 units of enzyme per gram of hemoglobin. The hemolysate thus prepared can also be used for the enzyme purification.

*Step 1. Elimination of Hemoglobin.* Ethanol–chloroform (2:1, v/v), 400–420 ml, cooled at −60°, was added at once to 1.2 liters of the ice-cold hemolysate (from 1 liter of blood, total activity 18,000–21,000 units) which was adjusted to pH 7.3 with 1 $M$ KOH. After stirring for 20 min in an ice bath, the mixture was centrifuged (15,000 $g$, 20 min), and the hemoglobin precipitate was discarded. Three liters of ethanol, cooled to −25°, were added to the supernatant fluid under stirring. The temperature was always kept lower than 0°. After 20 min, the precipitate that appeared was collected by centrifugation (15,000 $g$, 15 min). The precipitate was dissolved in 50–100 ml of 10 m$M$ phosphate buffer (KH$_2$PO$_4$–Na$_2$ HPO$_4$), pH 7.0, and dialyzed overnight against the same buffer. When larger quantities of chloroform–ethanol were added to the hemolysate, larger losses (35–40%) of the enzyme activity occurred, although the elimination of hemoglobin was more complete with such treatment.

*Step 2. Carboxymethyl-Sephadex Treatment.* The dialyzed enzyme solution (100–150 ml) was centrifuged, and CM-Sephadex (3 g, dry weight), which was washed with 1 $M$ acetic acid, washed with water, adjusted to pH 7.0 with 2 $N$ NaOH and then filtered through the Büchner funnel, was added to the supernatant liquid in small portions under stirring, maintaining the pH at 6.5 with 2 $N$ NaOH. The mixture was placed on a Büchner funnel and was washed with 200 ml of 10 m$M$ phosphate buffer at pH 7.0. The enzyme adsorbed on the CM-Sephadex was eluted with 250 ml of 0.1 $M$ phosphate buffer, pH 7.0, and concentrated by vacuum dialysis.

*Step 3. Further Elimination of Hemoglobin by DEAE-Sephadex Column Chromatography.* If the concentrated enzyme solution (20–30 ml) was contaminated with only small amounts of hemoglobin (faintly pink), it was unnecessary to carry out this step. Otherwise, the enzyme solution was dialyzed against 10 m$M$ Tris chloride, pH 8.0, and placed on a DEAE-Sephadex column (2.5 × 10 cm), equilibrated with 10 m$M$ Tris chloride, pH 8.0. The enzyme was eluted from the column by the same buffer, while hemoglobin remained adsorbed on the ion exchanger. The eluate was adjusted to pH 7.0 with 1 $M$ acetic acid and concentrated by vacuum dialysis to about 20 ml.

*Step 4. Sephadex G-75 Gel Filtration.* The enzyme solution was applied to a Sephadex G-75 column (2.5 × 100 cm), equilibrated with 10 m$M$ phosphate buffer, pH 7.5. Phosphoglycerate kinase was eluted before the high protein peak which contained carbonic anhydrase activity. The major phosphoglycerate kinase fractions were pooled and concentrated by vacuum dialysis.

*Step 5. Carboxymethyl-Sephadex Column Chromatography.* The concentrated enzyme solution (about 20 ml) was dialyzed against 10 m$M$ phosphate buffer, pH 7.0, and placed onto a CM-Sephadex column (2 $\times$ 30 cm), equilibrated with the same buffer. The enzyme was eluted with a linear gradient of buffer concentration from 10 m$M$ to 0.1 $M$. The gradient was produced by adding 500 ml of 0.1 $M$ buffer into a mixing chamber which contained 500 ml of 10 m$M$ buffer. The flow rate was 50 ml/hr. The major phosphoglycerate kinase fractions were pooled (total activity 10,000–12,000 units) and concentrated by vacuum dialysis to about 5 ml.

*Step 6. Crystallization.* The concentrated enzyme solution was centrifuged and a small amount of precipitate was discarded. The supernatant liquid was placed in a dialysis bag (1 cm diameter) and dialyzed against 0.1 $M$ phosphate buffer, pH 7.0, containing ammonium sulfate. The concentration of ammonium sulfate was increased stepwise from 30% saturation to 65% saturation over a period of 24 hr by adding saturated ammonium sulfate solution. The enzyme started to crystallize at about 35% saturation of ammonium sulfate; more than 95% of the enzyme was crystallized by the procedure. The enzyme can be recrystallized by repeating the procedure. Overall yield of the crystalline enzyme was about 15–18 mg from 1 liter of blood.

## Properties

*Enzymic Properties.*[3] The specific activity of crystalline human phosphoglycerate kinase measured in the reaction mixture specified in the text was 650–700 units/mg at 25°. The specific activity was about 1000 at 35°. $Mg^{2+}$ or $Mn^{2+}$ at a concentration of 5 m$M$ to 10 m$M$ is essential to the enzyme reaction. Higher $M^{2+}$ or $Mn^{2+}$ concentration (50 m$M$) slightly inhibited the enzyme reaction. The crystalline enzyme has no measurable activity of carbonic anhydrase, triosephosphate isomerase, glycerol-3-phosphate dehydrogenase, glyceraldehyde-3-phosphate dehydrogenase, pyruvate kinase, or adenylate kinase.

The Michaelis constant ($K_m$) for D-3-phosphoglycerate was 1.1 m$M$ under the assay condition specified. The rate of the enzyme reaction as a function of the concentration of ATP (or Mg ATP) did not fit the usual Michaelis-Menten relationship. At higher concentrations of ATP, the $K_m$ was estimated to be about 1 m$M$, and at lower concentrations it was estimated to be 0.37 m$M$. The optimal pH was 7.2 to 9.0, having a truncated pH-activity profile. Phosphate buffer inhibited the reaction.

The enzyme in the crude hemolysate is stable at neutral pH (pH 6.5–8.0) at 4° for at least a week, but is more rapidly inactivated at acidic pH (lower than pH 6.0) and alkaline pH (higher than pH 8.5). The purified crystalline enzyme, dissolved at a low concentration in a buffer of neutral pH, is rapidly inactivated, i.e., more than 10% inactivation per 24 hr at 4°. When bovine serum albumin was added to the enzyme solution at a final concentration of 10 mg/ml, the enzyme activity remained unchanged for at least a week at 4°. Freeze-drying caused inactivation of the enzyme.

*Molecular Properties.*[3] The molecular weight of the enzyme is 49,600 (by sedimentation equilibrium method). The sedimentation constant $(s_{20,w})$ is 3.35 S at a protein concentration of 0.6%. The enzyme could not be dissociated into smaller subunits by treatments which have caused dissociation of various other proteins. The enzyme has $N$-acetylserine as $NH_2$-terminal and isoleucine as COOH-terminal. The enzyme has the following amino acid composition: Asp 50, Thr, 19, Ser 27, Glu 38, Pro 21, Gly 45, Ala 45, CysH 11, Val 45, Met 14, Ile 20, Leu 44, Tyr 4, Phe 17, Lys 44, His 6, Arg 12, Trp 4.

Human phosphoglycerate kinase of other tissues should be identical to that of red blood cells, since a single structural gene located in the X chromosome produces the enzyme.[4]

[4] S.-H. Chen, L. A. Malcolm, A. Yoshida, and E. R. Giblett, *Amer. J. Human Genet.* **23**, 87 (1971).

# [26] Glycerol Kinase

By JEREMY W. THORNER

$$\text{Glycerol} + \text{ATP} \xrightarrow{\text{Mg}^{2+}} \text{glycerol-3-P} + \text{ADP}$$

In many microorganisms, glycerol kinase serves as the first enzyme in a pathway for the utilization of glycerol as a carbon and energy source.[1] The availability of regulatory mutants of *Escherichia coli* which produce glycerol kinase constitutively[2] permits the isolation of large amounts of crystalline enzyme.[3,4] The preparation of crystalline glycerol kinase from other sources has been described in this series previously.[5]

[1] J. W. Thorner and H. Paulus, *in* "The Enzymes" (P. D. Boyer, ed.), 3rd ed., p. 487f. Academic Press, New York, 1973.
[2] N. R. Cozzarelli, W. B. Freedberg, and E. C. C. Lin, *J. Mol. Biol.* **31**, 371 (1968).
[3] S. Hayashi and E. C. C. Lin, *J. Biol. Chem.* **242**, 1030 (1967).
[4] J. W. Thorner and H. Paulus, *J. Biol. Chem.* **246**, 3885 (1971).
[5] C. Bublitz and O. Wieland, this series, Vol. 5, p. 354; E. P. Kennedy, *ibid.*, p. 476.

## Assay Methods

*Principle.* Two sensitive radiochemical assays for the enzyme have been developed recently. Glycerol kinase activity can be measured either by the glycerol-dependent conversion of [γ-³²P]ATP to an acid-stable phosphate ester[4] (Method A) or by the formation from [¹⁴C]glycerol of glycerol-3-P which can be adsorbed to disks of DEAE-cellulose filter paper[6] (Method B). Although cell extracts might be expected to contain interfering activities such as adenylate kinase, glycerol-3-P dehydrogenase, D-triokinase, and phosphatases, both methods give reproducible results even with crude extracts of *E. coli* and other bacterial species.[7,8]

Spectrophotometric determinations of glycerol kinase activity, in which the formation of products is coupled to other enzymic reactions, have been reported previously.[5,9]

## Method A

### Reagents

2 m$M$ glycerol–4 m$M$ MgCl₂–0.1 $M$ triethanolamine hydrochloride, pH 7.0

[γ-³²P]ATP, 0.1 $M$ (20,000–50,000 cpm/μmole), neutralized with KOH

Bovine serum albumin, 10 mg/ml

HClO₄, 1 $N$

Ammonium molybdate, 80 m$M$

Triethylamine, 0.2 $M$

Bovine serum albumin, 0.1 mg/ml, in 10 m$M$ glycerol–20 m$M$ triethanolamine hydrochloride, pH 7.0 (enzyme diluent)

*Procedure.* Conical centrifuge tubes, containing 0.5 ml of glycerol–MgCl₂–triethanolamine hydrochloride, 0.01 ml of [γ-³²P]ATP, 0.1 ml of bovine serum albumin, and 0.34 ml of H₂O, are placed in a constant-temperature bath at 25° for 2 min, and then reaction is initiated by the addition of 0.05 ml of an appropriate dilution of glycerol kinase, to give a final volume of 1 ml. After 15 min of incubation at 25°, the reaction is terminated by immediate addition of 1 ml of cold HClO₄. The tubes

[6] E. A. Newsholme, J. Robinson, and K. Taylor, *Biochim. Biophys. Acta* **132**, 338 (1967).

[7] J. W. Thorner, Doctoral Dissertation, Harvard University, 1972.

[8] D. P. Richey and E. C. C. Lin, *J. Bacteriol.* **112**, 784 (1972); *ibid.* **114**, 880 (1973).

[9] P. B. Garland and P. J. Randle, *Nature (London)* **196**, 987 (1962).

are covered with marbles and placed in a boiling water bath for 40 min to convert the remaining $[\gamma^{32}P]$ATP to $^{32}P_i$, while leaving the $[^{32}P]$glycerol-3-P essentially unhydrolyzed. After cooling to ambient temperature, 0.5 ml of ammonium molybdate and 0.25 ml of triethylamine are added and mixed in thoroughly. The orthophosphate precipitates as a bright yellow complex,[10] whereas $[^{32}P]$glycerol-3-P remains in solution. The precipitate is removed by centrifugation and a sample (1 ml) of the supernatant solution is withdrawn for scintillation counting in 10 ml of Bray's scintillation fluid.[11] The experimental values are corrected by subtracting blank values (less than 10 nmoles) obtained from incubations without enzyme.

*Comments.* $[\gamma\text{-}^{32}P]$ATP can be obtained commercially or can be synthesized enzymically by the method of Glynn and Chappell.[12] Neither carrier glycerol-3-P nor carrier $P_i$ is needed in the assay procedure. It is best not to include a thiol in the incubation mixture since it will reduce the yellow phosphomolybdate complex to the soluble blue phosphomolybdous complex, thereby raising the blank values. Under conditions where less than 20% of the added ATP is consumed (when the final concentration of crystalline enzyme is below 0.4 $\mu$g/ml), the assay is a linear function both of the amount of enzyme added and of time, and thus provides an accurate measurement of initial rates.[7]

*Method B*

*Reagents*

20 m$M$ $MgCl_2$–0.5 $M$ triethanolamine hydrochloride, pH 7.0
Gelatin, 1 mg/ml
$[1,3\text{-}^{14}C]$Glycerol, 0.1 m$M$ (10,000–20,000 cpm per nmole)
ATP, 0.1 $M$, neutralized with KOH
Glycerol, 2 $M$
Glycerol, 0.1 $M$
HCl, 0.1 $N$
Gelatin, 0.1 mg/ml, in 1 m$M$ 2-mercaptoethanol–20 m$M$ triethanolamine hydrochloride, pH 7.0 (enzyme diluent)

*Procedure.* Small test tubes, containing 0.05 ml of $MgCl_2$–triethanolamine hydrochloride, 0.2 ml of $[1,3\text{-}^{14}C]$glycerol, 0.05 ml of gelatin, 0.185 ml of $H_2O$, and 0.01 ml of an appropriate dilution of glycerol kinase, are placed in a constant-temperature bath at 25° for 2 min, and

[10] Y. Sugino and Y. Miyoshi, *J. Biol. Chem.* **239**, 2360 (1964).
[11] G. A. Bray, *Anal. Biochem.* **1**, 279 (1960).
[12] I. M. Glynn and B. J. Chappell, *Biochem. J.* **90**, 147 (1964).

then reaction is initiated by the addition of 0.005 ml of ATP, to give a final volume of 0.5 ml. After 10 min of incubation at 25°, the reaction is terminated by immediate addition of 0.5 ml of cold 2 $M$ glycerol, followed by thorough mixing. Samples (0.05 ml) from each tube are spotted on disks (2.3 cm diameter) of DEAE-cellulose filter paper. After a few minutes, the disks are transferred to a perforated basket submerged in about 500 ml of 0.1 $M$ glycerol and are washed for 30 min in two more changes of glycerol solution. Essentially all the [$^{14}$C]glycerol-3-P is retained by the filters, while the excess radioactive glycerol is removed. The disks are then placed in vials containing 1 ml of HCl, and 10 ml of Bray's scintillation fluid are added for scintillation counting. The experimental values are corrected by subtracting blank values (less than 5 pmoles) obtained from incubations without enzyme.

*Comments.* Gelatin is used in this procedure since commercial preparations of bovine serum albumin contain glycerol. [$^{14}$C]Glycerol should be treated batchwise before use with microgranular DEAE-cellulose to remove any radioactive anionic impurities. The filter paper disks, once wetted, are rather fragile and should not be subjected to vigorous stirring during the washing procedure. Retention of [$^{14}$C]glycerol-3-P by the disks is unaffected by concentrations of ATP from 1 m$M$ to 10 m$M$ at either 1 or 10 m$M$ MgCl$_2$.[7] Under conditions where less than 20% of the added glycerol is consumed (when the final concentration of crystalline enzyme is below 4 ng/ml), the assay is a linear function both of the amount of enzyme added and of time, thus providing an accurate estimate of initial rates.[7]

*Definition of Unit and Specific Activity.* One unit of glycerol kinase activity is defined as the amount of enzyme which catalyzes the formation of 1 $\mu$mole of glycerol-3-P per minute under the conditions of Assay Method A. Under the conditions of Method B, nearly the same amount of product is formed by 1 unit of enzyme.

The specific activity of a preparation is defined as the units of glycerol kinase activity present per milligram of protein.

## Purification Procedure

*Reagents*

Culture medium (in grams per liter): casein acid hydrolyzate (20), K$_2$HPO$_4$(7), KH$_2$PO$_4$ (1.8), (NH$_4$)$_2$SO$_4$ (1), and MgSO$_4$ (0.5), at about pH 7.0

Standard buffer (buffer A): 10 m$M$ glycerol–1 m$M$ EDTA–1 m$M$ 2-mercaptoethanol–20 m$M$ triethanolamine hydrochloride, pH 7.0

Bovine pancreatic deoxyribonuclease I

6,9-Diamino-2-ethoxyacridine lactate (available from Winthrop
   Laboratories under the trade names Ethodin or Rivanol)

Ammonium sulfate (enzyme grade)

DEAE-cellulose (Cellex D from Bio-Rad)

Sephadex G-200, 40–120 $\mu$ (from Pharmacia), treated by the method
   of Kawata and Chase[13] to remove fines

Recrystallizing buffer (buffer B): 10 m$M$ glycerol–1 m$M$ EDTA–1
   m$M$ 2-mercaptoethanol–0.1 $M$ potassium phosphate, pH 7.0

*Growth of Bacteria.* A starter culture of *Escherichia coli* K10 strain
72,[2] constitutive for the expression of glycerol kinase due to a deletion
of the *glp* R locus and also lacking alkaline phosphatase, is grown in
2 liters of medium and is used to inoculate 100 liters of the same growth
medium in a Fermacell fermentor (New Brunswick Scientific), maintained at 37° with aeration at about 5 cubic feet per minute. Since synthesis of glycerol kinase is also subject to catabolite repression,[2,14] this
medium allows maximum production of the enzyme. The pH of the culture rises above 8 after 12 hr of growth. After 2 more hours, the culture
is transferred to chilled reservoir from which the cells are harvested
by a refrigerated Sharples supercentrifuge. The cells are washed at 4°
with buffer A. The total yield of cells is 600–700 g (wet weight), which
can be used immediately for the purification of glycerol kinase or kept
frozen at −15° without appreciable effect on the yield of enzyme.

*Preparation of Crude Extract.* This and all subsequent operations are
carried out at 4°. Cells (745 g) are suspended in 2 volumes of buffer
A and disrupted by passage through a Manton-Gaulin submicron disperser at 9000 psi in the presence of a few milligrams of bovine pancreatic
deoxyribonuclease I. The homogenate is first centrifuged for 20 min at
48,000 $g$ to remove cell debris, and then at 100,000 $g$ for 1 hr. The final
supernatant solution is the crude extract.

*Removal of Nucleic Acids.* The crude extract is treated with 6,9-di-
amino-2-ethoxyacridine lactate (about 0.2 g per gram of protein) which
is added dropwise with stirring as a 2.5% solution in buffer A, and the
suspension is stirred overnight. The heavy yellow precipitate which forms
is removed by centrifugation at 30,000 $g$ for 30 min and discarded.

*Ammonium Sulfate Fractionation.* The supernatant solution, after removal of nucleic acids, is fractionated by the addition of solid ammonium
sulfate. The material which precipitates between 24 and 32 g of ammo-

[13] H. Kawata and M. W. Chase, *J. Chromatogr.* **30**, 469 (1967).
[14] N. Zwaig and E. C. C. Lin, *Biochem. Biophys. Res. Commun.* **22**, 414 (1966);
   B. DeCrombrugghe, R. L. Perlman, H. Varmus, and I. Pastan, *J. Biol. Chem.*
   **244**, 5828 (1969).

nium sulfate added per 100 ml of the solution is collected by centrifugation at 30,000 $g$ for 20 min, redissolved in about one-tenth of the original volume in buffer A containing 50% glycerol, and stored at −15° overnight.

*DEAE-Cellulose Chromatography.* The ammonium sulfate fraction is diluted 20-fold with buffer A containing 0.1 $M$ KCl and applied over the course of 36 hr to a column (4 × 40 cm) of 500 ml of DEAE-cellulose (0.69 mEq/g) that was packed under pressure and equilibrated with buffer A containing 0.1 $M$ KCl.[15] The column is washed with 250 ml of buffer A containing 0.1 $M$ KCl, and then the protein is eluted with a linear gradient produced by 1650 ml of buffer A containing 0.1 $M$ KCl in the mixing chamber and 1650 ml of buffer A containing 0.25 $M$ KCl in the reservoir, followed by buffer A containing 0.25 $M$ KCl. Enzyme activity emerges at about 0.2 $M$ KCl, concomitant with a large protein peak.

*Concentration by Ammonium Sulfate Precipitation.* The peak fractions from the DEAE-cellulose chromatography are pooled, and solid ammonium sulfate (50 g/100 ml of solution) is added with gentle stirring. The material which precipitates after 30 min of stirring is collected by centrifugation for 20 min at 30,000 $g$ and redissolved in buffer A to yield a volume of 40 ml.

*Sephadex G-200 Chromatography.* The ammonium sulfate concentrate is passed through a column (5.6 × 130 cm) containing 2.8 liters of Sephadex G-200 that was equilibrated with buffer A. Protein is eluted with buffer A, and enzyme activity emerges from the column just after the void volume, coincident with the major protein peak.

*Concentration by Ammonium Sulfate Precipitation.* Protein in the pooled peak fractions from Sephadex G-200 chromatography is precipitated and redissolved in buffer A as described above, to yield a volume of 38 ml.

*Crystallization.* Buffer A saturated with ammonium sulfate (about 10 ml) is added very slowly with rather vigorous stirring to the concentrated protein solution until it becomes just slightly turbid. The solution is clarified by centrifugation at 16,000 $g$ for 20 min and allowed to stand for 2 days at 0°. If no crystals form during this time, crystallization is initiated either by the careful addition of buffer A saturated with ammonium sulfate, one or two drops at a time over the course of several days, or by seeding with crystals of glycerol kinase from another preparation. After 2 days, large and small needles with very little amorphous material appear. The crystals are separated from the mother liquor by

---

[15] At lower ionic strength, a significant fraction of the enzyme becomes irreversibly bound to the anion exchanger. On the other hand, at 0.1 $M$ KCl the sample must be applied to the column quite slowly; otherwise incomplete adsorption results.

PURIFICATION OF GLYCEROL KINASE

| Fraction | Volume (ml) | Total protein (g) | Total activity (units) | Specific activity (units/mg) | Yield (%) |
|---|---|---|---|---|---|
| Crude extract | 1585 | 45.5 | 55,700 | 1.2 | — |
| Ethodin supernatant | 1670 | 36.5 | 80,000 | 2.2 | (132) |
| (NH₄)₂SO₄ fraction | 171 | 15.0 | 59,000 | 3.9 | 106 |
| DEAE-cellulose eluate | 982 | 1.8 | 37,700 | 20.8 | 68 |
| (NH₄)₂SO₄ concentrate | 40 | 1.5 | 37,300 | 24.9 | 67 |
| Sephadex G-200 eluate | 403 | 1.1 | 33,300 | 29.1 | 60 |
| (NH₄)₂SO₄ concentrate | 38 | 0.85 | 35,200 | 41.3 | 63 |
| Crystals[a] | — | 0.82 | 34,000 | 41.2 | 61 |

[a] Combined crystals of two crops, see Purification Procedure.

centrifugation at 2000 $g$ for 5 min, suspended in buffer A saturated with ammonium sulfate, and stored at 0°. If the first crop of crystals does not represent the majority of the total units of activity, the remaining activity in the mother liquor is precipitated by slowly adding an equal volume of buffer A saturated with ammonium sulfate. The precipitate is collected by centrifugation and resuspended in buffer B. Buffer B saturated with ammonium sulfate is added dropwise with stirring until a faint haze appears in the concentrated protein solution. The solution is clarified as before and left to stand several days at 0°. Crystallization is then initiated by seeding with crystals from the first crop, and the solution left to stand for another 5 days at 0°. The second crop, mostly rectangular rods and plates, is collected as before and stored at 0° in buffer B saturated with ammonium sulfate. Only a few percent of the initial activity should remain in the mother liquor after the second crop.

The results of this purification procedure are summarized in the table. Electrophoresis in polyacrylamide gels and equilibrium ultracentrifugation indicate that this preparation is homogeneous.[4]

### Physical Properties

*Size and Subunit Structure.* The native enzyme has a molecular weight of 220,000 and is composed of four identical subunits of 55,000 molecular weight.[4]

*Composition.* The protein contains a net excess of acidic amino acids, glutamate being the only carboxyl-terminal amino acid.[4] The ultraviolet spectrum of the protein exhibits a distinct shoulder at 290 nm indicating a relatively high content of tryptophan.[4] At 280 nm, glycerol kinase has an absorption coefficient of 1.4 liter $g^{-1}cm^{-1}$.[16]

[16] J. W. Thorner and H. Paulus, *J. Biol. Chem.* **248**, 3922 (1973).

All cysteine residues are present in the reduced form; however, only 12 of the 20 sulfhydryl groups of the native enzyme are accessible for reaction with 5,5'-dithiobis(nitro-2-benzoic acid),[4] while in the presence of glycerol just two sulfhydryl groups can be titrated with this reagent.[16]

*Stability.* Glycerol kinase from *E. coli* is most stable at neutral pH, and the presence of glycerol, sulfhydryl compounds, and EDTA markedly enhances the stability of the enzyme in solution.[3,16] The enzyme has been stored at 0° for several years, without significant loss of activity, as a suspension of crystals in buffer B saturated with ammonium sulfate.[3,7]

## Catalytic Properties

*Substrate Specificity.* Crystalline glycerol kinase from *E. coli* also catalyzes the phosphorylation of dihydroxyacetone and L-glyceraldehyde.[3] However, the affinity of the enzyme for these compounds is very much less than that for glycerol; on the other hand, the phosphorylation of dihydroxyacetone is more rapid.[3] In the presence of D-glyceraldehyde, the enzyme effects a greater than stoichiometric release of ADP from ATP, but orthophosphate rather than D-glyceraldehyde-3-P is formed as the other product. Apparently, the hydrated form of this triose is phosphorylated at position 1 to yield an unstable intermediate that decomposes to D-glyceraldehyde and $P_i$.[3]

With respect to the phosphoryl group donor, the *E. coli* enzyme exhibits a high degree of specificity. Only ATP, and no other ribonucleoside triphosphate, can be utilized.[3] Thus, the enzyme has been used as a reagent for selectively removing ATP in the presence of other nucleoside triphosphates.[17] Other possible analytical uses for glycerol kinase have been described previously.[5]

The true substrate of the enzyme is apparently MgATP complex, and reaction is absolutely dependent on added divalent cation.[3,7] At equivalent concentrations, $Mg^{2+}$ gives a rate more than 3-fold higher than that observed with $Mn^{2+}$. $Ca^{2+}$ will not substitute.

*Kinetic Parameters.* At pH 7.0 and 25°, the apparent $K_m$ values of the enzyme for glycerol, dihydroxyacetone, L-glyceraldehyde, and D-glyceraldehyde are 10 $\mu M$, 0.5 m$M$, 3 m$M$, and 0.5 m$M$, respectively.[3,16] Increasing pH or decreasing temperature raises the apparent $K_m$ for glycerol.[7]

Even at saturating glycerol concentration and in the presence of excess $Mg^{2+}$, substrate saturation curves for ATP are not hyperbolic, and yield double reciprocal plots with limiting slopes that indicate two apparent $K_m$ values for MgATP, one in the range 0.08–0.1 m$M$ and the

[17] W. Wickner and A. Kornberg, *Proc. Nat. Acad. Sci. U.S.* **70**, 3679 (1973).

other between 0.4 and 0.5 m$M$.[16] At pH 9.5 or in 0.4 $M$ KCl, this "biphasic" behavior is less pronounced, but not eliminated.[7]

*Effect of Ionic Conditions.* The pH optimum of the enzyme is 9.5.[3,16] At pH 7.0, about half-maximal activity is observed.[3,16] At neutral pH, the activity of glycerol kinase is stimulated 30% by 0.4 $M$ KCl, while LiCl, NaCl, and $NH_4Cl$ have no effect.[16]

*Activators and Inhibitors.* Aside from the effects of ionic conditions, no specific activators of glycerol kinase have been found.

At neutral pH, glycerol kinase from *E. coli* is subject to allosteric inhibition by fructose 1,6-bisphosphate.[16,18] The inhibition is noncompetitive with respect to both substrates, with an apparent $K_i$ of 0.5 m$M$.[16] At pH 9.5 or in the presence of 0.4 $M$ KCl, LiCl, NaCl, or $NH_4Cl$, sensitivity of the enzyme to inhibition by fructose 1,6-bisphosphate is greatly reduced.[16,18] Glycerol kinase is also subject to product inhibition. The inhibition by ADP is mainly competitive with respect to MgATP (apparent $K_i = 0.5$ m$M$) and noncompetitive with respect to glycerol.[7,16] Glycerol-3-P is a weak competitive inhibitor with respect to glycerol (apparent $K_i = 2$ m$M$).[3,7,16] This pattern of product inhibition is consistent with an ordered reaction mechanism with glycerol as the first substrate to bind to the enzyme.

The enzyme is very sensitive to inactivation by $p$-hydroxymercuriphenyl sulfonate and other sulfhydryl group-modifying reagents.[7,16]

*Thermodynamics.* The reaction catalyzed by glycerol kinase is essentially irreversible, with a $\Delta G^{o'} = -5.1$ kcal/mole, corresponding to an equilibrium constant of greater than $10^3$.[19] The turnover number of the enzyme at pH 7.0 and 25° is 11,600 min$^{-1}$, with a $Q_{25°}^{37°}$ of about 1.5.[7]

At saturation about 4 moles of glycerol bind per mole of enzyme, with a dissociation constant 10 $\mu M$.[16] The dissociation constant of the enzyme-glycerol complex corresponds to a free energy of glycerol binding of $-6.8$ kcal/mole. Since the $K_m$ for glycerol is known to increase with decreasing temperature, the enthalpy of glycerol binding must be positive (about $+6$ kcal/mole as calculated from Van't Hoff plots).[7] Consequently, the entropy of glycerol binding must also have a positive value (about $+45$ cal mole$^{-1}$ deg$^{-1}$). Since the binding of glycerol per se will involve an entropy loss, glycerol binding must be accompanied by some other reaction in which considerable entropy is gained, possibly a conformational change in the protein.

At saturation about 4 moles of fructose 1,6-biphosphate, the allosteric inhibitor, bind per mole of enzyme, with a dissociation constant 60 $\mu M$.[16]

[18] N. Zwaig and E. C. C. Lin, *Science* **153**, 755 (1966).

[19] A. L. Lehninger, "Biochemistry," p. 304. Worth Publ. New York, 1970.

## [27] Pyruvate Kinase of *Bacillus licheniformis*

*By* F. WILLIAM TUOMINEN and ROBERT W. BERNLOHR

Phosphoenolpyruvate + ADP + H$^+$ → pyruvate + ATP

The pyruvate kinase ATP:pyruvate phosphotransferase, (EC 2.7.1.40) of *Bacillus licheniformis* is an example of an AMP-activated, ATP-inhibited glycolytic enzyme that can exist in a number of kinetically differentiable activity states.[1,2] Unless proper precautions are exercised with this enzyme during extract preparation, purification, and assay, the formation of inactive enzyme states will result in low or erroneous specific activity values, gross losses of activity, and misleading kinetic data. The methodology presented here emphasizes those details that allow the experimentalist to understand and deal with the difficulties inherent in this highly permutable pyruvate kinase.

*Culture and Harvest of Cells.* The organism used is the genetically stable (with respect to colonial morphology), rough colony-forming strain of *B. licheniformis* A-5. Spore stocks are prepared by inoculating 0.5 liter of 0.4% (w/v) peptone and 0.1% yeast extract contained in a 2.8-liter Fernbach flask with either vegetative or spore seed that has been derived from an isolated colony. Greater than 95% of the microscopically recognizable cells exist as free spores after 24 hr of incubation at 37° on a rotary shaker.

The minimal medium used routinely for the production of vegetative cells contains (per liter) 65 mmoles of $K_2HPO_4$–$KH_2PO_4$ (pH 7.2), 0.8 mmole of $MgSO_4$, 0.02 mmole of $MnCl_2$, 0.5 mmole of $CaCl_2$, 15 mmoles of $NH_4Cl$, and 15 mmoles of glucose. The glucose, $NH_4Cl$, $MgSO_4$, $MnCl_2$, and $NH_4Cl$ are added from sterile, concentrated stocks to autoclaved $K_2HPO_4$–$KH_2PO_4$ to yield a nutritionally complete, unprecipitated growth medium.

Cultures are started by adding approximately $10^{10}$ spores to 1 liter of minimal medium supplemented with 0.5 m$M$ L-alanine. Cultures are aerated by shaking on a Gyrotory shaker (New Brunswick Scientific Co.) at 37°. When cultures reach a turbidity of 150–170 Klett units (540 nm), 1 liter is transferred to aerated fermentation vessels of a New Brunswick Scientific Co. fermentor (Model FS-307) at 38°. Each vessel contains 10 liters of sterile, prewarmed medium.

Vegetative cells are harvested for enzyme studies from volumes of cul-

[1] F. W. Tuominen and R. W. Bernlohr, *J. Biol. Chem.* **246**, 1732 (1971).
[2] F. W. Tuominen and R. W. Bernlohr, *J. Biol. Chem.* **246**, 1746 (1971).

ture up to 3 liters by centrifugation (13,000 $g$ for 10 min at 0°) in a Sorvall RC2-B centrifuge after being cooled by the addition of ice Vol umes greater than 3 liters are harvested in a Lourdes continuous flow centrifuge (12,000 rpm) after the addition of ice. Cells are washed with a 0.1 $M$ Tris·HCl (pH 7.6) buffer that contains 10 m$M$ MgCl$_2$ and 0.1 $M$ KCl. The washed-cell pellets are either immediately used to prepare cell-free extracts or stored at −20°.

   *Preparation of Cell-Free Extracts.* Cells are suspended in a 0.1 $M$ Tris·HCl (pH 7.6) buffer that contains 10 m$M$ MgCl$_2$ and 0.1 $M$ KCl (Buffer TMK) on the basis of 1.0 ml buffer per gram, wet weight. Cells are broken by sonic disruption when the suspension volume is less than 5 ml and in a French pressure cell (at 12,000 psi) when larger volumes are employed. Sonic disruption is performed in a MSE sonicator (Measuring and Scientific Equipment, Ltd., London) at 110 V and 2 A with a 1.9-cm probe at 4–10°. The total time of sonic disruption is 3 min, i.e. six, 30-sec bursts that are interrupted by 30-sec cooling periods (ice bath). This sonic disruption time (3 min) is twice that required to release a maximum amount of pyruvate kinase from a suspension of late exponential phase cells.

   *Assay System.* Pyruvate kinase is assayed by employing the linked lactic dehydrogenase (EC 1.1.1.27) reaction as modified from Bücher and Pfleiderer.[3] The standard assay system (1.0 ml) contains 50 $\mu$moles imidazole-HCl (pH 7.1, the optimum), 50 $\mu$moles of KCl (also an optimum concentration), 7 $\mu$moles of MgCl$_2$, 2 $\mu$moles of tricyclohexylammonium phosphoenolpyruvate (PEP) (TCHA-PEP), 4 $\mu$moles NaADP (pH 7.1), 0.12 $\mu$mole NaNADH and 40 $\mu$g lactic dehydrogenase. Reactions are initiated by the addition of 2–20 $\mu$l of enzyme preparation to the reaction mixture at 30° in a 10-mm quartz cuvette. The time course of NADH oxidation is followed at 340 nm in a Zeiss PMQII spectrophotometer connected to a Sargent Model SRL recorder. Initial rates are converted to micromoles of pyruvate formed per minute by dividing the change in absorbance per minute by 6.22. All rates are corrected for blank rates obtained by the omission of ADP. One unit of activity is defined as the amount of pyruvate kinase yielding 1 $\mu$mole of pyruvate per minute in the standard assay system.

   Freshly prepared solutions of NaADP (P-L Biochemicals, Inc.), pH 7.1, and TCHA-PEP (Sigma) are used for kinetic studies and stored on ice for up to 16 hr. Stock solutions of 0.1 $M$ NaADP (pH 7.1) and 0.1 $M$ TCHA-PEP, for use in the standard assay system, are stored for up to 2 weeks at −60°. The concentrations of the PEP and ADP solutions are determined by assaying PEP and ADP in a modified standard assay

---

[3] T. Bücher and G. Pfleiderer, this series, Vol. 1, p. 435.

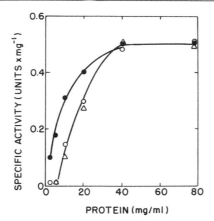

PROTEIN (mg/ml)

Fig. 1. Effect of dilution on the apparent specific activity of pyruvate kinase in crude extracts. A crude extract (80 mg of protein per milliliter) was prepared in 0.1 $M$ Tris·HCl, pH 7.6, containing 10 m$M$ MgCl$_2$ and 0.1 $M$ KCl at 22°. Aliquots of this extract were diluted to the indicated levels in the same buffer at 22°. Samples of each dilution were assayed in the standard assay system at 1 min (●——●), 2 hr (○——○), and 4 hr (△——△) postdilution. The amount of extract protein was approximately the same in all assays.

mixture containing 0.25 $\mu$mole NADH, 0.5 unit rabbit muscle pyruvate kinase and 0.1–0.2 $\mu$mole of either PEP or ADP. The change in absorbance after the addition of the pyruvate kinase is divided by 6.22 to give the $\mu$moles of PEP or ADP in the volume of solution assayed. Beef heart lactic dehydrogenase (LDH) is obtained as a crystalline suspension in 55% saturated (NH$_4$)$_2$SO$_4$. This preparation is diluted 5-fold in 50 m$M$ Tris·HCl, pH 7.5 and dialyzed against three changes of the same Tris·HCl buffer at 4°. The dialyzed preparation is diluted another 2-fold in glycerol and stored at 4°. The final LDH stock solution contains 4 mg/ml LDH (350 IU/mg) and 50% (v/v) glycerol. NaNADH (P-L Biochemicals, Inc.) stock solutions (8–15 m$M$) are stored at 2–4° in 10 m$M$ Na(HCO$_3$-CO$_3$) buffer, pH 10.5, in accord with published recommendations.[4]

*Determination of Specific Activity in Crude Extracts.* Crude extracts must contain greater than 40 mg per milliliter of protein in order to obtain meaningful pyruvate kinase specific activities. The reason for this precaution against greater dilution is illustrated in Fig. 1. The data presented in Fig. 1 show that pyruvate kinase in crude extracts is rapidly inactivated by dilution of these extracts to less than 40 mg/ml of protein at 22°. In addition to being sensitive to dilution at 22°, pyruvate kinase

[4] O. H. Lowry, J. V. Passonneau, and M. K. Rock, *J. Biol. Chem.* **236**, 2756 (1961).

is reversibly inactivated at 0°.[1] As the extent of dilution is increased, at 0°, both the amount of residual activity and reversibility of inactivation decrease. However, no cold inactivation can be detected in parallel experiments which employ warm pipettes (22°), rather than ice-cold pipettes, for removal of samples (2–10 $\mu$l) from the dilutions incubated near 0°. These results suggest that the undiluted enzyme is rapidly reactivated during the resident time (about 20–30 sec) in the pipettes.

The specific activity of pyruvate kinase in freshly prepared crude extracts, containing greater than 40 mg/ml of protein, has been found to be independent of the stage of growth (on glucose) or sporulation.[1] Similar results are observed when extracts are dialyzed against a buffer containing $KP_i$, $MgCl_2$ and PEP. The specific activities generally range between 0.55 and 0.65 unit per milligram of protein. The specific activity present in crude extracts of exponential-phase cells grown on malate range from 0.53 to 0.61 unit per milligram of protein. Values of 0.50 and 0.59 (unit per milligram of protein) have been observed with exponential-phase cells grown on pyruvate.

*Purification.* The otherwise labile pyruvate kinase activity of *B. licheniformis* is stabilized throughout the following purification scheme by avoiding extensive dilution of the enzyme and by the inclusion of a stabilizing ligand mixture (PEP, $P_i$, and Mg ions) in the supporting buffers when dilution is unavoidable. Protein concentrations through the penultimate purification step are determined by the method of Lowry *et al.*[5] The protein concentrations of the most highly purified enzyme preparations are estimated spectrophotometrically at 280 nm by employing an $E_{1cm}^{0.1\%} = 0.65$ which is the constant obtained by Hunsley and Suelter for crystalline yeast pyruvate kinase.

Thirty grams of late exponential-phase cells are suspended at 0° in 45 ml of 0.1 $M$ Tris·HCl (pH 7.6) that contains 10 m$M$ $MgCl_2$ and 0.1 $M$ KCl. Cells are ruptured by two passes through a French pressure cell at 12,000 psi. The extract is sonicated for 2 min in order to reduce the viscosity and centrifuged (0°) at 78,000 $g$ for 2 hr in a Spinco Model L centrifuge. The upper two-thirds of the resulting supernatants are combined to yield 40 ml of a solution that contains 2100 mg protein and 1200 units of pyruvate kinase activity. This supernatant fluid is designated as S78 (see the table).

The pH of S78 is adjusted to 8.0, at 4°, by the addition of 1 $M$ $NH_4OH$ from a capillary-tipped pipette. The extract is diluted 2-fold by the dropwise addition of a saturated solution (490 g/l; pH 8.0 at

[5] O. H. Lowry, N. J. Rosebrough, A. L. Farr, and R. J. Randall, *J. Biol. Chem.* **193**, 265 (1951).

[6] J. R. Hunsley and C. H. Suelter, *J. Biol. Chem.* **244**, 4815 (1969).

4°) of $(NH_4)_2SO_4$ at 4°, stirred another 20 min, and then centrifuged for 1 hr at 15,000 $g$. Sufficient powdered $(NH_4)_2SO_4$ is then added slowly to the supernatant until 0.8 saturation is achieved, based on an observed saturation of 490 g/liter at 4°. After 20 min of additional stirring and 1 hr of centrifugation, a precipitate is obtained which contains proteins precipitating between 0.5 and 0.8 saturation. The precipitate is then dissolved in 0.1 $M$ $KP_i$, pH 6.5 (4°), to a final volume of 34 ml. This solution, designated $(NH_4)_2SO_4$ I, contains 1300 mg of protein and 1100 units of pyruvate kinase activity.

A saturated solution of $(NH_4)_2SO_4$ (pH 6.5, 4°) is added, with stirring at 4°, to the above fraction until 0.30 saturation is achieved. After 20 min, the suspension is centrifuged for 1 hr at 15,000 $g$ and 4°. Saturated $(NH_4)_2SO_4$ is added to the resulting supernatant fluid to a concentration of 0.55 saturation. The precipitate (0.30–0.55 saturation) is then sedimented by 1 hr of centrifugation and dissolved in 40 m$M$ $KP_i$ (pH 7.0 at 22°) buffer that contains 5 m$M$ $MgCl_2$ and 3 m$M$ PEP. This produces 25 ml of a solution, designated $(NH_4)_2SO_4$ II, that contains 410 mg of protein and 660 units of pyruvate kinase activity.

The $(NH_4)_2SO_4$ II fraction is added to a 5 × 12 cm hydroxyapatite (Bio-Gel-HT Bio-Rad Laboratories, Richmond, California) column at 4°, that has been previously washed with 2 liters of 40 m$M$ $KP_i$ buffer (pH 7.0), 5 m$M$ $MgCl_2$, and 2 m$M$ PEP. Two-milliliter fractions are colinto the gel, the column is washed with 370 ml of a buffer that contains 40 m$M$ $KP_i$ (pH 7.0 at 22°), 5 m$M$ $MgCl_2$, and 2 m$M$ PEP. The bulk of the pyruvate kinase activity is eluted from the column by 210 ml of a solution that contains 75 m$M$ $KP_i$ (pH 7.0), 5 m$M$ $MgCl_2$ and 2 m$M$ PEP. The fractions collected between 380 and 480 ml of effluent are pooled and precipitated by adding solid $(NH_4)_2SO_4$ to 0.9 saturation. The precipitate is sedimented by centrifugation as before and suspended in 0.9 saturated $(NH_4)_2SO_4$. The suspension contains 17 mg of protein and 300 units of pyruvate kinase. This preparation is stable for at least 5 months when stored at 4°.

The $(NH_4)_2SO_4$ precipitate of the pooled, hydroxyapatite fractions (380–480 ml) is sedimented by centrifugation as before and dissolved in 20 m$M$ $KP_i$ (pH 7.0), containing 5 m$M$ $MgCl_2$ and 2 m$M$ PEP, to a final volume of 1 ml. This solution is passed through a 1.7 × 65 cm, Bio-Gel $A_{0.5m}$ (Bio-Rad) column at a flow rate of 10 ml/hr after the column is equilibrated (at 4°) with a buffer that contains 20 m$M$ $KP_i$ (pH 7.0), 5 m$M$ $MgCl_2$, and 2 m$M$ PEP. Two-milliliter fractions are collected after discarding the first 60 ml of eluent which represented the void volume as determined with Blue-Dextran 2000 (Pharmacia). The first six fractions contain a total of 210 units of pyruvate kinase and an

PURIFICATION OF PYRUVATE KINASE FROM *Bacillus licheniformis*

| Fraction | Protein (mg) | Units (μmoles/min) | Specific activity (units/mg) | Yield (%) |
|---|---|---|---|---|
| S78 | 2100 | 1200 | 0.570 | |
| (NH₄)₂SO₄ I | 1300 | 1100 | 0.845 | 92 |
| (NH₄)₂SO₄ II | 410 | 660 | 1.62 | 55 |
| Hydroxyapatite | 17 | 300 | 17.6 | 25 |
| Agarose | 1.0 | 210 | 210 | 17 |

estimated protein content of 1 mg. Each fraction is diluted 2-fold with glycerol and stored at −20°. No loss of pyruvate kinase activity is observed over a 5-month period under these conditions.

A summary of the purification scheme is presented in the table, which shows a yield of approximately 1 mg (17%) of pyruvate kinase at a specific activity of 210 units/mg. This specific activity compares favorably with that of the crystalline yeast enzyme.[6] This preparation yielded one protein band when a 50-μg aliquot was subjected to polyacrylamide electrophoresis according to the method described by Maizel.[7] A significant loss of activity is observed in a comparable purification scheme which does not include PEP during the hydroxyapatite and agarose chromatography steps.

The rate of catalysis is proportional to the concentration of both purified and unpurified enzyme. ADP, PEP, and Mg(II) are absolutely required for catalysis by the purified enzyme. The purified pyruvate kinase preparation contains less than 0.08 unit of adenylate kinase[8] and less than 0.04 unit of ATPase[9] per milligram.

*The Kinetic Detection of Different Activity States.* The progress curves shown in Fig. 2 illustrate that the order in which the reaction components are mixed together has a dramatic effect on the rate of catalysis by pyruvate kinase. Maximum activities and linear progress curves are observed when reactions are initiated with enzyme (curve a, Fig. 2), ADP (curve b), or by the simultaneous addition of ADP and PEP (curve d). However, 80–90% lower initial rates and nonlinear progress curves are observed when reactions are initiated by PEP addition (curve c). A comparison of curves c, d, and e demonstrates that such depressed initial rates are due to the inactivation of enzyme diluted in the absence

[7] J. V. Maizel, *in* "Fundamental Techniques in Virology" (K. Habel and N. P. Salzman, eds.), p. 334. Academic Press, New York, 1969.

[8] Assayed according to H. U. Bergmeyer, *in* "Methods of Enzymatic Analysis," p. 989. Academic Press, New York, 1963.

[9] Assayed by the adenylate kinase assay system, modified by excluding AMP.

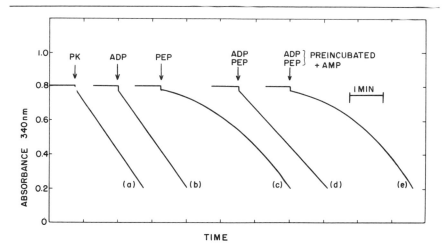

FIG. 2. Reaction time course and the detection of catalytically inactive enzyme states. Progress curves are shown for the standard assay system containing 0.05 unit of 50-fold purified pyruvate kinase. The method for initiating the reaction was the variable employed in this experiment. (a) Reaction initiated by the addition of enzyme; (b) initiated by ADP 2 min after addition of enzyme; (c) initiated by PEP 2 min after the addition of enzyme; (d) initiated by the simultaneous addition of ADP and PEP 2 min after the addition of enzyme; (e) initiated by the simultaneous addition of ADP and PEP 2 min after the addition of enzyme to a reaction system containing 60 $\mu M$ AMP. All operations were conducted at 30°.

of PEP and the presence of the low levels (1–2%) of AMP commonly found contaminating commercial ADP preparations. The increase in rate in reactions initiated with PEP shown in Fig. 2 is due to a reactivation process dependent upon PEP and Mg(II) ions.[1]

The nature of the inactivation and reactivation processes can be revealed by employing the following kinetic method. First, enzyme is inactivated by dilution in the absence of PEP and presence of AMP. Second, assays containing various amounts of inactive enzyme are initiated with PEP to produce curves resembling curves c and e in Fig. 2. The gain in activity, determined from reaction rate approximations over successive 10-, 20-, or 30-sec intervals, is then plotted against time. Finally, a rate of activity gain is calculated for each level of enzyme used. Such rates of reactivation for the *B. licheniformis* pyruvate kinase plotted against the log of inactive enzyme concentration yield a reaction order of two.[1] Thus, it is reasoned that the reactivation dependent on Mg(II) and PEP involves the association of the subunits that are produced by the dissociation induced by AMP and dilution.

A second inactive enzyme state can also be detected kinetically.[1] This state can be produced by diluting the enzyme in buffers lacking all ligands. This form cannot be directly reactivated by PEP and Mg(II). This inactive state can, however, be converted to the activatable state via a time-dependent reaction facilitated by AMP. Thus, incubation in the presence of AMP will, with time, yield the subunits that are in turn subject to the Mg + PEP-mediated association to the fully active enzyme state.

Studies similar to those described above have established that the complete stabilization of the active enzyme state at 0–37° can be achieved in the presence of 2 m$M$ PEP, 0.5 m$M$ AMP, and 5 m$M$ Mg(II) ions.[1] Ten millimolar $P_i$ (an inhibitor of catalysis) can be substituted for AMP (an activator of catalysis) in this protecting ligand mixture. Partial stabilization of the active enzyme can be achieved by 3 m$M$ Mg ATP, 50 $\mu M$ of either NAD or NADH, 50% glycerol, or 0.5 $M$ sucrose.

*Kinetic Characterization.* Many of the ligands that influence stability also effect the rate of catalysis of the pyruvate kinase of *B. licheniformis.*[2] Unlike the time-dependent effects on enzyme stability, the effects of ligands on the rate of catalysis occur almost instantaneously. Unfortunately, estimations of initial velocity are made over relatively lengthy periods of time (0.5–2 min) during which substantial losses of enzyme activity can occur. Such losses of activity become significant and unpredictable when the effect of varying (subsaturating) concentrations of a stabilizing ligand on initial velocity is being assessed. Therefore, the stability of the enzyme during assay must be followed closely in order to establish whether an apparent effect on the rate of catalysis might not be due to a hidden effect on enzyme stability.

Enzyme stability during initial velocity determinations can be assessed conveniently by employing the following general method. Initial rates at different concentrations of the variable ligand are determined from a continuous recording of the reaction time course over a 40-sec interval beginning 10 sec after the addition of enzyme. After the 40-sec rate determination, enzyme stability in the different reaction mixtures is estimated by adding (approximately 60 sec after the addition of enzyme) an amount of the variable ligand that is sufficient to yield a constant and saturating concentration. The rate of catalysis for the ligand-saturated system is then determined over a 60-sec interval. This new rate is expressed as a percentage of the initial rate that is observed when a ligand-saturated reaction mixture is initiated directly by the addition of enzyme. The application of these methods with PEP as the variable ligand (substrate) is illustrated in Fig. 3.

The experiment illustrated in the upper portion of Fig. 3 shows apparently sigmoid saturation curves for PEP and an apparent activation of catalysis by AMP. However, the stability data (inset) suggest that

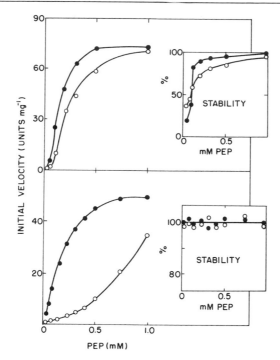

FIG. 3. Relationship between initial velocity and phosphoenolpyruvate (PEP) concentration in the presence and in the absence of AMP and ATP. The assay mixtures (1.0 ml) for the upper illustration contained 50 μmoles of imidazole-HCl (pH 7.1), 50 μmoles of KCl, 7 μmoles of MgCl₂, 0.25 μmole NADH, 40 μg of lactate dehydrogenase, 0.5 μg of pyruvate kinase (115 units/mg), 2 μmoles ADP and a variable concentration of PEP in the absence (○) and in the presence (●) of 1 μmole of AMP. The assay mixtures for the lower illustration were as above but for the presence of 2.8 μmoles MgATP, 0.8 μmole ADP, and 1.0 μg pyruvate kinase (PK). Initial rates were estimated from a continuous recording of the reaction time course over a 40-sec interval beginning 10 sec after the addition of enzyme. After the 40-sec rate determination, the stability of PK in the different reaction mixtures was estimated by adding sufficient PEP to yield a final concentration of 2 mM and obtaining a new reaction rate. This new rate is expressed as a percentage of the initial rate observed when a reaction, containing 2 mM PEP, was initiated by the addition of enzyme.

both the sigmoid nature of the saturation curves and the stimulation by AMP are caused by a variation of enzyme stability. In contrast, the sigmoidal nature of the PEP saturation curve in the presence of ATP and the activating influence of AMP are clearly established by the experiment illustrated in the lower portion of Fig. 3. The complete stabilization of the enzyme is achieved here at all concentrations of PEP by the stabilizing effect of ATP.

The kinetic characterization of the pyruvate kinase of *B. licheniformis*, conducted with a constant assessment of the effects of reaction conditions on enzyme stability, reveals the following properties.[2] Catalysis by this enzyme proceeds optimally at pH 7.0–7.4. The initial rate of catalysis is modulated by substrate activation (both PEP and ADP), activation by AMP, and inhibition by ATP, $P_i$ and carbamyl phosphate. Positive cooperatively is manifested by PEP, ADP, $P_i$ and ATP in the presence of Mg(II) ions. AMP relieves the inhibition by ATP, yielding hyperbolic saturation curves for both substrates. The apparent substrate $K_m$ values for the fully activated enzyme, 0.2 m$M$ for PEP, and 0.7 m$M$ for ADP, in the presence of Mg(II) are independent of the concentration of the second substrate. In the presence of Mn(II) as the required divalent cation, the $K_m$ for PEP is 7-fold lower than that obtained with the Mg(II)-activated enzyme. Finally, the inhibition of catalysis by ATP in the presence of Mn(II) is not reversed by AMP and does not effect the hyperbolic nature of the PEP saturation curve observed under these conditions. Thus, the K-type[10] allosteric properties manifested in the presence of Mg(II) are obliterated when Mn(II) is employed as the activating divalent cation.

[10] J. Monod, J. Wyman, and J.-P. Changeux, *J. Mol. Biol.* **12**, 88 (1965).

# [28] Isozymes of Pyruvate Kinase from the Grassfrog[1]

*By* L. E. FLANDERS, L. H. SCHLOEN, and H. J. SALLACH

Phospho*enol*pyruvate + ADP ↔ pyruvate + ATP

The occurrence of multiple forms of pyruvate kinase in different tissues of the rat was first reported by Tanaka *et al.*[2,3] These investigators detected from one to four isozymes in a given tissue as did Susor and Rutter.[4] Early work with human tissues demonstrated three different isozymes.[5] On the other hand, work in this laboratory[6] demonstrated that

[1] This work was supported by Research Contract No. AT(11-1)-1631 from the U.S. Atomic Energy Commission and by National Institutes of Health Grant No. NS10287.

[2] T. Tanaka, Y. Harano, H. Morimura, and R. Mori, *Biochem. Biophys. Res. Commun.* **21**, 55 (1965).
[3] T. Tanaka, Y. Harano, F. Sue, and H. Morimura, *J. Biochem. (Tokyo)* **62**, 71 (1967).
[4] W. A. Susor and W. J. Rutter, *Biochem. Biophys. Res. Commun.* **30**, 14 (1968).
[5] R. H. Bibley, P. Stanzel, R. T. Jones, J. P. Campas, and R. D. Koler, *Enzyme Biol. Clin.* **9**, 10 (1968).
[6] L. H. Schloen, J. R. Bamburg, and H. J. Sallach, *Biochem. Biophys. Res. Commun.* **36**, 823 (1969).

five isozymes of pyruvate kinase are found in a given cell type, e.g., the unfertilized frog egg, as well as in certain tissues of the adult grassfrog, *Rana pipiens*. Subsequent studies carried out in a number of laboratories have demonstrated multiple forms of pyruvate kinase in a variety of tissues from a number of different species (for further discussion and references, the reader is directed to recent articles[7-9] on this subject). On the basis of electrophoretic, kinetic, chromatographic, and immunological data,[7,8] there appear to be three unique forms of pyruvate kinase in animal systems; using the nomenclature of Tanaka and associates,[7,8] these are type L (major liver form), type $M_1$ (the only form in skeletal muscle), and type $M_2$ (predominant fetal form but found in certain adult tissues). Unique kinetic properties that have been reported for the three isozymes include the following. Although there are quantitative differences, both types L and $M_2$ show cooperativity with respect to phospho-*enol*pyruvate (PEP) concentration, inhibition by ATP or alanine, and activation by fructose 1,6-diphosphate (FDP) which lowers the apparent $K_m$ and shifts PEP kinetics from sigmoidal to hyperbolic. However, the two types differ in that inhibition of type L by either ATP or alanine is reversed by the addition of FDP whereas that of type $M_2$ is not. Both types appear to exist in sensitive and desensitive forms with respect to the effects of modulators. On the other hand, type $M_1$ displays hyperbolic kinetics with respect to PEP concentration either in the presence or in the absence of FDP. Each of the three isozymes can be distinguished on the basis of their electrophoretic mobilities. Type L has been shown to be under dietary and hormonal control.[2]

With respect to the isozymes of pyruvate kinase in the grassfrog, present data[10,11] indicate that the liver (major form), cardiac and skeletal muscle isozymes, the purification of which is described below, are similar in their properties to types L, $M_2$, and $M_1$ of the rat, respectively.

## Assay Method

*Principle.* The assay method is essentially that described by Bücher and Pfeiderer[12] and the reader is referred to a discussion of the spectrophotometric assay in which the pyruvate kinase reaction is coupled with that of lactate dehydrogenase (Vol. 1 [66]).

[7] K. Imamura and T. Tanaka, *J. Biochem. (Tokyo)* **71**, 1043 (1972).
[8] K. Imamura, K. Taniuchi and T. Tanaka, *J. Biochem. (Tokyo)* **72**, 1001 (1972).
[9] J. Osterman, P. J. Fritz, and T. Wuntch, *J. Biol. Chem.* **248**, 1011 (1973).
[10] L. E. Flanders, J. R. Bamburg, and H. J. Sallach, *Biochim. Biophys. Acta* **242**, 566 (1971).
[11] L. H. Schloen and H. J. Sallach, unpublished observations.
[12] T. Bücher and G. Pfleiderer, this series, Vol. 1 [66].

*Reagents.* For routine assays, the composition of the assay mixture is: 50 mM Hepes (*N*-2-hydroxyethylpiperazine-*N'*-2-ethanesulfonic acid) buffer, pH 7.4; 0.46 mM ADP; 0.77 mM PEP; 0.15 mM NADH; 10 mM MgCl$_2$; 100 mM KCl; 5 mM dithiothreitol; and 5 units of lactate dehydrogenase.[13] When FDP is included in the incubation system, it is added as an aqueous solution of the tetracyclohexylammonium salt to a final concentration of 0.5 mM; the same activation is achieved with the sodium salt.

*Procedure.* The reagents are pipetted into a cuvette (1-cm light path) which is placed in the spectrophotometer for temperature equilibration at 15°. The temperature is very important for cardiac and liver enzymes and should not exceed 15°; the muscle enzyme can be assayed at room temperature. The reaction is initiated by the addition of pyruvate kinase (25–90 units)[14] and is followed by measuring the change in absorbancy at 340 nm with a Beckman DU-Gilford 2000 recording spectrophotometer. The control cuvette contains all components except PEP.

*Definition of Unit and Specific Activity.* One unit is defined as that amount of enzyme that produces a change in absorbancy of 0.001 per minute under the above conditions. Specific activity is defined as the number of units per milligram of protein as determined by the method of Lowry *et al.*[15] with bovine serum albumin as the standard or from the absorbancy at 280 nm according to the method of Warburg and Christian.[16]


## Purification Procedures

The following general conditions are used unless stated otherwise. All steps are carried out at 0–4°. Fractionations with ammonium sulfate are made by the slow addition, with stirring, of the calculated amount of the solid salt. The resulting suspensions are equilibrated with stirring for 15 min prior to centrifugation. Centrifugations are carried out at 14,000 *g* for 30 min. All ammonium sulfate residues are dissolved in the

---

[13] The lactate dehydrogenase (Sigma, Type II) is in an ammonium sulfate suspension. A 1:250 dilution of the stock suspension into 25 mM tris(hydroxymethyl)-aminomethane buffer, pH 7.4, prior to use in the assay is sufficient both to perform the coupling of pyruvate to lactate and dilute out the ammonium sulfate present.

[14] The pyruvate kinase can be assayed directly from any of the purification steps except the ammonium sulfate fractionation; dialysis is used to eliminate the ammonium sulfate or dilution greater than 1:250 as with the lactate dehydrogenase.

[15] O. H. Lowry, N. J. Rosebrough, A. L. Farr, and R. J. Randall, *J. Biol. Chem.* **193**, 265 (1951).

[16] O. Warburg and W. Christian, *Biochem. Z.* **310**, 384 (1941). See also this series Vol. 3 [73].

appropriate buffer for the given purification step and are dialyzed against the same buffer until free of ammonium ions as measured by Nessler's reagent. CM- and DEAE-Sephadex ion exchange resins are used in column chromatography and are pretreated according to the directions supplied by Pharmacia. The final equilibration is carried out in the buffer to be used in the given chromatographic step. Protein in effluent solutions from columns is detected by absorbancy at 280 nm. Enzymic solutions are concentrated by ultrafiltration (Diaflo ultrafiltration cell, Amicon Corp., Lexington, Massachusetts).

Grassfrogs, *Rana pipiens*, are purchased from Mogel-Ed Corp., Oshkosh, Wisconsin.

## Purification of Skeletal Muscle Pyruvate Kinase[10]

*Step 1. Crude Extract.* One gram of muscle from the hind legs is homogenized in a Waring Blendor for 1 min in 5 ml of homogenizing buffer consisting of 0.25 $M$ sucrose, 25 m$M$ Tris·HCl, 2.5 m$M$ EDTA, and 1 m$M$ 2-mercaptoethanol (pH 7.5). Cellular debris is removed by centrifugation and discarded.

*Step 2. First Ammonium Sulfate Fractionation.* Ammonium sulfate (45 g/100 ml) is added to the supernatant solution from the above step. The 0–45% ammonium sulfate residue is removed by centrifugation and, to the resulting supernatant solution, ammonium sulfate (15 g/100 ml) is added. The 45–60% ammonium sulfate fraction is recovered by centrifugation and can be stored at −20° for 6 months with minimal loss in activity.

*Step 3. CM-Sephadex Chromatography.* The ammonium sulfate precipitate from the above step is dissolved in 0.04 imidazole acetate buffer, pH 7.4, containing 1 m$M$ EDTA and 10 m$M$ 2-mercaptoethanol, using a volume of buffer which results in a protein concentration of 10 mg/ml. The dialyzed fraction is then applied to a CM-Sephadex column which is equilibrated with the above buffer; a bed volume of 1 liter of resin is used for each 0.8 g of protein. After the protein solution has run into the column bed and the latter is washed with 50 ml of buffer, a linear grad·ent of 0 to 0.15 $M$ KCl, dissolved in column buffer, is applied. A total volume of 2 liters is used for a resin bed volume of 1 liter. The fractions comprising the peak of pyruvate kinase activity are pooled and concentrated by ultrafiltration to give a protein concentration of 2.5 mg/ml.

*Step 4. Second Ammonium Sulfate Fractionation.* The concentrated enzyme solution from the above step is dialyzed for 48 hr against a solution of ammonium sulfate containing 52 g/100 ml to precipitate pyruvate

TABLE I
PURIFICATION DATA FOR CARDIAC AND SKELETAL MUSCLE PYRUVATE KINASES

| Fraction | Total units ($\times 10^6$) | Total protein (mg) | Yield (%) | Specific activity ($\times 10^4$) |
|---|---|---|---|---|
| Skeletal muscle[a] | | | | |
| 1. Crude extract | 70 | 5000 | — | 1.4 |
| 2. First ammonium sulfate fraction | 60 | 800 | 86 | 7.5 |
| 3. CM-Sephadex chromatography[b] | 30 | 45 | 43 | 67.0 |
| 4. Second ammonium sulfate fraction | 28 | 32 | 40 | 87.0 |
| 5. DEAE-Sephadex chromatography[b] | 22 | 17 | 31 | 133.0 |
| Cardiac muscle[c] | | | | |
| 1. Crude extract | 0.525 | 100 | — | 0.52 |
| 2. Ammonium sulfate fraction | 0.210 | 21 | 40 | 1.0 |
| 3. CM-Sephadex chromatography | | | | |
| P-I | 0.065 | 1.65 | 12 | 4.0 |
| P-II | 0.077 | 1.70 | 15 | 4.6 |

[a] Reproduced from L. E. Flanders, J. R. Bamburg, and H. J. Sallach, *Biochim. Biophys. Acta*, **242**, 569 (1971) (Table I).

[b] This step is quite variable with respect to yield. Recovery of 100% of the activity can occasionally be obtained.

[c] Data are for 16 g of frog heart. Values in step 3 are theoretical from the total of step 2; in practice, only 72,000 units from step 2 are run on a CM-Sephadex column (1.8 $\times$ 42 cm). The total yield for step 3 from step 2 is 68%.

kinase. A ratio of one volume of enzyme solution to 50 volumes of ammonium sulfate solution is used. (The resulting suspension is a convenient storage form for the enzyme; no loss in activity is observed for a period of over 1 month.) The precipitated protein is recovered by centrifugation, and the supernatant solution is discarded.

*Step 5. DEAE-Sephadex Chromatography.* The ammonium sulfate residue from the above step is dissolved in 40 mM imidazole acetate buffer, pH 7.4, containing 1 mM EDTA, 10 mM 2-mercaptoethanol, and 0.5 mM FDP to give a protein concentration of 5–7 mg/ml (approximate dissolution volume is about two-thirds that of the volume of the suspension centrifuged in the above step). After dialysis against the same buffer, the enzyme solution is applied to a DEAE-Sephadex column (150 ml bed volume per 30 mg of protein) equilibrated with the same buffer. The column is washed with 10 ml of buffer, then the enzyme is eluted with a linear gradient of 0 to 0.3 M KCl, dissolved in column buffer; a total volume of 300 ml is used for a resin bed volume of 150 ml. The fractions containing pyruvate kinase are pooled and concentrated via ultrafiltration to give a final protein concentration of 4–8 mg/ml. This fraction can be stored at 4° for several weeks with no loss in activity. This fraction is judged to be homogeneous as determined by sedimentation velocity, gel electrophoresis in 1% sodium dodecyl sulfate, double diffusion precipitin reaction and immunoelectrophoresis.[10]

The results of a typical purification are summarized in Table I.

*Purification of Cardiac Pyruvate Kinase*[10]

*Step 1. Crude Extract.* Pyruvate kinase in cardiac tissue is extracted by homogenization in the same buffer as that employed above for the skeletal muscle enzyme except that a tissue to buffer ratio of 1:2 (w/v) is used.

*Step 2. Ammonium Sulfate Fractionation.* Ammonium sulfate (18 g/100 ml) is added to the supernatant solution from the above step. The 0–18% ammonium sulfate precipitate is removed by centrifugation and, to the resulting supernatant solution, is added 20 g of ammonium sulfate per 100 ml of solution. The ammonium sulfate precipitate is recovered by centrifugation and is retained.

*Step 3. CM-Sephadex Chromatography.* The ammonium sulfate fraction from the above step is dissolved in 40 m$M$ imidazole acetate buffer, pH 7.4, containing 1 m$M$ EDTA and 10 m$M$ 2-mercaptoethanol and dialyzed against the same buffer until free of ammonium ions. For CM-Sephadex chromatography, the enzyme solution (100 mg of protein per 100 ml bed volume of resin) is applied to the column and is then eluted with a 1 liter linear gradient of 0 to 0.15 $M$ KCl (dissolved in column buffer). At the termination of the KCl gradient, column buffer containing 0.15 $M$ KCl is passed over the column until all pyruvate kinase activity is eluted. The heart pyruvate kinase elutes as two peaks; the first (P–I) starts near the end of the gradient and the second (P–II) elutes with the 0.15 $M$ KCl-column buffer wash applied at the end of the gradient.[17] (These two individual fractions can be rechromatographed separately on CM-Sephadex columns under conditions identical to those employed above. Pyruvate kinase in P–II upon rechromatography is eluted from the second column in its original position whereas 62% of that originally in P–I is eluted in the position of P–II when rechromatographed with the remainder of the activity eluting in its original position. It can be estimated that the separated fractions used for the second chromato-

[17] For a typical run, 72,000 units of fraction II are placed on a column 1.8 × 42 cm. Flow rate is 25 ml/hr, and fraction volumes are 2.5 ml. Under these conditions, peak I begins eluting around fraction 100, ends at about fraction 130, at which point peak II begins to elute and is eluted after fraction 170; the KCl gradient ends at about fraction 125.

graphic experiments contained no more than 10% of the other fraction. As outlined below, these two fractions have different kinetic properties.) The results of a typical purification are summarized in Table I.

### Purification of Liver Pyruvate Kinase

There are five different forms of pyruvate kinase in frog liver. The major isozyme is the most rapidly migrating anodal band, and the procedure reported below is for the purification of this isozyme. The method[11] is a modification of the procedure of Staal et al.[18] used for the purification of pyruvate kinase from human erythrocytes and utilizes the property of the enzyme to bind specifically to blue dextran.[18,19]

Step 1. Crude Extract. Liver (66 g) is homogenized with two volumes (w/v) of the 50 m$M$ imidazole acetate buffer, pH 7.4, containing 80 m$M$ KCl, 10 m$M$ 2-mercaptoethanol, and 1 m$M$ EDTA, in a Waring Blendor at the low setting for 2 min. The homogenate is centrifuged at 14,000 $g$ for 20 min, and the resulting supernatant solution is then centrifuged at 100,000 $g$ for 30 min. The residue is discarded.

Step 2. Ammonium Sulfate Fractionation. Ammonium sulfate (30 g/100 ml) is added to the supernatant solution from the above step. The precipitate is removed by centrifugation and discarded. To the resulting supernatant solution, 20 g of ammonium sulfate per 100 ml of original solution is added. The precipitated protein is recovered by centrifugation and this 30–50% ammonium sulfate residue is stored at −15°.

Step 3. Gel Filtration of Enzyme–Blue Dextran Complex. One-third of the ammonium sulfate fraction from the above step is dissolved in 10 ml of 5 m$M$ potassium phosphate buffer, pH 6.8, containing 5m$M$ magnesium sulfate and 4 m$M$ 2-mercaptoethanol. (The volume of buffer used is one-tenth of the volume from which the ammonium sulfate fraction was prepared.) The resulting solution is dialyzed against 1 liter of the same buffer, with 30-min changes, until free of ammonium ions. After dialysis, any insoluble protein is removed by centrifugation and, to the resulting supernatant solution, Blue Dextran 2000[20] is added (final concentration = 0.5%). After equilibration for 10 min, any insoluble material is removed by centrifugation. The clear supernatant solution (12–13 ml) is applied to a Sephadex G-200 column (6 × 45 cm) equilibrated with the same buffer as used in the dialysis. The column is

[18] G. E. J. Staal, J. F. Koster, H. Kamp, L. Van Milligen-Boersma, and C. Veeger, Biochim. Biophys. Acta **227**, 86 (1971).
[19] R. Haeckel, B. Hess, W. Lauterborn and K. H. Wüster, Hoppe-Seyler's Z. Physiol. Chem. **349**, 699 (1968).
[20] The Blue Dextran 2000 (Pharmacia) is suspended in the same buffer (100 mg/ml) and is shaken at room temperature for several hours before use.

washed with the same buffer (flow rate = 35–40 ml/hr) and 10-ml fractions are collected. Pyruvate kinase is eluted together with the blue dextran. Fractions which are blue in color, not green, are pooled (110–120 ml).

*Step 4. First DEAE-Sephadex Chromatography.* The pooled fractions from the above step are applied to a DEAE-Sephadex column (5 × 8 cm) that is equilibrated with 0.2 $M$ potassium phosphate buffer, pH 7.4, containing 50 m$M$ 2-mercaptoethanol and 1 m$M$ EDTA. The column is washed with the same buffer. Under these conditions, the complex is dissociated and the bulk of the blue dextran is adsorbed to the gel whereas the enzyme is not and elutes in the main protein fraction. The enzyme is precipitated from this fraction by the addition of 55 g of ammonium sulfate per 100 ml of solution. The 0–55% ammonium sulfate residue is recovered by centrifugation and retained.

*Step 5. Second Gel Filtration.* The ammonium sulfate precipitate from the above step is dissolved in 6 ml of 0.2 $M$ potassium phosphate buffer, pH 6.8, containing 4 m$M$ 2-mercaptoethanol and 5% ammonium sulfate. Any insoluble material is removed by centrifugation prior to the application of the solution to a Sephadex G-200 column (4.7 × 90 cm) equilibrated with the above buffer. The column is washed with the same buffer (flow rate = 25–30 ml/hr) and 10 ml fractions are collected. Fractions with enzyme activity are pooled, concentrated by ultrafiltration (1–2 mg protein/ml) and enzyme activity is precipitated by the addition of 60 g of ammonium sulfate per 100 ml of solution. The 0–60% ammonium sulfate precipitate is recovered by centrifugation and is stored at −15°. In processing all of 30–50% ammonium sulfate residues from step 2, steps 3–5 are repeated two additional times and the 0–60% ammonium sulfate fractions from step 5 are pooled prior to proceeding to the next step.

*Step 6. Chromatography on Hydroxyapatite.* The ammonium sulfate fractions from the above step are dissolved in a minimal volume of 5 m$M$ potassium phosphate buffer, pH 6.5, containing 10 m$M$ 2-mercaptoethanol and 1 m$M$ EDTA and dialyzed against the same buffer. The sample is then applied to a hydroxyapatite column (1 × 15 cm) which has been equilibrated with the same buffer. For elution of the enzyme, a two-chamber linear gradient apparatus containing 200 ml of solution per chamber is used. The mixing vessel contains the above buffer and the reservoir contains 0.1 $M$ potassium phosphate buffer, pH 6.8, containing 10 m$M$ 2-mercaptoethanol and 1 m$M$ EDTA. The flow rate of the column is 45–50 ml/hr and 5-ml fractions are collected. Fractions containing enzyme activity are pooled.[21]

---

[21] Pyruvate kinase appears in the first protein eluting from the column; hence, enzyme in the first fractions has the highest specific activity. In practice, the last third to fourth of the activity peak is usually discarded.

## TABLE II
PURIFICATION DATA FOR LIVER PYRUVATE KINASE

| Fraction | Total units ($\times 10^6$) | Total protein (mg) | Yield (%) | Specific activity ($\times 10^4$) |
|---|---|---|---|---|
| 1. Crude extract | 2.00 | 6670 | — | 0.3 |
| 2. Ammonium sulfate fraction | 1.84 | 1935 | 91.8 | 0.95 |
| 3. Gel filtration of enzyme-blue dextran complex | 2.05 | 603 | 102.0 | 3.40 |
| 4. First DEAE-Sephadex chromatography | 2.00 | 372 | 100.0 | 5.40 |
| 5. Second gel filtration | 1.57 | 116 | 78.0 | 13.58 |
| 6. Chromatography on hydroxyapatite | 0.49 | 10.8 | 24.4 | 45.22 |
| 7. Second DEAE-Sephadex chromatography[a] | 0.26 | 1.2 | 13.0 | 220.00 |

[a] Low recoveries at this stage are due to: (a) separation of other isozymes; and (b) lability of the enzyme to concentration by ultrafiltration at this point.

*Step 7. Second DEAE-Sephadex Chromatography.* The pooled fractions from the above step are dialyzed against 25 volumes of 93 m$M$ imidazole acetate buffer, pH 7.4, containing 10 m$M$ 2-mercaptoethanol and 1 m$M$ EDTA, with hourly changes for 4 hr. The dialyzed solution is then applied to a DEAE-Sephadex column (1.5 $\times$ 15 cm) that has been equilibrated with the same buffer. The column is eluted with a linear gradient (150 ml of the above buffer in the mixing vessel and 150 ml of the same buffer containing 0.125 $M$ KCl in the reservoir), and fractions of 4 ml are collected. Under these conditions, the main peak of pyruvate kinase activity usually elutes around fraction numbers 50–55.[22] Fractions containing activity are pooled and concentrated (1–2 mg protein/ml). The purified enzyme preparation is free of all of the other four isozymes found in liver as determined by zone electrophoresis.[6] The enzyme is stabilized by 50% glycerol at this point.

The results of a typical preparation are summarized in Table II.

## Properties of the Isozymes [6,10,11]

The electrophoretic mobilities of cardiac, skeletal muscle, and liver (major form) isozymes of pyruvate kinase are different. Immunological cross-reactivity of the antibody to the skeletal muscle enzyme with the

[22] The conditions used in this chromatographic step resolve the chief isozyme from the four other isozymes found in liver. The main liver isozyme is the last to be eluted. Therefore, any pyruvate kinase activity eluting in earlier fractions, or as an ascending shoulder on the main peak, is discarded.

cardiac isozyme, but not with the purified liver isozyme, is observed. The antibody to the purified liver isozyme does not cross-react with either the cardiac or skeletal muscle isozymes. On the basis of gel filtration studies, all three isozymes appear to have approximately the same molecular weight.

A detailed description of the kinetic and other properties of the cardiac and skeletal muscle enzymes can be found in Flanders *et al.*[10] Briefly, the skeletal muscle enzyme exhibits Michaelis-Menten kinetics and is not affected by any of the modifiers demonstrated for regulatory-type pyruvate kinases from other sources. On the other hand, the cardiac isozyme is activated by PEP and FDP and is inhibited by L-alanine, but the extent to which the activity of the enzyme is altered by these modifiers is dependent upon the isolation procedure. Of the two fractions obtained upon CM-Sephadex chromatography, P–I and P–II (see above), only the P–II fraction exhibits allosteric kinetics indicative of a regulatory isozyme. With time (24 hr), the P–II fraction becomes desensitized to the effects of modifiers but can be regenerated by passage over CM-Sephadex as originally run. Although there are quantitative differences, the purified liver isozyme is similar kinetically to the cardiac isozyme in that it is subject to activation by FDP and PEP and is inhibited by L-alanine. In addition, two different fractions of liver pyruvate kinase, analogous to those obtained with the cardiac enzyme, can be obtained by CM-Sephadex chromatography. With respect to the effects of modulators on the amphibian cardiac and liver isozymes, the basis for the occurrence and/or interconversion of sensitive and insensitive forms is unknown. Other studies with hepatic pyruvate kinase from mammalian sources[23-25] indicate that the amount of FDP bound to the enzyme can be one explanation although similar changes have been observed in the presence of organic solvents.[8]

[23] B. Hess and C. Kutzbach, *Hoppe Seyler's Z. Physiol. Chem.* **352**, 453 (1971).
[24] M. G. Irving and J. F. Williams, *Biochem. J.* **131**, 287 (1973).
[25] M. G. Irving and J. F. Williams, *Biochem. J.* **131**, 303 (1973).

## [29] Pyruvate Kinase from Yeast (*Saccharomyces cerevisiae*)

*By* ANN AUST, SHYUN-LONG YUN, and C. H. SUELTER

$$H^+ + \text{phosphoenolpyruvate} + ADP \xrightarrow{\text{Mg}^{2+}, \text{K}^+} ATP + \text{pyruvate}$$

## Assay Method

*Principle.* The reaction catalyzed by pyruvate kinase may be continuously followed by measuring uptake of protons with a pH stat,[1] decrease in absorption at 230 nm due to loss of phosphoenolpyruvate (PEP)[2] or by coupling with lactic dehydrogenase to measure reduction of pyruvate with the oxidation of NADH.[3] The latter method, because of its most widespread use, is described in detail.

*Procedure.* The reaction mixture contains, per milliliter: $(CH_3)_4N$ cacodylate pH 6.2, 100 μmoles; $MgCl_2$, 24 μmoles; KCl, 100 μmoles; tricyclohexylammonium PEP, 10 μmoles[4]; Tris ADP, 10 μmoles; tetracyclohexylammonium FDP, 1 μmole; NADH, 0.16 μmole; and lactic dehydrogenase, 33 μg. After the reaction mixture is equilibrated at the desired temperature of 30°, a blank rate arising from pyruvate kinase contamination of the coupling lactic dehydrogenase is determined. The reaction is initiated by addition of 5–10 μl of pyruvate kinase solution containing 10–20 μg/ml. The initial rate of reaction is corrected for the blank rate.

*Definition of Unit and Specific Activity.* One unit of pyruvate kinase activity is the amount of enzyme that catalyzes the transformation of 1 μmole of substrate per minute under the conditions specified. Throughout the purification procedure and until the first cellulose phosphate column at pH 6.5, protein is determined by the biuret[5] method. After the cellulose phosphate column at pH 6.5, the extinction coefficient determined for yeast pyruvate kinase $OD_{280}^{0.1\%} = 0.51$ is used.[6]

## Purification Procedure

The purification procedure is a modification of that previously described.[7] All operations except for the lysis are completed at room tem-

[1] F. J. Kayne and C. H. Suelter, *J. Amer. Chem. Soc.* **87**, 897 (1965).
[2] N. G. Pon and R. J. L. Bondar, *Anal. Biochem.* **19**, 272 (1967).
[3] T. Bücher and G. Pfleiderer, this series, Vol. 1, p. 435 (1953).
[4] Two micromoles of PEP per milliliter may be used for general routine assays.
[5] E. Layne, this series, Vol. 3, 450 (1957).
[6] A. Aust, S. L. Yun, and C. H. Suelter, unpublished observation.
[7] J. R. Hunsley and C. H. Suelter, *J. Biol. Chem.* **244**, 4815 (1969).

PURIFICATION OF PYRUVATE KINASE FROM *Saccharomyces cerevisiae*

| Fraction | Volume (ml) | Total protein (mg) | Total units | Specific activity (units/mg) | Yield (%) |
|---|---|---|---|---|---|
| Initial lysate, fraction I[a] | 1070 | 24,055 | 245,000 | 10 | 100 |
| (NH$_4$)$_2$SO$_4$ fraction 40–55% | 195 | 12,135 | 226,400 | 18.6 | 92 |
| Phosphocellulose chromatography, pH 6.5 | 500 | 2,485 | 173,929 | 70 | 71 |
| DEAE-phosphocellulose chromatography, pH 7.5 | 520 | 529 | 132,260 | 250 | 54 |
| Phosphocellulose chromatography, pH 6.5 | 80 | 127 | 43,200 | 340 | 18 |
| Sephadex G-100 | 25 | 110 | 37,400 | 340 | 15 |

[a] This represents the yield from 1 pound of yeast.

perature. The results of a typical procedure are summarized in the table.

*Buffers and Strategy.* The following buffers are used: pH 6.5 buffer consists of 10 m$M$ sodium phosphate, 5 m$M$ EDTA, 5 m$M$ 2-mercapto-ethanol, and 25% glycerol adjusted to pH 6.5; pH 7.5 buffer is identical in composition to the pH 6.5 buffer but adjusted to pH 7.5; and (NH$_4$)$_2$SO$_4$ pH 6.5 buffer consists of 10 m$M$ sodium phosphate, 5 m$M$ EDTA, 5 m$M$ 2-mercaptoethanol, 25% glycerol, and (NH$_4$)$_2$SO$_4$ at the appropriate concentrations adjusted to pH 6.5.

The purification once initiated should be completed as soon as possible to prevent proteolytic degradation. If it is necessary to stop at some step, it is recommended that the enzyme be precipitated with 55% (NH$_4$)$_2$SO$_4$ and stored as an (NH$_4$)$_2$SO$_4$ suspension, at room temperature.

*Lysis of Yeast.* One pound of yeast (Budweiser bakers' yeast) is crumbled into 800 ml of distilled-deionized water at 4°C and stirred until suspension is complete. The resulting suspension is made 1 m$M$ in diiso-propylfluorophosphate (DFP),[8] and passed twice through a precooled

[8] One gram of DFP is dissolved in 49 ml of dried isopropanol (dried over Fisher type 5 A molecular sieve or sodium sulfate), giving an approximate 0.1 $M$ solution. This stock solution was stored at −20°. Condensation of water vapor in the stock solution should be avoided.

Whenever DFP is added, it is to a final concentration of 1 m$M$. To avoid denaturation of protein, the DFP stock solution (0.1 $M$) is diluted 10-fold in the protein solvent before addition to rapidly stirring protein solutions. Because of the toxic effects of DFP, the user is advised to take cognizance of the following precautions:

a. All operations with the pure liquid and concentrated solutions (greater than 1 m$M$) are to be done in a hood with good air flow. The user must wear gloves

Manton-Gaulin homogenizer at 8000–8500 psi. The homogenized suspension is immediately adjusted to pH 7.0 with 4 $N$ KOH and treated with DFP as before. The extract is then made 40 m$M$ in $CaCl_2$ (4.4 g/liter), and stirred thoroughly.[9] Immediately after dissolution of $CaCl_2$, 1.0 $M$ potassium phosphate (pH 7.0) is added to a final concentration of 0.1 $M$ and the solution allowed to stand for 1 hr at room temperature with occasional stirring. The clear yellow supernatant obtained after centrifugation at 14,000 $g$ for 30 min is fraction I. All remaining steps in the procedure are accomplished at room temperature.

*Ammonium Sulfate Fractionation.* To fraction I is added 243 g/liter solid ammonium sulfate, slowly, with constant stirring. After dissolution of ammonium sulfate, the mixture is adjusted to pH 6.2 with 4 $N$ KOH and allowed to remain for 1 hr with constant stirring. The supernatant obtained after centrifugation at 14,000 $g$ for 30 min is fraction II.

To fraction II is added 97 g/liter of solid ammonium sulfate, slowly, with constant stirring. After dissolution, the mixture is adjusted to pH 6.2 with 4 $N$ KOH, allowed to remain for 1 hr with constant stirring and centrifuged at 14,000 $g$ for 30 min. The pellet is suspended in pH 6.5 buffer to give a final volume of near 200 ml and dialyzed overnight against 2 liters of pH 6.5 buffer containing 1 m$M$ DFP. A precipitate at this stage must be removed by centrifugation. The dialyzate was diluted with pH 6.5 buffer to a final volume of 1500 ml to give fraction III. This dilution reduces the $(NH_4)_2SO_4$ concentration to near 10 m$M$ and protein concentration to near 10 mg/ml and is required to achieve adsorption of enzyme to the cellulose phosphate pH 6.5 column described in the next step.

Fraction III has a hazy white appearance which precipitates slowly on standing. This requires that the fraction be applied to the cellulose phosphate pH 6.5 column immediately after dilution. The precipitation is markedly enhanced if the precipitated fraction II is dialyzed by gel

---

at all times and take precautions not to contaminate clothing. Polyvinyl gloves are recommended.

b. All contaminated glassware is placed in 0.5 $N$ NaOH for a period of at least 24 hr for complete hydrolysis.

c. One-millimolar solutions are not allowed to come in contact with skin.

d. The following accounts of the chemistry and toxicity of fluorophosphate are available:

L. S. Goodman and A. Gelman, "The Pharmacological Basis of Therapeutics," 3rd ed., p. 441. Macmillan, New York, 1969.

B. C. Saunders, "Some Aspects of the Chemistry and Toxic Action of Organic Compounds Containing Phosphorus and Fluorine." Cambridge Univ. Press, London and New York, 1957.

[9] J. A. Illingworth, *Biochem. J.* **129**, 1119 (1972).

permeation chromatography. The nature of the precipitate is not known.

*Cellulose Phosphate Column pH 6.5.* Cellulose phosphate (1.1 mEq/g[10]) is washed as previously described.[7] The cellulose is suspended in pH 6.5 buffer, and the slurry is adjusted to pH 6.5 directly, using a pH meter. The cellulose is suction-filtered dry on a Büchner funnel, resuspended in the pH 6.5 buffer, and degassed. A column (5 × 20 cm) is poured and prior to use washed with at least 1 column volume of buffer. The pH of the eluate must be 6.5; if not, wash with additional buffer. After application of fraction III at an optimum flow rate of 500 ml/hour, the column is washed with pH 6.5 buffer to remove excess protein. Pyruvate kinase is eluted with the 0.1 $M$ $(NH_4)_2SO_4$ pH 6.5 buffer. All fractions containing eluted protein are pooled to give fraction IV, and dialyzed overnight against 10 volumes of pH 7.5 buffer.[11]

*DEAE-Cellulose Phosphate Column pH 7.5.* Cellulose phosphate (1.1 meq/g) is prepared as described earlier except pH 7.5 buffer is used. A column, 4 by 45 cm, is poured to a height of 20 cm. DEAE-cellulose (0.8 meq/g) is washed as previously described[7] and suspended in the pH 7.5 buffer. The slurry was adjusted to pH 7.5, then suction-filtered dry, resuspended in the pH 7.5 buffer, and degassed. The DEAE-cellulose was poured directly onto the pH 7.5 cellulose phosphate to form a combined DEAE-cellulose–cellulose phosphate column with a total height of 40 cm. Prior to use, the column is washed with 1 column volume of pH 7.5 buffer. If the pH of the eluate is not 7.5, wash with an additional volume of buffer. Fraction IV after adjustment to pH 7.5, is applied to the combined DEAE-cellulose–cellulose phosphate column followed by at least 1 column volume of pH 7.5 buffer. Pyruvate kinase is not adsorbed to this

---

[10] Cellulose phosphate with a lower binding capacity will not function satisfactorily for this procedure using the column sizes as prescribed. To maintain binding capacity, the washing should be completed rapidly and the resin stored at pH 6–8. Cellulose phosphate which has been subjected 6 or more times to the wash cycle appears to lose a significant percentage of its binding capacity. Please be aware that as this manuscript is being prepared, we were advised that the cellulose phosphate (Schwarz-Mann Mannex-P High Capacity) used for the development of this procedure, is no longer available. However, we have found that Brown Co. makes a cellulose phosphate of high capacity (greater than 1.1 meq/g) which may be satisfactorily substituted.

[11] Dialysis of fraction IV may be accomplished by passage over Sephadex G-25 or G-100. Passage over G-100 would serve an additional purpose of removing a contaminating protease activity which copurifies with the enzyme. To proceed, the enzyme is first precipitated by addition of 662 g of ammonium sulfate per liter. The precipitate obtained after centrifugation at 14,000 $g$ for 30 min is dissolved in a minimum volume of pH 7.5 buffer and applied to the Sephadex column equilibrated with the pH 7.5 buffer. To achieve separation of the contaminating protease, no more than 10 ml of solution are applied to a 5 × 90 cm G-100 column.

column, and thus the eluate is collected fractionally until all protein is eluted. Those fractions having a specific activity greater than 200 are pooled to give fraction V.

*Cellulose Phosphate Column, pH 6.5.* The cellulose phosphate resin (1.1 mEq/g) is prepared in the pH 6.5 buffer as described above and poured into a 2.5 cm column to a height of 28 cm. Prior to use, the column is washed with an additional volume of pH 6.5 buffer. If the pH of the eluate is not 6.5, wash with additional buffer.

Fraction V, after adjusting to pH 6.5 with 3 $N$ acetic acid, is applied to the cellulose phosphate column. After application of sample, the column is washed with pH 6.5 buffer to remove unadsorbed protein. Pyruvate kinase is eluted with a 600-ml linear $(NH_4)_2SO_4$ gradient (0.01 to 0.2 $M$) in pH 6.5 buffer, and fractions are collected fractionally. The peak of the pyruvate kinase activity elutes at 70 m$M$ $(NH_4)_2SO_4$. Those fractions having pyruvate kinase with specific activities greater than 340 are pooled (fraction VI).[12] The pooled fraction is dialyzed against 90% saturated ammonium sulfate prepared in 10 m$M$ sodium phosphate, pH 6.5, 5 m$M$ EDTA, 5 m$M$ 2-mercaptoethanol, to precipitate all enzyme.

The enzyme obtained at this stage migrates as a single band during sedimentation by analytical ultracentrifugation and electrophoresis on polyacrylamide gels. However, a solution of the protein in 0.1% sodium dodecyl sulfate containing 10 m$M$ 2-mercaptoethanol incubated for 12 hr at room temperature gives several bands on SDS polyacrylamide gels after electrophoresis, consistent with a protease contamination. The contaminating protease was removed by the Sephadex G-100 column.

*Sephadex G-100 Column.* The precipitated enzyme from fraction VI is sedimented by centrifugation at 27,000 $g$ for 20 min and dissolved in a minimum volume of pH 6.5 buffer. One milliliter is applied to a Sephadex G-100 column, 1.6 × 70 cm, and eluted with the pH 6.5 buffer. The fractions containing pyruvate kinase with specific activities greater than 340, and free of protease contamination as determined by SDS polyacrylamide gel electrophoresis, are pooled and dialyzed against 90% saturated ammonium sulfate prepared in 10 m$M$ sodium phosphate, pH 6.5, 5 m$M$ EDTA, 5 m$M$ 2-mercaptoethanol to precipitate enzyme.

Yeast pyruvate kinase is stored as a suspension at 2–4° in 90% saturated ammonium sulfate and retains nearly full activity for at least 1 month. The specific activity of the purified enzyme is 320–360 units/mg at 30°. The ratio of the specific activity determined with 10 m$M$ PEP in the presence and in the absence of fructose 1,6-diphosphate should be 1.

[12] If most of enzyme obtained at this stage has a specific activity less than 340 units/mg, a failure of the combined DEAE-cellulose phosphate pH 7.5 column is indicated.

## Properties

*Homogeneity and Molecular Properties.* The enzyme prepared by this procedure is homogeneous by the following criteria: the protein sediments as a single symmetrical peak during analytical ultracentrifugation, equilibrium sedimentation data indicates no heterogeneity, and a single band is obtained after electrophoresis on polyacrylamide gel prepared with 5 different concentrations of acrylamide.[6] The absorbancy ratio $A_{280nm}/A_{260nm}$ of the pure enzyme is greater than 1.7. The molecular weight of the enzyme determined by the sedimentation equilibrium method is 209,400 ± 1700. Electrophoresis of the protein in sodium dodecyl sulfate on polyacrylamide gel and comparison of its migration rate with those of standard proteins according to the method of Osborn and Weber[13] gives a subunit molecular weight of 50,000–52,000 consistent with a tetrameric subunit protein. The preparation from *Saccharomyces carlsbergensis*[14,15] has a molecular weight of 190,000 and also consists of 4 subunits.

The previous preparation of yeast (*S. cerevisiae*) pyruvate kinase had a molecular weight of 167,000 which dissociated in guanidine HCl to give subunits of 42,000–46,000 molecular weight.[16] These data in light of the present results suggest that the previous preparation had undergone partial proteolytic breakdown. The evidence available to date indicates that the preparation described in this chapter is native to yeast.[17] However, the techniques that have been used are not sufficient to eliminate a microheterogeneity.

The enzyme is cold labile[6] as described for an earlier preparation.[16] Fructose 1,6-diphosphate enhances the rate of inactivation both at 0° and at 23°. Glycerol (25%) and/or 0.1 $M$ KCl and 20 m$M$ MgCl$_2$ protects against the FDP-induced and cold inactivation.[19] The enzyme

[13] K. Weber and M. Osborn, *J. Biol. Chem.* **244**, 4406 (1969).

[14] P. Roschlau and B. Hess, *Hoppe-Seyler's Z. Physiol. Chem.* **335**, 435 (1972).

[15] H. Bischofberger, B. Hess, and P. Roschlau, *Hoppe-Seyler's Z. Physiol. Chem.* **352**, 1139 (1971).

[16] R. T. Kuczenski and C. H. Suelter, *Biochemistry* **9**, 4032 (1970).

[17] The presence of a contaminating protease at each step of the purification suggests a possible proteolytic degradation. However, the following evidence does not support this degradation. (1) Attempts to inhibit the contaminating protease by DFP at the stage just prior to the Sephadex G-100 column have failed, suggesting that the protease does not react with DFP or that it exists as a zymogen or as an enzyme-inhibitor complex[18]; SDS may activate the zymogen or disrupt the complex. (2) No protease activity can be detected in the absence of SDS. (3) The enzyme migrates as a single sharp band on polyacrylamide gels. (4) The enzyme in an initial extract of yeast has the same mobility as purified enzyme on polyacrylamide gels prepared from 5 different concentrations of acrylamide.

[18] J. F. Lenney and J. M. Dalbec, *Arch. Biochem. Biophys.* **129**, 407 (1969).

[19] R. T. Kuczenski and C. H. Suelter, *Biochemistry* **9**, 939 (1970).

contains 20 sulfhydryl residues; 8 are buried in the interior of the folded native enzyme, and 12 are exposed to the solvent as determined by reaction of native enzyme with 5,5'-dithiobis (2 nitrobenzoic acid).[6]

*Kinetic Properties.* Yeast pyruvate kinase is an allosteric enzyme exhibiting sigmoid substrate velocity saturation curves for PEP which are hyperbolic in the presence of fructose 1,6-diphosphate. The $K_m$ values at pH 6.2 and 30° in the presence of 1 mM FDP for PEP, ADP, $K^+$, and $Mg^+$ are 99 $\mu M$, 0.16 mM, 50 mM, and 2 mM, respectively. Reaction of the 12 exposed sulfhydryl residues with 5,5'-dithiobis(2-nitrobenzoic acid) results in a marked increase in the $K_m$ for PEP as noted by an increase in the ratio of the specific activity determined in the presence of FDP to that determined in the absence of FDP at 10 mM PEP. A ratio of 1 is found for native enzyme; ratios as high as 14 have been observed for modified enzyme.

# [30] Human Erythrocyte Pyruvate Kinase

*By* G. E. J. STAAL, J. F. KOSTER, and C. VEEGER

Phosphoenolpyruvate + ADP → pyruvate + ATP

Pyruvate kinase (EC 2.7.1.40) is a key enzyme in glycolysis. Human erythrocytes rely most exclusively on glycolysis to fulfill their energy requirements. Pyruvate kinase deficiency is, after glucose-6-phosphate dehydrogenase deficiency, the most common erythrocyte metabolic error. The deficiency comprises a heterogeneous group of disorders characterized by both qualitative and quantitative enzymic abnormalities.[1]

It is well accepted that different forms of pyruvate kinase exist. Erythrocyte pyruvate kinase resembles very much the liver L type, but differences do exist. An important difference between these two enzymes is that the hepatic enzyme is under hormonal and dietary control, whereas the erythrocyte enzyme is not.[2]

## Assay Method

Pyruvate kinase activity is measured at 25° by coupling the system with lactate dehydrogenase, according to the method of Bücher and Pfleiderer.[3] The oxidation of NADH is followed at 340 $\mu$m.

[1] P. Boivin, "Enzymopathies," Vol. 1, p. 190. Masson, Paris, 1971.

[2] T. Tanaka, Y. Harano, F. Sue, and H. Morimura, *J. Biochem. (Tokyo)* **62,** 71 (1967).

[3] T. Bücher and G. Pfleiderer, this series, Vol. 1, p. 435.

*Reagents.* All the compounds used for the determination of the activity are dissolved in the Tris·HCl buffer

 Tris·HCl, buffer 0.2 $M$ pH 7.2
 Tricyclohexylammonium PEP, 60 m$M$
 ADP, 60 m$M$
 KCl, 1 $M$
 MgSO$_4$, 0.3 $M$
 NADH, 2.7 m$M$
 Lactate dehydrogenase (2 mg/ml), dialyzed overnight at 4° against 0.2 $M$ Tris·HCl, pH 7.2.

*Procedure.* The activity is determined at 25° in 0.2 $M$ Tris·HCl (pH 7.2) in a final volume of 3 ml containing 5 m$M$ PEP, 5 m$M$ ADP, 65 m$M$ KCl, 20 mM MgSO$_4$, 0.02 ml dialyzed lactate dehydrogenase, 90 $\mu M$ NADH, and enzyme.

*Definition of Unit.* The activity is expressed as $\mu$moles NADH oxidized per minute. The specific activity is defined as $\mu$moles NADH oxidized per minute per milligram of protein. The protein content is determined by the method of Lowry *et al.*[4] using crystalline bovine serum albumin as a standard.

## Purification

*Reagents*

 Buffer A: phosphate buffer 10 m$M$, pH 6.8 containing 1 m$M$ $\beta$-mercaptoethanol and 1 m$M$ $\epsilon$-aminocaproic acid. To prevent microbial growth chloramphenicol and amphotericin B are added.
 Buffer B: phosphate buffer, 0.1 $M$, pH 6.8 containing 1 m$M$ $\beta$-mercaptoethanol, 1 m$M$ $\epsilon$-aminocaproic acid, chloramphenicol, and amphotericin B
 Buffer C: phosphate buffer, 0.1 $M$, pH 6.8, containing 2 m$M$ MgSO$_4$, 2 m$M$ $\beta$-mercaptoethanol, 1 m$M$ $\epsilon$-aminocaproic acid, chloramphenicol, and amphotericin B
 Buffer D: 5 m$M$ phosphate buffer, pH 6.8, containing 5 m$M$ MgSO$_4$, 2 m$M$ $\beta$-mercaptoethanol, chloramphenicol, and amphotericin B
 Buffer E: phosphate buffer, 0.2 $M$, pH 6.8, containing 5% $(NH_4)_2SO_4$ (w/v), 1 m$M$ $\beta$-mercaptoethanol, 1 m$M$ $\epsilon$-aminocaproic acid, chloramphenicol, and amphotericin B

[4] O. H. Lowry, N. J. Rosebrough, A. L. Farr, and R. J. Randall, *J. Biol. Chem.* **193**, 265 (1951).
[5] G. E. J. Staal, J. F. Koster, H. Kamp, L. van Milligen-Boersma, and C. Veeger, *Biochim. Biophys. Acta* **227**, 86 (1971).

Buffer F: phosphate buffer, 0.2 $M$, pH 6.8, containing 5% $(NH_4)_2SO_4$ (w/v) and 50% glycerol (v/v)

DEAE-Sephadex A-50, capacity 3.5 $\pm$ 0.5 meq/g, particle size 40–120 $\mu$m; Sephadex G-200, particle size 40–120 $\mu$m.

Blue Dextran 2000

Concentration of protein solutions with Diaflo ultrafiltration cells from Amicon (pressure about 20 psi) with XM-50 membranes.

*Procedure*

*Step 1.* About 0.5 liter of erythrocytes is mixed with an equal volume of water containing 1 m$M$ $\beta$-mercaptoethanol. The erythrocytes are hemolyzed by freezing and thawing (3$\times$). Approx. 1 liter of hemolyzate is obtained.

*Step 2.* To the hemolyzate are added 2 liters of DEAE-Sephadex equilibrated with buffer A. The mixture of DEAE-Sephadex and hemolysate is stirred during 1 hr at 4°. To remove the hemoglobin the mixture is placed on a Büchner funnel and repeatedly washed with buffer A, about 10 liters of buffer being used. The enzyme remains bound to the Sephadex, pyruvate kinase is eluted by washing the filter "cake" with 4 liters of buffer B.

*Step 3.* To the enzyme solution obtained in step 2, 22 g of $(NH_4)_2SO_4$ per 100 ml solution are added. After 12 hr the precipitate is collected by centrifugation (3000 $g$, for 30 min) and dissolved in 450 ml of 50 m$M$ phosphate buffer (pH 6.8).

*Step 4.* To the dissolved precipitate of step 3, 22 g $(NH_4)_2SO_4$ per 100 ml of enzyme solution are added. After 12 hr the precipitate is collected by centrifugation (10,000 g for 30 min) and suspended in 40 ml of buffer C.

*Step 5.* The dissolved precipitate of step 4 is put in an Erlenmeyer of 250 ml and heated during 2 min at 60° by stirring it gently in a water bath of 70°. After heating the solution is quickly cooled to 4°. This procedure is essential to remove much of the protein, while pyruvate kinase in this way is not inactivated.

*Step 6.* The supernatant of step 5 is concentrated to about 4 ml with the Amicon ultrafiltration cell. Blue Dextran 2000 is added (end concentration 0.5%). Any insoluble material is removed by centrifugation. The clear solution is placed on a Sephadex G-200 column (2.5 $\times$ 100 cm) equilibrated with buffer D. The pyruvate kinase elutes together with Blue Dextran 2000 as first described by Haeckel *et al.*[6] for the yeast

[6] R. Haeckel, B. Hess, W. Lauterborn, and K. H. Wüster, *Hoppe-Seyler's Z. Physiol. Chem.* **349,** 699 (1968).

PURIFICATION OF PYRUVATE KINASE FROM NORMAL HUMAN ERYTHROCYTES

| Step | Total volume (ml) | Total protein (mg) | Total activity (units) | Specific activity (units/mg protein) | Yield (%) | Cumulative purification (fold) |
|---|---|---|---|---|---|---|
| 1 | 1000 | $150 \times 10^3$ | 725 | $5 \times 10^{-3}$ | 100 | 1 |
| 2 | 4000 | $1.76 \times 10^3$ | 580 | $33 \times 10^{-2}$ | 80 | 66 |
| 3 | 450 | $1.66 \times 10^2$ | 500 | 3 | 69 | 600 |
| 4 | 40 | 87.6 | 456 | 5.2 | 63 | 1,040 |
| 5 | 40 | 26 | 300 | 11.5 | 40 | 2,300 |
| 6 | — | — | — | — | — | — |
| 7 | 2 | 0.38 | 58 | 150 | 8 | 30,000 |

enzyme. The enzyme–Blue Dextran complex is concentrated to 4 or 5 ml with the ultrafiltration cell.

*Step 7.* The concentrated pyruvate kinase–Blue Dextran solution is placed on a Sephadex G-200 column (2.5 × 100 cm) equilibrated with buffer E. Under these conditions pyruvate kinase is separated from the Blue Dextran. To stabilize the enzyme the active fractions (2 ml) are collected in tubes containing 1 ml 50% glycerol (w/v), mixed and further concentrated with the ultrafiltration cell to 1 or 2 ml. After concentration the enzyme is dialyzed overnight against 10 m$M$ Tris-maleic acid (pH 6.8) containing 50% glycerol (v/v). When the separation between pyruvate kinase and Blue Dextran is not complete, an additional DEAE-Sephadex column (1 × 3 cm) is used, equilibrated with buffer F; pyruvate kinase is eluted with the same buffer. Under these conditions Blue Dextran 2000 is adsorbed onto the DEAE-Sephadex while pyruvate kinase is not. The whole purification procedure is summarized in the table.

A quite identical isolation procedure has been reported by Blume *et al.*[7] with nearly the same results. Ching *et al.*[8] published a quite different method; however, the authors obtained the same purification and yield as with the method described above.

## Properties

*Molecular Weight.* With a Sephadex G-200 column (3.0 × 55 cm) according to the method of Andrews[9] a molecular weight of 205,000

[7] K. G. Blume, R. W. Hoffbauer, D. Busch, H. Arnold, and G. W. Löhr, *Biochim. Biophys. Acta* **227**, 364 (1971).
[8] J. C. Ching, M. B. Rittenberg, and J. A. Black, *J. Biol. Chem.* **247**, 7173 (1972).
[9] P. Andrews, *Biochem. J.* **96**, 595 (1968).

($\pm 5000$) was found. With the same method Blume *et al.*[7] found a value of 195,000 ($\pm 6000$). Ching *et al.*[8] reported a molecular weight of 225,400; this was measured with the sedimentation equilibrium method.

*Kinetics.* Erythrocyte pyruvate kinase belongs at physiological pH to the group of allosteric enzymes.[5] Positive cooperativity is observed at [ADP] = $\infty$ with respect to phosphoenol pyruvate ($h = 1.6$), no cooperativity is observed with ADP. In the presence of fructose 1,6-diphosphate as allosteric activator, the phosphoenolpyruvate-dependent rate curves change into hyperbolic curves. Phosphorylated hexoses and $PO_4^{3-}$ behave more or less like fructose 1,6-diphosphate. ATP acts as allosteric inhibitor. This kinetic behavior can be explained in a qualitative way with the model proposed by Monod *et al.*[10] and Rubin and Changeux.[11] This allosteric pattern disappears at pH 5.9. Kinetic data show that $V = 200$ $\mu$moles/min/mg, $K_m$ (ADP) = 0.6 m$M$ and $K_m$ (PEP) = 0.63 m$M$. In addition the reaction proceeds along a ternary complex mechanism.[5]

Recently it was found[12] that pyruvate kinase can be converted into an oxidized form by incubation with oxidized glutathione. The oxidized enzyme can be reduced again by incubation with mercaptoethanol. The oxidized form shows a lower affinity for the substrate phosphoenolpyruvate and for the allosteric effector fructose 1,6-diphosphate. The thermolability of the oxidized enzyme is markedly increased as compared to the freshly isolated enzyme. These kinetic and thermostability data are identical with those obtained with pyruvate kinase from most of the pyruvate kinase-deficient patients. Because combined defects of pyruvate kinase and glutathione reductase have been reported, it seems likely that an increased GSSG concentration in the red blood cell can be the primary effect causing alterations in pyruvate kinase.

[10] J. Monod, J. Wyman, and J. P. Changeux, *J. Mol. Biol.* **12**, 88 (1965).
[11] M. M. Rubin and J. P. Changeux, *J. Mol. Biol.* **21**, 265 (1966).
[12] T. J. C. van Berkel, J. F. Koster, and G. E. J. Staal, *Biochim. Biophys. Acta* in press.

# [31] Pyruvate, Orthophosphate Dikinase[1] from *Bacteroides symbiosus*

*By* Dorothy J. South and Richard E. Reeves

$$MgATP^{2-} + pyruvate^{1-} + P_i^{2-} \xrightarrow{(NH_4)^+} AMP^{2-} + MgPP_i^{2-}$$
$$+ \text{P-enolpyruvate}^{3-} + 2H^+$$

Pyruvate, phosphate dikinase functions in the gluconeogenic pathway in the leaves of tropical grasses[2] and in *Acetobacter xylinum*.[3] In *B. symbiosus* and in *Entamoeba histolytica* the enzyme seems to function in the glycolytic pathway, since pyruvate kinase has not been found in those organisms.[4] *Bacteroides symbiosus* dikinase has been employed in a spectrophotometric assay for pyrophosphate.[5] By virtue of its low $K_m$ for AMP, it is also useful in the assay of this metabolite.

## Assay Method

*Principle.* The dikinase has been assayed by following $H^+$ changes,[6] by phosphoenolpyruvate formation[2,3,7] ATP formation,[5] and pyruvate formation.[4-6] A spectrophotometric modification of the latter method is employed in this preparation. The method follows the rate of oxidation of NADH by pyruvate in the presence of lactic dehydrogenase.

*Reagents*

Imidazole·HCl buffer, 0.2 $M$, pH 6.8
NH₄Cl, 0.8 $M$
MgCl₂, 0.1 $M$
Inorganic pyrophosphate (PP$_i$), 50 m$M$, adjusted to pH 8.5 with HCl
Phosphoenolpyruvate, Na salt, 25 m$M$
Adenosine monophosphate (AMP), 40 m$M$
NADH, 10 m$M$
L-Lactate dehydrogenase, 1.25 mg/ml. Boehringer pig heart enzyme in ammonium sulfate suspension is centrifuged, and the pellet is dissolved in 1 m$M$ sodium EDTA, pH 7.0.

[1] EC 2.7.9.1.
[2] M. D. Hatch and C. R. Slack, *Biochem. J.* **106**, 141 (1968).
[3] M. Benziman and N. Eizen, *J. Biol. Chem.* **246**, 57 (1971).
[4] R. E. Reeves, R. A. Menzies, and D. S. Hsu, *J. Biol. Chem.* **243**, 5486 (1968).
[5] R. E. Reeves and L. K. Malin, *Anal. Biochem.* **28**, 282 (1969).
[6] R. E. Reeves, *Biochem. J.* **125**, 531 (1971).
[7] H. J. Evans and H. G. Woods, *Biochemistry* **10**, 721 (1971).

*Procedure.* To a quartz cuvette of 1 cm optical path, 3 mm inside width, is added 100 $\mu$l of imidazole buffer, 10 $\mu$l of ammonium chloride solution, 25 $\mu$l of magnesium chloride solution, 20 $\mu$l of phosphoenolpyruvate solution, 10 $\mu$l of AMP solution, 10 $\mu$l of lactate dehydrogenase solution, 0.005–0.02 unit of enzyme in dilute buffer, and water to a volume of 0.39 ml. After thermal equilibration of the cuvette at 30°, the reaction is started by addition of 10 $\mu$l of the pyrophosphate solution. The reaction is monitored at 30° and 340 nm using a Gilford recording spectrophotometer with a narrow aperture to avoid stray light effects. A control cuvette lacking added pyrophosphate serves to correct for NADH oxidase activity. This correction is negligible following the DEAE-column treatment (see below).

*Definition of Unit and Specific Activity.* One unit is defined as that amount of enzyme required to catalyze the oxidation of 1 $\mu$mole of NADH per minute under the above conditions. Specific activity is defined as units of enzyme per milligram of protein. Protein concentration is determined by the method of Lowry *et al.* [8,9]

*Application of Assay Method to Crude Homogenates.* When organisms containing 10–30 units per gram of fresh cells are employed assays can be made on lysozyme or sonic homogenates by using relatively small amounts of enzyme. Although such assays are of limited accuracy they are useful for determining whether the decryptified enzyme sediments with the pellet, and thus requires further treatment for its solubilization.

## Purification Procedure

*Growth.* The enzyme is obtained from the obligate anaerobe *Bacteriodes symbiosus* ATCC 14940. Somewhat greater yields are obtained from a culture long-established in this laboratory and from which a recent deposit in the American Type Culture Collection (ATCC 12829) has been made. Twenty-four hour growth at 37° in 16 × 125-mm screw-cap tubes should achieve 0.6 to 0.8 optical density unit before scale-up to larger quantities is attempted. The medium contains, in grams per liter: glucose, 10; Bactotryptone (Difco 0123-01), 20; yeast extract (Difco 0127-01), 2; $K_2HPO_4$ (anhydrous), 1.5; NaCl, 2.5; mercaptosuccinic acid (Eastman), 1.5 g; and NaOH to pH 7.0. The medium is autoclaved at 121° for 15 min and placed in screw-cap tubes or screw-cap Erlenmeyer flasks, which are filled to the neck after autoclaving. One 24-hr tube culture is used to inoculate the medium in a 125-ml flask in

[8] O. H. Lowry, N. J. Rosebrough, A. L. Farr, and R. J. Randall, *J. Biol. Chem.* **193**, 265 (1951).

[9] See this series, Vol. 3 [73].

the early morning. About 8 hr later this culture is used to inoculate freshly autoclaved medium in a 3-liter flask. After 16 hr at 37° the cells are harvested by centrifugation for 15 min at 1000 $g$. The cells are washed twice in the centrifuge with 0.15 $M$ NaCl and stored at $-20°$ until used. The yield of packed cells is about 3 g per liter of medium. In our experience the substitution of any other product for the Bactotryptone has resulted in diminished cell yields.

*Step 1. Crude Enzyme.* The frozen cell mass, consisting of 9–10 g of packed cells, is suspended in 100 ml of buffer at pH 7 containing 20 m$M$ imidazole HCl, 20 m$M$ NH$_4$Cl, and 5 m$M$ EDTA (sodium salt). The suspension is warmed to 35–37°. One milligram of lysozyme per gram of frozen cells is then added, and the suspension is kept at this temperature for 30 min, with stirring. The suspension is then centrifuged at 12,000 $g$ for 15 min, and the supernatant solution is saved. The pellet is resuspended in 50 ml of buffer and the lysozyme treatment and centrifugation is repeated. The pellet, which should now contain a large gelatinous white layer, is discarded. The combined supernatant solutions are precipitated with streptomycin.

One gram of streptomycin, as sulfate, is diluted with 20 ml of buffer and added dropwise, to the combined supernatants, with stirring. The flocculent white precipitate is removed by centrifugation. The supernatant is kept for step 2.

*Step 2. DEAE-Cellulose Column.* The DEAE-cellulose is prepared for use as follows: Twenty grams of DEAE-cellulose (Sigma, medium mesh) is allowed to sink in 1 liter of 1 $M$ NaOH. Then 750 g of cracked ice and 250 ml of concentrated HCl are added and the suspension is diluted to 4 liters with cold water. The cellulose is washed with water several times by decantation to remove free acid. To the cellulose slurry are added water and concentrated solutions of imidazole·HCl pH 6.4, NH$_4$Cl, and EDTA (sodium salt) to make a volume of 1 liter at final concentrations of 20, 20, and 1 m$M$, respectively. The suspension is adjusted to pH 6.4 with NaOH. Seven grams of the DEAE cellulose is used to pack a column, 16 cm $\times$ 5 cm$^2$. The supernatant from the streptomycin sulfate precipitation is passed through the column. The column is then washed with 75 ml of column buffer, which is a solution containing 20 m$M$ imidazole·HCl and 20 m$M$ NH$_4$Cl at pH 6.4. Elution is by a convex gradient obtained by allowing 0.4 $M$ NaCl in column buffer to drip into a reservoir containing 150 ml of column buffer. The flow is 2 ml/min and protein elution is monitored at 280 nm with a flow cell. The enzyme typically elutes after 200–240 ml of the gradient has entered the column. The fractions are assayed for enzyme activity and those for which the ratio units per ml/OD$_{280}$ exceeds 1 are pooled for the next step.

*Step 3. Hydroxyapatite Column.* Bio-Gel HT hydroxyapatite is used to prepare a column 19 cm × 2.5 cm². The column is washed with column buffer and the pooled fractions from the DEAE step are applied. Elution, at a rate of about 1 ml/min, is by means of a linear gradient obtained by allowing 125 ml of column buffer containing 0.4 $M$ potassium phosphate, pH 6.4, to run into 125 ml of column buffer without phosphate. The enzyme elutes after about 100 ml of the gradient has entered the column. It emerges with the principal protein peak. By judicious cutting of the fractions it is possible to get most of the enzyme activity in a 10–15 ml volume. This is then concentrated to 5 ml by vacuum dialysis in an S&S collodion membrane bag against a buffer containing 20 m$M$ imidazole·HCl, 20 m$M$ NH$_4$Cl, and 1 m$M$ EDTA, pH 7.0.

*Step 4. Bio-Gel Column.* Fifteen grams of Bio-Gel P-300, 50–100 mesh, is equilibrated with the above pH 7.0 buffer, and used to form a column 32 cm × 8 cm². After washing with the same buffer the concentrated enzyme from step 3 is applied to the column. Elution, at a rate of 0.7–1.0 ml/minute, is by the same pH 7.0 buffer. The enzyme elutes after 70–80 ml of buffer has entered the column. The fractions are assayed, and those for which the ratio of enzyme units per milliliter divided by OD$_{280}$ exceeds 15 are pooled. The resulting preparation, usually 25–30 ml, may be concentrated by vacuum dialysis. This preparation can be kept at 4° for several weeks without significant loss, or if 0.5 volume of glycerol is added it can be stored at −20° for several months without loss of activity.

*Comments about the Procedure.* Occasionally a large fraction of the enzyme activity sediments at the first centrifugation in step 1. When this occurs, the lysozyme treatment is repeated on the resuspended pellet. The final pellet should contain a large, fluffy, white, gelatinous layer over a small, compact layer.

Yields of enzyme at step 1 varied from 10 to 30 units per gram of packed cells. The supernatant from this step contains several yellow pigments, some of which pass directly through the DEAE-cellulose column. At step 2 a bright yellow substance, accompanied by the NADH oxidase activity, is eluted from the DEAE column immediately before the dikinase. Dull brownish-yellow materials appear in succeeding fractions, including those containing the enzyme. This color is removed at step 3. EDTA is omitted from the buffers used in step 3. In assaying fractions from the hydroxyapatite column EDTA is added to the assay mixture, to 1 m$M$ concentration. This reverses an inhibition presumably due to traces of calcium ion eluted from the hydroxyapatite.

The interfering activities of NADH oxidase, enolase, and adenylate

PURIFICATION OF PYRUVATE, ORTHOPHOSPHATE DIKINASE
FROM 9.5 G OF FROZEN CELLS

| Step | Enzyme (units) | Protein (mg) | Specific activity (units/mg protein) | Recovery (%) |
|---|---|---|---|---|
| 1. Crude enzyme | 300 | 1010 | 0.3 | (100) |
| 2. DEAE-cellulose | 220 | 75 | 3.4 | 73 |
| 3. Hydroxyapatite | 201 | 24 | 8.4 | 67 |
| 4. Bio-Gel | 138 | 11.5 | 12.0 | 46 |

kinase are removed by steps 2, 3, and 4, respectively. These activities amount to less than 0.01% that of the dikinase in the final product.

A typical purification is shown in the table.

## Properties

*Homogeneity.* Activity is proportional to absorbancy at 280 nm across the Bio-Gel enzyme peak. Enzyme from the Bio-Gel column gives a single band on SDS-acrylamide gel electrophoresis,[10] and its estimated molecular weight is 94,000. One milligram of assayed protein per milliliter gives an absorbancy of 1.42 at 280 nm.

*Inhibitors.* The dikinase is inhibited by oxalate with an apparent $K_i$ of 10 $\mu M$ under the conditions of the standard assay. ITP and CTP strongly inhibit the reaction with ATP in the direction of phosphoenolpyruvate formation. In the other direction no tested 5′-nucleoside monophosphate inhibited the reaction with AMP.[5] Calcium ions are strongly inhibitory.

*Specificity.* The dikinase is specific for the adenylate nucleoside mono- and triphosphates.

*Kinetic Values for Substrates and Activators.* Published kinetic values[5] for $K_m$ in the direction of phosphoenolpyruvate formation are $P_i$, 0.6 m$M$; ATP, 0.1 m$M$; and pyruvate, 80 $\mu M$. In the direction of pyruvate formation the values are AMP, 3.5 $\mu M$; $PP_i$, 0.1 m$M$; and phosphoenolpyruvate, 60 $\mu M$. Monovalent cations K$^+$ or NH$_4^+$ are required for activity. Their $K_a$ values are 20 and 2.5 m$M$, respectively.

*pH Optimum.* In the written direction the enzyme has optimum activity at pH 7.2–8.0. In the other direction optimum activity occurs at pH 6.4.

[10] See this series, Vol. 26 [1].

## [32] Pyruvate, Orthophosphate Dikinase from *Acetobacter xylinum*

*By* Moshe Benziman

$$\text{Pyruvate} + \text{ATP} + P_i \xrightleftharpoons{\text{Mg}^{2+}} \text{PEP} + \text{AMP} + PP_i$$

**Assay Methods**

*Method 1*

*Principle.* Pyruvate, orthophosphate dikinase activity in the forward direction (formation of PEP) is measured as the ATP-dependent formation of PEP from pyruvate and $P_i$. The PEP in deproteinized reaction mixtures is determined by the oxidation of NADH in the presence of pyruvate kinase and lactate dehydrogenase. To prevent the reverse reaction from occurring, assay systems are supplemented with exogenous pyrophosphatase or adenylate kinase. The carboxylation of the formed PEP to oxaloacetate, in crude extracts, is prevented by the addition of succinate, which strongly inhibits PEP-carboxylase activity.[1] This assay method is equally applicable to crude and purified preparations.

*Reagents*

Tris·$H_2SO_4$, 1 $M$, pH 8.2
Sodium pyruvate, 0.1 $M$
$MgCl_2$, 0.1 $M$
ATP, 0.05 $M$, pH 7.0
Sodium succinate, 0.1 $M$, pH 8.2
Potassium phosphate, 50 m$M$, pH 8.2
NADH, 10 m$M$
KCl, 1 $M$
ADP, 10 m$M$, pH 7.0
Pyrophosphatase, crystalline (200 units/ml)
Lactate dehydrogenase, crystalline (360 units/ml)
Pyruvate kinase, crystalline (300 units/ml)
Enzyme, diluted with 5 m$M$ Tris·$H_2SO_4$ buffer (pH 7.4) containing 1 m$M$ $MgCl_2$, 5 m$M$ EDTA, and 0.5 m$M$ dithiothreitol (TMED-buffer) to obtain 0.5–2.0 units (forward direction) per milliliter (for definition of units, see below)

[1] M. Benziman, *J. Bacteriol.* **98**, 1005 (1969).

*Procedure.* Reagents are pipetted into test tubes as follows: Tris·$H_2SO_4$ buffer, 0.10 ml; $MgCl_2$, 0.10 ml; Na succinate, 0.20 ml; ATP, 0.10 ml; K phosphate, 0.20 ml; pyrophosphatase, 0.01 ml; diluted enzyme preparation, 0.05 ml, and distilled water to make a final volume of 1.00 ml. ATP is omitted from the control tube. The tubes are equilibrated at 30° and the reaction is started by the addition of 0.02 ml of the pyruvate solution. After incubation for 5 min the mixtures are placed in a boiling water bath for 1 min, to stop the reaction and then cooled in ice. After centrifugation, 0.2-ml samples of the supernatant solutions are transferred to silica cuvettes (10 mm light path) containing 0.1 ml Tris·$H_2SO_4$ buffer, 0.05 ml $MgCl_2$, 0.05 ml KCl, 0.05 ml ADP, 0.05 ml NADH, 0.01 ml lactate dehydrogenase, and water to give a final volume of 1.0 ml. Absorbancy is measured at 340 nm. When absorbancy readings become constant, 0.01 ml of the pyruvate kinase suspension is added and the change in absorbancy is followed until no further change occurs (about 5 min). The difference between the initial and final absorbancies is used to calculate the amount of PEP present in the sample.

Under these conditions, the rate of PEP formation is linear with respect to time up to an incubation period of approximately 10 min, and directly proportional to the amount of enzyme up to 0.07 unit of pyruvate, orthophosphate dikinase.

### Method 2

*Principle.* The activity of pyruvate, phosphate dikinase in the reverse reaction (formation of pyruvate) is measured as the $PP_i$-dependent formation of pyruvate from PEP and AMP. This is assayed spectrophotometrically in the presence of lactate dehydrogenase, by measuring the rate of decrease in absorbancy at 340 nm concomitant with the oxidation of NADH. The oxidation of NADH in the absence of $PP_i$ is used as a control, and this absorbancy change is subtracted from the observed value for the complete system. With crude extracts, where the correction for the control is large, the assay is done in a stepwise manner. The dikinase reaction is allowed to proceed for 5 min in the absence of NADH, and lactate dehydrogenase and is stopped by heating the reaction mixture for 1 min in a boiling water bath. After centrifugation the pyruvate in the supernatant solution is determined with lactate dehydrogenase.

### Reagents

Imidazole·HCl, 0.5 $M$, pH 7.0
PEP (tricyclohexylammonium salt), 0.1 $M$, pH 7.0
AMP, 0.02 $M$, pH 7.0

Sodium pyrophosphate, 0.1 $M$, pH 7.0
MgCl$_2$ 0.1 $M$
NADH, 0.01 $M$
Lactate dehydrogenase, crystalline (360 units/ml)
Enzyme, diluted with TMED-buffer (pH 7.4) to obtain 0.1–1.0 unit
    (reverse direction) per milliliter (for definition of units, see
    below)

*Procedure.* Reagents are pipetted into silica cuvettes (10 mm light path) as follows: imidazole·HCl buffer, 0.10 ml; PEP, 0.02 ml; MgCl$_2$, 0.10 ml; AMP, 0.10 ml; NADH, 0.02 ml; diluted enzyme preparation, 0.05 ml; lactate dehydrogenase, 0.01 ml and water to make a final volume of 1.00 ml. A blank cuvette contains all components with the exception of NADH. Any changes in absorbancy at 340 nm, which may be caused by NADH-oxidase present in the enzyme preparation, are recorded for 1–2 min. The reaction is then started by the addition of 0.01 ml of the PP$_i$ solution and the linear rate of decrease in absorbancy at 340 nm is recorded thereafter. The control value is subtracted from the rate observed after the addition of PP$_i$. The reaction is carried out at 20–22°.

Pyruvate formation, when coupled to the lactate dehydrogenase system is linear with time and with enzyme concentration up to 0.05 unit of pyruvate, orthophosphate dikinase.

## Method 3

*Principle.* The pyruvate phosphate dikinase reaction in the reverse direction can also be assayed spectrophotometrically by measuring the rate of ATP formation as the increase in absorbancy at 340 nm in a reaction coupled with hexokinase and glucose-6-P dehydrogenase. This assay cannot be used with crude extracts because of their NADPH-oxidase activity. It can, however, be used satisfactorily after step 1 of the purification procedure.

### Reagents

Imidazole·HCl, 0.5 $M$, pH 7.0
PEP (tricycloammonium salt) 0.1 $M$, pH 7.0
AMP, 0.02 $M$, pH 7.0
Sodium pyrophosphate, 0.1 $M$, pH 7.0
MgCl$_2$ 0.1 $M$
NADP, 10 m$M$
Glucose, 0.1 $M$
Hexokinase, crystalline (400 units/ml)

Glucose-6-phosphate dehydrogenase (500 units/ml)

Enzyme, diluted with TMED-buffer (pH 7.4) to obtain 0.1–1.0 unit (reverse direction) per milliliter (for definition of units, see below)

*Procedure.* Reagents are pipetted into a silica cuvette (10-mm light path) as follows: imidazole·HCl buffer, 0.10 ml; PEP, 0.02 ml; MgCl₂, 0.1 ml; AMP, 0.10 ml; NADP, 0.03 ml; glucose, 0.05 ml; hexokinase, 0.01 ml; glucose-6-P dehydrogenase 0.01 ml; diluted enzyme preparation 0.05 ml, and water to a final volume of 1.00 ml. The reaction is started by adding 0.01 ml of the $PP_i$ solution and is followed by the increase in absorbancy at 340 nm. The reaction is carried out at 20–22°.

ATP formation, when coupled to the hexokinase-glucose-6-P dehydrogenase system, is linear with respect to time and is directly proportional to enzyme concentrations up to 0.05 unit of pyruvate, phosphate dikinase.

*Units.* Units of enzyme activity are based either on the forward or on the reverse reaction when carried out under the conditions described. When measured by the forward reaction, one enzyme unit is defined as the amount of enzyme that will catalyze the formation of 1 μmole of PEP per minute. When measured by the reverse reaction, one enzyme unit is defined as the amount that will catalyze the formation of 1 μmole of pyruvate or ATP per minute. Units for the forward and reverse reaction are not equivalent.

Specific activity is expressed as units per milligram of protein. Protein is determined by the colorimetric method of Lowry et al.[2] with crystalline bovine serum albumin as standard.

## Other Assay Methods

Various other assay methods for this enzyme have been reported. The reaction proceeding toward PEP synthesis has been followed spectrophotometrically by measuring the increase in absorbancy at 340 nm in a coupled system in which lactate, NAD, and lactate dehydrogenase were substituted for pyruvate.[3] This method, however, restricts the concentration of pyruvate in the assay mixture. The forward reaction has also been measured either spectrophotometrically by coupling with PEP-carboxytransphosphorylase and malate dehydrogenase,[4] or by measuring $^{14}CO_2$ incorporation into aspartate in the presence of PEP carboxylase,

[2] O. H. Lowry, N. J. Rosebrough, A. L. Farr, and R. J. Randall, *J. Biol. Chem.* **193**, 265 (1951).

[3] M. Benziman and A. Palgi, *J. Bacteriol.* **104**, 211 (1970).

[4] H. J. Evans and H. G. Wood, *Biochemistry* **10**, 721 (1971).

aspartate aminotransferase, and glutamate.[5] Enzyme activity in the reverse reaction has been assayed by measuring the formation of pyruvate determined colorimetrically as the dinitrophyenyl hydrazine derivative.[6] Titrimetric methods have been described based on hydrogen ion release in the forward reaction or on its uptake in the reverse reaction.[7]

## Purification Procedure

*Growth of Organism.* The organism used in this study was the cellulose-synthesizing *A. xyllinum* wild type employed in earlier investigations reported from this laboratory.[8] It was grown on succinate and harvested as described elsewhere in this series.[9]

All the following steps were carried out at 15–25°.

*Step 1. Preparation of Crude Extract.* Cells were suspended in 5 m$M$ Tris·$H_2SO_4$—1 m$M$ $MgCl_2$–5 m$M$ EDTA–0.5 m$M$ dithiothreitol buffer (TMED buffer) at pH 7.4, to give a final concentration of 30 mg (dry weight) of cells per milliliter. Portions (25 ml) of this suspension were treated for 15 min in a Raytheon Model DF 101 magnetorestrictive oscillator at 200 W and 10 kHz/sec. The sonic extract was then centrifuged at 18,000 $g$ for 15 min, and the precipitate was discarded.

*Step 2. High Speed Centrifugation.* The crude extract was centrifuged in a Beckman Model L-2 ultracentrifuge at 150,000 $g$ for 60 min. The resulting sediment was discarded.

*Step 3. DEAE-Cellulose Chromatography.* The high speed supernatant fraction was applied to a 1.5 × 20 cm DEAE-cellulose (Whatman DE-52) column, preequilibrated with TMED-buffer, pH 7.4. The column was eluted with 80 ml of 0.15 $M$ KCl in TMED buffer pH 7.4 at a rate of 90 ml/hr to remove undesired proteins. Dikinase activity is not eluted by this treatment. The concentration of KCl was next increased to 0.25 $M$, and 3-ml fractions were collected at a flow rate of 40 ml/hr. The peak of enzymic activity appeared after 15 ml of eluate were collected. Eighty-five percent of the original activity was recovered in three fractions, which were combined.

*Step 4. Ammonium Sulfate Fractionation.* Solid ammonium sulfate to 35% saturation (20.9 g/100 ml) was added to the combined fractions from the previous step. The salt was added slowly with stirring, and 15 min was allowed for precipitation. The precipitate was removed by cen-

[5] M. D. Hatch and C. R. Slack, *Biochem. J.* **106**, 141 (1968).
[6] Y. Milner and H. G. Wood, *Proc. Nat. Acad. Sci. U.S.* **69**, 2463 (1972).
[7] R. E. Reeves, *Biochem. J.* **125**, 531 (1971).
[8] M. Schramm and S. Hestrin, *J. Gen. Microbiol.* **11**, 123 (1954).
[9] M. Benziman, this series, Vol. 13 [22].

PURIFICATION PROCEDURE FOR PYRUVATE ORTHOPHOSPHATE DIKINASE

| Step and fraction | Volume (ml) | Protein (mg) | Activity (units)[a] | Specific activity[b] | Yield (%) |
|---|---|---|---|---|---|
| 1. Crude extract | 70 | 1000 | 19 | 0.019 | 100 |
| 2. Supernatant of high speed centrifugation | 70 | 450 | 18 | 0.040 | 95 |
| 3. DEAE-cellulose pooled fractions | 9 | 30 | 16 | 0.520 | 85 |
| 4. (NH₄)₂SO₄ precipitate (0.35 to 0.55 saturation) | 7 | 15 | 13 | 0.880 | 68 |

[a] One unit is the amount of enzyme catalyzing the formation of 1 $\mu$mole of PEP per minute under the conditions of Assay Method 1.
[b] Units per milligram of protein.

trifugation at 15,000 $g$ for 15 min and discarded. The supernatant solution was then brought to 55% saturation (12.9 g/100 ml), stirred, and centrifuged as before. The precipitate was dissolved in a small volume of TMED buffer pH 7.4. This fraction had a specific activity of 0.88 in the forward reaction and 0.36 in the reverse reaction when assayed under standard conditions. It represents a 46-fold purification over the crude extract in 68% yield as measured by activity in both the forward and the reverse directions. A summary of the purification procedure is presented in the accompanying table.

## Properties

The 46-fold purified kinase is free of pyruvate kinase, PEP carboxylase, adenylate kinase, enolase and contains only traces of ATPase and pyrophosphatase activities. Crude extracts lost 65% of their pyruvate, phosphate dikinase activity in about 5 hr when kept at 4°. The purified enzyme retained about 85% of its activity for at least 2 weeks when kept at room temperature.

*Stoichiometry.* One mole of PEP, AMP, and $P_i$ is formed for each mole of pyruvate, ATP, and $PP_i$ utilized in the reaction.

*Effect of pH.* The optimum pH for PEP synthesis is pH 8.2; for the reverse reaction it is pH 6.5.

*Ratio of Activities of the Forward to Reverse Reaction.* When the enzyme is assayed in both directions at 30° the ratio of the rate of the forward to the reverse reaction is 5.1 at pH 8.2 and 0.45 at pH 6.5.

*Specificity.* GTP, CTP, ITP, or TTP at 10 m$M$ concentrations cannot replace ATP in the reaction with pyruvate. Arsenate can be substituted for phosphate. At equimolar concentrations, however, the rate of PEP formation in the presence of arsenate is 65% of that obtained with phosphate. In the presence of 5 mM phosphate the further addition of 5 mM arsenate does not affect activity. GMP, CMP, and ADP (5 mM each) do not substitute in the reverse reaction for AMP.

*Metal Ion Requirement.* $Mg^{2+}$ is required for activity. $Mn^{2+}$ or $Ca^{2+}$ at 5 mM concentrations cannot replace $Mg^{2+}$ in either the forward or the reverse reaction.

*Substrate Affinity Constants.* The apparent $K_m$ values for the components of the system are as follows: Forward reaction, pyruvate, 0.2 mM; ATP, 0.4 mM; $P_i$, 0.8 mM; $Mg^{2+}$, 2.2 mM; reverse reaction, PEP, 0.1 mM; AMP, 1.6 $\mu M$; $PP_i$ 0.06 mM; $Mg^{2+}$, 0.87 mM.

*Inhibitors.* PEP formation is competitively inhibited by AMP with respect to ATP ($K_i = 0.2$ mM). ADP, GMP, and CMP at 1–3 mM concentration do not affect the forward reaction. $PP_i$ inhibits PEP formation in assay systems in which excess adenylate kinase is substituted for pyrophosphatase; 0.5 mM $PP_i$ giving 50% inhibition. PEP at concentrations up to 2 mM does not affect the forward reaction. The reverse reaction is competitively inhibited by ATP with respect to AMP ($K_i = 0.22$ mM). The reverse reaction is also inhibited to the extent of 65% by 2 mM pyruvate. $P_i$ up to concentration of 10 mM does not affect the reverse reaction. The enzyme is inhibited by $p$-hydroxymercuribenzoate. Within the range of enzyme activities used, preincubation at 25° for 5 min in the presence of 2 $\mu M$ PHMB brought about an inhibition of approximately 50%. The inhibition may be reversed by dithiothreitol (1 mM), or glutathione (5 mM). Potassium fluoride (50 mM) inhibits the forward and reverse reaction to the extent of 80% and 60%, respectively. The inhibition of the forward reaction by fluoride is not affected by raising the $P_i$ concentration to 50 mM.

*Physiological Function.*[10] Although the enzyme readily catalyzes the reverse reaction, the major physiological role of pyruvate, phosphate dikinase in *A. xylinum* is gluconeogenic in that it promotes the formation of PEP from pyruvate. Dikinase formation is induced on the transfer of glucose-grown cells of *A. xylinum* to succinate- or pyruvate-containing media. If such media are supplemented with glucose, fructose, or glycerol, enzyme formation is suppressed by 96%, 76%, and 67%, respectively. When glucose is added to cultures of cells growing on succinate, the growth of the organism is not affected, but is accompanied by a virtual cessation of dikinase synthesis. The changes in the pyruvate, phosphate dikinase activity of cells, are accompanied by parallel changes in the ability of cells to convert pyruvate into cellulose. Gluconeogenesis in this

[10] M. Benziman and N. Eizen, *J. Biol. Chem.* **246,** 57 (1971).

organism appears to be subject to control exerted on the synthesis of the dikinse. The reciprocal effect of AMP and ATP on the activity of the dikinase may have physiological implications in the regulation of gluconeogenesis and carbohydrate oxidation in *A. xylinum*.[11,12]

*Distribution.* In addition to cellulose-forming *A. xylinum* cells, the enzyme has been found in celluloseless mutants of this organism.[10] in photosynthetic grasses[5] in propionibacteria,[4,6] in *Endamoeba histolytica* and *Bacteroides symbiosus*.[13] Whereas in the former three sources the physiological role of the enzyme is considered to be gluconeogenic,[5,10,14] a function in a glycolytic direction has been ascribed to it in *E. histolytica* and *B. symbiosus*.[13]

[11] H. Weinhouse and M. Benziman, *Biochem. Biophys. Res. Commun.* **43**, 233 (1971).
[12] H. Weinhouse and M. Benziman, *Eur. J. Biochem.* **28**, 83 (1972).
[13] R. E. Reeves, R. A. Menzies, and D. S. Hsu, *J. Biol. Chem.* **243**, 5486 (1968).
[14] H. J. Evans, and H. G. Wood, *Proc. Nat. Acad. Sci. U.S.* **61**, 1448 (1968).

## [33] Pyruvate, Orthophosphate Dikinase[1] of *Bacteroides symbiosus* and *Propionibacterium shermanii*

*By* YORAM MILNER, GEORGE MICHAELS, and HARLAND G. WOOD

$$\text{Pyruvate} + \text{ATP} + \text{P}_i \underset{\longleftarrow}{\overset{\text{Mg}^{2+}, \text{NH}_4^+}{\longrightarrow}} \text{P-enolpyruvate} + \text{AMP} + \text{PP}_i \qquad (1)$$

### Assay Methods

*Principle.* The assay measures the formation of pyruvate in the reverse of Reaction (1) by the formation of the hydrazone derivative, which can be determined spectrophotometrically. This assay is used in our laboratory routinely coupled with controls without added AMP and occasionally without added pyrophosphate.

*Reagents*

A. Mixture A which contains (in $\mu$moles/ml): imidazole·HCl (Sigma) buffer (pH 6.7), 80; $(NH_4)_2SO_4$, 40; P-enolpyruvate (tricyclohexyl ammonium salt, Sigma), 4.8, and $MgCl_2$, 24. The mixture is prepared in 30-ml lots and kept frozen until used.

B. AMP solution (Sigma disodium salt), 60 m$M$, pH = 6.7. Stored frozen.

C. Sodium pyrophosphate buffer, 30 m$M$, pH = 6.7, kept at 4°.

[1] [ATP:pyruvate, orthophosphate phosphotransferase, EC 2.7.9.1].

D. Dinitrophenyl hydrazine, 0.1%, in 2 $N$ HCl. The reagent is stable for a few months at room temperature if kept in a brown bottle.

E. HCl, 2 $N$

F. NaOH, 10% solution

*Assay Procedure.* Properly diluted enzyme in 0.05 ml is added to test tubes and then 0.1 ml of pyrophosphate solution (C) and 0.05 ml of AMP solution (B) and the reaction is initiated by adding 0.3 ml of mixture A and mixing. Incubation is at 25° for 5–30 min. (Pyrophosphate or AMP is omitted and replaced by equivalent amounts of water in controls.)

The reaction is terminated by adding 0.3 ml of dinitrophenyl hydrazine reagent (D), mixing and after 5 min at 25°, 0.8 ml of NaOH is added, and then 3.4 ml water and the optical density is read in a colorimeter or spectrophotometer at 420 nm. A standard curve is constructed using all ingredients of the assay except enzyme and containing 0.05–0.35 $\mu$mole of sodium pyruvate (Sigma) freshly prepared in 0.01 $N$ HCl containing 1 m$M$ EDTA. A value of 130 Klett units (Klett-Summerson colorimeter filter 42) is usually observed per 0.1 $\mu$mole of pyruvic acid.

An assay also can be done by the dinitrophenyl hydrazine method without adding NaOH. Instead 0.2 ml 2 $N$ HCl solution is added and the optical density is read directly in a spectrophotometer at 415 nm after 10 min incubation. In this determination $\epsilon_{415} = 1.1 \times 10^7$ OD/mole per liter.

Various other determinations have been used. The forward reaction has been measured by coupling the reaction with carboxytransphosphorylase (EC 4.1.1.38) and malate dehydrogenase[2] and the back reaction has been measured by coupling with lactate dehydrogenase and DPNH.[2,3] A titrimetric assay has been employed[4] for assaying the dikinase in either direction.

An enzyme unit is the quantity of enzyme that produces 1 $\mu$mole of pyruvate per minute under the conditions given above.

### Purification of Pyruvate, Phosphate Dikinase

The enzyme has been found in *Propionibacterium shermanii*,[5] *Bacteroides symbiosus*,[6] and *Entamoeba histolytica*,[7] *Acetobacter xylinum*,[3]

[2] H. J. Evans and H. G. Wood, *Biochemistry* **10**, 721 (1971).
[3] M. Benziman and A. Palgi, *J. Bacteriol.* **104**, 211 (1970).
[4] R. E. Reeves, *Biochem. J.* **125**, 531 (1971).
[5] H. J. Evans and H. G. Wood, *Proc. Nat. Acad. Sci. U.S.* **61**, 1448 (1968).
[6] R. E. Reeves, R. A. Menzies, and D. S. Hsu, *J. Biol. Chem.* **243**, 5486 (1968).
[7] R. E. Reeves, *J. Biol. Chem.* **243**, 3202 (1968).

and in tropical grasses.[8] Purification of the enzyme from *Bacteroides symbiosus* and from *Propionibacterium shermanii* will be described here.

*Growth of Cells.* *Bacteriodes symbiosus* ATCC 14940[9] was kept at 4° in culture tubes and revitalized before use by 3–4 transfers at 37°. Media for all growth contained the following components:

Casein hydrolysate (peptone No. 50–General Biochemicals, Cleveland), 2%

  Yeast extract (microbiological grade—General Biochemicals, Cleveland), 0.2%

  $KH_2PO_4$ (anhydrous), 0.15%

  NaCl, 0.25%

  Mercaptosuccinic acid (Eastman Organic-Technical), 0.15%

  Glucose (Nutritional Biochemical Company), 2%

The mixture was brought to pH 7.0 with NaOH solution before sterilization.

*Preparation of Inoculum.* Three-liter portions of the medium were autoclaved in 6-liter Erlenmeyer flasks for 30 min, 122°, 15 psi. The contents of two flasks were combined into one flask and inoculated with 2–3 tubes of active culture and incubated at 37° for 15 hr. The flasks were stoppered with cotton plugs capped with aluminum foil.

*Preparation of Fermenter and Inoculation.* A Fermacell Model CF-130 Fermenter (120-liter capacity, New Brunswick Scientific) was used without the drive shaft and agitator. The fermenter containing 90 liters of distilled water was steam-sterilized for 90 min. It was then cooled to 37° by passing cold water through the jacket while 50 liters of helium was blown through the solution per hour at 10 psi. Then 6 liters of inoculum, 4 liters of sterile glucose solution, 9 liters of the other sterilized nutrients, and 50 ml of sterile antifoam solution (Dow Chemical FG-10) were introduced aseptically through a port in the lid. The passage of helium through the solution was increased to 100 liters/hr during these additions to decrease the chances of contamination while introducing materials through the port. The pH was maintained at a constant value by a peristaltic pump using 10 N NaOH and a pH control unit (Leeds and Northrop, 117143). The optimal pH for cell growth was found to be between 6.2 and 6.4, and after the second batch (Fig. 1) the pH was maintained in this range. During the growth of the cells 5 liters of helium per hour were bubbled through the medium to ensure anaerobiosis and proper agitation. The growth was followed by measurement of the turbidity in a Klett-Summerson colorimeter with a No. 66 filter. As soon

[8] M. D. Hatch and C. R. Slack, *Biochem. J.* **106**, 141 (1968).
[9] We wish to express our thanks to Dr. R. E. Reeves for the initial culture and valuable advice concerning its growth and purification of the enzyme.[4]

as the end of the logarithmic growth phase was reached, the medium was removed and cooled to 4° in a 200-liter stainless steel vat equipped with a cooling coil. About 10 liters of culture were left in the fermenter as an inoculum for the subsequent fermentation. The cooled culture was centrifuged in a Sharples centrifuge (40,000 rpm, 20 liters per hour). The cell mass was collected and stored at —15°, until used (within 2 weeks). Eighty liters of sterile distilled water supplied from a 200-liter sterilizer vat, 4 liters of sterile glucose solution, 9 liters of sterile nutrients, and 50 ml of antifoam solution were added as before. This fermentation process was repeated 7 times (Fig. 1).

One hundred Klett units are equivalent to 0.63 g of wet packed cells per liter. An average of 13 units of enzyme was recovered in the crude extract per gram of wet cells. The overall yield of cells was 4300 g wet weight from the 7 batches.

*Lysis of Cells and Streptomycin Treatment.* The frozen cells were divided into two approximately equal parts and treated on consecutive days as described below. After thawing at 37° with vigorous mechanical stirring in approximately 25 liters of buffer consisting of 20 m$M$ imidazole·HCl (pH 6.8) 75 m$M$ KCl, 2.5 m$M$ EDTA, and 3 m$M$ $\beta$-mercaptoethanol; lysozyme (Worthington) 1 mg per gram of wet cells was added in about 100 ml of the above buffer, and the stirring was continued for about 2 hr. Assay at various times during the lysozyme treatment indicated that complete lysis occurs after about 1 hr. Then per gram cells 0.1 g of streptomycin (Sigma) dissolved in several hundred milliliters of buffer was added over a period of 15–20 min, and the stirring was continued for 30 more minutes. The lysate was then placed in the cold

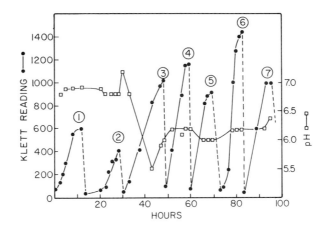

Fig. 1. Semicontinuous growth of *Bacteroides symbiosus* in a 120-liter fermenter.

TABLE I

SUMMARY OF THE PURIFICATION OF PYRUVATE, ORTHOPHOSPHATE
DIKINASE FROM *Bacteroides symbiosus*

| Step | Fraction | Total volume (liters) | Protein (g) | Total units $\times 10^{-4}$ | Specific activity[c] | Yield (%) |
|------|----------|------------------------|-------------|-------------------------------|------------------------|-----------|
| 1 | Lysate | 64.00 | 703[a] | 9.60 | 0.14 | 100 |
| 2 | Crude extract | 61.00 | 506[a] | 9.50 | 0.19 | 99 |
| 3 | First chromatography on DEAE cellulose | | | | | |
| | Fraction I | 9.5 | 24.6[b] | 6.00 | 2.43 | 62 |
| | Fraction II | 14.5 | 25.0[b] | 0.75 | 0.3 | 8 |
| 4 | Second chromatography on DEAE cellulose | | | | | |
| | Fractions D, E, F | 2.70 | 4.03[b] | 3.77 | 9.4 | 40 |
| | Fractions C, G | 2.70 | 3.23[b] | 1.61 | 5.0 | 17 |
| | Fractions B, H | 2.90 | 3.07[b] | .93 | 2.9 | 10 |
| 5 | Chromatography on Bio-Gel A 1.5 | 2.5 | 2.00[b] | 3.22 | 16.1 | 35 |
| 6 | Reverse $(NH_4)_2SO_4$ extraction | 1.3 | 0.96[b] | 2.41 | 25 | 25 |
| 7 | Chromatography on Bio-Gel A 0.5 | 1.5 | 0.55[b] | 1.92 | 35 | 20 |

[a] Protein was determined by biuret reaction because of presence of nucleic acids.
[b] Protein was determined by absorption at 280 nm assuming 1 OD equals 1 mg/ml (see comments on purification).
[c] Specific activity = units per milligram of protein.

room, and all subsequent purification steps were done at 4°. The lysate was centrifuged at 10,000 rpm for 20 min in the GSA head of a Sorvall RC-2B centrifuge at 0–5°, and the supernatant was centrifuged for 60 min under the same conditions for complete clarification of the extract. The cellular debris from the first centrifugation was resuspended in 10 liters of the buffer containing additional lysozyme and stirred overnight. It was combined with a similar fraction from the second half of the cells and then centrifuged. The three supernatant fractions from the 4.3 kg of bacteria were combined to give the crude extract fraction (see Table I).

*First DEAE-Cellulose Column.* The crude extract, which had a conductivity of 19 mS[10] (Radiometer, CD-8 conductivity meter at 25° setting), was pumped by air pressure (5–7 psi) from a 20-liter carboy directly onto a 12 × 105 cm plexiglass column containing 2 kg of DEAE-cellulose

[10] S (Siemens) = ohm$^{-1}$.

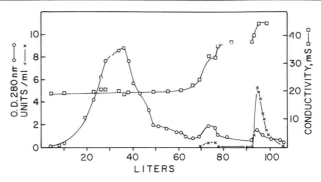

Fɪɢ. 2. Elution profile of pyruvate, orthophosphate dikinase from the first DEAE-cellulose column.

(fibrous, Schleicher and Schull, 0.8 meq/g) equilibrated with 20 m$M$ imidazole·HCl buffer (pH 6.8) containing 2.5 m$M$ EDTA, 0.6 m$M$ $\beta$-mercaptoethanol and brought to 20 mS with KCl, giving a concentration of 88 m$M$ KCl. (All subsequent purifications were done in the same buffer with varying conductivity obtained by variation of the concentration of KCl.) Some difficulty was encountered during the loading of the large amount of extract onto this column in that the flow rate became very slow. The column was therefore unpacked, washed with 30 liters of 20 mS buffer, and then repacked. Effluent fractions of 3–5 liters were collected manually and the protein, enzymic activity and conductivity were monitored. The column was washed with 50 liters of buffer containing enough KCl to bring the conductivity to 21 mS until the OD at 280 nm dropped to a low value, then with 22 liters of buffer containing KCl of 31.4 mS (160 m$M$ KCl), which gave a small protein peak, and finally with 24 liters of buffer with KCl of 44 mS (250 m$M$ KCl). The enzyme began to elute at 33 mS, and effluent from 115–126 liters was pooled, precipitated at 80% saturation,[11] with (NH$_4$)$_2$SO$_4$, centrifuged, then dissolved in 700 ml of buffer and dialyzed overnight at 4° against 20 liters of buffer containing no KCl. The specific activity was 2.43 (fraction 1, Table I). Some side portions (fraction 2) were not included in the main pool, but saved for purification at some future time. Figure 2 shows the elution profile of an identical column in which no difficulty was encountered. In this case, extract from a smaller batch of cells was used (1.7 kg). The specific activity of the resultant preparation was 2.08 compared to the 2.43 of Table I and the elution profile was quite similar, except that the enzyme emerged from the column at slightly higher conductivity.

[11] This series, Vol. I [1] Table I. No temperature correction was used.

*Second Chromatography on DEAE-Cellulose.* The dialyzate and then the gradient were pumped at a rate of 300 ml/hr by a syringe type pump (Milton-Roy, Minipump 196–47) onto a $5 \times 63$ cm Pharmacia column (K50/100) packed with microgranular DEAE-cellulose (Whatman's DE-32) equilibrated with 21 mS buffer (88 m$M$ KCl). The gradient was 10 liters of buffer with 48 mS of KCl flowing into 10 liters of buffer with 21 mS of KCl. Fractions of 20 ml each were collected by an LKB Fraction Collector. Portions of the effluent were pooled as soon as the assay was complete and precipitated in 80% saturated $(NH_4)_2SO_4$ to minimize dilution inactivation. Fractions designated D, E, and F had an average specific activity of 9.4, C and G, 5.0; and H 2.9 (Table I).

*Chromatography on Bio-Gel A 1.5.* The precipitates from D, E, and F and from C and G were dissolved in a minimal volume of buffer (usually 40–50 ml) and the enzyme was purified in aliquots of 1.5–2 g by layering on top of a $4.4 \times 158$ cm Bio-Gel column ($V_t = 2350$ ml and $V_0 = 900$ ml). The column was equilibrated and eluted with buffer containing 25 mS of KCl. The enzyme emerges at $V_e/V_0 = 1.6$–1.66 (enzyme eluted between 1400 and 1700 ml) and was precipitated immediately with 80% saturated $(NH_4)_2SO_4$. The average yield and specific activity from these columns is shown in Table I.

*Reverse Ammonium Sulfate Extraction on a Cellulose Column.* About 1 g of the ammonium sulfate precipitate was mixed with 20 g of cellulose powder (LKB) in 50–60 ml of buffer solution 80% saturated with $(NH_4)_2SO_4$. The suspension was placed on a $4 \times 30$ cm column of cellulose powder equilibrated with buffer 80% saturated with $(NH_4)_2SO_4$. After the suspension settled, a small amount of cellulose suspended in buffer 80% saturated with $(NH_4)_2SO_4$ was placed on top of the cellulose–enzyme layer and the column was connected to a linear gradient composed of 1500 ml of buffer, 30% saturated with $(NH_4)_2SO_4$ flowing into 1500 ml of buffer 80% saturated with $(NH_4)_2SO_4$.

The elution profile is shown in Fig. 3. Usually the enzyme eluted between 1700 and 2200 ml [between 70% and 60% saturated $(NH_4)_2SO_4$], and the fractions were pooled and precipitated with 80% saturated $(NH_4)_2SO_4$. The first fractions emerging had the highest specific activity enzyme, and the trailing lower activity was precipitated and fractionated again on the same column after combining trailing edges from several columns. Results summarized in Table I give the average purification and total yields from 4 columns.

*Gel Chromatography on Bio-Gel A 0.5.* Portions of the $(NH_4)_2SO_4$ precipitates from the above step were centrifuged at 10,000 cpm (Sorvall RC-2B), and the precipitates were dissolved in a minimal volume of buffer (usually 5–10 ml). About 365 mg of the dissolved enzyme was

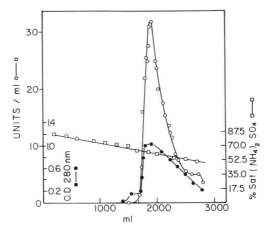

Fig. 3. Reverse ammonium sulfate extraction of pyruvate, orthophosphate dikinase from a cellulose column.

then chromatographed on a Pharmacia column packed with Bio-Gel
A 0.5 (200–400 mesh, $V_t$ = 1500 ml) equilibrated with buffer containing 18 mS KCl. The enzyme emerges at $V_e/V_o$ = 1.15–1.25. Since
at this stage all the contaminants are of smaller molecular weight than
the enzyme, the ascending limb of the elution profile emerges as pure
protein while the descending limb has to be rechromatographed on the
on the same column. Table I summarizes the average yield from all
runs performed, and the specific activity was more or less constant at
this stage.

### Comments about Purification and Stability of the Enzyme

*Protein Determination.* The assumption that 1.0 $OD_{280}$ = 1 mg/ml of
protein agreed with the value as calculated by the method of Warburg
and Christian.[12] However, this determination of protein probably is not
correct. Estimation of protein by the biuret method[13] (BSA standard)
and by the Kjeldahl method[14] both yielded values 30–35% higher than
the spectrophotometric method using enzyme preparations of steps 3, 5,
and 7 of Table I. Therefore, the true specific activity was probably 24–25
and the values given in Table I may be about 30% too large.

*Purity.* The enzyme of step 7, Table I, appears to be quite pure. A

[12] O. Warburg and W. Christian, *Biochem. Z.* **310**, 384 (1941).
[13] S. Zamenhof, this series, Vol. 3, p. 396.
[14] Digestion by method of C. Lang, *Anal. Chem.* **30**, 1692 (1958) and nesslerization
was done using Fischer Scientific reagent.

sharp symmetrical sedimenting single peak was obtained on analytical ultracentrifugation, and straight lines of log $[Y(r)-Y_0]$ vs $r^2$ on equilibrium sedimentation by meniscus depletion technique[15] at three different enzyme concentrations. The enzyme was found to be at least 95–98% pure by polyacrylamide gel disc electrophoresis.

*Stability.* The enzyme is very unstable, and it is advisable to work with highly concentrated solutions since the enzyme is sensitive to dilution, particularly at concentrations below 0.3 mg/ml. In one case at 0° (pH 6.8) with 0.06 mg/ml, 50% of the activity was lost in 3 hr. Purification of the enzyme with $\frac{1}{5}$ to $\frac{1}{6}$ the amount of cells used here and without mercaptoethanol gave much lower yields of enzyme of specific activity 12 (which appeared to be 95% pure). Although the reason for the variation of the specific activities is not known, oxidation of essential sulfhydryl groups and "dilution death" may contribute to low specific activity and low yield of the enzyme. The pH optimum for stability is 5.5–7.0, and the enzyme is best preserved as a precipitate in 80% saturated $(NH_4)_2SO_4$ at 0–4° or in solutions with greater than 5 mg/ml bovine serum albumin, the former being better. Enzyme kept as a precipitate displayed full activity after 6 months. Freezing and thawing causes inactivation of the enzyme, and formation of an insoluble precipitate.

*Sedimentation Coefficient and Molecular Weight.* By the sedimentation velocity technique using fresh enzyme, a single peak was obtained, $s_{20,w} = 8$ S. However, with diluted enzyme (partially inactivated) a second peak was observed with a sedimentation constant of $s_{20,w} = 5$ S. Likewise, enzyme treated with pHMB gives two peaks with a rough correlation between the loss of activity by pHMB treatment and amounts of 5 S species appearing in the ultracentrifuge.[16] Molecular weight determination by chromatography on Bio-Gel A 1.5 or Sephadex G-200 gave values of 150,000[16] and equilibrium sedimentation using meniscus depletion in the Yphantis cell technique[15] at three different enzyme concentrations gave a molecular weight of 160,000 (assuming $\bar{v} = 0.74$).[16] Molecular weight determination in SDS acrylamide gels[17] gave a single band migrating as a 75,000 molecular weight species.[16] The enzyme therefore appears to be composed of two subunits of similar molecular weight.

*pH Optimum.* The pH optimum for the forward reaction is 7.0 to 7.8 and for the backward reaction 6.4 to 6.7.

*Specificity and Kinetic Constants.* The only nucleoside monophosphate that was found to substitute for AMP is deoxy AMP and at about 20% of the activity of AMP.[4] No substitute was found for ATP. Oxalate,

[15] D. A. Yphantis, *Biochemistry* 3, 297 (1964).
[16] G. Michaels, Y. Milner, and H. G. Wood, manuscript in preparation.
[17] K. Weber and M. Osborn, *J. Biol. Chem.* 244, 4406 (1969).

## TABLE II
### THE $K_m$'s (m$M$) FOR PYRUVATE, ORTHOPHOSPHATE DIKINASE FROM *Bacteroides symbiosus*

|            | Reeves[a,b] | Authors' laboratory[c] |
|------------|-------------|------------------------|
| ATP        | 0.1         | 0.05                   |
| $P_i$      | 0.6         | 0.8                    |
| Pyruvate   | 0.08        | 0.1                    |
| AMP        | 0.0035      | 0.015                  |
| $PP_i$     | 0.1         | 0.08                   |
| P-enolpyruvate | 0.06    | 0.1                    |

[a] See Reeves *et al.*[4,6,7]

[b] $K_m$'s for ATP, $P_i$, pyruvate were at pH 7.4 and for AMP, $PP_i$, and P-enolpyruvate at pH 6.8.

[c] All $K_m$ values were determined at pH 6.7.

oxaloacetate, or lactate do not substitute for pyruvate. Phosphoglycolate does not substitute for P-enolpyruvate.

Kinetic constants for all the substrates recorded in the forward and back directions as determined in two laboratories are given in Table II. The ratio of the rates of the forward to back reactions was about 3.5 at pH = 6.7.

*Inhibitors.* The enzyme is inhibited by sulfhydryl reagents, such as pHMB, DTNB, NEM, but not iodoacetamide or iodoacetic acid. dATP, ITP, and CTP inhibit the enzyme[4] (probably competitively with ATP). $\alpha$-$\beta$ methylene ATP, $\beta$-$\gamma$ methylene ATP inhibit competitively with ATP. The methylene analog of pyrophosphate (methylene diphosphonate) inhibits competitively with respect to pyrophosphate. Phosphoglycolate inhibits competitively with respect to P-enolpyruvate.

*Mechanism.* The mechanism appears to involve three partial reactions.[4,18]

$$\text{(i) ATP} + \text{enzyme} \overset{\beta\gamma}{\rightleftarrows} \text{enzyme-PP} + \text{AMP}$$

$$\text{(ii) Enzyme-PP} \overset{\beta\gamma}{} + P_i \rightleftarrows \text{enzyme-P} \overset{\beta}{} + PP_i \overset{\gamma}{}$$

$$\text{(iii) Enzyme-P} \overset{\beta}{} + \text{pyruvate} \rightleftarrows \text{enzyme} + \text{P-enolpyruvate} \overset{\beta}{}$$

$$\text{Sum (iv) ATP} + P_i + \text{pyruvate} \rightleftarrows \text{AMP} + PP_i + \text{P-enolpyruvate}$$

The enzyme catalyzes the exchange reaction [$^{14}$C]AMP $\rightleftarrows$ ATP without involvement of other substrates, an ATP or P-enolpyruvate dependent

[18] Y. Milner, G. Michaels, and H. G. Wood, manuscript in preparation.

$[^{32}P]P_i \rightleftarrows PP_i$ exchange, and a $[^{14}C]$pyruvate $\rightleftarrows$ P-enolpyruvate exchange which is independent of other substrates. The rate of the slowest exchange reaction, the $P_i \rightleftarrows PP_i$ exchange, is about 30% of the overall rate of reaction. The exchange rate might be a minimal value because of mutual $P_i$, $PP_i$ inhibition. The mechanism with three partial reactions demands a Tri (Uni Uni) Ping-Pong steady-state kinetic pattern. Preliminary kinetic studies with the enzyme from B. symbiosus have verified this pattern and with the enzyme from the propionibacteria have been fully established.[18-20] Finally, the most crucial test for this mechanism is the direct demonstration of the phosphoryl-enzyme and pyrophosphoryl-enzyme species. These forms have been isolated by gel chromatography with the enzyme from B. symbiosus and demonstrated to contain 4 moles of $^{32}P$ from $[^{32}P]$P-enolpyruvate (reaction iii) and four moles of $^{32}P$ from $[^{32}P]PP_i$ (reaction ii) per mole of enzyme of 150,000 molecular weight.[18]

Metal Requirements. The enzyme has an absolute requirement for divalent metal. The relative activities[16] are: $Mg^{2+}$, 100; $Mn^{2+}$, 30; and $Co^{2+}$, 45 in the backward reaction. In addition, a monovalent cation is required with the following relative activities: $NH_4^+$, 100; $K^+$, 70; $Rb^+$, 25; and $Tl^+$, 94.[16]

Equilibrium Constant. The equilibrium constant is strongly influenced by the pH[6], $K_{eq} = 1.75 \times 10^{-4}$ (pH = 6.51), $8.76 \times 10^{-4}$ (pH = 7.0), $1 \times 10^{-2}$ (pH = 7.53).

## Purification of Pyruvate, Orthophosphate Dikinase of *Propionibacterium shermanii*

Growth of P. shermanii and the initial purification steps were the same as those reported for preparation of oxaloacetate transcarboxylase.[21] These steps involve grinding the cells in an Eppenbach colloid mill, adsorption on DEAE-cellulose, and batch elution and chromatography on cellulose phosphate. The transcarboxylase adsorbs, but the dikinase does not adsorb, on cellulose phosphate in 50 m$M$ phosphate buffer pH = 6.8. Table III summarizes these steps. Usually, 600-g lots of wet packed cells were processed. One gram of glycerol grown cells contains 1.5–3 units of enzyme. All purification procedures were done at 4°.

Step 4. $(NH_4)_2SO_4$ Fractionation. The enzyme from step 3 was brought to 35% saturation with solid $(NH_4)_2SO_4$. The suspension was centrifuged at 10,000 rpm for 30 min (Sorvall RC-2B). The precipitate was discarded

[19] Y. Milner and H. G. Wood, Fed. Proc. Fed. Amer. Soc. Exp. Biol. 31, 452 (1972).
[20] Y. Milner and H. G. Wood, Proc. Nat. Acad. Sci. U.S. 69, 2463 (1972).
[21] This series, Vol. 13, p. 215.

TABLE III
Summary of Purification of Pyruvate, Orthophosphate Dikinase
from *Proprionibacterium shermanii*

| Step | Volume (ml) | Total units | Total protein (mg) | Specific activity[a] | Yield (%) |
|---|---|---|---|---|---|
| 1. Crude extract | 600 | 598 | 15,890 | 0.037[b] | 100 |
| 2. DEAE-cellulose 50 m$M$ P$_i$ eluent | 800 | 518 | 5,180 | 0.10[b] | 87 |
| 3. Cellulose phosphate 50 m$M$ P$_i$ eluent | 600 | 492 | 2,860 | 0.181[c] | 82 |
| 4. 35–65% (NH$_4$)$_2$SO$_4$ | 120 | 375 | 1,500 | 0.25[c] | 63 |
| 5. Chromatography on DEAE-cellulose | 360 | 58 | 105 | 0.55[c] | 10 |
| 6. Chromatography on TEAE-cellulose | 5 | 13 | 11 | 1.2[c] | 2.2 |
| 7. Chromatography on Sephadex G-200 | 25 | 7 | 3.32 | 2.1[c] | 1.2 |

[a] Specific activity: micromoles per milligram of protein per minute.

[b] Protein was determined by the biuret method [S. Zamenhof, this series, Vol. 3, p. 396].

[c] Protein was determined by the method of O. Warburg and W. Christian [*Biochem. Z.* **310,** 384 (1941)] and agreed within 15% with the biuret method [S. Zamenhof, this series, Vol. 3, p. 396].

and the supernatant solution was brought to 65% saturation with solid (NH$_4$)$_2$SO$_4$ and treated as before. The precipitate was dissolved in a small amount of 10 m$M$ potassium phosphate buffer (pH = 6.65) plus 1 m$M$ EDTA and dialyzed for 7 hr against 3 liters of 75 m$M$ potassium phosphate buffer (pH 6.65) containing 1 m$M$ EDTA.

*Step 5. DEAE-Cellulose Chromatography.* The dialyzate was placed on a 3 × 40 cm microgranular DEAE-cellulose column preequilibrated with the dialysis buffer. The enzyme was eluted with a linear gradient of 2 liters of 0.3 $M$ phosphate buffer (pH = 6.65) containing 1 m$M$ EDTA flowing into 2 liters of the dialysis buffer. Fractions of 10 ml were collected at the rate of 200 ml/hr. The enzyme was eluted by about 0.25 $M$ phosphate and was precipitated in 75% saturated (NH$_4$)$_2$SO$_4$.

*Step 6. TEAE-Cellulose Chromatography.* The ammonium sulfate precipitate from the preceding step was centrifuged, and the precipitate was dissolved in 100 ml of dialysis buffer and then placed on a 2.3 × 10 cm TEAE-cellulose column preequilibrated with 0.12 $M$ phosphate buffer (pH = 6.65) containing 1 m$M$ EDTA. The enzyme was eluted by a linear gradient of phosphate buffer consisting of 250 ml of 0.45 $M$ buffer flowing into 250 ml of 0.15 $M$ buffer, both at pH = 6.7 and containing

1 m$M$ EDTA; 5-ml fractions were collected at a rate of 50 ml per hour. The enzyme eluted in 0.24 $M$ phosphate (determined by conductivity measurements) and was precipitated by 75% saturation with $(NH_4)_2SO_4$.

*Step 7. Chromatography on Sephadex G-200.* The precipitate from the previous step was dissolved in about 1.0 ml of Tris-malate buffer, 20 m$M$ (pH 6.7) containing 0.5 m$M$ EDTA and was placed on a 1.2 × 100 cm column of Sephadex G-200 equilibrated in Tris-malate buffer as above. The column was washed with the same buffer and 1.0-ml fractions were collected at a rate of 8 ml/hr. The enzyme emerged at $V_e/V_0 = 1.15$–$1.25$ and was pooled and held at −20°. Results of the purification are given in Table III.

*Comments on the Purification and Stability of the Enzyme.* The enzyme is very unstable, especially when highly purified, and this leads to low yields. The reason for this instability is not clear. The enzyme from propionibacteria is stable upon freezing and thawing in contrast to the enzyme from *B. symbiosus;* it could be kept for at least a half a year frozen without major loss. The enzyme is essentially devoid of ATPase, phosphatase, adenylate kinase, or pyrophosphatase activity after purification to specific activity 0.6. It is, however, not pure and shows on acrylamide gel electrophoresis at pH = 8.0 about 4–5 bands with one main band containing about 70% of the stained proteins (Amido black staining).

*Molecular Properties.* The purified enzyme upon sedimentation in an analytical ultracentrifuge showed two peaks, one at $s_{20,w} = 7.0$ S and one at $s_{20,w} = 4.9$ S. Occasionally enzyme has been obtained without loss of activity during the preparation and it consisted primarily of the 7 S species.[2] By gel filtration on Sephadex G-200, the molecular weight has been estimated to be approximately 150,000.

*pH Optima.* For stability, the pH optimum is 6.5–6.7; for activity in the formation of pyruvate, 6.8; and for P-enolpyruvate, 6.5–7.5.

*Specificity and Kinetic Constants.* Of the nucleotides, only AMP and ATP were found to be active as substrates. Kinetic constants were ATP, 40 $\mu M$; AMP, 15 $\mu M$; $P_i$, 1 m$M$; $PP_i$, 0.1 m$M$; pyruvate, 0.1 m$M$; P-enolpyruvate, 40 $\mu M$; $Mg^{2+}$, 3 m$M$ and $NH_4^+$, 2 m$M$.[2] The ratio of the rate of the backward/forward reactions was 3.0 at pH 6.7.

*Inhibitors.* The enzyme is inhibited by sulfhydryl reagents, such as pHMB, NEM, iodoacetate, and heavy metals. $Ca^{2+}$ and $Fe^{2+}$ inhibit, probably through competition with $Mg^{2+}$. GTP and especially GMP inhibit competitively with respect to ATP and AMP. $\alpha$-$\beta$-Methylene ATP and $\beta$-$\gamma$-methylene ATP inhibit competitively with respect to AMP and ATP, and methylene diphosphonate inhibits competitively with respect to pyrophosphate and phosphate.[18] Bromopyruvate inhibits, in high ionic strength competitively with respect to P-enolpyruvate (unpublished results). Product inhibition between AMP;ATP, $P_i$:$PP_i$, and pyruvate·P-enolpyruvate was competitive; all other combinations were noncompetitive.

*Mechanism.* The enzyme from *P. shermanii* possesses a mechanism similar to that described for the enzyme from *B. symbiosus.* Exchange patterns are in accord with the three separate partial reactions.[19,20] The exchange reactions give Ping-Pong kinetics and the steady-state kinetics suggest a Tri (Uni, Uni) Ping-Pong mechanism. The rates of exchange in micromoles per minute per unit of enzyme in the forward direction, whether performed separately or in an equilibrium mixture were: $P_i \rightleftarrows PP_i$, 0.3; AMP $\rightleftarrows$ ATP, 0.35, and pyruvate $\rightleftarrows$ P-enolpyruvate, 5.0. The enzyme-P and enzyme-PP forms have been isolated[20] with 1 mole covalently bound per mole of enzyme (molecular weight 150,000). The inhibition patterns indicate the existence of three subsites for the reaction, where each substrate–product pair occupies a single binding site. The pyrophosphoryl group probably is attached to a histidyl residue, located centrally, thus permitting transfer between the three substrate sites.

# [34] Pyruvate, $P_i$ Dikinase from Leaves

*By* M. D. HATCH *and* C. R. SLACK

$$\text{Pyruvate} + \text{ATP} + P_i \xrightleftharpoons{\text{Mg}^{2+}} \text{phosphoenolpyruvate} + \text{AMP} + PP_i$$

## Assay Method A

*Principle.* Phosphoenolpyruvate (PEP) formation is measured as acid-stable radioactivity incorporated from $NaH^{14}CO_3$ into aspartate via the coupling enzymes PEP carboxylase and aspartate aminotransferase.[1] The instability of oxaloacetate necessitates its conversion to aspartate. This method offers the advantage of rapid and simultaneous assay of a number of enzyme samples of widely varying activity.

*Reagents*

Stock reagent mixture. Each day a mixture of stock reagents of the required volume is prepared by mixing 2 volumes of water and 1 volume each of 0.25 $M$ Tris·HCl buffer (pH 8.3), 0.1 $M$ dithio-

[1] M. D. Hatch and C. R. Slack, *Biochem. J.* **106**, 141 (1968).

threitol, 0.1 $M$ MgSO$_4$, 0.1 $M$ glutamate, 0.1 $M$ sodium pyruvate, 2.5 m$M$ ATP, 30 m$M$ K$_2$HPO$_4$, and 50 m$M$ NH$_4$Cl (see Sugiyama[2] for stimulation by NH$_4^+$), 0.1 $M$ NaH$^{14}$CO$_3$ (approximately $10^8$ dpm/ml. NaH$^{14}$CO$_3$ should be added last and the mixture stored at 0°. Fresh solutions of sodium pyruvate should be prepared daily.

Coupling enzyme mixture. 2.5 units/0.1 ml of PEP carboxylase and 5.0 units/0.1 ml of aspartate aminotransferase in 50 m$M$ Tris·HCl buffer, pH 8.3, containing 5 m$M$ dithiothreitol and 1 m$M$ MgSO$_4$. Samples of partially purified PEP carboxylase[3] are recovered by centrifuging suspensions of protein stored in (NH$_4$)$_2$SO$_4$ solution. The redissolved PEP carboxylase may be activated as much as 5-fold by incubating at 25° for 90 min.

Enzyme. Samples of enzymes prepared with the conditions and precautions described below and containing 0.01–5 units/ml (defined below).

*Procedure.* To the bottom of small test tubes add 25 $\mu$l of enzyme samples and 10 $\mu$l of the mixture of PEP carboxylase and aspartate aminotransferase. Duplicate tubes should be prepared for each enzyme sample for two assay periods. Place the tubes in a water bath at 30° and then start the reaction by adding 0.1 ml of the stock reagent mixture and rapidly mixing the contents. Stop the reactions after periods of 5 min and 10 min by adding 25 $\mu$l of 5 $N$ HCl and rapidly mixing the contents. A convenient procedure for counting acid-stable radioactivity is to pipette 50 $\mu$l of the reactions onto 2.5 cm Whatman glass-fiber disks, which are then thoroughly dried in a stream of warm air and counted by a liquid scintillation procedure. The exact specific activity of the NaH$^{14}$CO$_3$ should be determined by diluting samples of the stock reagent mixture with 0.1 $M$ Hyamine hydroxide in ethanol and counting samples of this mixture directly with the same scintillant mixture.

*Application to Crude Extracts.* This assay procedure has proved suitable for filtrates of leaf extracts prepared under the conditions described below. However, particularly where activity is low, it is advisable to remove small molecular weight components from extracts by treatment on a column of Sephadex G-25 (equilibrated with 20 m$M$ Tris·HCl, pH 8.2, containing 2 m$M$ MgSO$_4$ and 5 m$M$ dithiothreitol) and to run separate controls with either orthophosphate or pyruvate omitted from

[2] T. Sugiyama, *Biochemistry* **12**, 2862 (1973).
[3] J. Lowe and C. R. Slack, *Biochim. Biophys. Acta* **235**, 207 (1971). Suitable preparations can also be obtained during the purification of pyruvate, P$_i$ dikinase by eluting the Hypatite C column with 100 m$M$ potassium phosphate after pyruvate, P$_i$ dikinase has emerged (see Purification Procedure).

the reaction. We use a 5-ml column of Sephadex G-25, apply about 1.5 ml of extract, and collect about 0.7 ml of the mid-part of the emerging protein band, coveniently defined by the green chlorophyll–protein complex. The rate of the reverse reaction of pyruvate kinase is very low but an additional check for any contribution of this activity would be provided by measuring activity insensitive to dithiol-binding agents (see Properties section).

## Assay Method B

*Principle.* PEP formation is measured spectrophotometrically at 340 nm via the oxidation of $NADH_2$ in the presence of a coupling system containing PEP carboxylase and malate dehydrogenase.[4]

*Reagents*

Stock reagent mixture. Each day a mixture of stock reagents is prepared containing 1 volume each of 50 m$M$ dithiothreitol, 0.1 $M$ $MgSO_4$, 50 m$M$ $NaHCO_3$, 0.25 $M$ Tris·HCl (pH 8.3), 15 m$M$ sodium pyruvate, 25 m$M$ $K_2HPO_4$, and 50 m$M$ $NH_4Cl$. The sodium pyruvate solution should be freshly prepared.
$NADH_2$, 5 m$M$
ATP, 50 m$M$
Malate dehydrogenase, 0.8 unit/10 $\mu$l in 20 m$M$ Tris·HCl (pH 8.3)
PEP carboxylase, 0.3 units/10 $\mu$l (see Method A above)
Enzyme. Less than 0.02 unit/25 $\mu$l (see Method A above)

*Procedure.* To a 1-ml cuvette (1-cm light path) add 0.6 ml of the stock reagent mixture, 25 $\mu$l of $NADH_2$, 10 $\mu$l of malate dehydrogenase, 10 $\mu$l of PEP carboxylase, 25 or 50 $\mu$l of enzyme, and water to give a total volume of 0.975 ml. Mix and check that the absorbance change at 340 nm is zero. Start the reaction with 25 $\mu$l of 50 m$M$ ATP.

*Application to Crude Extracts.* We have preferred the assay described in Procedure A. However, Procedure B should be suitable if the enzyme extract is centrifuged and treated on a Sephadex G-25 column (see Procedure A). Lactate dehydrogenase would interfere, but its activity is negligible or absent in leaves.

## Reverse Direction Assays

*Principle.* Two spectrophotometric procedures have been described.[4] The one outlined below measures pyruvate formation as $NADH_2$ oxidation at 340 nm in the presence of lactate dehydrogenase.

[4] T. J. Andrews and M. D. Hatch, *Biochem. J.* **114,** 117 (1969).

*Reagents*

Stock reagent mixture. A mixture of stock reagents is prepared each day containing 1 volume each of 0.25 $M$ Tris·HCl buffer (pH 8.3), 50 m$M$ MgSO$_4$, 50 m$M$ dithiothreitol, 10 m$M$ phosphoenolpyruvate, 10 m$M$ AMP and 0.5 $M$ NH$_4$Cl (see Sugiyama[2] for NH$_4^+$ requirement).
NADH$_2$, 5 m$M$
Sodium pyrophosphate, 50 m$M$
Lactate dehydrogenase, 2 units/10 $\mu$l in 20 m$M$ Tris·HCl (pH 8.3)

*Procedure.* To a 1-ml cuvette, add 0.6 ml of the stock reagent mixture, 25 $\mu$l of NADH$_2$, 10 $\mu$l of lactate dehydrogenase, enzyme, and water to give a total volume of 1 ml. Check that there is no absorbance change at 340 nm. Start the reaction by adding 10 $\mu$l of 50 m$M$ pyrophosphate.

*Application to Crude Extracts.* The high pyrophosphatase activity in leaves containing pyruvate, $P_i$ dikinase[5] and other enzymes including pyruvate kinase and PEP carboxylase plus malate dehydrogenase would complicate the use of this procedure for crude extracts.

*Definition of Enzyme Units and Specific Activity.* One unit of enzyme is defined as the amount that forms 1 $\mu$mole of product in 1 min at 30°. Specific activity is expressed as units per milligram protein measured by the procedure of Lowry *et al.*[6]

## Sources and Extraction of Enzyme

The only plants in which the enzyme has been detected are those utilizing C$_4$-pathway photosynthesis.[5] Crude enzyme extracts are prepared by grinding leaves (with sand) in a mortar with 2–4 volumes (w/v) of 50 m$M$ Tris·HCl buffer, pH 8.3, containing 2 m$M$ MgSO$_4$ and 5 m$M$ dithiothreitol and then filtering the homogenate through miracloth. The enzyme is inactivateed in the dark *in vivo*[7,8] so that leaves should be illuminated until extraction commences. Alternatively, the enzyme can be activated after extraction.[8] Some additional activation is often seen after extraction of preilluminated leaves. The reader is referred to the section on properties of the enzyme for further details on other factors affecting enzyme activity in leaves and extracts.

[5] M. D. Hatch and C. R. Slack, *in* "Progress in Phytochemistry" (L. Reinhold and Y. Liwschitz, eds.), Vol. 2, pp. 35–106. Wiley (Interscience), New York, 1970.
[6] O. H. Lowry, N. J. Rosebrough, A. L. Farr, and R. J. Randall, *J. Biol. Chem.* **193**, 265 (1951).
[7] C. R. Slack, *Biochem. Biophys. Res. Commun.* **30**, 483 (1968).
[8] M. D. Hatch and C. R. Slack, *Biochem. J.* **112**, 549 (1969).

## Purification Procedure

*Step 1. Extraction.* The following procedure has proved suitable for maize and sugar cane leaves.[1,4] One hundred grams of leaves are extracted by blending in a Sorvall OmniMixer for 2 min with 400 ml of 50 m$M$ Tris·HCl (pH 8.2) containing 50 m$M$ mercaptoethanol, 6 m$M$ MgSO$_4$, 1 m$M$ EDTA, and 2 m$M$ pyruvate. Extract at 0° but conduct all subsequent steps except (NH$_4$)$_2$SO$_4$ treatments at about 22°. The extract obtained by filtering the homogenate through muslin is centrifuged at 10,000 $g$ for 5 min. At this stage further activation of the enzyme can be achieved by incubating this extract at 25° with K$_2$HPO$_4$ added to give a final concentration of 2 m$M$.[8]

*Step 2. Ammonium Sulfate.* To the 10,000 $g$ supernatant is added a saturated solution of (NH$_4$)$_2$SO$_4$ (5°, pH 7.0) to give 40% saturation. After standing for 20 min the precipitated protein is removed by centrifugation at 10,000 g for 5 min and discarded. Additional saturated (NH$_4$)$_2$SO$_4$ solution is then added to give 53% saturation, and after 20 min the precipitated protein is recovered by centrifugation at 10,000 $g$ for 5 min and dissolved in 30 ml of 50 m$M$ Tris·HCl, pH 8.2, containing 50 m$M$ 2-mercaptoethanol, 6 m$M$ MgSO$_4$, 1 m$M$ EDTA, and 2 m$M$ pyruvate. This buffer mixture is used for all subsequent steps and will be referred to as the standard buffer mixture.

*Step 3. Sephadex G-200.* The redissolved (NH$_4$)$_2$SO$_4$ precipitate is applied to a 380 ml (4 × 30 cm) column of Sephadex G-200 previously washed with the standard buffer mixture, and eluted at a rate of 1 ml/min. Activity appears in a band of approximately 50 ml that commences to emerge about 30 ml after the void volume (previously determined with dextran blue).

*Step 4. Hypatite C.* The fraction from Sephadex G-200 should be immediately applied to a 32 ml column (1.5 × 19 cm) of Hypatite C (calcium phosphate gel, Clarkson Chemical Co, Williamsport, Pennsylvania), previously washed with the standard buffer mixture until the eluate is pH 7.4. After the enzyme has entered, successively elute the gel at a flow rate of 1 ml/min with 30 ml of the standard buffer mixture and then with two 45-ml lots of the standard buffer mixture containing first 30 m$M$ K$_2$HPO$_4$ and then 40 m$M$ K$_2$HPO$_4$. Elution of the enzyme coincides with the emergence of the buffer containing 40 m$M$ potassium phosphate. Active fractions are pooled, and the protein is precipitated by adding 2 volumes of saturated (NH$_4$)$_2$SO$_4$ solution. The enzyme was stable in this form when stored at 0°.

*Step 5. Sephadex G-200.* The protein suspension in (NH$_4$)$_2$SO$_4$ solution is recovered by centrifugation, dissolved in 3.5 ml of a standard

PURIFICATION OF PYRUVATE, $P_i$ DIKINASE FROM MAIZE LEAF

| Step and fraction | Total units | Conc. of protein (mg/ml) | Specific activity (units/mg protein) |
|---|---|---|---|
| 1. Crude extract | 210 | 7.2 | 0.065 |
| 2. 40–53% $(NH_4)_2SO_4$ | 180 | 15 | 0.25 |
| 3. First Sephadex G-200 | 108 | 8.6 | 0.28 |
| 4. Hypatite C | 30 | 8.6 | 0.76 |
| 5. Second Sephadex G-200 | 23 | 1.1 | 1.2 |

buffer, and applied to a 180 ml column (5 cm$^2$ × 36 cm) of Sephadex G-200 previously equilibrate with the standard buffer mixture. The enzyme is eluted from the column at a flow rate of 0.5 ml/min, appearing in a band of about 18 ml, 15 ml after the void volume. It should be noted that the use of superfine G-200 at this stage, combined with slower flow rates, removes traces of pyrophosphatase that otherwise remains.[4] Enzyme obtained from this treatment remains active for long periods when the protein is precipitated by adding 2 volumes of saturated $(NH_4)_2SO_4$ solution and stored at 0°. If there is insufficient protein in these fractions it may be necessary to add bovine serum albumin to a concentration of 2 mg/ml and additional solid $(NH_4)_2SO_4$ to give a 75% saturated solution.

An example of the purification obtained by this procedure is given in the table. The procedure provides a preparation free of enzymes that react with any of the substrates or products of pyruvate, $P_i$ dikinase. Purification of the maize leaf enzyme to homogeneity has recently been reported.[2]

## Properties

*Activity in Leaves.* The enzyme appears to be confined to plants with C$_4$-pathway photosynthesis[5] and crassulacean acid metabolism.[9] It is inactivated in darkened leaves but rapidly regains activity when leaves are illuminated.[7,8] The quantity of enzyme in leaves is reduced manyfold when plants are grown at lower light intensities.[10]

*Activation and Inactivation in Vitro and Stability.* Both thiols and $Mg^{2+}$ are essential for maintaining activity in solution.[1,8] Particularly

[9] M. Kluge and C. B. Osmond, *Naturwissenschaften* **58**, 414 (1971).
[10] M. D. Hatch, C. R. Slack, and T. A. Bull, *Phytochemistry* **8**, 697 (1969).

without $Mg^{2+}$, inactivation appears to be irreversible. Activation of the inactive enzyme isolated from darkened leaves requires a thiol, $Mg^{2+}$, orthophosphate (and also pyruvate with some species) as well as a protein component present in crude extracts.[8] AMP is an antagonist of phosphate in this reaction. In air, ADP mediates a rapid but reversible inactivation of enzyme in the presence of a thiol.[8] Enzyme inactivated in this way has the same requirements for activation as described above for the inactive enzyme isolated from darkened leaves. The enzyme is inactivated at 0° even in the presence of a thiol and $Mg^{2+}$; activity is rapidly recovered by rewarming to 20–30°.[1,2] High concentrations of pyruvate (30 m$M$) prevent inactivation at 0° (C. R. Slack, unpublished results). The enzyme retains activity for many months as a precipitate kept at 0° in a 66% saturated solution of $(NH_4)_2SO_4$.

*Activators and Inhibitors.* Sugiyama[2] has reported that $NH_4^+$ marginally increases the activity of the maize enzyme in the forward direction. However, in the reverse direction activity is increased severalfold with maximum rates at about 50 m$M$ $NH_4Cl$. The enzyme is inhibited by thiol-binding agents such as $p$-chloromercuribenzoate and also reagents such as $\gamma$($p$-arsenophenyl)-$n$-butyrate and arsenate plus 2,3-dimercaptoethanol that strongly bind with vicinal dithiol groups.[8] The rate of loss of activity seen without thiols is rapidly accelerated by diethyldithiocarbamate and its oxidized derivative tetraethylthiuram disulfide. However, activity lost in the presence of these reagents is completely recovered by the subsequent addition of a thiol.[8]

All products of the reaction in the forward direction have some inhibitory effect. Using Method B for assay, 0.5 m$M$ $PP_i$ and 0.5 m$M$ AMP cause 54% and 52% inhibition, respectively.[4] Inhibition by AMP is competitive with respect to ATP. Pyruvate and $P_i$ inhibit the reverse direction rate and inhibition by these compounds is competitive with PEP and $PP_i$, respectively.[4]

*Specificity.* Activity with UTP, GTP, and CTP is only 1–3% of that with ATP.[1] $MnCl_2$ and $CaCl_2$ will not substitute for $MgCl_2$. The enzyme operates with arsenate in place of orthophosphate and the reaction is effectively irreversible under these conditions.

*Kinetic Properties.* Apparent $K_m$ values of 0.11 m$M$ for pyruvate, 90 $\mu M$ for ATP and 0.5 m$M$ for $P_i$ were obtained for the maize enzyme at pH 8.3.[1] Values of 0.25 m$M$ for pyruvate, 20 $\mu M$ for ATP, and 1.5 m$M$ for $P_i$ have been reported for reactions run at pH 7.5 and containing 25 m$M$ $NH_4Cl_2$.[1] $Mg^{2+}$ is apparently required as such, since there is little activity when the ratio of $Mg^{2+}$ to ATP was less than one.[1] Maximum activity was observed with 10 m$M$ $Mg^{2+}$ when ATP was added at a concentration of 2.5 m$M$. For the reverse direction, apparent $K_m$ values at pH

8.3 for sugar cane pyruvate, P$_i$ dikinase are 0.11 m$M$ for PEP, 40 $\mu M$ for PP$_i$ and less than 4 $\mu M$ for AMP.[4] Very similar values have been reported for the maize enzyme.[2]

With the sugarcane enzyme the maximum velocity in the forward direction at pH 8.3 was found to be 5–6 times that in the reverse direction.[4] However, Sugiyama[2] reports the forward rate for the maize enzyme to be only slightly greater at pH 7.5 provided NH$_4^+$ ions are added to stimulate the reverse reaction. In the former studies reverse direction reactions would have contained only 2–5 m$M$ (NH$_4$)$_2$SO$_4$ added with coupling enzymes.

*Equilibrium Constant.* Studies with the bacterial enzyme[11] have shown that the equilibrium favors the reverse direction but that the $K_e$ approaches unity as the pH is increased. For the direction toward PEP formation, the apparent $K_e$ was 0.5 at pH 8.4. A value of 0.005 at pH 7.2 for the maize enzyme[2] was in reasonable agreement with the value recorded for the bacterial enzyme at this pH.

*Mechanism and Physical Properties.* Pyrophosphate produced in the forward reaction is derived from orthophosphate and the $\gamma$-phosphate of ATP, and the phosphate of PEP from the $\beta$-phosphate of ATP.[4] Studies of isotope exchange reactions[4] indicate a sequential mechanism for the addition of ATP and P$_i$ and a "Ping-Pong" mechanism for the addition of pyruvate and release of PEP as shown in the following equations:

$$\text{Enzyme} + \text{ATP} + \text{P}_i \rightleftharpoons \text{enzyme-P} + \text{AMP} + \text{PP}_1$$
$$\text{Enzyme—P} + \text{pyruvate} \rightleftharpoons \text{enzyme} + \text{PEP}$$

Sedimentation analysis indicates a molecular weight for the purified maize enzyme of 387,000.[2] This enzyme is dissociable into four subunits of MW 94,000.

[11] R. E. Reeves, R. A. MacKenzie, and D. S. Hsu, *J. Biol. Chem.* **243**, 5486 (1968).

# Section II

# Aldolases

# [35] Fructose-diphosphate Aldolase from Lobster Muscle

*By* ARABINDA GUHA

Fructose 1,6-diphosphate $\rightleftarrows$ dihydroxyacetone phosphate
$+$ D-glyceraldehyde 3'-phosphate

Fructose-1,6-diphosphate aldolase (FDP-aldolase) which catalyzes the cleavage of fructose 1,6-diphosphate to dihydroxyacetone phosphate (DHAP) and D-glyceraldehyde 3-phosphate (GAP) is widely distributed in nature.[1] FDP-aldolase from lobster tail muscle has been shown to form a Schiff-base intermediate with substrate[2] and is thus similar to other vertebrate and plant FDP-aldolases and is a Class I aldolase.[3] The physicochemical and catalytic properties of invertebrate FDP-aldolase[4,5] are very similar. The fact that hybridization between FDP-aldolase from *Drosophila melanogaster* and FDP-aldolase from rabbit or calf brain[6] yields isoenzymes proves that interspecies hybridization is possible between FDP-aldolases from vertebrates and invertebrates. Thus aldolases from various species seem to be homologous.

## Assay Method

*Principle.* The assay is based on the oxidation of NADH by dihydroxyacetone phosphate in a coupled system containing triosephosphate isomerase and $\alpha$-glycerophosphate dehydrogenase.[7]

*Reagents*

Buffer-EDTA solution: A stock solution is prepared 1.0$M$ triethanolamine and 0.2 $M$ in EDTA; the pH is adjusted to 7.4. The stock solution is stored at 4° and diluted 20-fold with distilled water at room temperature before use.

FDP-Na (Boehringer), 0.1 M pH 7.5

NADH-Na (Boehringer) 10 mg/ml in 0.001 $N$ NaOH (approxi-

[1] D. E. Morse and B. L. Horecker, *Advan. Enzymol. Relat. Areas Mol. Biol.* 31, 125, 1968.
[2] A. Guha, C. Y. Lai, and B. L. Horecker, *Arch. Biochem. Biophys.* 147, 692 (1971).
[3] See this series, Vol. 9 [87].
[4] L. Levenbook, A. C. Bauer, and H. Shigematsu, *Arch. Biochem. Biophys.* 157, 625 (1973).
[5] P. J. Anderson, I. Gibbons, and R. N. Perham, *Eur. J. Biochem.* 11, 503 (1969).
[6] O. Brenner-Holzach and F. Leuthardt, *Helv. Chim. Acta* 54, 2809 (1971).
[7] E. Racker, *J. Biol. Chem.* 167, 843 (1947).

mately 10 m$M$) stored frozen at —16°. The frozen solution may be kept indefinitely.

$\alpha$-Glycerophosphate dehydrogenase–triosephosphate isomerase mixed crystals (Bohringer) 10 mg/ml.

*Assay Mixture.* To 9.75 ml of diluted buffer-EDTA solution is added 0.1 ml of FDP solution, 0.15 ml of NADH solution, and 0.01 ml of $\alpha$-glycerophosphate dehydrogenase–triosephosphate isomerase crystal suspension. This assay mixture is prepared fresh and kept at room temperature, but unused portions may be stored for several days at 4° and warmed to room temperature before use.

*Procedure.* One milliliter of assay mixture is measured into a cuvette with a 1-cm light path and placed in a spectrophotometer with the cuvette chamber thermostated at 25°. The enzyme is diluted with the buffer-EDTA solution to contain about 0.25–1.0 unit of activity per milliliter; 10–20 $\mu$l of diluted enzyme are added to the assay mixture in the cuvette, and the mixture is rapidly stirred. The rate of oxidation of NADH is measured at 340 nm on a recording spectrophotometer. The reaction rate is expressed as micromoles of FDP cleaved per minute, with $6.22 \times 10^3$ as the molar absorbancy coefficient of NADH.[8] In this assay each mole of FDP cleaved results in the oxidation of 2 moles of NADH.

*Definition of Unit.* One unit of enzyme activity is defined as the amount required to catalyze the cleavage of 1 $\mu$mole of FDP per minute, under the conditions described above.

*Protein Determination.* In the initial stages of purification of lobster muscle aldolase, the protein concentrations are determined by the turbidometric method of Bücher.[9] After isolation of the enzyme, the absorbance of protein ($E_{1\,cm}^{0.1\%} = 1.12$) at 280 nm, as determined by the method of Walsh and Brown,[10] is used for determination of protein concentration.

## Purification Procedure

All procedures were carried out at 4° or in an ice bath.

*Step 1. Extraction.* Tail muscles from fresh lobsters from a local fish market are dissected out. Tail muscle (200 g) obtained from 4500 g of fresh lobster is minced in a meat grinder. The minced tissue is suspended in 1200 ml of 1 m$M$ EDTA, pH 7.0, and stirred for 15 min. The suspension is strained through 4 layers of cheesecloth. The muscle brei is extracted two more times with 600 ml of the same solution. The combined, strained extracts (2300 ml) contained $1.41 \times 10^4$ units of aldolase activity (crude extract).

---

[8] B. L. Horecker and A. Kornberg, *J. Biol. Chem.* **175**, 385 (1948).
[9] T. Bücher, *Biochim. Biophys. Acta* **1**, 292 (1947).
[10] K. A. Walsh and J. R. Brown, *Biochim. Biophys. Acta* **58**, 596 (1962).

*Step 2. First Ammonium Sulfate Fractionation.* To the crude extract, 590 g of solid $(NH_4)_2SO_4$ is added. The resulting suspension is stirred for 20 min and then centrifuged at 16,000 $g$ for 45 min in an RC-2B Sorvall centrifuge. To the supernatant solution (2400 ml), 295 g of solid $(NH_4)_2SO_4$ is added and the suspension is stirred for 30 min. The precipitate is collected by centrifugation at 16,000 $g$ as above, and the precipitate is dissolved in 1 m$M$ sodium phosphate buffer, pH 6.5, 1 m$M$ in $\beta$-mercaptoethanol. The solution (125 ml) is dialyzed against 3 changes of 4 liters each of the same buffer (first ammonium sulfate fraction, 150 ml).

*Step 3. Chromatography on DEAE-Cellulose.* The dialyzed enzyme solution is applied to a column (3.5 $\times$ 70 cm) of DEAE-cellulose which has been equilibrated with 1 m$M$ sodium phosphate, pH 6.5, 1 m$M$ in $\beta$-mercaptoethanol. The column is washed at a flow rate of 40 ml/hr with 150 ml of the same buffer. The enzyme is eluted with 2 m$M$ sodium phosphate buffer, pH 6.5, 1 m$M$ in $\beta$-mercaptoethanol. Fractions of 10 ml each are collected and analyzed for aldolase activity. Fractions containing aldolase activity (fractions 11–30) are pooled (DEAE-cellulose eluate, 200 ml).

*Step 4. Second Ammonium Sulfate Fractionation.* To the DEAE-cellulose eluate, 59.2 g of $(NH_4)_2SO_4$ are added and the suspension is stirred for 30 min. The slight turbidity formed is removed by centrifugation for 20 min at 36,800 $g$. To the supernatant solution, saturated $(NH_4)_2SO_4$ solution ($\sim$4°) is added to give 0.6 saturation and the pH is adjusted to pH 7.5. The suspension is stirred for 1 hr. The precipitate obtained after centrifugation is dissolved in 0.1 $M$ Tris-Cl buffer, pH 7.5, 1 m$M$ in EDTA and 1 m$M$ in $\beta$-mercaptoethanol. The enzyme solution is dialyzed against the same buffer. (Second ammonium sulfate fraction, 50 ml).

*Step 5. Crystallization.* The dialyzed enzyme solution is adjusted to a protein concentration of about 4 mg/ml. To this solution (100 ml), 100 ml of saturated ammonium sulfate solution pH 7.0, is slowly added with continuous stirring. The solution is centrifuged for 20 min at 3700 $g$. To the supernatant 80% $(NH_4)_2SO_4$ solution pH 7.0 is added dropwise until visible turbidity appears. The suspension is stirred gently and crystals appeared after 2 hr. Additional 80% $(NH_4)_2SO_4$ solution is added to raise the saturation to 0.54 and the stirring is continued for another 48 hr. The specific activity of the crystalline enzyme is 14–16 units/mg. Recrystallization does not increase the specific activity. Crystalline lobster muscle aldolase could be stored in 0.54 saturated $(NH_4)_2SO_4$ at 4° without appreciable loss of activity over a period of 6 months. The results of a typical preparation are summarized in Table I.

## TABLE I
### PURIFICATION OF ALDOLASE FROM LOBSTER MUSCLES

| Step | Volume (ml) | Total protein (mg) | Total activity (units) | Specific activity (units/mg protein) | Yield (%) |
|---|---|---|---|---|---|
| 1. Crude extract | 2300 | 11,000 | 14,100 | 1.28 | 100 |
| 2. First ammonium sulfate fraction | 150 | 2,200 | 10,600 | 4.82 | 75 |
| 3. DEAE-cellulose column eluate | 200 | 530 | 7,420 | 14 | 52 |
| 4. Second ammonium sulfate fraction | 50 | 414 | 5,800 | 14.5 | 41 |
| 5. Crystallization | 200 | 318 | 4,600 | 14.8 | 33 |

*Notes on the Purification Procedure.* Exhaustive dialysis of FDP-aldolase from the first ammonium sulfate is necessary since the enzyme is eluted from the column by increasing the ionic strength from 1 m$M$ sodium phosphate to 2 m$M$ sodium phosphate. The enzyme is eluted as the first protein peak from the column. Considerable activity is lost in the trailing edge of the peak.

Occasionally higher concentrations of ammonium sulfate are required (greater than 0.54 saturation) for the appearance of the visible turbidity in the crystallization step, but this does not interfere with the crystal formation.

## Properties

*Homogeneity.* Lobster muscle aldolase purified by this procedure is found to be homogeneous in disc gel electrophoresis, Sephadex gel filtration, and ultracentrifugation.

*Molecular Properties.* The molecular weight of native lobster muscle aldolase is 160,000, as determined by sedimentation equilibrium, sedimentation velocity, and quantitative Sephadex gel filtration.[2] The molecular weight of lobster aldolase is the same as that of mammalian aldolase[1] and other vertebrate aldolases[11,12] and considerably larger than the value of 120,000 reported for spinach aldolase.[13]

The amino acid composition of lobster muscle aldolase, calculated on

[11] C. Y. Lai and C. Chen, *Arch. Biochem. Biophys.* **144**, 467 (1971).
[12] S. M. Ting, C. L. Sia, C. Y. Lai, and B. L. Horecker, *Arch. Biochem. Biophys.* **144**, 485 (1971).
[13] G. Rapoport, L. Davis, and B. L. Horecker, *Arch. Biochem. Biophys.* **132**, 286 (1969).

TABLE II
AMINO ACID COMPOSITION OF LOBSTER MUSCLE—RESIDUES PER MOLECULE OF
MW 160,000

| Amino acid | Lobster muscle aldolase | Amino acid | Lobster muscle aldolase |
|---|---|---|---|
| Lysine | 136 | Alanine | 132 |
| Histidine | 36 | Cysteine | 12 |
| Arginine | 56 | Valine | 100 |
| Aspartic acid | 136 | Methionine | 20 |
| Threonine | 92 | Isoleucine | 64 |
| Serine | 80 | Leucine | 124 |
| Glutamic acid | 156 | Tyrosine | 56 |
| Proline | 52 | Phenylalanine | 40 |
| Glycine | 114 | Tryptophan | 20 |

the basis of a molecular weight of 160,000, is shown in Table II. The overall amino acid composition of lobster muscle aldolase is very similar to those of rabbit muscle and rabbit liver aldolases.[14,15] The most striking differences are lower contents of cysteine and proline and the higher amounts of methionine and aromatic amino acids relative to the values for rabbit muscle aldolase.

The COOH-terminal amino acid of lobster muscle aldolase is tyrosine and it is released readily by digestion with carboxypeptidase A; no residues other than tyrosine are released during prolonged digestion or by higher concentrations of carboxypeptidase A. The release of 4 moles of tyrosine per mole of enzyme confirmed the tetrameric structure of lobster muscle aldolase, and hydrazinolysis after carboxypeptidase digestion shows that the COOH-terminal sequence is -Ala-Tyr, like that of rabbit muscle aldolase.

The substrate-binding site isolated from tryptic digests contains 28 amino acid residues and its sequence is very much like that of the peptides isolated from other aldolases.[2]

*Catalytic Activity.* The crystalline lobster muscle aldolase shows specific activities with fructose 1,6-diphosphate as substrate of 14–16 units/mg. The specific activity with fructose 1-phosphate is 0.14 unit/mg, and the activity ratio FDP:FIP is about 50.

Removal of the COOH-terminal tyrosine residue by carboxypeptidase A digestion leads to loss of activity toward fructose 1,6-diphosphate, but little or no change in the ability of the enzyme to cleave fructose 1-phos-

[14] C. Y. Lai, *Arch. Biochem. Biophys.* **128**, 202 (1968).
[15] R. W. Gracy, A. G. Lacko, and B. L. Horecker, *J. Biol. Chem.* **244**, 3913 (1969).

phate. It is interesting to note that the loss of activity toward fructose 1,6-diphosphate does not parallel the removal of tyrosine residue. When 80% of this activity is lost, less than half of the tyrosine have been released. On the other hand, loss of activity toward fructose 1-phosphate is observed only during release of the last 2 tyrosine residues. This suggests a high degree of interaction between subunits of lobster muscle aldolase that has not been observed for rabbit muscle aldolase.[2]

# [36] Fructose-diphosphate Aldolase from Blue-Green Algae[1,2]

By JAMES M. WILLARD and MARTIN GIBBS

$$\text{Fructose 1,6-diphosphate} \xrightleftharpoons{Me^{2+}+RSH} \text{dihydroxyacetone phosphate} + \text{D-glyceraldehyde 3-phosphate}$$

Rutter[3] proposed two broad classes of fructose-1,6-diphosphate (FDP) aldolases: Type I aldolases not requiring a divalent metal, strongly inhibited by mercurials, and unaffected by chelating agents and $K^+$; type II aldolases requiring a divalent metal, slightly inhibited by mercurials, strongly inhibited by chelating agents, and stimulated by $K^+$. Variants within each aldolase type have been proposed.[4,5]

The FDP aldolase of blue-green algae is a type II aldolase.[3,5-7] Unlike most other type II aldolases, the enzyme from the blue-green algae, *Anacystis nidulans*, is unaffected by $K^+$, has an absolute requirement for cysteine, and $Fe^{2+}$ stimulates the rate obtained with cysteine alone.[8]

## Assay Methods

*Principle.* The assay of aldolase is based on the estimation of triose phosphate formed with FDP cleavage. Two methods of assay can be employed depending upon the specific assay conditions and both yield com-

[1] Fructose-1,6-biphosphate D-glyceraldehyde-3-phosphate-lyase, EC 4.1.2.13.
[2] This work was assisted by grants from the National Science Foundation and the United States Atomic Energy Commission.
[3] W. J. Rutter, *Fed. Proc., Fed. Amer. Soc. Exp. Biol.* 23, 1248 (1964).
[4] W. J. Rutter, B. M. Woodfin, and R. E. Blostein, *Acta Chem. Scand.* 17, Suppl. 1, 226 (1963).
[5] J. M. Willard and M. Gibbs, *Plant Physiol.* 43, 793 (1968).
[6] C. Van Baalen, *Nature (London)* 206, 193 (1965).
[7] J. M. Willard, M. Schulman, and M. Gibbs, *Nature (London)* 206, 195 (1965).
[8] J. M. Willard and M. Gibbs, *Biochim. Biophys. Acta* 151, 438 (1968).

parable results: the colorimetric method of Sibley and Lehninger[9] and the spectrophotometric method of Wu and Racker.[10] Specifically, the colorimetric method should be used whenever there is appreciable NADH oxidase activity or when additions result in UV-absorbing complexes. The spectrophotometric assay can be used in the presence of cysteine and iron only if both are preincubated at least 5 min prior to initiation of the reaction with FDP.[11]

The rate with either assay is linear with time and proportional with enzyme.

### Reagents for Assay

Tris(hydroxymethyl)aminomethane·HCl (Tris·HCl) (Sigma), 0.2 $M$, pH 7.6

FDP-Na, 50 m$M$ (Sigma)

Hydrazine sulfate (Eastman Chemicals), 0.56 $M$, pH 7.5

Cysteine-HCl·H$_2$O (Pfanstiehl Laboratories), 40 m$M$ pH 7.5, (prepared just before use)

Fe(NH$_4$)$_2$(SO$_4$)$_2$·6 H$_2$O (Mallinckrodt Chemicals), 20 m$M$

Trichloroacetic acid, 10% (w/v) (Fisher Chemicals)

2,4-Dinitrophenylhydrazine (Eastman Chemicals) (0.1% in 2 $N$ HCl)

NADH (Sigma), 10 m$M$

Triosephosphate isomerase and $\alpha$-glycerolphosphate dehydrogenase (Boehringer and Soehne).

*Colorimetric Assay.* The assay is performed at 37° in 2.5-ml reaction mixtures and contains 0.5 ml of Tris·HCl, 0.25 ml hydrazine sulfate, 0.2 ml FDP, 0.5 ml cysteine·HCl, 0.125 ml Fe (NH$_4$)$_2$(SO$_4$)$_2$ and aldolase. The aldolase is incubated 10 min to achieve maximum activity after which FDP is added to initiate the reaction. Incubations are terminated at the end of 20 min by the addition of 2.0 ml cold trichloroacetic acid. Blanks consist of adding FDP after the acid and controls with no enzyme present are run in all cases where cofactors or inhibitors are employed. Following chromagen development[9] the absorption at 540 nm is determined in 1-cm cuvettes with a Beckman DU spectrophotometer.

*Spectrophotometric Assay.* This is the method of choice and may be

---

[9] J. A. Sibley and A. L. Lehninger, *J. Biol. Chem.* **177**, 859 (1949).

[10] R. Wu and E. Racker, *J. Biol. Chem.* **234**, 1029 (1959).

[11] The intense red-violet color strongly absorbing at 340 nm slowly disappears as cysteine stoichometrically reduces ferric ions to the ferrous state. A. E. Martell and M. Calvin, *in* "Chemistry of the Chelate Compounds," p. 384. Prentice-Hall, Englewood Cliffs, New Jersey, 1956.

employed provided (a) the NADH oxidase activity is low and (b) a 5-min preincubation occurs when using cysteine and iron.

Assays are performed at either 26° or 37° employing the coupling system of triosephosphate isomerase and $\alpha$-glycerolphosphate dehydrogenase. In a final volume of 1.0 ml the reaction mixture contains 0.2 ml of Tris·HCl, 9.6 EU of triosephosphate isomerase, 0.24 EU of $\alpha$-glycerolphosphate dehydrogenase, 0.02 ml of NADH, 0.1 ml of FDP, 0.2 ml of cysteine HCl, 0.05 ml of $Fe(NH_4)_2(SO_4)_2$, and aldolase. Reference cuvettes lack NADH while control cuvettes lack FDP or aldolase. The reaction is initiated by addition of FDP or aldolase. NADH oxidation at 340 nm is followed in 1-cm cuvettes with a Beckmann DU spectrophotometer equipped with a Gilford Model 2000 multiple absorbance recorder. A molecular extinction coefficient for NADH of $6.22 \times 10^6$ $cm^2/mole$ is employed.[12] The temperature is controlled by use of a Haake Model F circulator.

*Units.* Units are expressed as micromoles of FDP cleaved per minute at a specified temperature of 26° or 37°, and specific activities are expressed in units per milligram of protein. Protein is measured spectrophotometrically[13] in purified preparations and colorimetrically[14] at 750 nm in crude preparations using crystalline bovine serum albumin as standard. The two methods yield similar results provided the 280/260 ratio is greater than 0.9.

## Purification of Anacystis Aldolase[8]

A typical purification obtained in the different steps is summarized in the table. All procedures are carried out at 4° unless otherwise stated.

*Source of Enzyme. Anacystis nidulans* (Richt.) was obtained from Dr. J. Meyer at the University of Texas, Austin. Ten-liter cultures were grown at 30° in 12-1 Florence flasks in a medium[15] containing, per liter, sodium citrate·$2H_2O$, 0.165 g; $MgSO_4·7H_2O$, 0.122 g; $KNO_3$, 1 g; $K_2HPO_4$, 1.0 g; $Ca(NO_3)_2$ $4H_2O$, 0.025 g, 1 ml of 1% (w/v) $FeSO_4$-EDTA solution[16] and 1 ml of Hoagland and Arnon $A_5$ solution.[17] Cells were

---

[12] B. L. Horecker and A. Kornberg, *J. Biol. Chem.* **175**, 385 (1948).

[13] O. Warburg and W. Christian, *Biochem. Z.* **310**, 384 (1941); see this series, Vol. 3 [73].

[14] O. H. Lowry, N. J. Rosebrough, A. L. Farr, and R. J. Randall, *J. Biol. Chem.* **193**, 265 (1951).

[15] W. A. Kratz and J. Myers, *Amer. J. Botany* **42**, 282 (1955).

[16] L. Jacobson, *Plant Physiol.* **26**, 411 (1951).

[17] D. R. Hoagland and D. I. Arnon, *Calif. Agr. Exp. Sta. Circ.* **347** (1938). 1 ml of solution contains 2.86 mg of $H_3BO_3$; 1.81 mg of $MnCl_2·H_2O$; 0.11 mg of $ZnCl_2$; 0.079 mg of $CuSO_4·5H_2O$; 0.03 mg of $Na_2MoO_4·2H_2O$.

SUMMARY OF PURIFICATION PROCEDURE FOR *Anacystis* ALDOLASE

| Fractionation step | Total activity[a] (units) | Specific activity (units/mg protein) | Recovery (%) |
|---|---|---|---|
| Cell-free extracts: sum of 3 lots | 64 | 0.13 | 100 |
| 35–75% $(NH_4)_2SO_4$: sum of 3 lots | 38 | 0.58 | 60 |
| Stored and pooled 35–75% $(NH_4)_2SO_4$ | 38 | 0.58 | 60 |
| Calcium phosphate gel and 35–80% $(NH_4)_2SO_4$ eluate | 33 | 2.1 | 51 |
| Combined Sephadex G-25 fractions | 31 | 2.3 | 48 |
| Combined DEAE-cellulose fractions | 20 | — | 30 |
| 80% $(NH_4)_2SO_4$ fraction off DEAE-cellulose | 11 | 128 | 17 |

[a] Aldolase activity was determined by the colorimetric assay method at 37° with 8 m$M$ cysteine and 1 m$M$ $Fe^{2+}$ present.

grown under continuous fluorescent illumination of 500 foot-candles with a constant supply of 1% $CO_2$/99% air. Five-day-old cultures yield 0.5 to 1.0 g wet weight of cells per liter.

*Step 1. Preparation of Cell-Free Extracts.* Two grams wet weight of *Anacystis* cells were suspended in a final volume of 15 ml of 50 m$M$ Tris·HCl (pH 7.6). After deposition in a 25-ml Rosette cell the suspension was deaerated 4 min with $N_2$, then sonicated under $N_2$ at full power on a Branson Model S-75 sonifier. After centrifugation at 10,000 $g$ the supernatant fraction constituted the cell-free extract. The aldolase of this step is extremely unstable (80% loss in 10 hrs at 4°, 50% loss in 2 days at −15°) and the following step 2 must be performed immediately for maximal yield.

*Step 2. Fractionation with Ammonium Sulfate.* Freshly prepared cell-free extract (29 ml) is brought to 35% saturation with 7.2 g solid $(NH_4)_2SO_4$ and centrifuged at 10,000 $g$. The resultant pellet is discarded and the supernatant is brought to 75% saturation with a further 8.2 g $(NH_4)_2SO_4$. After centrifugation the 75% supernatant is discarded and the pellet is suspended in 4 ml of deaerated (by $N_2$) 50 m$M$ Tris·HCl (pH 7.6). This 35–75% $(NH_4)_2SO_4$ fraction is stable for several months when stored under $N_2$ at −15°. This remarkable stability in $(NH_4)_2SO_4$ allows for the accumulation of sufficient material for subsequent purification.

*Step 3. Calcium Phosphate Treatment and Ammonium Sulfate Elution.* Calcium phosphate gel suspension, 61 ml in 50 m$M$ Tris·HCl (pH 7.6, containing 16.5 mg of gel dry weight per milliliter) is added to 10 ml of three pooled 35–75% $(NH_4)_2SO_4$ fractions (66.25 mg protein). After 10 min of magnetic mixing, the suspension is centrifuged and the supernatant is discarded. The calcium phosphate gel pellets are extracted twice with 30 ml each of 35% saturated (248.5 g/liter) $(NH_4)_2SO_4$. This $(NH_4)_2SO_4$ solution is then raised to 80% saturation with 13.6 g of solid $(NH_4)_2SO_4$ and centrifuged; the resultant precipitate is suspended in 4.5 ml of 50 m$M$ Tris·HCl (pH 7.6) and recentrifuged.

*Step 4. Chromatography on DEAE-Cellulose.* This step completely separates aldolase from the molecularly similar biliprotein, phycocyanin, the latter constituting 40% of the total soluble protein, having an isoelectric point of 4.5–5.0 and a molecular weight of 138,000.[18]

The 35–80% $(NH_4)_2SO_4$ fraction (5.3 ml) is passed through a 1.5 × 15 cm Sephadex G-25 column previously equilibrated with 50 m$M$ Tris (pH 7.6). Fractions of 2.3 ml are collected at 0.8 ml/minute. Fractions 6 through 11 are pooled, and 13.5 ml are applied directly to a 1.5 × 23 cm DEAE-cellulose column equilibrated with 50 m$M$ Tris·HCl (pH 7.6). Fractions of 4.6 ml are then collected at 1.9 ml/min. Stepwise gradient elution with NaCl solutions of 50, 100, 150, 200, and 350 m$M$ concentrations is performed and aldolase and phycocyanin (reflected by its absorption at 615 nm) located. The aldolase in fractions 43 through 57 (67 ml) is concentrated by addition of 37.8 g solid $(NH_4)_2SO_4$ to yield 80% saturation. The resultant precipitate after centrifugation is suspended in a final 3.0 ml of 50 m$M$ Tris-HCl (pH 7.6). The final preparation has a specific activity of 128 representing a 980-fold purification.

## Properties

*Apparent Molecular Weight.* When determined by the sucrose density gradient method,[19] the *Anacystis* aldolase migrates with muscle lactic acid dehydrogenase[20] and therefore has an apparent molecular weight of 137,000. This molecular weight was obtained with both catalytically inactive and activated samples.

*Purity.* The purified aldolase appears homogeneous when subjected to polyacrylamide disc gel electrophoresis.[21] Aldolase activity corresponds exactly to the single protein band obtained.[22] The purified enzyme contained no NAD- or NADP-linked glyceraldehyde-3 phosphate dehydrogenase activity and, on a unit of activity basis, contained about 0.1% triosephosphate isomerase.

[18] C. O'hEocha, *in* "Physiology and Biochemistry of Algae" (R. A. Lewin, ed.). p. 421. Academic Press, New York, 1962.

[19] R. G. Martin and B. N. Ames, *J. Biol. Chem.* **236**, 1372 (1961).

[20] Generously provided by Dr. N. O. Kaplan.

[21] B. J. Davis, *Ann. N. Y. Acad. Sci.* **121**, 321 (1964).

[22] J. M. Willard, Ph.D. Thesis, Cornell Univ., Ithaca, New York, 1967.

*pH Optimum.* Like other type II aldolases, the *Anacystis* enzyme exhibits a sharp pH optimum at pH 7.6.[8] No difference is seen when either activating and assaying at the same pH or activating at pH 7.6, then assaying at varying pH.

*Cofactor Requirements.* The *Anacystis* aldolase exhibits maximal activity only when both 8 m$M$ cysteine and 0.1 m$M$ $Fe^{2+}$ is present. Use of either alone elicits no activity. A 10-min preincubation is required for maximum activity. Unlike other type II aldolases, $K^+$ is without effect. When assayed in the presence of cysteine, $Fe^{3+}$ and $Mn^{2+}$ could replace $Fe^{2+}$. In the presence of $Fe^{2+}$, reduced glutathione and thiolycolate yield rates 20% and 58%, respectively, those obtained with cysteine. $\beta$-Mercaptoethanol and BAL were uneffective.

*Evidence for Metal in Anacystis Aldolase.* When excess aldolase is assayed in the presence of cysteine alone, the rate obtained is 3% that obtained with both cysteine and $Fe^{2+}$ present. This rate with cysteine alone is totally inhibited by $o$-phenanthroline and 2,2′-bipyridine suggesting the presence of a tightly bound metal, possibly iron.

*Specificity and Kinetics.* The *Anacystis* enzyme exhibits a high degree of specificity for FDP ($K_m = 0.16$ m$M$).

The enzyme cleaved 4 m$M$ sedoheptulose-1,7-P ($K_m = 10$ m$M$) at a rate 59% of that obtained with FDP. No activity was noted with fructose-1-P, ribulose 1,5-diphosphate, 2-keto,3-deoxyphosphogluconate, sorbose 1-phosphate, sorbose 1,6-diphosphate, rhamnulose 1-phosphate, or fuculose 1-phosphate, each at 4 m$M$ concentration. The turnover number for the *Anacystis* enzyme is 5200 compared to 6900 obtained for the yeast enzyme.[23] It appears that type II, metal-requiring aldolases have 10-fold lower affinities for FDP than do type I, nonmetal-requiring, aldolases, but their turnover numbers are twice those of the type I aldolases.

*Distribution.* Type II aldolase activity for which there is an absolute requirement for cysteine with $Fe^{2+}$ stimulating such activity has been demonstrated in cell-free extracts of the following blue-green algae: *Anabaena variabilis*, *Plectonema* sp., *Anabaenopsis* sp., and *Nostoc muscorum.*[5]

A type II aldolase with cofactor requirements identical to those of the yeast enzyme, i.e., maximal activity results with cysteine, $Zn^2$, and $K^+$ present,[3,23] has been obtained from the flexibacterium, *Saprospira thermalis.*[8]

Russell and Gibbs[24,25] have demonstrated that autotrophically grown *Chlamydomonas mundana* possesses a type I aldolase whereas heterotro-

[23] O. C. Richards and W. J. Rutter, *J. Biol. Chem.* **236**, 3177 (1961).
[24] G. K. Russell and M. Gibbs, *Biochim. Biophys. Acta* **132**, 145 (1967).
[25] G. K. Russell and M. Gibbs, *Plant Physiol.* **41**, 885 (1966).

phically grown cells possess only a type II aldolase.[24,25] Similarly, autotrophic *Euglena gracilis* and *Chlorella pyrenoidosa* possess type I aldolases, and heterotropic, dark grown cells, type II aldolases.[5] The type II aldolase of such dark-grown *Euglena* has been shown to be of cytoplasmic and not chloroplastic origin.[26] It appears that in the leaves of higher plant and green algae the type I aldolase participates in both photosynthesis and cellular respiration. In those cells which lose their ability to photosynthesize, a type II enzyme functions in cellular respiration. The red alga *Chondrus crispus* and the golden brown algae *Ochromonas danica* possess type I and II aldolases.[5]

Seeds and the etiolated leaves and cotyledons of plants appear to possess a variant of the type I aldolase, i.e., they lack mercurial inhibition. Subsequent illumination results in normal type I aldolase formation.[5]

[26] Y. Mo, B. G. Harris, and R. W. Gracy, *Arch. Biochem. Biophys.* **157**, 580 (1973).

## [37] Fructose Bisphosphate Aldolase from Spinach

*By* B. L. HORECKER

Fructose 1,6-bisphosphate $\rightleftharpoons$ dihydroxyacetone phosphate
$$+ \text{D-glyceraldehyde 3-phosphate}$$

Fructose-1,6-bisphosphate (Fru-P$_2$) aldolase in higher plants has been shown to form a Schiff-base intermediate with the substrate and therefore belongs to the class I aldolases.[1] The enzyme has been purified from peas,[2-4] cactus,[5] and spinach chloroplasts[6,7] and whole spinach leaves.[8] The procedure described is that of Fluri *et al.*[8] as modified by Davis.[9]

There have been some suggestions that plant cells may contain two forms of Fru-P$_2$ aldolase. The green alga *Chlamydamonas* has been reported to contain a class II aldolase as well as a class I aldolase, the latter probably playing a role in photosynthetic carbon dioxide fixation.[10]

[1] See this series, Vol. 9 [87].
[2] P. K. Stumpf, *J. Biol. Chem.* **176**, 233 (1948).
[3] C. Hatz and F. Leuthardt, *Hoppe Seyler's Z. Physiol. Chem.* **348**, 354 (1967).
[4] J. M. Willard, Thesis dissertation, Cornell University, Ithaca, New York, 1967.
[5] G. G. Sanwal and P. S. Krishnan, *Enzymologia* **23**, 249 (1961).
[6] K. Brooks and R. S. Criddle, *Arch. Biochem. Biophys.* **117**, 650 (1966).
[7] G. Jacobi, *Z. Pflanzen Physiol.* **56**, 262 (1967).
[8] R. Fluri, T. Ramasarma, and B. L. Horecker, *Eur. J. Biochem.* **1**, 117 (1967).
[9] J. C. Davis, Thesis dissertation, Albert Einstein College of Medicine, New York, N.Y., 1970.
[10] G. K. Russell and M. Gibbs, *Biochim. Biophys. Acta* **132**, 145 (1967).

However, the aldolase isolated from higher plants appears to be a class I aldolase,[8,11] and evidence for a class II aldolase in angiosperms is lacking. It is of interest that the purified enzymes isolated from chloroplasts[6] and from whole leaves[8] showed significant differences in molecular weight and amino acid composition. However, the cellular localization and heterogeneity of aldolase in higher plants remains an open question.

The procedure described here is a modification of that employed by Fluri *et al.*[8] for the isolation of aldolase from whole mature spinach leaves. A major difficulty with the original procedure was the large variation in activity and stability of the enzyme in the crude extracts, which also showed large seasonal variation. To some extent these variations could be reduced by allowing the spinach leaves to soak overnight in cold water (2°) before extraction, but even under these conditions, some preparations must be discarded.

## Assay Method

*Principle.* The assay is based on the oxidation of NADH by dihydroxyacetone phosphate (DHAP) in a coupled system containing triose-phosphate isomerase, and $\alpha$-glycerophosphate dehydrogenase.[12]

*Reagents*

TEA-EDTA solution: A stock solution is prepared containing 0.8 $M$ triethanolamine and 0.2 $M$ EDTA. The pH of this solution should be adjusted to 7.5 with NaOH or HCl as required. The stock solution is stored at 2° and diluted 20-fold with distilled water at room temperature before use. The diluted solution may be kept at room temperature for several weeks.

Fructose bisphosphate-Na (Boehringer) 0.1 $M$, pH 7.5

NADH-Na (Boehringer) 10 mg/ml in 0.001 $N$ NaOH (approximately 10 m$M$). Stored frozen at −16°.

$\alpha$-Glycerophosphate dehydrogenase–triosephosphate isomerase mixed crystals (Boehringer), 10 mg/ml.

Assay mixture: To 9.7 ml diluted TEA-EDTA solution add 0.2 ml Fru-$P_2$ solution, 0.1 ml NADH solution, and 0.01 ml of $\alpha$-glycerophosphate dehydrogenase–triosephosphate isomerase crystal suspension. This assay mixture is prepared fresh and kept at room temperature, but unused portions may be stored for several days at 2° and warmed to room temperature before use.

---

[11] C. Hatz and F. Leuthardt, *Biochim. Biophys. Acta* **139**, 460 (1967).
[12] E. Racker, *J. Biol. Chem.* **167**, 843 (1947).

*Procedure.* One ml of assay mixture is measured into a 1.0-ml cuvette with a 1.0-ml light path. The enzyme is diluted in TEA-EDTA solution to contain about 0.3–1.0 unit/ml, and 10–20 $\mu$l of diluted enzyme is added to the solution in the cuvette and mixed rapidly. The cuvette is then placed in a spectrophotometer with the cuvette chamber thermostated at 25°, and the change in absorbance at 340 nm is measured. The reaction rate is calculated from the linear portion of the curve. The reaction rate is expressed as micromoles of Fru-P$_2$ cleaved per minute, with $6.22 \times 10^3$ as the molar absorbancy coefficient of NADH.[13] In this assay each mole of Fru-P$_2$ cleaved results in the oxidation of 2 moles of NADH.

*Definition of Unit and Specific Activity.* One unit of enzyme activity is defined as the amount that will catalyze the cleavage of 1 $\mu$mole of Fru-P$_2$ per minute, under the conditions described above. Protein concentrations are determined by the method of Bücher,[14] or in purified fractions after gel filtration, from the absorbance at 280 nm. At pH 7.4 the specific absorbance at this wavelength, determined with a sample of purified enzyme dialyzed against distilled water, a portion of which was dried at 105°C and weighed, was 1.35 absorbance units for a solution containing 1 mg of enzyme per milliliter.


## Purification Procedure

All procedures are carried out at 4° or in an ice bath.

*Step 1. Extraction.* Fresh spinach from the local market is destemmed and stored in water at 4° until used, usually overnight. The extract is prepared by adding 800 g of drained leaves to 800 ml of cold deionized water in a large Waring Blendor. The blender is turned on for short periods until all the leaves are minced, then it is run at high speed for 1 min to obtain a smooth homogenate. Four such extracts are combined and filtered through 4 layers of wet cheesecloth. This removes air bubbles and fibrous material. Five liters of extract are recovered (crude extract).

*Step 2. Acidification.* The extract is stirred and adjusted to pH 5.0 with 4 N acetic acid, and the solution is kept at 4° for 45 min. It is then centrifuged for 15 min at 9000 rpm in a Sorvall RC2 centrifuge, using the GSA rotor. The supernatant (4.6 liters) is decanted (acid fraction).

*Step 3. First Ammonium Sulfate Fractionation.* To the acid fraction, 1.29 kg of solid (NH$_4$)SO$_4$ is added, with rapid stirring. The precipitate is collected by centrifugation at 9000 rpm as above, and the supernatant solution is retained (see note below). The precipitate is suspended in 120 ml of 20% cold (2°) saturated ammonium sulfate solution, and the

[13] B. L. Horecker and A. Kornberg, *J. Biol. Chem.* **175**, 385 (1948).
[14] T. Bücher, *Biochim. Biophys. Acta* **1**, 292 (1947).

suspension is centrifuged for 15 min at 9000 rpm. The residue is again extracted with 30 ml of 20% saturated ammonium sulfate solution, the suspension centrifuged and the two extracts, totaling 150 ml, combined (ammonium sulfate fraction I). This solution is frozen and stored at −16°C.

*Step 4. Chromatography on DEAE-Sephadex.* The frozen solution is thawed and passed through a column of Sephadex G-25 (3.5 cm × 60 cm) which has previously been equilibrated with 50 mM phosphate buffer, pH 7.4, using the same buffer to wash the protein through the column. For subsequent preparations the column is washed with 0.1 M acetic acid and then equilibrated with 50 mM phosphate buffer, pH 7.4.

The fractions containing enzyme are then applied to a column of DEAE-Sephadex (3.5 cm × cm) which has been equilibrated with 50 mM phosphate buffer, pH 7.4, at a flow rate of 60 ml/hr. The column is then washed with 50 ml of the same buffer and the enzyme eluted with a linear gradient of 0–0.4 M KCl in the same buffer in a total volume of 1 l. Fractions of 10 ml each are collected and analyzed for aldolase activity. The enzyme is eluted between 0.15 and 0.25 M KCl. Colored material is removed, and remains tightly bound to the column. The most active fractions (usually numbers 56–62, 70 ml) are pooled (DEAE eluate).

*Step 5. Second Ammonium Sulfate Fractionation.* The DEAE eluate is treated with an equal volume of saturated ammonium sulfate solution and adjusted to pH 7.4. The suspension is kept at 2° for 10 min and centrifuged at 9000 rpm for 15 min. The precipitate is discarded and to the supernatant solution is added the same volume as before of saturated ammonium sulfate solution. The suspension is centrifuged, and the precipitate is dissolved in 0.1 M TEA buffer, pH 7.8, containing 10 mM EDTA, to a final volume of 10 ml (ammonium sulfate fraction II). This fraction was stored at −20°.

*Notes on the Purification Procedure.* The type of spinach available in the local market is subject to large seasonal variations, affecting the yield of aldolase in the extract and first ammonium sulfate fraction. Except for the period from mid-June to mid-August, acceptable yields are obtained. During the summer months, the total activity after the first ammonium sulfate fractionation (step 3) may be only half that shown for the preparation in the table. A corresponding decrease in protein content is usually observed. Occasionally higher concentrations of ammonium sulfate are required to precipitate the enzyme in step 3, and the supernatant solution should not be discarded until the final fraction has been assayed. If necessary, additional fractions are collected by the further addition of ammonium sulfate.

In DEAE-Sephadex chromatography (step 4), a small peak contain-

PURIFICATION OF ALDOLASE FROM SPINACH LEAVES

| Step | Fraction | Volume (ml) | Total protein (mg) | Total activity (units) | Specific activity (units/mg protein) | Yield (%) |
|------|----------|-------------|--------------------|-----------------------|--------------------------------------|-----------|
| 1 | Extract | 5000 | 45,000 | 6200 | 0.13 | — |
| 2 | Acid fraction | 4600 | 5,500 | 4200 | 0.72 | 68 |
| 3 | Ammonium sulfate fraction I | 150 | 1,250 | 2400 | 1.25 | 39 |
| 4 | DEAE eluate | 70 | — | 1300 | — | — |
| 5 | Ammonium sulfate fraction II | 10 | 120 | 810 | 6.50 | 17 |

ing aldolase activity is consistently eluted before the main peak, at approximately 0.1 $M$ KCl. This material, of low specific activity, is usually discarded; except for this difference in chromatographic behavior its properties are similar to those of the bulk of the enzyme.

## Properties

*Homogeneity.* Preparations at the final stage of purification (step 5) show a single major protein band in disc gel electrophoresis, despite considerable variation in specific activity (see below). Some preparations show evidence of faint minor bands accounting for less than 5% of the total protein.[9] The preparations were also found to be homogeneous in equilibrium sedimentation.[15]

*Molecular Properties.* The molecular weight of native spinach aldolase is 120,000, as determined by sedimentation velocity.[8,15,16] The same molecular weight has been reported for aldolase isolated from spinach chloroplasts.[6] Spinach leaf aldolase is composed of four subunits. Dissociation by maleylation[15,16] yielded subunits with a molecular weight of 30,000. The COOH-terminal residue is tyrosine, but the number of residues released by digestion with carboxypeptidase is variable. Preparations of high specific activity appear to contain at least three COOH-terminal tyrosines, but the number may be as few as one with preparations showing low specific activity toward fructose bisphosphate. These preparations also contain COOH-terminal lysine, suggesting modification by proteolytic digestion during isolation of the enzyme[15] (see below).

[15] G. Rapoport, L. Davis, and B. L. Horecker, *Arch. Biochem. Biophys.* **132,** 286 (1969).

[16] C. L. Sia and B. L. Horecker, *Biochem. Biophys. Res. Commun.* **31,** 731 (1968).

Solutions of spinach leaf aldolase in 50 m$M$ Tris buffer may contain dimers having an apparent molecular weight of 235,000 and even larger aggregates. The higher molecular weight species are converted to monomers (molecular weight $\cong$ 120,000) by increasing the ionic strength.[9]

*Catalytic Activity.* The best preparations show specific activities with fructose 1,6-bisphosphate as substrate of 10–12 units/mg. The specific activity with fructose 1-phosphate is 0.4–0.5 unit/mg, and the activity ratio Fru-P$_2$/Fru-1-P with these preparations is about 25. Although some preparations may show lower specific activities toward Fru-P$_2$, the specific activity with Fru-1-P remains relatively constant, rarely falling below 0.3 units/mg. These differences are directly related to the number of COOH-terminal tyrosine residues. Preparations with high specific activity yield 3 or more tyrosines on digestion with carboxypeptidase A, while only one or two COOH-terminal tyrosine residues may be detected in preparation with low specific activity.[15]

Digestion with carboxypeptidase causes a marked decrease in activity toward fructose 1,6-bisphosphate, and a decrease in the Fru-P$_2$/Fru-1-P catalytic ratio, similar to that observed for rabbit muscle aldolase.[17,18] However, whereas the change in catalytic properties during digestion of rabbit muscle aldolase with carboxypeptidase is complete only when all the COOH-terminal tyrosine residues have been removed, the spinach enzyme loses all of its activity toward Fru-P$_2$ when only one tyrosine is removed, even though two or three intact subunits containing COOH-terminal tyrosine may still be present. This suggests a high degree of interaction between the subunits of spinach aldolase. This has not been observed for rabbit muscle aldolase, and may be related to the primarily gluconeogenic function of the spinach leaf enzyme. Similar cooperation of interaction between subunits during digestion with carboxypeptidase has been reported for fructose bisphosphate aldolase from lobster muscle.[19]

[17] E. R. Drechsler, P. D. Boyer, and A. G. Kowalsky, *J. Biol. Chem.* **234**, 2627 (1959).

[18] W. J. Rutter, *in* "The Enzymes," (P. D. Boyer, H. Lardy, and K. Myrbäck, eds.), 2nd ed., Vol. 5, p. 361 (1961).

[19] A. Guha, C. Y. Lai, and B. L. Horecker, *Arch. Biochem. Biophys.* **147**, 692 (1971).

# [38] Detection and Isolation of Mammalian Fructose-diphosphate Aldolases

By EDWARD E. PENHOET and WILLIAM J. RUTTER

Three distinct class I fructose-diphosphate aldolases have been purified in vertebrate tissues.[1] These aldolases, A, B, and C, have been purified from vertebrate skeletal muscle, liver, and brain, respectively.[2] Although functionally homologous, they have different catalytic and physical properties and different amino acid sequences, indicating that they are the products of three different genes.[3] Each of the enzymes contains 4 subunits, and hybrid tetrameric enzymes composed of the subunits of the parental types are found in many vertebrate tissues.[4] The kind and amount of aldolase present in a given tissue are strictly regulated, and change during embryological development.[5] These developmental transitions result in patterns of aldolase in the adult which are characteristic for specific organs and tissues.[4,6] In many cases, neoplasia in an adult tissue results in a reverse transition of aldolase complement to the fetal pattern.[7] Thus analysis of multiple aldolase forms can be useful both for studying the regulation of aldolases themselves in differentiation and as a marker for differentiation or dedifferentiation in other studies of ontogeny or oncogeny. A full definition of the aldolase present in a given tissue sample must include an analysis of the total catalytic activity of aldolase and an analysis of the enzyme forms present which are responsible for this activity. Methods for performing these determinations are presented below.

## Assay Method

*Preparation of Tissue Extracts.* Dissect the organ or tissue to be analyzed from a freshly killed animal and drop it into a beaker containing

[1] E. E. Penhoet, T. V. Rajkumar, and W. J. Rutter, *Proc. Nat. Acad. Sci. U.S.* **56**, 1275 (1966).

[2] E. E. Penhoet, M. Kochman, and W. J. Rutter, *Biochemistry* **8**, 4391 (1969).

[3] W. J. Rutter, T. V. Rajkumar, E. E. Penhoet, and M. Kochman, *Ann. N.Y. Acad. Sci.* **151**, 102 (1968).

[4] H. G. Lebherz and W. J. Rutter, *Biochemistry* **8**, 109 (1969).

[5] W. J. Rutter, and C. S. Weber, *in* "Developmental and Metabolic Control Mechanisms in Neoplasia" (D. N. Ward, ed.), pp. 195–218. Williams & Wilkins, Baltimore, Maryland, 1965.

[6] C. J. Masters, *Biochim. Biophys. Acta* **167**, 161 (1968).

[7] Y. Nordmann and F. Schapira, *Eur. J. Cancer* **3**, 247 (1967).

10 m$M$ Tris Cl, 1 m$M$ EDTA pH at 0°C. After the sample has cooled for a few minutes, remove and weigh it. Mince it into small pieces with dissecting scissors and suspend the pieces in 2 volumes (v/w) of the above buffer at 0°C. Pour this suspension into a glass homogenizer with a Teflon plunger and homogenize for about 10 strokes. Pour the resulting homoggenate into appropriate-sized ultracentrifuged tubes and spin in an ul-

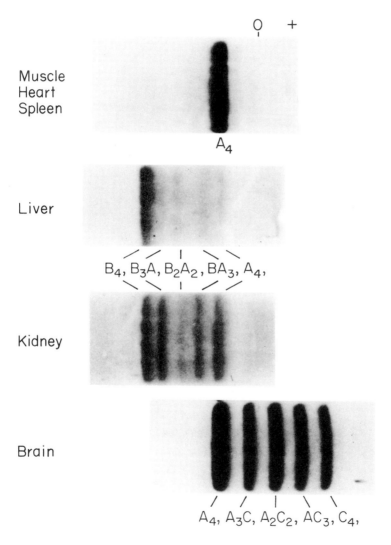

Fig. 1. Patterns of aldolase activity of adult rabbit tissues after cellulose acetate electrophoresis. Electrophoresis and activity staining for aldolase were performed as indicated by W. A. Susor, E. Penhoet, and W. J. Rutter, this series, Vol. 41 [15].

tracentrifuge for 60 min at 100,000 $g$ at 0–4°. After the centrifugation is completed, carefully remove the clear supernatant with a Pasteur pipette and keep at 0° for short-term storage (0–8 hr) or freeze at —20° for longer-term storage.

*Assay.* The type of aldolase present in a tissue extract can be qualitatively determined by electrophoresis of the extract followed by activity staining, as described in the accompanying paper.[8] Typical results of such analyses are presented in Fig. 1. A quantitative estimate of the total aldolase activity present in a tissue extract obtained as indicated above is determined by measuring the rate of cleavage of either fructose 1,6-diphosphate or fructose 1-phosphate according to the following reactions.

Fructose 1,6-diphosphate $\rightleftharpoons$ glyceraldehyde 3-phosphate
$$+ \text{ dihydroxyacetone phosphate}$$
Fructose 1-phosphate $\rightleftharpoons$ glyceraldehyde + dihydroxyacetone phosphate

Details of the assay are presented by Lebherz and Rutter.[8]

## Purification of Aldolases A, B, and C

For studies of aldolase enzymology it is necessary to obtain the various enzyme forms in a pure state. The methodologies presented below allow purification of these enzymes very rapidly and easily by affinity chromatography using phosphocellulose chromatographic columns and selective elution with the substrate FDP, a modification of the method originally utilized by Pogell.[9]

## General Methodology

All operations in the purification procedures are performed at 4° unless otherwise noted. Contact of aldolase preparations with the substrate is an integral part of the purification procedures, but the length of time of exposure to substrate should be minimized since it has been demonstrated that the enzymes are subject to slow inactivation in the presence of substrate.[10] Concentrations of purified aldolases are determined spectrophotometrically using extinction coefficients of 0.91, 0.89, and 0.88 optical density unit per milligram per milliliter, in a 1-cm light path for aldolases A, B, and C, respectively.

*Purification of Aldolase A.* A summary of the purification procedure is presented in Table I. A detailed description of the methods follows.

[8] H. G. Lebherz and W. J. Rutter, this volume [39].
[9] B. M. Pogell, *Biochem. Biophys. Res. Commun.* **7**, 225 (1962).
[10] B. M. Woodfin, *Biochem. Biophys. Res. Commun.* **29**, 288 (1968).

### TABLE I
#### PURIFICATION OF ALDOLASE A FROM RABBIT MUSCLE

| Purification steps | Total activity | Total protein (g) | Specific actiyity | Yield (%) |
|---|---|---|---|---|
| 1. 14,000 $g$ supernatant of crude extract | 78,700 | 60.26 | 0.13 | 100 |
| 2. 45–60% $(NH_4)_2SO_4$ precipitate | 55,900 | 11.8 | 4.8–5.2 | 71 |
| 3. Phosphocellulose chromatography | 54,400 | 3.2 | 17 | 69 |
| 4. Crystallization | 46,500 | 3.0 | 15–16 | 59 |

*Step 1. Crude Extract.* Anesthetize a young (6 months or so) rabbit by injecting 30 mg of a Nembutal solution per kilogram. Remove the skin of the rabbit, quickly remove the skeletal muscles, and cool in an ice bath. Grind the meat in a meat grinder and weigh the ground muscle. Suspend the ground muscle in 3 volumes (v/w) of 50 m$M$ Tris Cl, 5 m$M$ EDTA, 4 m$M$ 2-mercaptoethanol, pH 7.5. Homogenize this suspension in a commercial blender at low speed for 60 sec, and centrifuge at 14,000 $g$ for 30 min. Decant the supernatant solution for use in the following steps.

*Step 2. Ammonium Sulfate Fractionation.* Bring the opalescent supernatant from above to 45% saturation ammonium sulfate by adding 278 g of ammonium sulfate per liter of extract over a 1-hr period while maintaining the solution at approximately 2°. Let the suspension stand for 1 hr and then centrifuge at 14,000 $g$ for 1 hour. Remove the supernatant from this spin and discard the precipitate. Bring the supernatant to 60% saturation of ammonium sulfate by adding an additional 98 g per liter of supernatant. Adjust the pH of the suspension to 7.5 by adding an appropriate amount of 6 $N$ $NH_4OH$ and allow the solution to stand at least 2 hr before centrifuging at 14,000 $g$ for 1 hr. Discard the supernatant from this spin and dissolve the precipitate in 10 m$M$ Tris Cl, 1 m$M$ EDTA, pH 7.5. Adjust the protein concentration to approximately 40 mg/ml, and desalt the solution by passing it over a Sephadex G-25 column equilibrated with 10 m$M$ Tris Cl, 1 m$M$ EDTA, pH 7.5.

*Step 3. Phosphocellulose Chromatography.* Apply the desalted 45–60% ammonium sulfate fraction to a 3.7 $\times$ 100 cm phosphocellulose column equilibrated with 10 m$M$ Tris Cl, 1 m$M$ EDTA, pH 7.5. After applying the sample wash the column with 1–2 liters of 50 m$M$ Tris Cl, 5 m$M$ EDTA, pH 7.5 until the narrow red band visible on the column is eluted and the optical density of the effluent at 280 nm drops below 0.1. Proceed to elute the aldolase from the column with its substrate, fructose diphos-

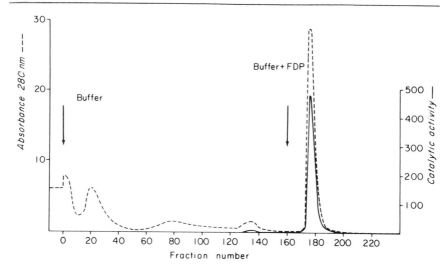

FIG. 2. Phosphocellulose chromatography of rabbit muscle aldolase. The 45–60% ammonium sulfate fraction obtained as described in the text, and equilibrated with 10 m$M$ Tris Cl, 1 m$M$ EDTA, pH 7.5 was applied to a 3.7 × 100 cm phosphocellulose column equilibrated with the same buffer. After the sample was applied, the column was washed with 50 m$M$ Tris Cl, 5 m$M$ EDTA, pH 7.5 until the red hemoglobin band was removed and the absorbance of the effluent at 280 nm fell below 0.1. The aldolase was then eluted with the same buffer to which 2.5 m$M$ FDP had been added. The volume of individual fractions collected was 10 ml.

phate, by washing the column with approximately 1 liter of 2.5 m$M$ fructose diphosphate in 50 m$M$ Tris Cl, 5 m$M$ EDTA, pH 7.5. This substrate wash elutes the aldolase A sharply with a maximum specific activity of 14–18 μmoles of FDP cleaved per minute per milligram of protein as shown in Fig. 2.

*Step 4. Crystallization.* Combine the peak fractions obtained from phosphocellulose chromatography above and dialyze against a solution which is 10 m$M$ Tris Cl, 1 m$M$ EDTA, 50% saturated ammonium sulfate, pH 7.5 (4° saturated, diluted 1.1 to achieve 90%). Crystals will form in the dialysis tubing after a period of 12–20 hr. Collect the crystals by centrifugation at 35,000 $g$ for 30 min. Resuspend the aldolase crystals in a small volume of the 50% saturated ammonium sulfate solution and store at 2–4° as a crystalline suspension for up to 6 months.

### Purification of Aldolase

A summary of the isolation of aldolase B is presented in Table II. With the following exceptions it is purified in the same manner as described above for aldolase A.

TABLE II

PURIFICATION OF ALDOLASE B FROM RABBIT LIVER

| Fraction | Volume (ml) | Activity (μmoles of FDP/min/ml) | Protein (mg/ml) | Total protein (mg) | Total activity | Specific activity | Yield (%) |
|---|---|---|---|---|---|---|---|
| 37,000 g supernatant | 950 | 0.84 | 28 | 26,300 | 790 | 0.032 | 100 |
| 45% supernatant | 1620 | 0.43 | 9 | 14,600 | 690 | 0.048 | 87 |
| 60% precipitate | 203 | 2.80 | 35 | 7,100 | 568 | 0.080 | 73 |
| Phosphocellulose fraction | 58 | 6.30 | 2.92 | 170 | 365 | 2.15 | 46 |

TABLE III

PURIFICATION OF THE ALDOLASE A-C SET FROM RABBIT BRAIN

| Fraction | Volume (ml) | Activity (μmoles of FDP/min/ml) | Protein (mg/ml) | Total protein (mg) | Total activity | Specific activity | Yield (%) |
|---|---|---|---|---|---|---|---|
| 37,000 g supernatant | 14,000 | 1.78 | 7.5 | 105,000 | 24,900 | 0.25 | 100 |
| 45% $NH_4SO_4$ supernatant | 14,800 | 1.47 | 4.1 | 60,700 | 21,700 | 0.36 | 87 |
| 60% $NH_4SO_4$ precipitate | 1,500 | 11.25 | 26 | 39,000 | 16,900 | 0.43 | 68 |
| Desalted | 2,300 | 8.0 | 17 | 39,000 | 18,400 | 0.47 | 74 |
| Cell $PO_4$ fraction | 403 | 20 | 3.5 | 1,410 | 8,060 | 5.77 | 32 |
| Total DEAE eluted material[a] | | | | | 5,210 | | 21 |

[a] Total activity: $A_1 = 720$, $A_3C = 1540$, $A_2C_2 = 1300$, $AC_3 = 980$ and $C_4 = 670$.

*Step 1. Preparation of a Crude Extract.* Remove the liters from fully fed, freshly killed rabbits and homogenize in 2 volumes of 0° 10 mM Tris Cl, 1 mM EDTA, pH 7.5, for 60 sec in a commercial blender, and centrifuge at 37,500 g for 60 min. It is important in this step to use only fresh livers as use of frozen rabbit livers gives a much lower yield of aldolase B with low specific activities. Take the supernatant from the 37,000 g spin and pass through Whatman No. 1 filter paper to remove excess lipid material.

*Step 2. Ammonium Sulfate Fractionation.* Obtain a 45 to 60% ammonium sulfate fraction as outlined above for aldolase A except use 0° saturated ammonium sulfate to obtain the fraction instead of solid ammonium sulfate, as in the preparation of aldolase A. As above desalt the fraction on G-25 and purify the aldolase by affinity chromatography on phosphocellulose.

*Step 3. Crystallization.* Dialyze the fractions of the phosphocellulose chromatography containing aldolase B against 10% saturated ammonium sulfate in 10 mM Tris-Cl, 1 mM EDTA, pH 7.5, for 2 hr. Then slowly raise the ammonium sulfate concentration to 55% by adding saturated ammonium sulfate to the dialysis vessel over a period of several hours. In about 12 hr homogeneous needle-type crystals will appear in the dialysis bag. The specific activity of aldolase B prepared by this method should be between 1.5 and 2.

## Purification of Aldolase C and the AC-Hybrids from Rabbit Brain

The fractionation procedure utilized for this purification is summarized in Table III. Details which differ from those presented above for aldolase A and B are presented below.

Place either fresh or freshly thawed frozen rabbit brains in 2.5 volumes of 10 mM Tris Cl, 1 mM EDTA, 0.1 mM phenylmethane sulfonylfluoride, pH 7.5. Homogenize for 10 sec at low speed in a commercial blender and then homogenize for 30 sec in a Sorvall omnimixer at 40,000 rpm. Centrifuge the resulting homogenate at 37,500 g for 1 hr and remove the supernatant for use in the succeeding steps.

*Step 1. Ammonium Sulfate Fractionation and Phosphocellulose Chromatography.* These steps are identical to those presented for aldolase A except that the phosphocellulose column is washed with 1 column volume of the sample buffer before the aldolases are eluted with 28 mM Tris Cl, 5 mM EDTA, 2.5 mM FDP, pH 7.5. The brain aldolases A4, A3C, A2C2, AC3, and C4 are eluted in a single peak with a specific activity of approximately 6. The enzymes are then purified further and resolved from each other by ion exchange chromatography.

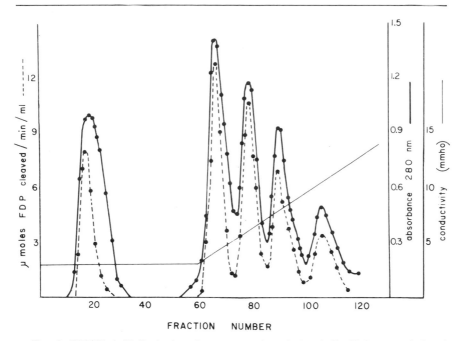

Fig. 3. DEAE A-50 Sephadex chromatography of the A-C aldolase set isolated from rabbit brain. Aldolases A-C eluted from phosphocellulose were equilibrated with 50 m$M$ Tris Cl, 4 m$M$ EDTA, 200 m$M$ sucrose, 5 m$M$ $\beta$-mercaptoethanol, pH 8.0. This elution was then applied to a 0.8 $\times$ 50 cm DEAE-Sephadex A-50 column equilibrated with the same buffer. Elution was carried out with a linear sodium chloride gradient (0 to 0.4 $M$ in the same buffer); 2-ml fractions were collected. Aldolase A was present in the breakthrough peak, and the order of elution beyond this was A$_3$C, A$_2$C$_2$, AC$_3$, C$_4$.

Step 2. DEAE-Sephadex Chromatography. Combine the aldolase containing fractions from the phosphocellulose column above, and concentrate to 10–20 mg/ml using ultrafiltration. Then pass them over a Sephadex G-25 column equilibrated with 50 m$M$ Tris Cl, 3 m$M$ EDTA, 200 m$M$ sucrose, and 5 m$M$ $\beta$-mercaptoethanol, pH 8.0. Apply the desalted solution to a DEAE-Sephadex A-50 column (2.5 $\times$ 50 cm) equilibrated with the same buffer and elute the aldolases with a linear sodium chloride gradient (0 $\rightarrow$ 0.35 $M$ NaCl) in 50 m$M$ Tris Cl, 3 m$M$ EDTA, 200 m$M$ sucrose, and 5 m$M$ $\beta$-mercaptoethanol, pH 8. As shown in Fig. 3, the various members of the AC hybrid set are resolved from each other by this procedure and elute from the column in the order A4, A2C2, AC3, and C4. Combine the fractions in each peak and concentrate to approximately 10 mg/ml by ultrafiltration and then precipitate by dialyzing against 55% saturated ammonium sulfate containing 1 m$M$ EDTA, pH 7.5.

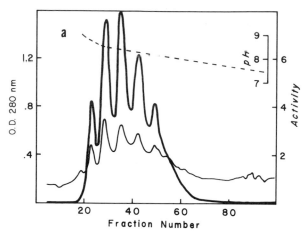

FIG. 4. Isoelectric resolution of crystalline rabbit muscle aldolase A of specific activity 16, electrofocused on an ampholine gradient (pH 7 to 9).

## Purity of the Aldolase Preparations

To determine purity of the aldolase preparations the samples may be subjected to polyacrylamide disc gel electrophoresis or analyzed by other traditional methods of protein chemistry such as ultracentrifugation, etc. To determine the efficacy of the resolution of the various members of the A-C set from each other in the purification from brain tissue perform the qualitative aldolase assay using cellulose acetate electrophoresis and activity staining which was presented in an earlier section of this manuscript. Purified, crystalline aldolase A preparations give rather diffuse bands in native polyacrylamide gels. This phenomenon is apparently due to an inherent heterodispersity of this protein, as shown in the studies of Susor, Kochman, and Rutter,[10a] who demonstrated that crystalline aldolase A preparations could be resolved into 5 peaks by electrofocusing (Fig. 4). These peaks behaved as a 5-membered hybrid set formed by combination of two different subunits in tetrameric molecules. This possibility was confirmed by the isolation of the two subunits and recombination to form the 5-membered set. Peptide maps confirmed a minor difference in the primary structure (probably the alteration of a single amino acid). This suggested the possibility of two separate aldolase A alleles in rabbits, or the degradation, for example by deamination *in situ*, or during the isolation procedure. The results of subsequent studies by Horecker and colleagues support the latter alternative.[11]

[10a] W. A. Susor, M. Kochman, and W. J. Rutter, *Science* **165**, 1260 (1969).
[11] C. Y. Lai, C. Chen, and B. L. Horecker, *Biochem. Biophys. Res. Commun.* **40**, 461 (1970).

This heterogeneity has little if any effect on general catalytic properties of the molecules. However, detailed studies of catalysis or chemistry should probably be carried out with an enzyme composed of a single subunit. The "native" aldolase A is probably the species of pI 8.2.

## [39] The Class I (Schiff Base) Fructose-diphosphate Aldolase of *Peptococcus aerogenes*

*By* HERBERT G. LEBHERZ and WILLIAM J. RUTTER

Fructose 1,6-diphosphate → dihydroxyacetone phosphate
+ D-glyceraldehyde 3-phosphate
Fructose 1-phosphate → dihydroxyacetone phosphate + D-glyceraldehyde

The fructose-1,6-diphosphate (FDP) aldolases found in biological systems can be assigned to one of two main classes.[1] Class I aldolases, as exemplified by the mammalian muscle enzyme, catalyze the reversible cleavage of FDP and fructose 1-phosphate (F-1-P) via the formation of a Schiff base intermediate between substrate and a lysyl residue at the active site of the enzyme.[2] In contrast, class II aldolases, as exemplified by the yeast enzyme, do not cleave F-1-P nor do they participate in a Schiff base complex with the substrate (FDP). Rather, these aldolases are metalloenzymes; they require divalent cations, usually $Zn^{2+}$ or $Fe^{2+}$, for activity.[3,4] The two aldolase types can further be distinguished on the basis of their catalytic and molecular properties.[5-7]

Until recently the class I aldolases were thought to be restricted to "higher organisms"; namely, animals, plants, protozoa, and algae. Class II aldolases were found in fungi and *all* prokaryotic cells tested.[1,6] *Euglena*[6] and *Chlamydomonas*[8] contain both aldolase types.

We have recently shown that the aldolase isolated from the bacterium *Peptococcus aerogenes*[9] ATCC No. 14963 (also called *Micrococcus aero-*

[1] W. J. Rutter, *Fed. Amer. Soc. Exp. Biol.* **23**, 1248 (1964).
[2] B. L. Horecker, P. T. Rowley, E. Grazi, T. Cheng, and O. Tchola, *Biochem. Z.* **338**, 36 (1963).
[3] R. D. Kobes, R. T. Simpson, B. L. Vallee, and W. J. Rutter, *Biochemistry* **8**, 585 (1969).
[4] W. J. Rutter and K. H. Ling, *Biochim. Biophys. Acta* **30**, 71 (1958).
[5] O. C. Richards, and W. J. Rutter, *J. Biol. Chem.* **236**, 3177 (1961).
[6] W. J. Rutter and W. E. Groves, *in* "Taxonomic Biochemistry, Physiology and Serology" (C. A. Leone, ed.), p. 417. Ronald Press, New York, 1964.
[7] W. E. Groves, Ph.D. Thesis, University of Illinois, Urbana, Illinois, 1962.
[8] G. K. Russell, and M. Gibbs, *Biochim. Biophys. Acta* **132**, 145 (1967).
[9] R. S. Breed, W. G. D. Murray, and N. R. Smith, "Bergey's Manual of Determinative Bacteriology," 7th Ed. Williams & Wilkins, Baltimore, Maryland, 1957.

*genes*) in contrast to the aldolases of other prokaryotes, is a class I enzyme.[10,10a] A method for culturing *P. aerogenes* and the procedure for isolating the aldolase in pure form are described here. In addition, we compare the properties of this enzyme with those of the well-characterized rabbit (class I) and yeast (class II) aldolases.

## Assay Method

*Principle.* Limiting amounts of aldolase are incubated in an appropriate buffer with excess amounts of substrate, NADH, $\alpha$-glycerolphosphate dehydrogenase, and triose-phosphate isomerase. The cleavage of substrate generates triose phosphates, which are then quantitatively converted to $\alpha$-glycerol phosphate with the concomitant oxidation of stoichiometric amounts of NADH to NAD.[11] This latter conversion is followed spectrophotometrically by measuring the rate of decrease in absorbance at 340 nm which results from the oxidation of NADH. The rate of NADH oxidation is calculated using the molar extinction coefficient of NADH at 340 nm. The quantity of NADH oxidized is directly related to the micromoles of substrate cleaved in the aldolase reaction. Two micromoles of NADH are oxidized for each micromole of FDP cleaved since the two triose phosphates formed are both utilized, via the mediation of triosephosphate isomerase. Only 1 $\mu$mole of NADH is oxidized for each micromole of F-1-P cleaved since only 1 $\mu$mole of triose phosphate is formed in this reaction.

*Reagents*

Glycylglycine buffer, 0.1 $M$, pH 7.5
D-Fructose 1,6 diphosphate, trisodium salt (99% pure, Sigma), 50 m$M$
D-Fructose 1-phosphate, dimonocyclohexylammonium salt (99% pure, Sigma), 0.1 $M$
$\alpha$-Glycerolphosphate dehydrogenase–triosephosphate isomerase, mixed crystals (Sigma or Boehinger), 10 mg/ml
NADH (99% pure, Sigma)

*Procedure.* Enzyme "reaction mix" contains the following: 20 ml of glycylglycine buffer, 4 mg of NADH, and 0.05 ml of.$\alpha$-glycerolphosphate dehydrogenase–triosephosphate isomerase. Reactions are routinely initiated by the addition of 5–50 $\mu$l of enzyme solution, containing

[10] H. G. Lebherz and W. J. Rutter, *J. Biol. Chem.* **248**, 1650 (1973).
[10a] H. G. Lebherz, Ph.D. Thesis, University of Washington, Seattle, Washington (1970).
[11] E. Racker, *J. Biol. Chem.* **167**, 843 (1947).

0.005–0.02 unit, to 1-ml glass cuvettes (1 cm light path) which contain "reaction mix" and either 0.05 ml of FDP or 0.1 ml of F-1-P solutions (final volume = 1 ml). Alternatively, large volumes of "reaction mix" containing FDP may be prepared fresh each day. The decrease in $A_{340}$ with time is determined at 25° in a spectrophotometer which is equipped with a water jacket maintained at 25°. Since the molar extinction coefficient of NADH at 340 nm is $6.22 \times 10^6$ $cm^2$ mole,[12] absorbance changes of 12.44 and 6.22 OD units correspond to the cleavage of 1 $\mu$mole of FDP and 1 $\mu$mole F-1-P, respectively. The rate of cleavage of FDP and F-1-P in micromoles per minute can then be calculated.

Crude extracts show appreciable "activity" in the absence of added substrate. This substrate independent oxidation of NADH is detected by deleting the substrate from the reaction mixtures. This endogenous rate is then subtracted from the values obtained with assays containing substrate. Corrections of this kind are necessary only when assaying crude enzyme preparations.

*Definition of a Unit and Specific Activity.* A unit of aldolase is defined as that amount of enzyme which catalyzes the cleavage of 1 $\mu$mole of substrate per minute at 25° under the above conditions of assay. The specific activity is defined as the number of activity units per milligram of protein. Protein concentrations of crude extracts and partially purified enzyme preparations are estimated by measuring the absorbance at 260 nm and 280 nm according to the method of Warburg and Christian.[13] Protein concentrations of purified enzyme solutions are determined by measuring the absorbance at 280 nm using the extinction coefficient ($E_{288}$ 1 mg/ml, 1 cm = 0.58) of this enzyme at 280 nm[10] (i.e., protein concentration in mg/ml = $A$ 280/0.58).

## Growth of *P. aerogenes* Cells

*Peptococcus aerogenes* strain 228 cells are cultured anaerobically at 37° in glutamate-supplemented media.[14] This medium is prepared as follows: 49 g of Bacto peptone (Difco Laboratories), 49 g of Bacto yeast extract (Difco), 25 g of sodium glutamate, 6.5 g of sodium thioglycolate, 130 ml of 2 $M$ potassium phosphate, pH 7.0, is dissolved in tap water and diluted to 6500 ml. The medium is then divided into two 200-ml volumes (in 250-Erlenmeyer flasks) and two 3000-ml volumes (in 4000-ml Erlenmeyer flasks) followed by sterilization in the autoclave at 120° for 20 min. After cooling to 37°, cultures are initiated by adding 30 ml of

---

[12] A. Kornberg and B. L. Horecker, *in* "Biochemical Preparations (E. E. Snell, ed.), Vol. 3, p. 27. Wiley, New York, 1953.

[13] O. Warburg and W. Christian, *Biochem. Z.* **310**, 348 (1942).

[14] H. R. Whiteley, *J. Bacteriol.* **74**, 324 (1957).

inocula[15] to each of the 250-ml flasks. After 20 hr, these cultures are added to the two 4000-ml flasks and the cultures grown for an additional 20 hr. Afterward, the combined cultures are added to a 100-liter fermenter (New Brunswick Fermacell) containing sterile media comprised of 1.5 lb of peptone, 1.5 lb of yeast extract, 0.5 lb of sodium glutamate, 1400 ml of 2 $M$ potassium phosphate buffer, and tap water to 100 liters. The fermenter cultures are grown at 37° for 20–24 hr with mild stirring (50 rpm), and the cells are harvested with the aid of a Sharples continuous-flow centrifuge (10,000 rpm); 150–200 g of packed cells are routinely obtained from each fermenter culture. The cells are washed by suspension in distilled water and, after centrifugation, the cell paste is stored at —20° until used.

### Purification Procedure

All procedures are performed at 0° to 4° unless otherwise noted.

*Step 1. Crude Extract.* Frozen cells, 1000 g, are suspended in 1000 ml of Buffer I (10 m$M$ Tris·HCl, 1 m$M$ EDTA, 1 m$M$ 2-mercaptoethanol, pH 7.5); 1200 ml of 120 $\mu$l glass beads (previously washed in 6 $N$ HCl) are added to the suspension, and the cells are disrupted in a Gifford-Wood colloid mill ($\frac{1}{30}$ inch gap) at 5–10°C. Cell suspensions are routinely disrupted for 50 min, longer grinding times did not result in a substantial increase in activity or in preparations with higher specific activity. Volumes of 1000–1500 ml are conveniently accommodated by the mill at one time. After grinding, the disrupted cell suspension is decanted from the glass beads and the beads are washed with 2000 ml of buffer I. The suspension and wash are combined and adjusted to 6 m$M$ MgCl₂ by the addition of the appropriate volume of 1 $M$ MgCl₂; then 6 mg of bovine pancreatic DNase (Calbiochem, B grade) are added per 1000 ml of suspension.[16] The crude extract is then centrifuged at 16,000 $g$ for 90 min in the Sorvall centrifuge and the supernatant used for further purification.

*Step 2. Ammonium Sulfate Fractionation.* The following procedure was chosen for the large-scale fractionation of step I supernatant with ammonium sulfate. The supernatant from step I (4050 ml) is brought to 50% saturation by the slow addition of 312 g of ammonium sulfate per 1000 ml with constant stirring. The pH is adjusted to 7.5 by addition of 7 $N$ NH₄OH, and the precipitate is allowed to form overnight. The

---

[15] Generously supplied by Dr. H. R. Whiteley, Department of Microbiology, University of Washington, Seattle, Washington.

[16] DNase treatment is essential to obtain good separation of soluble and insoluble cell components and, later, to achieve good separation of ammonium sulfate precipitate and supernatant fractions.

precipitated protein is sedimented by centrifugation at 13,000 $g$ for 90 min and discarded. The supernatant (4308 ml) is brought to 65% saturation by the slow addition of 100 g of ammonium sulfate per 1000 ml; after 3 hr, the precipitated protein is collected by centrifugation and discarded. The supernatant (4280 ml) is brought to 100% ammonium sulfate saturation, the pH is adjusted to 5.5 with 7 $N$ acetic acid, and the precipitate is allowed to form overnight at 23°.[17] The precipitated proteins are then collected by centrifugation and used for further purification.

*Step 3. DEAE-Sephadex $A_{25}$ Ion Exchange Chromatography.* The material from step 2, even after extensive dialysis, is very viscous; consequently, since DEAE $A_{25}$, unlike $A_{50}$, has a high flow rate, it was selected for the first ion exchange chromatography. The material from step 2 is dissolved in approximately 400 ml of buffer I and extensively dialyzed against buffer I which contains 0.15 $M$ NaCl. When the conductivity of the dialyzed solution approximates that of the buffer, the solution is centrifuged at 11,000 $g$ for 30 min to remove denatured proteins. The supernatant is then diluted to a final protein concentration of 10 mg/ml or less, and applied to a DEAE-Sephadex $A_{25}$ column (4.5 × 50 cm). The Sephadex column is prepared as follows[3] DEAE-Sephadex beads are suspended in buffer I at room temperature, and the slurry is adjusted to pH 7.5 by the addition of 1 $M$ Tris base. The beads are allowed to swell for 3 days at room temperature and then the Sephadex is equilibrated with buffer I containing 0.15 $M$ NaCl at 4° by washing several times with this buffer. pH adjustments, when necessary, are made with 1 $M$ Tris-base or 1 $M$ HCl. After the column is poured, one column volume of buffer is passed through the column before the sample is applied. After sample application, the column is washed with 1000 ml of buffer I and is then developed with a 0.15 $M$ to 0.50 $M$ NaCl linear gradient (4000 ml total). Fractions of 15–20 ml are collected, and fractions containing aldolase activity are pooled (1660 ml) and concentrated to approximately 200 ml by ultrafiltration through Diaflo PM-10 membranes.

*Step 4. DEAE-Sephadex $A_{50}$ Ion Exchange Chromatography.* The enzyme solution from step 3 is adjusted to pH 6.0 with 7 $N$ acetic acid and diluted with water until the conductivity of the sample is equal to that of buffer II (10 m$M$ sodium malonate, 1 m$M$ 2-mercaptoethanol, 0.2 $M$ NaCl, pH 6.0, conductivity = 8 millimho). The sample (approximately 300 ml) is then applied to a DEAE-Sephadex $A_{50}$ column (4.5 × 68 cm), which is prepared as follows: The Sephadex beads are suspended in buffer II, and the slurry is adjusted to pH 6.0 with 1 $N$

---

[17] The lower pH and higher temperatures of this last fractionation facilitates precipitation of aldolase activity from the saturated ammonium sulfate solution.

NaOH. After 3 days at room temperature, the Sephadex is equilibrated with buffer II at 4° and, when necessary, pH adjustments are made with 1 $N$ NaOH or 1 $M$ malonic acid. After pouring the column, one column volume of buffer is passed through the column before application of the sample. After sample application, the column is developed with a 0.2 $M$ to 0.4 $M$ NaCl linear gradient (4000 ml total); 15–20 ml fractions are collected, and fractions containing aldolase activity are pooled (525 ml) and concentrated to 25 ml by ultrafiltration.

*Step 5. Sephadex G-100 Gel Filtration.* Sephadex G-100 suspended in buffer I and allowed to equilibrate for 4 days at room temperature. The Sephadex is then washed several times at 4° with buffer I, poured into the column and washed with one column volume of buffer. The sample from step 4 is applied and the column developed with buffer I; 8–10-ml fractions are collected, and fractions containing aldolase activity are pooled (133) and concentrated to 13 ml by ultrafiltration. The sample is rechromatographed on the same G-100 column to remove minor protein contaminants, and the fractions containing constant specific activity are pooled. The final preparation is concentrated to a protein concentration of approximately 5 mg/ml, divided into aliquots, and stored frozen at −20°C until used.

With the use of this procedure, *P. aerogenes* aldolase is purified to essential homogeneity with about 70% recovery of activity.[10] The protein migrates as a single component upon electrophoresis and gel filtration. The total purification is about 1500-fold when protein measurements are made according to the method of Warburg and Christian.[13] However, when the protein concentration of the purified enzyme is mea-

TABLE I
PURIFICATION OF *Peptococcus aerogenes* ALDOLASE

| Procedure | Volume (ml) | Total activity (units) | Total protein (mg) | Specific activity (units/ mg) | Purification (fold)[a] | Recovery (%) |
|---|---|---|---|---|---|---|
| 16,000 $g$ supernatant | 4050 | 1445 | 146,000 | 0.010 | — | 100 |
| 50% $(NH_4)_2SO_4$ supernatant | 4308 | 1422 | 106,800 | 0.013 | 1.3 | 98.5 |
| 65% $(NH_4)_2SO_4$ supernatant | 4280 | 1284 | 51,500 | 0.025 | 2.5 | 89 |
| 65–100% $(NH_4)_2SO_4$ precipitate | 400 | 1231 | 12,400 | 0.10 | 10 | 85.2 |
| Dialysis | 550 | 1154 | 9,620 | 0.115 | 11.5 | 79.9 |
| DEAE $A_{25}$ Chromatography | 1660 | 1079 | 1,776 | 0.61 | 61 | 74.6 |
| DEAE $A_{50}$ Chromatography | 525 | 1046 | 273 | 3.83 | 383 | 72.4 |
| Sephadex G-100 gel filtration | 133 | 1026 | 69 | 14.8 | 1480 | 71.0 |

[a] Specific activity relative to that of step 1.

sured by $A_{280}$[10] using the calculated extinction coefficient for the enzyme, the total purification is somewhat less, i.e., 800-fold.

A typical purification is summarized in Table I.

## Properties

*Catalytic Mechanism.* *Peptococcus aerogenes* aldolase is a class I aldolase and is distinct from the aldolases of fungi and prokaryotes previously studied. This enzyme has no metal requirement for activity. The enzyme is not inhibited by high concentrations (10 m$M$) of EDTA or $o$-phenanthroline.[10] Furthermore, the enzyme does not contain bound $Zn^{2+}$,[10] the metal commonly found in class II aldolases.[1] Like the aldolases of higher organisms, this aldolase is inactivated by incubation with dihydroxyacetone phosphate and NaBH$_4$. A covalent bond between [$^3$H]dihydroxyacetone phosphate and a lysyl residue of the enzyme is found on treatment with NaBH$_4$.[10] Lysine has been identified as the substrate binding site in other aldolases of both animal[2,18] and plant [19] origin. By analogy with the aldolases of higher organisms, it is proposed that the reversible cleavage of FDP and F-1-P mediated by this aldolase proceeds via the formation of a Schiff base between C-2 of dihydroxyacetone phosphate and the $\epsilon$-NH$_2$ group of a lysyl residue at the active site of the enzyme.[2] The enzyme contains one catalytically active site per enzyme subunit as judged by the stoichiometric incorporation of [$^3$H]dihydroxyacetone phosphate upon reduction of the enzyme substrate Schiff base with NaBH$_4$, and the concomitant inactivation of catalytic activity.

*Substrate Specificity.* Purified *P. aerogenes* aldolase has a specific activity of about 8 $\mu$moles of FDP cleaved per minute per milligram of protein,[10] which is nearly identical with that of rabbit aldolase C. Class II aldolases (Table II) exhibit much higher specific activity. In common with other class I aldolases, this enzyme can use F-1-P as substrate. The $K_m$ values with FDP and F-1-P as substrate are essentially identical with those of rabbit aldolases B and A, respectively (Table II). The $V_{max}$ values (units/subunit) with FDP and F-1-P are quite similar to those of rabbit aldolases C and B, respectively (Table II).

*pH Optima.* The broad pH optima for FDP cleavage and the narrow optimum for F-1-P cleavage (pH 7.5) are similar to those of rabbit aldolases A and C.[10,20]

*Inactivation.* All class I aldolases so far studied are irreversibly inactivated by incubation with low levels of carboxypeptidase A.[1,7] In con-

[18] Z. E. Swartz and B. L. Horecker, *Arch. Biochem. Biophys.* **115**, 407 (1966).
[19] R. Fluri, J. Rasnasarma, and B. L. Horecker, *Eur. J. Biochem.* **1**, 117 (1967).
[20] E. E. Penhoet, M. Kochman, and W. J. Rutter, *Biochemistry* **8**, 4396 (1969).

## TABLE II
### CATALYTIC PROPERTIES OF FRUCTOSE-DIPHOSPHATE ALDOLASES

| Property | Class 1 (rabbit)[a] | | | Class I (Peptococcus aerogenes)[b] | Class II (yeast)[c] |
|---|---|---|---|---|---|
| | A | B | C | | |
| Specific activity ($\mu$moles FDP cleaved/min/mg protein) | 15–16 | 1–2 | 6–10 | 7–9 | 85–100 |
| $K_m$ FDP | $3 \times 10^{-6}$ | $1 \times 10^{-6}$ | $3 \times 10^{-6}$ | $1 \times 10^{-6}$ | $3 \times 10^{-4}$ |
| $K_m$ F-1-P | $1 \times 10^{-2}$ | $3 \times 10^{-4}$ | $4 \times 10^{-3}$ | $1 \times 10^{-2}$ | — |
| $V_{max}$ FDP/subunit | 725 | 63 | 250 | 280 | 4500 |
| $V_{max}$ F-1-P/subunit | 29 | 63 | 31 | 79 | — |
| $V_{max}$ FDP/$V_{max}$ F-1-P | ~25[b] | 1 | 8–10 | ~3 | — |
| pH optima (FDP) | 6.6–8.5 | 7.2–7.6 | 6.5–8.5 | 6.5–8.5 | ~7.2 |
| Inactivation by carboxypeptidase A | 98% | 54% | 93% | 0 | 0. |

[a] R. Fluri, J. Rasnasarma, and B. L. Horecker, *Eur. J. Biochem.* **1**, 117 (1967).
[b] H. G. Lebherz and W. J. Rutter, *J. Biol. Chem.* **248**, 1650 (1973).
[c] O. C. Richards, Ph.D. Thesis, University of Illinois, Urbana, Illinois, 1960.

trast, the *P. aerogenes* enzyme is insensitive to carboxypeptidase A even at concentrations 100 times greater than those required to achieve rapid inactivation of the rabbit enzymes[10] (Table II). Rabbit aldolase A is inactivated by treatment with the sulfhydryl reagent 5,5'-dithiobis- '2-nitrobenzoic acid (DTNB). This inactivation can be prevented by FDP,[21] and it has been proposed that one SH group per enzyme subunit is essential for catalytic activity.[22,23] In contrast, *M. aerogenes* aldolase is not inactivated nor is there sulfhydryl group modification upon treatment of the enzyme with DTNB.[10]

## Molecular Properties

*Amino Acid Composition. Peptococcus aerogenes* and rabbit muscle aldolases contain similar amounts of many amino acids.[24] However, the *P. aerogenes* enzyme contains considerably lower levels of proline, histidine, and threonine and higher levels of methionine as compared to the rabbit enzyme.[24] Of particular interest is the detection of no more than 0.3 residue of cysteine per enzyme subunit by amino acid analysis, alkyla-

[21] P. J. Anderson and R. N. Perham, *Biochem. Z.* **117**, 291 (1970).
[22] D. E. Morse and B. L. Horecker, *Advan. Enzymol.* **31**, 125 (1968).
[23] B. L. Horecker, *Fed. Eur. Biochem. Soc. Symp.* **19**, 181 (1970).
[24] H. G. Lebherz, R. A. Bradshaw, and W. J. Rutter, *J. Biol. Chem.* **248**, 1660 (1973).

tion under strongly denaturing conditions, or by reaction with DTNB.[10,24] Consequently, it appears that sulfhydryl groups are not involved in the catalytic function of this enzyme, in contrast to the proposed involvement of a sulfhydryl group in catalysis by the mammalian aldolases.[22,23]

*Amino Terminal Amino Acid.* *Peptococcus aerogenes* aldolase has a methionine at the amino terminus,[2] in contrast to proline at the amino terminus of the rabbit[22,25] and other vertebrate[26] aldolases.

*Carboxy-Terminal Amino Acid.* The purified enzyme has a lysyl residue at the COOH terminus,[24] in contrast to tyrosine at the COOH terminus of all animal[18,26,27] and plant[19] aldolases so far studied. The inactivation of FDP cleavage activity of other class I aldolases by carboxypeptidase A digestion occurs simultaneously with the release of the COOH tyrosyl residue.[27] Consequently, this tyrosine has been implicated to function in the accelerated cleavage of FDP as compared with F-1-P seen with class I aldolases.[22,23,27] Because lysine is at the COOH terminus of the *P. aerogenes* enzyme, COOH-terminal tyrosyl residues cannot be involved in the accelerated cleavage of FDP observed with this enzyme.[10]

*Primary Structure at the Active Site.* The tryptic peptide containing the lysyl residue which participates in Schiff-base formation with substrate has been isolated and partially characterized.[24] The peptide is 28 amino acids long, the same length as the corresponding peptides of rabbit aldolases A and B.[22] The position of 6 residues has been determined and alignment of the remaining 22 residues to obtain maximum homology to the rabbit A peptide results in identity at 13 of the 28 positions.[24] Thus, there could be a maximum of 48% homology between the primary structures of the two peptides.

*Subunit Structure.* As determined by high speed sedimentation-equilibrium ultracentrifugation in neutral buffers and in solutions of 6 $M$ guanidine·HCl, *P. aerogenes* aldolase has a molecular weight of approximately 33,000 and is composed of a single polypeptide chain.[10] All other aldolases studied are oligomeric enzymes. The rabbit aldolase A, B, C (MW = 160,000)[28,29] and spinach aldolase[30] (MW = 120,000) are all composed of four subunits. The class II aldolase of yeast (MW = 80,000) is a dimer.[31]

[25] W. J. Rutter and P. Edman, unpublished observations.
[26] C. E. Cuban and L. F. Hass, *J. Biol. Chem.* **246**, 6807 (1971).
[27] E. R. Drechsler, P. D. Boyer, and A. G. Kowalsky, *J. Biol. Chem.* **234**, 2627 (1959).
[28] K. Kawokara and C. Tandord, *Biochemistry* **5**, 1578 (1966).
[29] E. Penhoet, M. Kochman, R. R. Valentine, and W. J. Rutter, *Biochemistry* **6**, 2940 (1967).
[30] G. Rapoport, L. Davis, and B. L. Horecker, *Arch. Biochem. Biophys.* **132**, 286 (1969).
[31] C. E. Harris, R. D. Kobes, D. C. Teller, and W. J. Rutter, *Biochemistry* **8**, 2442 (1969).

## Distribution

Prior to the initial report that *P. aerogenes* contains a class I aldo-lase,[32] all prokaryotic cells studied (at least 50 species) were found to contain class II aldolases.[1,7,32] Even a bacterium which has certain nutri-tional and metabolic similarities to *P. aerogenes* (*Acidaminococcus fermentans* strain vr-2[33]) was found to contain a class II aldolase.[10] More recently, Class I aldolases have been detected in two additional bacteria. Kaklij and Nadkarni[34] have shown that the aldolase activity of *Lactobacillus casei* is insensitive to EDTA and is inactivated by incubation with dihydroxyacetone phosphate and NaBH$_4$. Stribling and Perham[35] have isolated both a class I and a class II aldolase from *Escherichia coli* cells grown on pyruvate or lactate growth media; only the class II aldolase is detected in cells grown on glucose. Both the *L. casei*[36] and *E. coli*[35] Class I aldolases have some properties distinct from those of the *P. aerogenes* enzyme. In addition, both aldolases are presumably oligom-eric since they have molecular weights greater than 100,000.

[32] H. G. Lebherz and W. J. Rutter, *Biochemistry* **8**, 109 (1969).
[33] M. Rogosa, *J. Bacteriol.* **98**, 756 (1969).
[34] G. S. Kaklij and G. B. Nadkarni, *Arch. Biochem. Biophys.* **140**, 334 (1970).
[35] D. Stribling and R. N. Perham, *Biochem. J.* **131**, 833 (1973).
[36] G. B. Nadkarni, and G. S. Kaklij, *Proc. Int. Congr. Biochem. 9th 1973*. Stockholm, Sweden, p. 62 (1973).

# [40] 2-Keto-3-deoxy-6-phosphogluconic Aldolase from *Pseudomonas putida*

By R. H. HAMMERSTEDT, H. MÖHLER, K. A. DECKER, DIANA ERSFELD, and W. A. WOOD

2-Keto-3-deoxy-6-phosphogluconate $\rightleftarrows$ pyruvate + D-glyceraldehyde 3-phosphate

2-Keto-3-deoxy-6-phosphogluconate (KDPG) aldolase, along with 6-phosphogluconate dehydrase, are the unique enzymes of the Entner-Doudoroff[1] pathway for glucose and gluconate dissimilation in pseudo-monads and similar bacteria. Because the aldolase has an unusual tri-meric arrangement of subunits, relatively large amounts of enzyme have been prepared in crystalline form, using the procedure described below, for physical and chemical studies including X-ray crystallography and sequencing.

[1] N. Entner and M. Doudoroff, *J. Biol. Chem.* **196**, 853 (1952).

**Assay Method**[2]

The assay utilizes the continuous measurement of pyruvate formation as the disappearance of NADH in the presence of excess lactic dehydrogenase and NADH.

*Reagents*

NaNADH, 10 m$M$

Imidazole buffer, 0.5 $M$, pH 8.0

KDPG, 50 m$M$. Ten milligrams of BaKDPG[3] is dissolved in 0.25 ml of $H_2O$ with the aid of Dowex 50 ($H^+$). The mixture is then filtered on a micro sintered-glass filter. The filtrate should be acid with Congo red. If not, add more resin and repeat. The pH is raised to 6.0 with NaOH and the volume brought to 0.5 ml.

Lactic dehydrogenase heart muscle (crystals in ammonium sulfate) 0.1 ml of NADH, 0.2 ml of imidazole buffer, 0.02 ml of lactic dehydrogenase, and 0.7 ml of water are added to form a reagent stock solution.

The assay is performed in microcuvettes (3 $\times$ 25 mm, 1-cm light path, 0.5-ml capacity) by adding 0.01 ml of KDPG, 0.08 ml of reagent stock solution, and enzyme and water to a volume of 0.15 ml. The enzyme is added last to start the reaction. The decrease in absorbance at 28° is followed at 340 nm in a Gilford spectrophotometer equipped with a pinhole mask. The rate is linear with time and enzyme concentration over a wide range of velocities.

One unit of activity produces 1 $\mu$mole of pyruvate per minute under the above conditions. Protein was measured by the method of Lowry *et al.* or at 280 nm using bovine serum albumin as a standard. The ratios of dry weight, $A_{280}:A_{260}$ and Lowry values are 1.0:1.0:1.33; the $E_{1\%}$ at 280 nm in 0.1 $N$ NaOH is 8.6.[4]

*Enzyme Source. Pseudomonas putida*, ATCC 12633 is grown in a medium consisting of 2% potassium gluconate as the carbon source, 0.6% $(NH_4)_2HPO_4$, 0.3% $K_2HPO_4$. 0.1% $MgSO_4\cdot7\ H_2O$, 0.15% citric acid, and 0.0005% $FeCl_3$. A solution of 0.1% $FeCl_3$ in 0.1 $N$ HCl is added at a rate of 5 ml per liter of media. The citrate and $MgSO_4\cdot7\ H_2O$ are combined first so that the resulting chelation retards magnesium-ammonium phosphate formation. The pH should be 7.0 $\pm$ 0.1. The organism is kept on slants of the same medium except that 1.0% K-gluconate and 1.5% Difco Bacto agar were added. The slants were prepared bimonthly.

[2] H. P. Meloche and W. A. Wood, *J. Biol. Chem.* **239**, 3511 (1964).

[3] This series, Vol. 9 [12].

[4] D. C. Robertson, R. H. Hammerstedt, and W. A. Wood, *J. Biol. Chem.* **246**, 2075 (1971).

The organism is grown in a New Brunswick Fermacell 130-liter batch fermentor (100 liters of media) at 28° with an aeration rate of 255 liters (9 ft³) per minute of air (previously passed through an incinerator unit at 350°) and agitation at 200 rpm. Foam is controlled by an addition of 300 ml of a 4:1 dilution of Antifoam A (Dow Corning) sterilized separately and added with the inoculum. The inoculum is prepared by making a series of transfers, allowing 12 hr for growth of each as follows: 10 ml of medium is inoculated from a slant and this is added to 100 ml of medium used to inoculate five Fernbach flasks, each containing 1 liter of medium. The Fernbach flasks provide the inoculum for 100 liters in the fermentor. At all stages of inoculum preparation, growth is with continuous shaking. Growth is allowed to proceed in the fermentor for 11–15 hr until a 10:1 dilution of cells gives an optical density of 1.10 at 660 nm. The cells are chilled rapidly to 4° and collected in a Sharples AS-12 continuous-flow centrifuge. The cells are stored either in the frozen or lyophilized state and fractionated within 3 months. The average yield is 1800 g (wet weight). The cell paste must be pink-red and have a putty-like consistency.

### Purification Procedure

The procedure reported below contains only minor modifications of one reported previously.[5] All steps are conducted at 0–4°.

*Crude Extract.* Eight-hundred grams of frozen cells were thawed overnight at 4° and then were suspended in 800 ml. Two milligrams of DNase (Sigma) were added, and the mixture was passed twice through a Manton-Gaulin laboratory homogenizer at 7000 psi. The homogenate now at 15–18° is quickly cooled to 4° by adding ice equivalent to an additional 800 ml of $H_2O$.

*Acid Treatment.* Slowly, 4 $N$ HCl was added to the crude extract with vigorous stirring to 0.2 $N$ HCl final concentration. After 15 min, the precipitate was removed by centrifugation and discarded. The pH of the supernatant should be 3 or lower.

*Ammonium Sulfate Precipitation.* Solid ammonium sulfate was added to 0.8 $M$ concentration. After centrifugation, additional ammonium sulfate was added to 1.25 $M$. Both precipitates, recovered by centrifugation, were discarded. The clear yellow supernatant was adjusted to 1.9 $M$ with solid ammonium sulfate. The solution was stirred for an additional 30 min and then centrifuged. The precipitate was dissolved in 10 m$M$ $KPO_4^-$ buffer, pH 6, and adjusted to pH 6 with solid $KHCO_3$. Purification steps

[5] R. H. Hammerstedt, H. Möhler, K. A. Decker, and W. A. Wood, *J. Biol. Chem.* **246**, 2069 (1971).

to this point are performed in 1 day so that the dialysis can run overnight.

*Dialysis.* The 1.25–1.9 $M$ ammonium sulfate fraction was desalted by dialysis overnight against several 1-liter portions of 10 m$M$ KPO₄ buffer, pH 6. The precipitate present at the end of dialysis was removed by centrifugation.

*Calcium Phosphate Gel Treatment.* The dialyzed fraction was diluted with an equal volume of cold distilled water and calcium phosphate gel was added at the rate of 1–2 mg dry weight per 10 units of activity. This amount is required to adsorb 90% of the activity after 10–15 min. The gel was recovered by centrifugation, washed twice with 0.10 $M$ KPO₄⁻, pH 6. The activity was eluted by stirring the gel with 0.01 KPO₄⁻ buffer, pH 6 (1 ml per 250 units of activity) for 1 hr. One treatment was sufficient to recover the activity.

*Ammonium Sulfate Precipitation.* Solid ammonium sulfate was added to the calcium phosphate gel eluate to 2.5 $M$; the solution was stirred for 30 min and then centrifuged. The pellet was then dissolved in 0.10 $M$ KPO₄⁻, pH 6, to a protein concentration of 12–15 mg/ml, and the solution was centrifuged to remove any turbidity.

*Crystallization.* The ammonium sulfate fraction was brought to 0.5 $M$ by the careful addition of finely powdered ammonium sulfate. The solution was then centrifuged and a solution of ammonium sulfate (saturated at 0°) was added to the supernatant until a faint turbidity was observed. The crystallization was complete at 4° overnight. The crystals were collected by centrifugation and stored in 2.5 $M$ ammonium sulfate or dissolved in 0.1 $M$ KPO₄⁻, pH 6.0. The calcium phosphate gel, ammonium sulfate, and crystallization steps were carried out in the second day.

TABLE I

PURIFICATION OF KDPG ALDOLASE FROM *Pseudomonas putida*[a]

| Step | Volume (ml) | Total activity (units $\times 10^{-3}$) | Specific activity (units/mg protein) | Purification (fold) | Recovery (%) |
|---|---|---|---|---|---|
| Crude extract | 2400 | 50.6 | 0.17 | — | 100 |
| Acid precipitation | 1900 | 38.3 | 0.62 | 3.6 | 73 |
| Ammonium sulfate precipitate | 60 | 26.3 | 26.0 | 150.0 | 52 |
| Dialysis | 80 | 23.4 | 36.6 | 210.0 | 46 |
| Calcium phosphate gel eluate | 60 | 16.4 | 194.0 | 1140.0 | 32 |
| Ammonium sulfate precipitate | 12 | 14.7 | 280.0 | 1650.0 | 29 |
| Crystals | 10 | 13.3 | 330.0 | 1940.0 | 26 |

[a] All assays were run at 28°C. The data are based on 800 g of cell paste.

This procedure results in a 1940-fold purification with a 26% yield (*ca.*, 40 mg) as crystals (Table I).

## Properties

Crystalline KDPG aldolase is stable for months at 4° to —15°. It is essentially pure, based on the following criteria: there was no change of specific activity upon recrystallization, and one symmetrical peak was observed in sedimentation velocity centrifugation; one major component and a minor component amounting to no more than 5% of the protein was observed by disc gel electrophoresis at pH 8.3, 7.0, and 4.5.[5]

The characteristics of KDPH aldolase have been summarized in detail elsewhere.[6] Physical,[5] chemical,[4] and X-ray[7] crystallographic data establish that the native molecule is a trimer of molecular weight 72,000 composed of identical subunits of MN 24,000. Large (0.1 mm) crystals belonging to the cubic system form at pH 3.5 in ammonium sulfate. In this form, the space group is P2₁3. This arrangement has 4 trimeric units in the unit cell.

An active site peptide for 50 amino acids has been sequenced.[8,9] The region around the Schiff base lysine residue is very basic with the structure, Arg-Arg-Phe-Lys-. Two sulfhydryl groups are on the surface in that they are readily accessible to Ellman's reagent and are not necessary for activity. Two others are inaccessible to Ellman's reagent.[10] In addition to the Schiff base lysine residue, one other lysine residue, probably in the active site, is indirectly involved in activity but not binding of the substrate.[11] There is one histidine residue per polypeptide chain. Its role in catalysis is unclear; however, photooxidation destroys the histidine residue and activity at the same rate.[12] From inactivation studies with bromopyruvate by Meloche,[13–17] a cysteine residue has been found in the

[6] W. A. Wood, *in* "The Enzymes" (P. D. Boyer, ed.), Vol. 7, 3rd ed., p. 281. Academic Press, New York, 1972.

[7] R. L. Vandlen, D. L. Ersfeld, A. Tulinsky, and W. A. Wood, *J. Biol. Chem.* **248**, 2251 (1973).

[8] D. C. Robertson, W. W. Altekar, and W. A. Wood, *J. Biol. Chem.* **246**, 2084 (1971).

[9] D. Tsay and W. A. Wood, unpublished data, 1973.

[10] H. Möhler, K. A. Decker, and W. A. Wood, *Arch. Biochem. Biophys.* **151**, 251 (1972).

[11] L. R. Barran and W. A. Wood, *J. Biol. Chem.* **246**, 4024 (1971).

[12] L. R. Barran and W. A. Wood, unpublished data (1970).

[13] H. P. Meloche, *Biochem. Biophys. Res. Commun.* **18**, 277 (1965).

[14] H. P. Meloche, *Biochemistry* **6**, 2273 (1967).

[15] H. P. Meloche, *Biochemistry* **9**, 5050 (1970).

[16] H. P. Meloche, M. A. Luczak, and J. R. Wurster, *J. Biol. Chem.* **247**, 4186 (1972).

[17] H. P. Meloche, *J. Biol. Chem.* **248**, 6945 (1973).

TABLE II
KINETIC AND PHYSICAL PARAMETERS

| Characteristics | Data |
|---|---|
| *Native enzyme* | |
| Sedimentation coefficient $S^{\circ}_{20,w}$ | $4.35$ S $\pm 0.06$ S |
| Diffusion coefficient $D^{\circ}_{20,w}$ | $5.60 \pm 0.10 \times 10^{-7}$ cm$^2$ sec$^{-1}$ |
| Intrinsic viscosity | $3.2 \pm 1.0$ cm$^3$ g$^{-1}$ |
| Molecular weight[a] | |
| Miniscus depletion | $73,300 \pm 2,000$ |
| Low speed | $73,000-74,000$ |
| Disc gel | $72,000-78,000$ |
| *Subunit* | |
| Partial specific volume | $0.745, 0.735$ |
| Molecular weight[a] | |
| Low speed | $23,000-24,000$ |
| Disc gel | $23,000-24,000$ |
| Amino acid composition[b] | $24,012$ |
| $K_{eq}^2$ | $0.62$ to $1.87 \times$ m$M$ |
| $K_m$ KDPG[b] | $73 \mu M$ |
| Turnover number | |
| Cleavage[c] | $17,250$ ($28°$) |
| Schiff base formation[d] | $38,000-45,000$ |
| Proton exchange[d,e] | $20,000$ ($28°$)$-33,000$ |
| Oxaloacetate decarboxylation[f] | $860$ |

[a] R. H. Hammerstedt, H. Möhler, K. A. Decker, and W. A. Wood, *J. Biol. Chem.* **246**, 2069 (1971).

[b] D. C. Robertson, R. H. Hammerstedt, and W. A. Wood, *J. Biol. Chem.* **246**, 2075 (1971).

[c] H. P. Meloche and W. A. Wood, *J. Biol. Chem.* **239**, 3511 (1964).

[d] I. A. Rose and E. L. O'Connell, *Arch. Biochem. Biophys.* **118**, 758 (1967).

[e] L. R. Barran and W. A. Wood, *J. Biol. Chem.* **246**, 4024 (1971).

[f] J. M. Ingram and W. A. Wood, *J. Biol. Chem.* **241**, 3256 (1966).

active site region and a carboxyl group has been implicated in catalysis.

The aldolase catalyzes four reactions at measurable rates, cleavage of KDPG, Schiff base formation between a lysine $\epsilon$-NH$_2$ group and the carbonyl group of substrates and analogs,[18-21] proton exchange between pyruvate and H$_2$O,[20,22] and oxaloacetate decarboxylation.[20] The aldolase is highly specific for cleavage. 2-Keto-3-deoxy-gluconate is cleaved at

[18] E. Grazi, H. P. Meloche, G. Martinez, W. A. Wood, and B. L. Horecker, *Biochem. Biophys. Res. Commun.* **10**, 4 (1963).

[19] J. M. Ingram and W. A. Wood, *J. Biol. Chem.* **240**, 4146 (1965).

[20] J. M. Ingram and W. A. Wood, *J. Biol. Chem.* **241**, 3256 (1966).

[21] I. A. Rose and E. L. O'Connell, *Arch. Biochem. Biophys.* **118**, 758 (1967).

[22] H. P. Meloche, *Anal. Biochem.* **38**, 389 (1970).

0.1% the rate with KDPG[18]. No other substrates are cleaved, including KDP galactonate[11] and 2-keto-6-phosphogluconate.[23] On the other hand, specificity for Schiff base formation, as judged by inactivation in the presence of carbonyl compounds and borohydride, is very low.[20] Table II lists some of the kinetic and physical parameters.

[23] E. W. Frampton and W. A. Wood, *J. Biol. Chem.* **236**, 2571 (1961).

# [41] L-Rhamnulose-1-phosphate Aldolase[1]

By TEH-HSING CHIU, KAREN L. EVANS, and DAVID S. FEINGOLD

L-Rhamnulose 1-phosphate = dihydroxyacetone phosphate + L-lactaldehyde

## Assay Method

*Principle.* L-Rhamnulose-1-phosphate aldolase activity is measured by determining the rate of dihydroxyacetone phosphate formation from L-rhamnulose 1-phosphate. Dihydroxyacetone phosphate is determined with $\alpha$-glycerophosphate dehydrogenase (EC 1.1.1.8) and NADH. Decrease in absorbance at 340 nm indicates the oxidation of NADH and corresponding reduction of DHAP.

*Reagents.* L-Rhamnulose 1-phosphate, 60 m$M$, is prepared as described previously.[2] Either the crystalline dicyclohexylammonium or the sodium salt is used.

Tris·HCl buffer, 50 m$M$, pH 7.5
KCl, 2.0 $M$
NADH, freshly prepared in above buffer, 0.13 mg/ml
$\alpha$-Glycerophosphate dehydrogenase, 10 mg/ml

Since L-rhamnulose-1-phosphate aldolase is inhibited by traces of some divalent cations, all assay components should be as pure as possible. Glass-distilled deionized water should be used for preparation of solutions, and all solutions be stored in carefully cleaned polyethylene containers.

*Procedure.* All reagents are tempered at 30° before the assay. The following are then added to a 0.5-ml cuvette with a 1.0-cm light path: 20 $\mu$l of L-rhamnulose 1-phosphate, and 0.38 ml of buffer containing 0.052 mg of NADH, 20 $\mu$moles of KCl, and 0.012 mg of $\alpha$-glycerophosphate dehydrogenase. A negligible volume of enzyme (i.e., 2–5 $\mu$l) containing

[1] T. H. Chiu and D. S. Feingold, *Biochemistry* **8**, 98 (1969).
[2] D. F. Fan, M. Sartoris, T. H. Chiu, and D. S. Feingold, this series Vol. 41. [19].

10–20 activity units/ml is added, and the contents are mixed without delay. Continuous absorbance measurements are made at 340 nm and 30° in a spectrophotometer equipped with a thermostatted cuvette holder until a linear rate of absorbance decrease is obtained (1–2 min). The oxidation of 1 $\mu$mole of NADH (15.5 absorbance units) is equivalent to the cleavage of 1 $\mu$mole of substrate under the assay conditions described.

*Definition of Unit and Specific Activity.* One unit of L-rhamnulose 1-phosphate aldolase is defined as the amount of enzyme which catalyzes the cleavage of 1 $\mu$mole of L-rhamnulose 1-phosphate per minute under the assay conditions described. Specific activity is defined as units of enzyme activity per milligram of protein. Protein concentration may be determined spectrophotometrically by the method of Waddell.[3] The concentration of pure enzyme may be determined as follows: absorbance 280 nm/1.04 = mg protein per milliliter (determined in a 1-cm light path cuvette).[4]

*Purification Procedure.* L-Rhamnulose-1-phosphate aldolase can be prepared either from a strain of *Escherichia coli* inducible for the enzyme or from a mutant strain of the organism, constitutive for L-rhamnulose-1-phosphate aldolase. The purification procedure is the same in either case.

*Growth of Cells.* When *E. coli* K40, *inducible,* is used, *E. coli* K40 (derived from K12) is grown in a medium which contains the following (grams per liter): $K_2HPO_4$, 7.0; $KH_2PO_4$, 3.0; $(NH_4)_2SO_4$, 1.0; $MgSO_4 \cdot 7 H_2O$, 0.1; L-rhamnose, 2.0. The sugar is sterilized by filtration and added aseptically to the sterile salts solution. To prepare a starter culture, 100 ml of medium (contained in a 300-ml flask) is inoculated from agar slants (prepared from the same medium with 2% agar added) and incubated at 37° on a rotary shaker until a Klett reading of approximately 200 (Filter No. 42) is attained. A 2-liter flask containing 1 liter of medium is inoculated with the starter and shaken at 37° until the Klett reading is in the vicinity of 200 (about 12 hr). The contents of the flask are used to inoculate 9 liters of medium contained in a New Brunswick 10-liter MicroFerm fermentor. Growth is at 37° with 1.5 ft³/min aeration and 200 rpm agitation. After 12–16 hr, the culture is in late log phase. The culture is rapidly cooled to 5–10°, and cells are collected by centrifugation and washed twice with ice-cold 0.85% NaCl. The cells are frozen and stored at −4°. Approximately 30 g of packed cells is obtained. When *E. coli* constitutive for L-rhamnulose 1-phosphate aldolase is used, the cells are maintained, grown, and harvested as described above, in the following medium (grams per liter): $KH_2PO_4$, 3; $K_2HPO_4$, 7; $MgSO_4 \cdot 7$

---

[3] J. J. Waddell, *J. Lab. Clin. Med.* **48,** 311 (1956).
[4] N. B. Schwartz and D. S. Feingold, *Bioinorg. Chem.* **1,** 233 (1972).

$H_2O$, 0.1; $(NH_4)_2SO_4$, 1; glycerol, 5; and casamino acids, 15. The yield is approximately 100 g of packed cells.

Three different buffer solutions are used in this preparation. Buffer I is 20 m$M$ potassium phosphate, pH 7.0, 50 m$M$ in 2-mercaptoethanol (2-ME). Buffer II is 5 m$M$ potassium phosphate, pH 7.0, 10 m$M$ in 2-ME. Buffer III is 5 m$M$ potassium phosphate, pH 7.0, 50 m$M$ in 2-ME. The 2-ME is added to the solutions just before use. Unless otherwise stated, all operations are carried out at 0–4°. All centrifugations are performed at 40,000 $g$ for 20 min.

*Preparation of Cell-Free Extracts.* One hundred grams of frozen cells are suspended in 200 ml of cold 20 m$M$ potassium phosphate buffer, pH 7.0, and allowed to stand with occasional stirring until thawed. The cells are then disrupted in an ice-cooled 150-ml Rosett cell with the Branson sonifier Model S-110 (Branson Instruments, Inc., Stamford, Connecticut). Seven or eight 1-min periods of sonication are used, at a 10 A output. The suspension is centrifuged, and the supernatant is retained. This supernatant is the crude extract.

*$MnCl_2$ Treatment.* To the supernatant is added 0.1 volume of 0.5 $M$ $MnCl_2$. The mixture is held at 0–4° for 12–14 hr and then clarified by centrifugation.

*$(NH_4)_2SO_4$ Fractionation I.* 2-ME is added to the supernatant obtained in the previous step, to a final concentration of 50 m$M$. Solid $(NH_4)_2SO_4$ is added with stirring to 30% saturation (167 g/liter). After it has stood for 10 min, the suspension is centrifuged; the pellet is discarded. Solid $(NH_4)_2SO_4$ is added to the supernatant to 40% saturation (59 g/liter). The precipitate is collected by centrifugation and dissolved in a volume of buffer II equal to one-third that of the crude extract.

*Acetone Fractionation.* The solution is placed in an ice–salt bath at —15° and stirred with a magnetic stirrer; 0.89 volume of cold (—20°) acetone is added slowly. The mixture is centrifuged immediately at —20°, and the precipitate is discarded. To the supernatant solution is added 0.61 additional volume of cold acetone; the suspension is centrifuged, and the supernatant is discarded. The well-drained precipitate is dissolved in sufficient buffer I to bring the protein concentration to less than 25 mg/ml (approximately 15–20 ml).

*Sephadex G-150 Chromatography.* The enzyme solution is immediately loaded onto a column (8.5 × 30 cm) of Sephadex G-150 equilibrated with buffer II, which is also used for elution. Six-milliliter fractions are collected, and each is tested for activity. Active fractions are pooled (70–80 ml).

*DEAE-Sephadex A-50 Fractionation.* The pooled active fractions are fractionated on a DEAE-Sephadex A-50 (3.75 × 55 cm) column equilibrated with buffer II. The enzyme solution is loaded onto the column, which is then washed with 50 ml of buffer. The column is eluted with an increasing gradient of NaCl obtained with 200 ml of buffer II in a mixing chamber and 500 ml of the same buffer, 0.5 $M$ in NaCl, in a reservoir. Six-milliliter fractions are collected and those containing enzyme activity are pooled (approximately 100–120 ml); 2-ME is added to a final concentration of 50 m$M$.

$(NH_4)_2SO_4$ *Fractionation II.* Solid $(NH_4)_2SO_4$ is added to the pooled active fractions to 60% saturation (371 g/liter). After standing for 15 min, the suspension is centrifuged, the supernatant solution is discarded, and the well-drained precipitate is suspended in 5 ml of buffer III, 35% saturated with $(NH_4)_2SO_4$. The precipitate is removed by centrifugation, drained, and suspended in 2 ml of buffer III, 30% saturated with $(NH_4)_2SO_4$. After centrifugation the precipitate is suspended in 2 ml of buffer III, 28% saturated with $(NH_4)_2SO_4$. The precipitate so obtained is recovered by centrifugation, dissolved in 2 ml of buffer III, and held at 0–4° for 18 hr. Insoluble material is removed by centrifugation.

*Crystallization.* The clear supernatant from the preceding step is brought to 25% saturation with solid $(NH_4)_2SO_4$ (137 g/l) and held at 0–4° for 10 min. Additional $(NH_4)_2SO_4$ (119 g/l) is added to 45% saturation, whereupon rapid crystallization occurs. After an additional 10 min, the crystals are removed by centrifugation and dissolved in 2 ml of buffer III.

PURIFICATION OF L-RHAMNULOSE-1-PHOSPHATE ALDOLASE

| Purification step | Volume (ml) | Protein (mg/ml) | Specific activity (units/mg) | Total units | Purification (fold) | Recovery (%) |
|---|---|---|---|---|---|---|
| MnCl$_2$ | 237 | 10 | 0.52 | 1244 | 1 | 100 |
| (NH$_4$)$_2$SO$_4$ fractionation I | 85 | 15 | 0.86 | 1094 | 1.7 | 96 |
| Acetone fractionation | 20 | 19 | 2.5 | 1000 | 4.8 | 80 |
| Sephadex G-150 chromatography | 70 | 5 | 2.1 | 735 | 4.0 | 68 |
| DEAE-Sephadex chromatography | 106 | 3 | 2.6 | 848 | 5.0 | 69 |
| (NH$_4$)$_2$SO$_4$ fractionation II | 2 | 45 | 4.4 | 400 | 8.4 | 32 |
| Crystallization | 2 | 40 | 5.4 | 430 | 10.2 | 33 |
| Recrystallization | 2 | 25 | 6.0 | 300 | 11.0 | 24 |

*Recrystallization.* Solid $(NH_4)_2SO_4$ is added to 25% saturation (137 g/l) and the enzyme is allowed to crystallize at 0–4° for 10 min. The crystals are removed by centrifugation, drained well, dissolved in a desired volume of buffer II, and stored in a well stoppered vessel at 0–4°. A typical purification is summarized in the table.

## Properties

*Specificity.*[1] L-Rhamnulose-1-phosphate aldolase cleaves only those ketose 1-phosphates which have the D configuration at C-3 and the L configuration at C-4 in their Fischer projection formulas. The products are dihydroxyacetone phosphate and an aldehyde. L-Rhamnulose 1-phosphate, D-sorbose 1-phosphate, and L-xylulose 1-phosphate are cleaved. However, the following are unaffected by the enzyme: D-fructose 1,6-diphosphate, L-fuculose 1-phosphate, D-ribulose 1,5 diphosphate, 6-deoxy-L-sorbose 1-phosphate, L-sorbose 1-phosphate, L-sorbose 1,6 diphosphate, D-fructose 1-phosphate, L-sorbose 6-phosphate, and D-fructose 6-phosphate. In the condensation reaction, only those isomers are formed from dihydroxyacetone phosphate and appropriate aldehydes which have the configuration D at C-3 and L at C-4. Dihydroxyacetone phosphate yields L-rhamnulose 1-phosphate with L-lactaldehyde, D-sorbose 1-phosphate with D-glyceraldehyde, L-xylulose 1-phosphate with glycolaldehyde, 5-deoxy-L-xylulose 1-phosphate with acetaldehyde, D-erythrulose 1-phosphate with formaldehyde, and 6-deoxy-D-sorbose 1-phosphate with D-lactaldehyde. The relative rates of formation of these compounds under the experimental conditions employed are 100, 50, 40, 15, 10, and 10, respectively. Neither L-glyceraldehyde, D-glyceraldehyde 3-phosphate, nor propionaldehyde condenses with dihydroxyacetone phosphate.

*Equilibrium and Kinetic Constants.* The equilibrium constant for the reaction L-rhamnulose 1-phosphate ⇌ dihydroxyacetone phosphate + L-lactaldehyde is 83 $\mu M$ at pH 7.5.[1] $K_m$ values for L-rhamnulose 1-phosphate and dihydroxyacetone phosphate are 0.2 m$M$, for L-lactaldehyde, 0.1 m$M$.

*Turnover Number.* The turnover number of the enzyme is 2300 moles of substrate split per minute per mole of enzyme at 37°.

*pH Optimum.* The pH optimum of L-rhamnulose-1-phosphate aldolase is 7.5, with activity decreasing sharply on either side of this value. At pH 6.3 and 9.0, activity is 50% of the maximum.

*Inhibitors and Activators.*[1] Like other class II aldolases, L-rhamnulose-1-phosphate aldolase is activated by, and may have an absolute requirement for, a specific monovalent cation. In increasing order of efficacy, $Na^+$, $Cs^+$, $NH_4^+$, $Rb^+$, and $K^+$ activate the enzyme. $K^+$ stimulates the enzyme activity at least 30-fold; $K_a$ for KCl is 6 m$M$. Thus, assay mixtures are 50 m$M$ in KCl. Li$^+$ is an inhibitor strictly competitive with K$^+$; $K_i$ = 6 m$M$.

The aldolase is stabilized by sulfhydryl compounds. Mercurials such as p-mercuribenzoic acid or mercuric chloride are reversible inhibitors.

The substrate analog L-rhamnitol 1-phosphate is an inhibitor competitive with substrate. This was the only effective inhibitor of a wide variety of glycoses and glycose phosphates tested.

*Inhibition by Chelating Agents.*[4] Enzyme activity is inhibited by chelating agents such as 1,10-phenanthroline, 8-hydroxyquinoline-5-sulfonic acid, α,α-dipyridyl, and ethylenediaminetetraacetic acid. The inhibition by 1,10-phenanthroline is strictly competitive with substrate. The inhibition by EDTA is instantly relieved by addition of Co$^{2+}$, Zn$^{2+}$, Mn$^{2+}$, or Ni$^{2+}$. The inhibitions observed are consistent with the presence of a zinc ion in the protein. Upon prolonged dialysis against EDTA, the activity and metal content of the enzyme are markedly reduced.

*Molecular Properties.*[1,4] Crystalline L-rhamnulose-1-phosphate aldolase is a homogeneous protein when examined by polyacrylamide gel and cellulose polyacetate electrophoresis, immunoelectrophoresis, immune double diffusion, isoelectric focusing, gradient density centrifugation, analytical ultracentrifugation, or thin-layer gel chromatography. The molecular weight is of the order of 140,000 as determined by gel chromatography and density gradient centrifugation. The isoelectric point is 5.0. The crystalline protein contains 2 gram atoms of zinc per mole of enzyme and consists of 4 identical subunits.

*Stability.* The enzyme can be stored in 50 m$M$ 2 ME at pH 7.0 and 0–4° for at least 3 months without loss of activity.

# [42] 2-Keto-3-deoxy-L-arabonate Aldolase

*By* RICHARD L. ANDERSON and A. STEPHEN DAHMS

2-Keto-3-deoxy-L-arabonate → pyruvate + glycolaldehyde
2-Keto-3-deoxy-D-fuconate → pyruvate + D-lactaldehyde

This enzyme functions in the metabolism of L-arabinose[1] and D-fucose[2] in pseudomonad MSU-1

## Assay Method[2]

*Principle.* The continuous spectrophotometric assay measures the rate of pyruvate formation by coupling the reaction to lactate dehydrogenase.

[1] A. S. Dahms and R. L. Anderson, *Biochem. Biophys. Res. Commun.* **36,** 809 (1969).
[2] A. S. Dahms and R. L. Anderson, *J. Biol. Chem.* **247,** 2238 (1972).

With lactate dehydrogenase in excess, the rate of 2-keto-3-deoxy-L-arabonate or 2-keto-3-deoxy-D-fuconate cleavage is equivalent to the rate of NADH oxidation, which is measured by the absorbance decrease at 340 nm.

### Reagents

$N$-2-Hydroxyethylpiperazine-$N'$-2-ethanesulfonate (HEPES buffer), 0.3 $M$, pH 7.8
2-Keto-3-deoxy-D-fuconate,[3] 0.25 $M$
MnCl$_2$, 0.10 $M$
NADH, 14 m$M$
Crystalline lactate dehydrogenase

*Procedure.* The following are added to a cuvette with a 1.0-cm light path: 0.05 ml of buffer, 0.02 ml of 2-keto-3-deoxy-D-fuconate, 0.01 ml of MnCl$_2$, 0.005 ml of NADH, a nonlimiting amount of lactate dehydrogenase, a limiting amount of 2-keto-3-deoxy-L-arabonate aldolase, and water to a volume of 0.15 ml. The reaction is initiated by the addition of the aldolase. A control cuvette minus substrate measures NADH oxidase activity, which must be subtracted from the total rate. The rates are conveniently measured with a Gilford multiple-sample absorbance recorder. The cuvette compartment should be thermostatted at 25°. Care should be taken to confirm that the rates are proportional to the 2-keto-3-deoxy-L-arabonate aldolase concentration.

*Definition of Unit and Specific Activity.* One unit is defined as the amount of enzyme that catalyzes the cleavage of 1 $\mu$mole of 2-keto-3-deoxy-D-fuconate per minute. Specific activity is in terms of units per milligram of protein. Protein is conveniently measured by the method of Warburg and Christian,[4] but with crude extracts and fractions high in nucleic acid content, the biuret method[5] may be used.

### Purification Procedure[2]

The enzyme was purified from pseudomonad MSU-1.[6] It was grown as described previously[7] except that the carbon source was L-arabinose instead of D glucose. The preparation of cell extracts was also as de-

---

[3] 2-Keto-3-deoxy-D-fuconate may be prepared as described by A. S. Dahms and R. L. Anderson, *J. Biol. Chem.* **247**, 2233 (1972).
[4] O. Warburg and W. Christian, *Biochem. Z.* **310**, 384 (1941).
[5] A. G. Gornall, C. J. Bardawill, and M. M. David, *J. Biol. Chem.* **177**, 751 (1941).
[6] This organism is obtainable from the American Type Culture Collection as strain ATCC 27855.
[7] See Vol. 41 [34].

PURIFICATION OF 2-KETO-3-DEOXY-L-ARABONATE ALDOLASE

| Fraction | Volume (ml) | Total protein (mg) | Total activity (units) | Specific activity (units/mg protein) | $A_{280}:A_{260}$ |
|---|---|---|---|---|---|
| Cell extract | 600 | 16,260 | 521 | 0.032 | 0.62 |
| Protamine sulfate supernatant | 840 | 13,520 | 520 | 0.038 | 0.87 |
| (NH₄)₂SO₄ precipitate | 82 | 3,450 | 450 | 0.140 | 1.20 |
| Sephadex G-200 | 100 | 210 | 114 | 0.545 | 1.38 |
| Heat step | 100 | 51 | 82.6 | 1.63 | 1.59 |

scribed,[7] except that the cells were suspended in a solution of 0.14 m$M$ 2-thioethanol and 0.1 $M$ $N,N$-bis(2-hydroxyethyl)glycine (Bicine buffer), pH 7.4. Except for the heat step, the following procedures were performed at 0° to 4°. A summary of the purification procedure is given in the table.

*Protamine Sulfate Treatment.* To a cell extract containing 0.2 $M$ ammonium sulfate was added an amount of 2% (w/v) protamine sulfate solution (pH 7.0) to give a concentration of 0.33%. After 30 min, the precipitate was removed by centrifugation and discarded.

*Ammonium Sulfate Fractionation.* The protein in the supernatant from the protamine step was fractionated by the addition of crystalline ammonium sulfate. The protein precipitating between 40 and 60% saturation was collected by centrifugation and was dissolved in 0.1 $M$ Bicine–0.14 m$M$ 2-thioethanol (pH 7.4).

*Chromatography on Sephadex G-200.* The above fraction was chromatographed on a column (6 × 60 cm) of Sephadex G-200 equilibrated with 50 m$M$ Bicine–0.14 m$M$ 2-thioethanol (pH 7.4). Fractions (13 ml) were collected during elution with the same buffer, and those which contained the highest specific activity were combined.

*Heat Step.* MgCl₂ was added to the above fraction to a concentration of 0.6 m$M$. The solution was heated at 55° for about 8 min and then quickly cooled and clarified by centrifugation. 2-Keto-3-deoxy-L-arabonate aldolase in the supernatant was 50-fold purified with an overall recovery of 16%.

## Properties[2]

*Substrate Specificity.* Of nine 2-keto-3-deoxyaldonic acids and related compounds tested, only 2-keto-3-deoxy-L-arabonate ($K_m = 1.8$ m$M$) and

2-keto-3-deoxy-D-fuconate ($K_m$ = 2.9 m$M$) served as substrates for this aldolase. Compounds that were not cleaved (<1% of the rate on 2-keto-3-deoxy-D-fuconate) at 10 m$M$ concentrations were 2-keto-3-deoxy-D-galactonate and its 6 phospho ester, 2-keto-3-deoxy-D-gluconate and its 6-phospho ester, $N$-acetylneuraminate, 2-keto-4-hydroxy-DL-glutarate, and 2-keto-3-deoxy-D-arabonate.

*pH Optimum.* Aldolase activity as a function of pH is maximal in the pH range of about 7.8 to 8.3.

*Metal Requirement.* The aldolase requires a divalent metal for activity when assayed in the presence of ethylenediaminetetraacetate (EDTA). With 0.33 m$M$ EDTA, the relative activities in the presence of various 1.3 m$M$ metal salts were as follows: $MnCl_2$, 100; $CoCl_2$, 95; $MgCl_2$ or $MgSO_4$, 54; $NiCl_2$, 45; and none, 0.

*Effect of Thiols and Thiol Group Reagents.* The purified enzyme was insensitive to thiols and thiol group reagents during assay, as none of the following compounds affected the reaction velocity: 3 m$M$ 2-thioethanol, 3 m$M$ reduced glutathione, 3 m$M$ dithiothreitol, 2 m$M$ iodoacetate, and 2 m$M$ $p$-chloromercuribenzoate. However, the enzyme was stabilized by 2-thioethanol during purification.

*Stability.* The purified enzyme (in 0.1 $M$ Bicine–0.14 m$M$ 2-thioethanol) lost 70% of its activity in 3 weeks when stored at −20°. When it was lyophilized in the same buffer, it was stable for at least 1 month.

### Equilibrium Constants

The equilibrium constant for cleavage of 2-keto-3-deoxy-L-arabonate at pH 7.4 and 28° is about 0.37 m$M$.[1] The equilibrium constant for cleavage of 2-keto-3-deoxy-D-fuconate at pH 7.5 and 28° is about 0.12 m$M$.[2]

## [43] D-4-Deoxy-5-ketoglucarate Hydro-lyase (Decarboxylating)

*By* STANLEY DAGLEY and ROGER JEFFCOAT

D-4-Deoxy-5-ketoglucarate → 2,5-diketovalerate + $H_2O$ + $CO_2$

When D-glucaric acid serves as carbon source for growth of Enterobacteriaceae or Pseudomonadaceae the initial product of catabolism is D-4-deoxy-5-ketoglucarate, but different metabolic pathways are used beyond this point. In *E. coli* or *K. aerogenes* the compound undergoes aldol

[1] H. J. Blumenthal and D. C. Fish, *Biochem. Biophys. Res. Commun.* 11, 239 (1963).
[2] R. Jeffcoat, H. Hassall, and S. Dagley, *Biochem. J.* 115, 969 (1969).

fission.[1] whereas in *Pseudomonas acidovorans*[2] and other pseudomonads[3] it is converted into 2,5-diketovalerate which is then oxidized to 2-ketoglucarate.[2] Formation of 2,5-diketovalerate occurs with the simultaneous liberation of carbon dioxide, catalyzed by a single enzyme, D-4-deoxy-5-ketoglucarate hydro-lyase.[4]

## Assay Method

*Principle.* The decrease in concentration of substrate is measured by oxidizing deoxyketoglucarate with periodate to yield formylpyruvic and glyoxylic acids. The chromophore formed by the reaction of thiobarbituric and formylpyruvic acids[5] is measured at 546 nm.

*Reagents*

  Potassium phosphate buffer, 0.1 $M$, pH 7.2
  D-4-Deoxy-5-ketoglucarate, 10 m$M$
  Metaphosphoric acid, 3 $M$
  Sodium arsenite, 2.0% in 0.5 $M$ HCl
  Periodic acid, 25 m$M$ in 62 m$M$ H$_2$SO$_4$
  2-Thiobarbituric acid, 0.3%

*Procedure.* Solutions may be incubated at 20° in 5-inch test tubes. To 2.3 ml of phosphate buffer, pH 7.2 is added 0.5 ml (5 $\mu$moles) of deoxyketoglucarate and 0.2 ml of enzyme. The mixture is incubated for 10 min, the reaction stopped by adding 0.2 ml of metaphosphoric acid, and if a precipitate is formed it is removed by centrifugation. To 0.25 ml of solution is added 0.25 ml of periodic acid and the mixture allowed to stand for 20 min at 20°, when excess of periodic acid is removed by adding 0.5 ml of sodium arsenite. The solution develops a pink color on addition of 2.0 ml of thiobarbituric acid when heated at 100° for 10 min, and the extinction of the chromophore is measured at 546 nm. The concentration of D-4-deoxy-5-ketoglucarate remaining in the original incubation mixture is calculated using an extinction coefficient of 60,000.[6]

*Definition of Unit and Specific Activity.* One unit of enzyme is defined as that amount of enzyme that catalyzes the removal of 1 $\mu$mole of D-4-deoxy-5-ketoglucarate per minute under these conditions. Specific activity is expressed as units per milligram of protein.

*Preparation of D-4-Deoxy-5-ketoglucarate.* The preparation of this compound from D-glucarate (Sigma Chemical Co. Ltd., London S.W. 6)

[3] P. W. Trudgill and R. Widdus, *Nature (London)* **211**, 1097 (1966).
[4] R. Jeffcoat, H. Hassall, and S. Dagley, *Biochem. J.* **115**, 977 (1969).
[5] A. Weissbach and J. Hurwitz, *J. Biol. Chem.* **234**, 705 (1959).
[6] H. J. Blumentahl, D. C. Fish, and T. Jepson, this series, Vol. 9 p. 56.

by the action of the glucarate hydro-lyase from *Escherichia coli* has been described.[6] In the present instance it is convenient to use, as a source of glucarate hydro-lyase, a portion of the crude extract of *Pseudomonas acidovorans* from which D-4-deoxy-5-ketoglucarate hydro-lyase is to be purified. This extract is taken through protamine sulfate treatment, and fraction II of the table is brought to 30% saturation (17.6 g/100 ml) with ammonium sulfate. The precipitate is then dissolved in phosphate buffer, pH 7.2, to give a glucarate hydro-lyase preparation which requires the addition of 2 m$M$ magnesium sulfate for activity. The solution of deoxyketoglucarate prepared by using this enzyme is taken through the isolation procedure previously described[6] to obtain a lyophilized powder. This material is dissolved in water, brought to pH 7, and held at 100° for 8 min to ensure the hydrolysis of any lactonized deoxyketoglucatate.

### Purification Procedure

*Growth of Culture and Preparation of Crude Extracts.* Pseudomonas *acidovorans* (ATCC 17455) is grown with forced aeration at 30° in media adjusted with sodium hydroxide to pH 7.2 containing, per liter, 2 g of potassium dihydrogen phosphate, 1 g of ammonium sulfate, 1 g of potassium hydrogen glucarate, and 0.4 g of magnesium sulfate. The harvested cells are washed once by suspension in phosphate buffer, pH 7.2, centrifuged, and the packed cells are frozen in a Hughes bacterial press at —20°. They are crushed and then taken up in phosphate buffer, pH 7.2, and the extract is clarified by centrifugation at 26,000 $g$ for 45 mins. The extract (fraction I in the table) contains about 30 mg of protein per milliliter when each 1 g of crushed cell paste is suspended in 2 ml of buffer. Sonic disintegration may also be used, with the same ratio of cells to buffer and exposure for 4 min at 0° to the output of a MSE-Mullard ultrasonic disintegrator at an average frequency of 20 kHz per second.

*Protamine Sulfate Treatment.* To each 10 ml of extract at 4° is added, with stirring, 2 ml of 2% protamine sulfate and the precipitate is removed by centrifugation (fraction II in the table).

*Ammonium Sulfate Fractionation.* The extract is made 30% saturated (17.6 g/100 ml) with powdered ammonium sulfate added with stirring at 4° over a period of 15 min, allowed to stand for 30 min and centrifuged at 30,000 $g$ for 20 min and is used as previously described, to prepare deoxyketoglucarate. The supernatant solution is brought to 50% saturation (12.7 g/100 ml) with ammonium sulfate, the precipitate removed by centrifuging and discarded, and the protein precipitating at 75% saturation is collected, dissolved in 50 m$M$ phosphate buffer, pH 7.2, and

PURIFICATION OF D-4-DEOXY-5-KETOGLUCARATE HYDRO-LYASE

| Fraction | Total protein (mg) | Enzyme (total units) | Specific activity (units/ mg) | Yield (%) | Purification (fold) |
|---|---|---|---|---|---|
| I. Crude extract | 3500 | 700 | 0.20 | 100 | 1 |
| II. Protamine sulfate treatment | 3130 | 754 | 0.23 | 107 | 1.2 |
| III. 50–75% Ammonium sulfate | 328 | 595 | 1.8 | 85 | 9 |
| IV. DEAE-Cellulose | 70 | 300 | 5.9 | 43 | 29.5 |

dialyzed against the same buffer until free from ammonium ions (fraction III in the table).

*DEAE-Cellulose Chromatography.* The extract (about 30 mg of protein per milliliter) is applied to a column (1 × 14 cm) of DEAE-cellulose which has equilibrated with 50 mM phosphate buffer, pH 7.2, and is then eluted with a 0–0.15 M sodium chloride linear concentration gradient in the same buffer (total volume 200 ml). Protein is eluted throughout the first 50 fractions of 4 ml, but the enzyme is confined to tubes 30–35, which are eluted at 0.08–0.10 M sodium chloride. These solutions are pooled, brought to 80% saturation with ammonium sulfate, the precipitate dissolved in phosphate buffer, pH 7.2, and dialyzed. When the solution is frozen, the enzyme retains activity for several weeks at −15° (fraction IV in the table).

## Properties

*Effect of pH.* The pH optimum is at 7.0–7.3 with a relatively sharp decrease in activity below pH 6.5 and above pH 8.0. A published curve is incorrectly displaced by 1 pH unit.[4]

*Molecular Weight.* A value of 118,000 is given by sedimentation equilibrium.[7] Amino acid analysis indicates 1050 residues for this molecular weight, the most abundant being alanine (141) and leucine (103).[8] The protein is a dimer as shown by ultracentrifugation in 6 M guanidine hydrochloride.[7] The subunits of this dimer are composed of two nonidentical polypeptide chains having the same molecular weight, as shown by ultracentrifugation in 6 M guanidine hydrochloride plus 0.1 M β-mercaptoethanol,[8] and also by SDS gel electrophoresis[8] using the method of Weber and Osborn.[9]

[7] H. Hassall, R. Jeffcoat, and S. Dagley, *Biochem. J.* **114**, 78p (1969).
[8] R. Jeffcoat, unpublished observations.
[9] K. Weber and M. Osborn, *J. Biol. Chem.* **244**, 4406 (1969).

*Catalytic Action.* The substrate is bound to the enzyme in Schiff base formation with lysine,[4] 2 moles reacting with 1 mole of enzyme.[7] In agreement with the mechanism proposed to account for the fact that decarboxylation and dehydration occur simultaneously,[4] the product expected from a single-step decarboxylation, namely 4,5-dihydroxy-2-oxovalerate, is not attacked by the enzyme. No added cofactors are required; the enzyme does not attack, and is not inhibited by the tartaric acids or a range of $\alpha$-keto acids at 1 m$M$. Instant inhibition is found with $p$-chloromercuribenzene sulfonic acid, but iodoacetate and iodoacetamide inhibit only after prolonged incubation.

# [44] Deoxyribose-5-phosphate Aldolase from *Salmonella typhimurium*

*By* Patricia Hoffee

Deoxyribose 5-phosphate $\rightleftharpoons$ D-glyceraldehyde 3-phosphate + acetaldehyde

## Assay of Enzymic Activity

*Principle.* Enzyme activity can be measured either in the forward direction by a coupled spectrophotometric assay,[1] or in the reverse direction employing the diphenylamine method of Dische.[2,3] The spectrophotometric method is favored by the author because of its simplicity and accuracy. This is particularly true when dealing with high levels of enzyme such as those found in wild-type cells induced by deoxyribose or in uninduced constitutive cells.[4]

*Spectrophotometric Assay.* The production of acetaldehyde is followed by the change in absorbance at 340 nm produced by NADH in a coupled enzyme system with deoxyribose 5-phosphate as substrate and alcohol dehydrogenase as the indicator enzyme.

### Reagents

2-Deoxyribose 5-phosphate, 50 m$M$. Prepare from commercially obtained barium salt.[5] Appropriate amounts of the solid barium salt are dissolved in water and treated with 10% molar excess of

---

[1] E. Racker, *J. Biol. Chem.* **196**, 347 (1952).
[2] Z. Dische, *Mikrochemie* **8**, 4 (1930).
[3] W. E. Pricer and B. L. Horecker, *J. Biol. Chem.* **235**, 1292 (1960).
[4] J. Blank and P. Hoffee, *Mol. Gen. Genet.* **116**, 291 (1972).
[5] Obtained from Calbiochem, Los Angeles, California.

sodium sulfate. The barium sulfate precipitate is removed by centrifugation and washed with a small amount of water several times; the combined supernatant solutions are adjusted to the desired volume with water.

TEA-EDTA buffer. Triethanolamine-HCl buffer, 40 m$M$, pH 7.8, is made 1 m$M$ with respect to ethylenediaminetetraacetic acid.

NADH, 10 m$M$

Alcohol dehydrogenase, yeast, 30 mg/ml (the crystalline enzyme preparation is obtained from Boehringer Mannheim Corporation)

*Procedure.* To 0.97 ml of TEA-EDTA buffer in a quartz cuvette add 0.01 ml of NADH, 0.001 ml of alcohol dehydrogenase, 0.01 ml of deoxyribose 5-phosphate and deoxyribose-5-phosphate aldolase solution that gives a change in absorbance of 0.01–0.06 per minute. Measure the change in absorbance at 340 nm at 25°. A control cuvette without substrate is run to correct for any nonspecific oxidation of NADH.

*Definition of Unit and Specific Activity.* One unit of activity cleaves 1 $\mu$mole of deoxyribose 5-phosphate per minute. Specific activity is defined as units per milligram of protein. Protein is determined by the method of Lowry *et al.*[6] before the ammonium sulfate fractionations (see Purification) and by the method of Bücher[7] after ammonium sulfate extraction.

## Purification of Enzyme[8]

*Growth of Cells.* Wild-type *Salmonella typhimurium* (ATCC 15277): Inoculate 2-liter flasks containing 1 liter of casamino acid medium, buffered at pH 7.0 with phosphate, with 25 ml of an overnight culture. Incubate the flasks with shaking at 37° for 4 hr or until the cells are in log phase. Add sterile 2-deoxyribose to the cells to a final concentration of 0.2%. Incubate the cells at 37° with aeration for an additional 90 min and then rapidly chill and harvest. Wash the cell paste in 1 m$M$ EDTA and store frozen at −10°. *Salmonella* strains constitutive for deoxyribose 5-phosphate aldolase: Inoculate 2-liter flasks containing 1 liter of casamino acid medium, buffered at pH 7.0 with phosphate, with 25 ml of an overnight culture. Incubate the flasks with shaking at 37° overnight and harvest in the morning. Wash the cell paste as above and store frozen at −10°. The use of the constitutive strains is much more

[6] O. H. Lowry, N. J. Rosebrough, A. L. Farr, and R. J. Randall, *J. Biol. Chem.* **193**, 265 (1951).

[7] T. Bücher, *Biochim. Biophys. Acta* **1**, 292 (1947).

[8] This is a modified procedure based on work previously published: P. Hoffee, *Arch. Biochem. Biophys.* **126**, 795 (1968).

convenient since it does not require induction with 2-deoxyribose and the amount of enzyme obtained is about twice that amount found with the wild-type cells. The purification procedure described below is identi cal for both induced wild-type cells and the constitutive strains.

*Step 1. Preparation of Cell Extract.* Extract frozen cells by one of the following two methods: (a) Resuspend frozen cells in TEA-EDTA buffer to a 15% suspension and sonicate in a Branson sonifier. Keep the temperature of the mixture below 10° during sonication. Centrifuge the sonicate in a refrigerated Sorvall, Model RC-2 at 27,000 $g$ for 45 min and assay the supernatant fluid. (b) Grind 25 g of cells (wet weight) in a cold mortar with 50 g of Alumina A-301 (Bacteriological grade, Alcoa Chemicals, Bauxite, Arkansas). Take up the paste in 125 ml of TEA-EDTA buffer, centrifuge as above, and assay the supernatant fluid.

*Step 2. Protamine Step.* Make a 2% solution of protamine sulfate (Eli Lilly). To each milliliter of extract, add 0.2 ml of the protamine sulfate solution slowly with constant stirring. After 10 min at 0°, remove the precipitate by centrifugation at 12,000 $g$ for 10 min and assay the supernatant fluid.

*Step 3. Ammonium Sulfate Fractionation.* Treat the supernatant fluid from the protamine fraction with solid ammonium sulfate (199 mg/ml). After 30 min at 4°, centrifuge the suspension. Treat the supernatant fluid with more solid ammonium sulfate (123 mg/ml) to give 0.55 saturation. After 30 min, centrifuge the suspension. Treat the supernatant fluid with more solid ammonium sulfate (200 mg/ml). After 30 min, centrifuge the suspension, dissolve the precipitate in 5 m$M$ potassium phosphate buffer, pH 7.5, and dialyze overnight against 1000 volumes of the same buffer.

*Step 4. DEAE-Cellulose Column.* Apply the dialyzed solution to a DEAE-cellulose column (2.5 × 25 cm) equilibrated with 5 m$M$ potassium phosphate buffer, pH 7.5. Wash the column with the same buffer until no more 280-nm absorbing material is eluted. Deoxyribose-5-phosphate aldolase which binds to the column under these conditions is then eluted with a linear gradient of potassium phosphate buffer, 5 m$M$, pH 7.5, to 0.1 $M$, pH 6.5 in 2000 ml. The enzyme elutes from the column at about 25–30 m$M$ potassium phosphate buffer and is concentrated from dilute solution by precipitation with ammonium sulfate (600 mg/ml).

*Step 5. Crystallization.* Dissolve the ammonium sulfate precipitate in 5 m$M$ phosphate buffer, pH 7.5, and remove any insoluble material by centrifugation. To the clear supernatant fluid add solid ammonium sulfate slowly until the solution is slightly turbid. Store the solution in the cold for 2 days to allow crystallization. For the second crystallization, spin out the first crystals and dissolve in 5 m$M$ potassium phosphate, pH 7.5. Add solid ammonium sulfate until the solution becomes turbid.

PURIFICATION PROCEDURE FOR DEOXYRIBOSE-5-PHOSPHATE
ALDOLASE FROM *Salmonella typhimurium*[a]

| Fraction | Total volume (ml) | Units/ml | Total units | Protein (mg/ml) | Specific activity (units/ mg) | Re-covery (%) |
|---|---|---|---|---|---|---|
| Extract[b] | 98 | 46 | 4500 | 23 | 2.0 | — |
| Protamine fraction | 108 | 37 | 4000 | 14 | 2.6 | 89 |
| Ammonium sulfate extract | 10 | 275 | 2750 | 30 | 9.1 | 62 |
| DEAE-cellulose column + ammonium sulfate | 10 | 165 | 1650 | 3.5 | 47 | 37 |
| Crystals | 7.5 | 200 | 1500 | 3.6 | 55 | 33 |

[a] Data given for a constitutive mutant JB 3041 which has 2–3 times the enzyme level found for wild-type induced cells. Constitutive strains are available from the author or can be selected by the procedure described by J. Blank and P. Hoffee, *Mol. Gen. Genet.* **116,** 291 (1972).

[b] Extract prepared by sonication.

Store the solution in the cold until crystallization takes place (about 2–3 days). The final ammonium sulfate concentration should be about 55–60% saturation as determined by a conductivity meter.

The purification procedure is summarized in the table.

## Properties[8]

*Specificity.* Deoxyribose-5-phosphate aldolase is specific for 2-deoxyribose 5-phosphate. It does not cleave ribose 5-phosphate or 2-deoxyribose 1-phosphate. 2-Deoxyribose is split at $\frac{1}{200}$ the rate of 2-deoxyribose 5-phosphate.

*Stability.* The enzyme is quite stable when stored in ammonium sulfate solution at 4°. It is not stable to freezing and loses all activity when frozen and thawed.

*Effect of pH.* The enzyme has a broad optimum pH range from pH 7.3 to pH 8.4. The activity diminishes rapidly on either side of this range.

*Other Properties.* The apparent $K_m$ values for deoxyribose-5-phosphate aldolase are, deoxyribose 5-phosphate, 0.1 m$M$; D-glyceraldehyde 3-phosphate, 0.6 m$M$; and acetaldehyde, 3.5 m$M$. The molecular weight of deoxyribose-5-phosphate aldolase is about 57,000. The enzyme is composed of two identical subunits of molecular weight 28,500.

# [45] 2-Keto-4-hydroxyglutarate Aldolase from Bovine Liver[1]

By EUGENE E. DEKKER, RODGER D. KOBES, and SHARON R. GRADY

DL-2-Keto-4-hydroxyglutarate $\rightleftharpoons$ pyruvate + glyoxylate

## Assay

*Principle.* A number of methods for the assay of KHG-aldolase have been published. The most convenient procedures measure the products released as a result of KHG cleavage. Glyoxylate[2] or pyruvate[3] may be determined colorimetrically. In a spectrophotometric assay at 340 nm, pyruvate is reduced by NADH in the presence of added lactate dehydrogenase. Only purified fractions of the aldolase can be assayed by the latter method; details of this method have already been presented[4] which differ only slightly in quantities of materials used from the procedure followed in the author's laboratory.[5]

The condensation of glyoxylate with pyruvate can be followed in two ways. In one instance, the 2,4-dinitrophenylhydrazone derivatives of the substrates and KHG are prepared, separated by paper chromatography, extracted, and quantitated spectrophotometrically.[6] Solutions of crystalline derivatives of each compound serve as standards (these values are used in the author's laboratory—KHG: $\lambda_{max}$ 380 nm, $\epsilon$ $1.97 \times 10^4$ $M^{-1}cm^{-1}$; glyoxylate I + II: $\lambda_{max}$ 367, $\epsilon$ $2.18 \times 10^4$ $M^{-1}cm^{-1}$; pyruvate: $\lambda_{max}$ 367, $\lambda$ $2.14 \times 10^4$ $M^{-1}cm^{-1}$). Alternatively, the incorporation of [1-$^{14}$C]glyoxylate into KHG is measured.[7] After reaction, the incubation mixture is treated with hydrogen peroxide to decarboxylate any residual glyoxylate; the product is decarboxylated to [$^{14}$C]malate, which is counted. These two methods are very laborious and are generally useful only for special purposes. The procedure outlined here for the colorimetric determination of glyoxylate is simple, sensitive, and highly specific.

## Reagents

DL-2-Keto-4-hydroxyglutarate, 50 m$M$, pH 6.8. See Vol. 41 [27] for the preparation of DL-2-keto-4-hydroxyglutarate.

[1] Abbreviation used: KHG, 2-keto-4-hydroxyglutarate.
[2] E. E. Dekker and U. Maitra, *J. Biol. Chem.* **237**, 2218 (1962).
[3] F. B. Straub, *Hoppe-Seyler's Z. Physiol. Chem.* **244**, 117 (1936).
[4] This series, Vol. 17B [175D].
[5] R. D. Kobes and E. E. Dekker, *J. Biol. Chem.* **244**, 1919 (1969).
[6] K. Kuratomi and K. Fukunaga, *Biochim. Biophys. Acta* **78**, 617 (1963).
[7] R. G. Rosso and E. Adams, *J. Biol. Chem.* **242**, 5524 (1967).

Tris·HCl buffer, 1.0 $M$, pH 8.4

Glutathione (reduced), 50 mM

Metaphosphoric acid, 12%

Phenylhydrazine hydrochloride, 1% in water, made fresh daily (the commercial solid is recrystallized twice from 50% ethanol)

Potassium ferricyanide, 5%, made fresh daily

Sodium glyoxylate (commercial material is purified by precipitation as the bisulfite addition product, recrystallizing three times, and decomposing in acid solution).

*Procedure.* Assay mixtures contain 0.1 ml Tris·HCl buffer, 0.1 ml of glutathione, 0.1 ml of KHG, water, and aldolase in a final volume of 1.0 ml. After the mixtures are incubated for 20 min at 37°, the reaction is stopped by adding 0.4 ml of metaphosphoric acid. If necessary, any precipitate is removed by centrifugation. Controls should include a mixture lacking substrate and another lacking the enzyme sample. An aliquot of the supernatant solution is made to 1.0 ml with water, mixed with 1 drop of concentrated hydrochloric acid and 2 drops of 1% phenylhydrazine solution, and heated in a boiling water bath for exactly 2 min. After chilling for 8 min by immediate transfer to an ice water bath, the samples are diluted and mixed thoroughly with 1.2 ml of cold, concentrated hydrochloric acid and sufficient water to bring the volume to the mark in 5-ml calibrated Klett tubes. Finally, the solutions are mixed with 2 drops of cold 5% potassium ferricyanide solution; the intensity of the colored 1,5-diphenylformazancarboxylic acid[8] is measured in a Klett-Summerson colorimeter (No. 54 filter) exactly 10 min after this last addition. Standard solutions of glyoxylate (0.01–0.10 μmole) containing as much metaphosphoric acid as the unknown samples are carried through the same procedure.

*Definition of Unit and Specific Activity.* One unit of KHG-aldolase activity is defined as the amount of enzyme that catalyzes the formation of 1.0 μmole of glyoxylate per minute under the conditions of the assay. The calculations are based on the premise that 0.10 μmole of glyoxylate has an absorbance of 0.60 (300 Klett units). Specific activity is expressed as units of enzyme activity per milligram of protein.

## Purification Procedure

Bovine liver, obtained from a local abattoir immediately at the time of slaughter, is trimmed of connective tissue, chilled in ice, and frozen in 250-g portions. Liver stored in frozen state for about a month has

[8] D. N. Kramer, N. Klein, and R. A. Baselice, *Anal. Chem.* **31**, 250 (1959).

been found usable. Unless indicated otherwise, all steps are carried out at 0–4°. Protein is estimated by the method of Lowry et al.[9] with crystalline bovine serum albumin as a standard.

Step 1. *Preparation of Extract.* Frozen bovine liver (1000 g) is thawed overnight at 5–6°. The liver is then cut into small pieces, which are rinsed with 1 liter of cold 0.5 $M$ KCl and removed by filtering through cheesecloth. The tissue is homogenized in a precooled Waring Blendor (1 gallon capacity) with 1000 ml of 0.5 $M$ KCl and 3.3 ml of 30 m$M$ EDTA for 4 min at low speed and 1 min at high speed. The resulting thick suspension is diluted with 1 liter of cold 0.5 $M$ KCl, 20 ml of $M$ Tris·HCl buffer (pH 7.4), and 1.2 ml of 2-mercaptoethanol and stirred for 30 min. The crude homogenate is centrifuged at 20,000 $g$ for 30 min, and the precipitate is discarded.

Step 2. *Heat Treatment.* The supernatant fluid is placed in a 5-liter round-bottomed flask fitted with a stirrer and heated with constant stirring in a 95° water bath. As soon as the temperature of the extract reaches 70°, the flask is placed in an ice water bath and the solution stirred until the temperature approaches 5°. The suspension is diluted with 180 ml of $M$ Tris·HCl buffer (pH 7.4) and centrifuged at 20,000 $g$ for 30 min to remove the denatured protein.

Step 3. *Ammonium Sulfate Fractionation.* Crystalline ammonium sulfate is added with stirring to the supernatant fluid until the salt concentration is 38% saturation (235 g/liter). The mixture is stirred for an additional 45 min and then centrifuged at 20,000 $g$ for 45 min. The precipitate is dissolved in a maximal volume of 70 ml of 50 m$M$ Tris·HCl buffer (pH 7.4) containing 5 m$M$ 2-mercaptoethanol ("Tris buffer mixture") and the resulting solution frozen. Three individual kilograms of frozen liver, treated in the manner outlined, are combined at this point and dialyzed against 8 liters of Tris buffer mixture for 8 hr.

Step 4. *DEAE-Cellulose Chromatography.* A suspension of DEAE-cellulose in water is degassed (water pump) and used to pack a column (4.3 × 32 cm) under 5 pounds of pressure. The column is equilibrated with 2 liters of Tris buffer mixture, then dialyzed enzyme solution is applied. The cellulose column is first washed with 3 liters of Tris buffer mixture. The enzyme is eluted with a nonlinear gradient of potassium chloride (0 → 0.22 $M$) prepared by passing Tris buffer mixture containing 0.22 $M$ potassium chloride into a mixing reservoir (800 ml) of Tris buffer mixture. Fractions (20 ml) are collected at a flow rate of approximately 150 ml/hr. The fractions with highest aldolase activity (normally tubes 55–75) are pooled.

[9] O. H. Lowry, N. J. Rosebrough, A. L. Farr, and R. J. Randall, *J. Biol. Chem.* **193**, 265 (1951).

*Step 5. Calcium Phosphate Gel-Cellulose Chromatography.* A suspension of Whatman standard grade cellulose powder (20 g) in 200 ml of water is mixed thoroughly with 100 ml of a suspension of calcium phosphate gel (30 mg/ml) prepared by the method of Swingle and Tiselius.[10] The mixture is degassed (water pump) for 15 min, then poured as a slurry into a column and allowed to settle by gravity. The column is then equilibrated with 200 ml of 300 m$M$ Tris buffer mixture. The pooled eluates from step 4 are applied to this column. The column is washed with 100 ml of 0.3 $M$ Tris·HCl buffer (pH 7.4) plus 5 m$M$ 2-mercaptoethanol and the enzyme then eluted with the same Tris-mercaptoethanol mixture containing 4% ammonium sulfate. Fractions (10 ml) are collected and those with highest enzymic activity (usually tubes 5–11) are pooled. Crystalline ammonium sulfate is added with stirring to the pooled eluates until the salt concentration is 40% saturation (241 g/liter). The mixture is stirred for an additional 45 min and then centrifuged at 20,000 $g$ for 45 min. The precipitate is dissolved in 3.7 ml of 25 m$M$ Tris·HCl buffer (pH 7.4) containing 5 m$M$ 2-mercaptoethanol and dialyzed against 4 liters of the same buffer solution for 8 hr.

*Step 6. DEAE-Cellulose Chromatography in Ammonium Sulfate.*[11] DEAE-cellulose (Whatman DE-32, microgranular) is suspended several times in a solution containing 25 m$M$ Tris·HCl buffer (pH 7.4), 5 m$M$ 2-mercaptoethanol, and 30% saturated ammonium sulfate ("buffer-salt mixture"). After the suspension is degassed (water pump), a column (2.5 × 8.5 cm) is poured and equilibrated with the buffer–salt mixture. The enzyme solution from step 5 is diluted with 25 mM Tris·HCl buffer (pH 7.4) to a protein concentration of about 7.5 mg/ml and made 30% saturated with ammonium sulfate by adding the crystalline solid and stirring. The solution is stirred an additional 30 min, then applied to the DEAE-cellulose column. The cellulose is washed with the buffer–salt mixture (about 200 ml) until the effluent, monitored at 280 nm, shows complete removal of a protein peak. The aldolase is then eluted with a gradient (30% → 10%) of ammonium sulfate concentration. The gradient is made by continuous dilution of 150 ml of the initial buffer-salt mixture (containing 30% ammonium sulfate) with about 200 ml of a mixture containing 25 m$M$ Tris·HCl buffer (pH 7.4), 5 m$M$ 2-mercaptoethanol, and 10% saturated ammonium sulfate. Fractions (10 ml) are collected with the effluent monitored at 280 nm. KHG-aldolase activity is usually coincident with the observed protein peak. Fractions with highest enzymic activity (usually tubes 25–35) are pooled and concentrated with the use of an Amicon Diaflo ultrafiltration setup (PM-10 mem-

[10] S. M. Swingle and A. Tiselius, *Biochem. J.* **48**, 171 (1951).
[11] S. G. Mayhew and L. G. Howell, *Anal. Biochem.* **41**, 466 (1971).

PURIFICATION PROCEDURE FOR 2-KETO-4-HYDROXYGLUTARATE ALDOLASE
FROM BOVINE LIVER

| Step and fraction | Total volume (ml) | Total protein (mg) | Total units | Specific activity (units/ mg) | Re- covery (%) |
|---|---|---|---|---|---|
| 1. Crude extract | 7860 | 412,650 | 1140 | 0.0028 | 100 |
| 2. Heat-treated supernatant | 6250 | 57,813 | 953 | 0.017 | 84 |
| 3. Ammonium sulfate (0–38%) | 296 | 2,871 | 881 | 0.307 | 77 |
| 4. DEAE-cellulose eluate | 320 | 333 | 499 | 1.5 | 44 |
| 5. Calcium phosphate-cellulose eluate, ammonium sulfate (0–40%) | 5 | 76 | 370 | 4.9 | 32 |
| 6. DEAE-cellulose eluate in ammonium sulfate | 5 | 16 | 155 | 9.7 | 14 |

brane). The final enzyme solution (4–8 ml) is dialyzed for 8 hr against 4 liters of 25 m$M$ Tris·HCl buffer (pH 7.4) containing 5 m$M$ 2-mercapto-ethanol and stored at 5°.

Overall, the aldolase is purified several thousandfold; a summary of the steps and typical results is shown in the table. Purified enzyme samples show variable degrees of stability; in general, preparations are stable for at least 2 weeks whereas some have retained full activity for months. At times, a few minor protein bands are seen by polyacrylamide gel electrophoresis but otherwise the purified enzyme yields a single major band.

## Properties

KHG-aldolase has an average molecular weight of 120,000.[5] The D- and L-isomers of KHG are virtually equally effective as substrate[12] ($K_m$ values for DL-, L-, and D-KHG are 0.11 m$M$, 71 $\mu M$, and 0.14 m$M$, respectively, whereas the relative $V_{max}$ values are 100, 105, and 84, respectively).[13] The enzyme has a pH optimum of 8.8, no divalent metal ion requirement, and no absolute requirement for added thiol compounds although sulfhydryl-reacting reagents strongly inhibit activity.[5] The aldolase preparation is essentially specific for the cleavage of KHG; only 2-keto-4,5-dihydroxyvalerate, 2-keto-4-hydroxy-4-methylglutarate, 5-keto-4-deoxyglucarate, 2-keto-3-deoxy-6-phosphogluconate, and 2-keto-4-hydroxybutyrate are cleaved at 33%, 8%, 3%, 2%, and 1%, respectively, the rate of KHG-cleavage.[12] The aldolase is inactivated by

[12] R. D. Kobes and E. E. Dekker, *Biochim. Biophys. Acta* **250**, 238 (1971).
[13] R. S. Lane and E. E. Dekker, *Biochemistry* **11**, 3295 (1972).

borohydride when either KHG, pyruvate *or* glyoxylate is present, indicating the formation of Schiff-base intermediates.[14] One mole of each substrate is bound per mole of enzyme. The $N_6$-lysine derivatives have been isolated and identified with pyruvate and glyoxylate; competitive binding in the presence of borohydride suggests that all *three* substrates are bound at the same active-site lysyl residue. Tests (inactivation by borohydride) of analogs for their ability to form a Schiff-base intermediate indicate KHG-aldolase is highly specific for pyruvate, but not for glyoxylate; likewise, the enzyme is quite specific for analogs of KHG having a pyruvate-like structure on one end of the molecule.[12] Treatment of this aldolase with glyoxylate (or glyoxylate analogs) in the presence of cyanide irreversibly destroys all enzymic activity, and 1 mole of glyoxylate is bound per mole of enzyme[14]; the formation of an inactive aminonitrile has been demonstrated. This effect of cyanide is not seen with either pyruvate or KHG.

The purified enzyme is also an effective $\beta$-decarboxylase, catalyzing the $\beta$-decarboxylation of oxaloacetate at 50% the rate of KHG cleavage.[12,15] Tetranitromethane treatment of KHG-aldolase completely destroys aldolase and $\beta$-decarboxylase activities of the enzyme concomitant with the oxidation of four free sulfhydryl groups.[13]

The equilibrium constant for the reaction at 37° is 11 mM in Tris·HCl buffer (pH 8.4) and 1.32 mM in Krebs original Ringer phosphate buffer (pH 7.3); the difference is probably due to Tris forming a complex with glyoxylate shifting the equilibrium in the direction of KHG cleavage. This aldolase catalyzes a terminal step in hydroxyproline[16] and homoserine[17] catabolism in mammals.

[14] R. D. Kobes and E. E. Dekker, *Biochemistry* **10**, 388 (1971).
[15] R. D. Kobes and E. E. Dekker, *Biochem. Biophys. Res. Commun.* **27**, 607 (1967).
[16] U. Maitra and E. E. Dekker, *J. Biol. Chem.* **238**, 3660 (1963).
[17] R. S. Lane, A. Shapley, and E. E. Dekker, *Biochemistry* **10**, 1353 (1971).

# [46] 2-Keto-4-hydroxyglutarate Aldolase from *Escherichia coli*

*By* EUGENE E. DEKKER, HIROMU NISHIHARA, and SHARON R. GRADY

DL-2-Keto-4-hydroxyglutarate $\rightleftarrows$ pyruvate + glyoxylate

The KHG-aldolase present in mammalian liver (see this volume [45]) and in *Escherichia coli* are Class I aldolases; i.e., they function via Schiff-base mechanisms.[1] In mammals, KHG-aldolase catalyzes a specific reaction in the catabolism of hydroxyproline and homoserine whereas the function of this aldolase in *E. coli* is as yet uncertain.[2] A note[3] describing a 10-fold purification of KHG-aldolase from a soil bacterium grown on 2-ketoglutarate is the only other report on this aldolase in microorganisms.

## Assay Method

The most convenient method measures a product released as a result of KHG cleavage. The same assay described earlier (this volume [45], bovine liver KHG-aldolase), wherein glyoxylate formation is determined colorimetrically, is also used to estimate activity in extracts of *E. coli*. For the procedure and results described here, DL-KHG was used as substrate. Since KHG-aldolase from *E. coli* preferentially utilizes the L- rather than the D-isomer of KHG, somewhat higher specific activity values are obtained with L-KHG as substrate.

*Definition of Unit and Specific Activity.* One unit of KHG-aldolase activity is defined as the amount of enzyme that catalyzes the formation of 1.0 $\mu$mole of glyoxylate per minute under the conditions of the assay. Calculations are based on 0.10 $\mu$mole of glyoxylate having an absorbance of 0.60 (300 Klett units). Specific activity is expressed as units of enzyme activity per milligram of protein.

## Purification Procedure

*Growth of Bacterial Cells.* KHG-aldolase of *E. coli* appears to be a constitutive enzyme since changes in growth medium seem to have little effect on activity levels. The enzyme has been obtained in homogeneous form from cells grown on a medium of 2.5% nutrient broth.[2] The medium described here is now used since it presents less foaming problem and the cells obtained are also used as a source for other enzymes.

*Escherichia coli* K12 is grown aerobically (100-liter volumes) in a New Brunswick Scientific Fermacell Fermentor. The growth medium contains (in percentages, weight/volume): 0.84% $Na_2HPO_4$, 0.45% $KH_2PO_4$, 0.15% $NH_4Cl$, 0.03% $MgSO_4 \cdot 7H_2O$, 0.8% glycerol, and 0.25% casein hydrolysate. The solid components except $MgSO_4 \cdot 7H_2O$ are dissolved in 15 liters of water and the glycerol is added; this solution is

[1] See this series, Vol. 9 [87] for the distinction between class I and class II aldolases on the basis of the properties of fructose-diphosphate aldolases.
[2] H. Nishihara and E. E. Dekker, *J. Biol. Chem.* **247**, 5079 (1972).
[3] L. D. Aronson, R. G. Rosso, and E. Adams, *Biochim. Biophys. Acta* **132**, 200 (1967).

transferred to the fermentor and diluted to the appropriate volume. The magnesium sulfate dissolved in 250 ml of water is then added. The medium is sterilized for 30 min at 121°, then allowed to cool to 37°. A 4% (v/v) inoculum, prepared by growing cells for 11 hr in 2-liter batches on a rotary shaker at 37°, is added, and the cells allowed to grow for 12 hr at 37° with 4 cubic feet per minute of sterile air aeration and paddle agitation (120 rpm). The cells are harvested, washed by resuspending in cold 50 mM Tris·HCl buffer (pH 8.4), and finally removed by centrifuging for 20 min at 10,000 g at 4°. Approximately 1 kg of cells (wet weight) are obtained from 100 liters of medium; they can be stored as a frozen paste at −20° for several months without significant loss of enzyme activity. All purification steps are carried out at 0–4° unless otherwise specified. Protein is estimated by the method of Lowry et al.[4] with crystalline bovine serum albumin as a standard.

Step 1. Preparation of Extract. Cells (423 g, wet weight) are thawed overnight at 5–6° and then suspended in 1 liter of 50 mM potassium phosphate buffer (pH 7.4) containing 5 mM 2-mercaptoethanol and 0.1 mM EDTA ("buffer mixture"). This suspension is divided into two equal portions which are alternately sonicated with stirring until the temperature approaches 20° and then cooled in an ice–salt bath; a Branson Sonifier (Model MS2) at full power (setting of 8) is used. The temperature is kept below 20° at all times; total time of sonication is 4 hr for each portion of the suspension. The crude homogenates are combined, added to 690 ml of buffer mixture, and stirred overnight. The starting extract is obtained by centrifuging at 20,000 g for 30 min and discarding the precipitate.

Step 2. Ammonium Sulfate Fractionation. Crystalline ammonium sulfate is added with stirring to the supernatant fluid until the salt concentration is 50% saturation (303 g/liter); the pH of the solution is adjusted as necessary to 7.4 with 5 N NH₄OH. The mixture is stirred for an additional 2 hr and centrifuged at 20,000 g for 30 min; the precipitate is discarded. More crystalline ammonium sulfate is added with stirring until the salt concentration is 80% saturation (230 g additional per liter of original volume). This solution is stirred for 2 hr, after the pH is adjusted to 7.4, and subsequently centrifuged at 20,000 g for 30 min. The precipitate is dissolved in about 150 ml of buffer mixture, and the resulting solution dialyzed against two changes (8 liters each) of buffer mixture for a total of 12 hr.

Step 3. Heat Treatment. Two equal portions of the dialyzed solution are placed in 250-ml beakers and heated with stirring in a 90° water

[4] O. H. Lowry, N. J. Rosebrough, A. L. Farr, and R. J. Randall, J. Biol. Chem. 193, 265 (1951).

bath. As soon as the temperature of the solution reaches 80°, the beaker is placed in an ice water bath and the solution stirred until the temperature falls below 30°. The precipitate is removed by centrifuging at 20,000 $g$ for 30 min and discarded.

Step 4. *Protamine Sulfate Addition.* A 1% solution of neutralized protamine sulfate is added (0.041 ml per milligram of protein) dropwise with stirring to the supernatant fluid obtained in step 3. The turbid solution is stirred for an additional 20 min, centrifuged at 20,000 $g$ for 30 min, and the precipitate discarded.

Step 5. *Ammonium Sulfate Fractionation.* Crystalline ammonium sulfate is added slowly with stirring to the enzyme solution until the salt concentration is 52% saturation (317 g/liter); if necessary, the pH is adjusted to 7.4 with 5 $N$ NH$_4$OH. The mixture is stirred for an additional 2 hr, centrifuged at 20,000 $g$ for 30 min, and the precipitate discarded. More crystalline ammonium sulfate is added with stirring until the salt concentration is 80% saturation (230 g/liter of original volume). After the pH of the solution is adjusted to 7.4 with 5 $N$ NH$_4$OH, it is stirred for an additional 2 hr and then centrifuged at 20,000 $g$ for 30 min. The precipitate is dissolved in a minimal volume of buffer mixture and the resulting solution dialyzed against two changes (4 liters each) of buffer mixture for a total period of 12 hr. This solution is concentrated to about 10–15 ml with the use of an Amicon Diaflo ultrafiltration setup (PM-10 membrane).

Step 6. *Sephadex Filtration.* The fraction from step 5 is applied to a column (3.5 × 65 cm) of Sephadex G-100 previously equilibrated with 50 m$M$ Tris·HCl buffer (pH 8.0) containing 5 m$M$ 2-mercaptoethanol. The column is washed with the same buffer-mercaptoethanol mixture at a flow rate of approximately 25 ml/hr; 10-ml fractions are collected. Those fractions having highest aldolase activity (usually tubes 30–40) are pooled.

Step 7. *DEAE-Sephadex Chromatography.* The pooled fractions are finally applied to a column (3.0 × 27 cm) of DEAE-Sephadex A-50, previously equilibrated with 50 m$M$ Tris·HCl buffer (pH 8.0) containing 5 m$M$ 2-mercaptoethanol. The column is first washed with 2–3 column volumes of 50 m$M$ Tris·HCl buffer (pH 8.0) containing 5 m$M$ 2-mercaptoethanol and then with 2–3 column volumes of a mixture of 0.10 $M$ Tris·HCl buffer (pH 8.0) with 5 m$M$ 2-mercaptoethanol; 20-ml fractions are collected. A linear gradient (0.10 $M$ → 0.50 $M$; 250 ml of each solution) of Tris·HCl buffer (pH 8.0) containing 5 m$M$ 2-mercaptoethanol is used to elute the enzyme. Fractions (5 ml) are collected; those having highest enzymic activity (usually tubes 33–48) are pooled and concentrated as in step 5 to about 4–5 ml. The resulting enzyme sample is

PURIFICATION PROCEDURE FOR 2-KETO-4-HYDROXYGLUTARATE ALDOLASE
FROM *Escherichia coli*

| Step and fraction | Total volume (ml) | Total protein (mg) | Total units | Specific activity (units/ mg) | Re- covery (%) |
|---|---|---|---|---|---|
| 1. Crude extract | 1822 | 55,249 | 45.8 | 0.0008 | — |
| 2. Ammonium sulfate (50–80%) | 281 | 10,774 | 64.0 | 0.006 | 100 |
| 3. Heat-treated supernatant | 231 | 2,012 | 41.5 | 0.020 | 65 |
| 4. Protamine sulfate | 310 | 1,993 | 39.9 | 0.020 | 62 |
| 5. Ammonium sulfate (52–80%) | 10.8 | 1,382 | 53.5 | 0.039 | 84 |
| 6. Sephadex G-100 | 68.0 | 168 | 37.9 | 0.226 | 59 |
| 7. DEAE-Sephadex eluate | 4.0 | 13 | 29.5 | 2.27 | 46 |

dialyzed for 15 hr against 4 liters of 50 m$M$ potassium phosphate buffer (pH 7.4) containing 5 m$M$ 2-mercaptoethanol and stored at 5°. The presence or the absence of 2-mercaptoethanol in the final dialysis mixture seems to have little effect on the level of activity or the stability of the enzyme sample.

By this procedure, KHG-aldolase is purified over 2000-fold and migrates as a single protein band in polyacrylamide gel electrophoresis. The table shows a summary of the steps and typical results. Solutions of the pure enzyme show various degrees of stability. Most preparations are stable for at least a week and some have retained full activity for months. Others, however, lose about 50% of their initial activity after approximately a week and remain comparatively stable at that level.

## Properties

In contrast to the same enzyme from bovine liver,[5] KHG-aldolase from *E. coli* has an average molecular weight of 63,000 and preferentially utilizes the L-isomer of KHG as substrate ($K_m$ values for L-, D-, and DL-KHG are 2.3 m$M$, 25 m$M$, and 4.2 m$M$ respectively, whereas the relative $V_{max}$ values are 100, 19, and 84, respectively).[2] The enzyme has a pH optimum of 8.6 and no divalent metal ion requirement; sulfhydryl-reacting reagents only partially inhibit activity. Of a wide variety of compounds tested, only 2-keto-4-hydroxybutyrate is cleaved 8% relative to the rate of cleavage of DL-KHG at pH 8.1 and 25°. Like liver KHG-aldolase, the bacterial enzyme has the unusual property of forming a Schiff-base intermediate not only with KHG and pyruvate, but also with

[5] This volume [45].

glyoxylate; competition studies indicate that all three substrates are apparently bound at the same active site. Schiff-base formation (as determined by borohydride inactivation) is highly specific for pyruvate, but not for glyoxylate; 1 mole of either pyruvate or glyoxylate is bound per mole of enzyme. KHG-aldolase from *E. coli* like the liver enzyme, is irreversibly inactivated by cyanide only in the presence of glyoxylate (not with KHG or pyruvate), and under these conditions also 1 mole of glyoxylate is bound per mole of enzyme.

The aldolase catalyzes a typical hydrogen exchange half-reaction with pyruvate; chemical modification of either all five histidyl or all five cysteinyl residues in the molecule has essentially no effect on aldol cleavage or hydrogen exchange activities.[6] Whereas liver KHG-aldolase catalyzes the $\beta$-decarboxylation of oxaloacetate at 50% the rate of KHG cleavage, the pure bacterial aldolase is a better $\beta$-decarboxylase toward oxaloacetate than it is an aldolase toward KHG.[2]

[6] S. R. Grady and D. J. Dunham, *Fed. Proc., Fed. Amer. Soc. Exp. Biol.* **32**, 667Abs (1973).

# [47] Transaldolase

By O. Tsolas and L. Joris

Fructose-6-P + erythrose 4-P $\rightleftarrows$ sedoheptulose-7-P + glyceraldehyde 3-P

Transaldolase can most conveniently be purified from *Candida utilis*, a yeast rich in this enzyme. The activity in the yeast extracts can be separated into three fractions called isoenzymes I, II, and III, in their order of elution from DEAE-Sephadex columns.[1] The percentage distribution of the three isoenzymes varies from batch to batch of the commercial yeast preparation, isoenzyme I being almost always the major form. Dried cells yield about two-thirds of the activity found in frozen cells, and consequently, the latter are preferable as starting material.

Isoenzymes I and III are composed of similar or identical subunits, whereas isoenzyme II is a hybrid of the other two species, formed by exchange of their subunits ($\alpha\alpha + \beta\beta \rightleftarrows 2\ \alpha\beta$). The hybridization takes place under neutral conditions with an equilibrium constant of 0.7.[2,3] The preparation of this isoenzyme will not be presented here since it can be generated and prepared from purified isoenzymes I and III.[2]

[1] See this series, Vol. 9 [88].
[2] O. Tsolas and B. L. Horecker, *Arch. Biochem. Biophys.* **136**, 287 (1970).
[3] O. Tsolas and B. L. Horecker, *Arch. Biochem. Biophys.* **136**, 303 (1970).

Transaldolase has been purified from sources other than *Candida*. Purification methods have been reported from dried brewer's yeast[4] and from baker's yeast.[5] The activity from the latter source has been obtained in crystalline form. Recently a purification method for the activity in bovine mammary gland has also been reported.[6]

Purification procedures for isoenzymes I and III from dried or from frozen cells of *Candida* have been published earlier,[1,2] and a simple procedure for obtaining crystalline isoenzyme III from cells rich in this species was reported some time ago.[7] A procedure on a pilot-plant scale has also been published.[8] Recently, the procedure of preparation of isoenzymes I and III from frozen cells of *Candida* has been simplified and in addition to isoenzyme III, isoenzyme I has now been obtained in crystalline form. This procedure, together with the analytical chromatographic method for obtaining the relative amounts of each isoenzyme in various batches of yeast, will be described below.

## Assay Method

*Principle.* Glyceraldehyde 3-phosphate produced in the reaction is converted to dihydroxyacetone phosphate by triose-phosphate isomerase. This product is measured with DPNH and $\alpha$-glycerophosphate dehydrogenase. Under appropriate conditions the rate of oxidation of DPNH is proportional to the quantity of transaldolase present. The reaction is measured spectrophotometrically by the change in absorbance at 340 nm.[9]

*Reagents*

D-Fructose 6-phosphate (0.14 $M$). Dissolve 122 mg of D-fructose 6-phosphate, Na salt,[10] in water and adjust the volume to 2.0 ml.

DPNH (10 m$M$). Dissolve 10 mg of reduced diphosphopyridine nucleotide in 1.0 ml of $10^{-3}$ $N$ NaOH.

Erythrose 4-phosphate (10 m$M$). Prepare D-erythrose 4-phosphate by the oxidation of glucose 6-phosphate with lead tetraacetate.[11,12]

[4] B. L. Horecker and P. Z. Smyrniotis, *J. Biol. Chem.* **212**, 811 (1955).

[5] R. Venkataraman and E. Racker, *J. Biol. Chem.* **236**, 1876 (1961).

[6] E. Kuhn and K. Brand, *Biochemistry* **11**, 1767 (1972).

[7] S. Pontremoli, B. D. Prandini, A. Bonsignore, and B. L. Horecker, *Proc. Nat. Acad. Sci. U.S.* **47**, 1942 (1961).

[8] H. E. Blair, S. E. Charm, D. Wallace, C. C. Matteo, and O. Tsolas, *Biotechnol. Bioeng.* **12**, 321 (1970).

[9] P. T. Rowley, O. Tchola, and B. L. Horecker, *Arch. Biochem. Biophys.* **107**, 305 (1964).

[10] Boehringer Mannheim Corp., New York, New York.

[11] J. N. Baxter, A. S. Perlin, and F. J. Simpson, *Can. J. Biochem. Physiol.* **37**, 199 (1959).

[12] See this series, Vol. 9 [6].

Store the final product as a solution (10 m$M$ at pH 2–3) in the cold. Commercial preparations of erythrose 4-phosphate may also be used.

$\alpha$-Glycerophosphate dehydrogenase-triosephosphate isomerase. A mixture of the two enzymes (10 mg/ml) is commercially available.[10]

TEA-EDTA buffer. Triethanolamine-EDTA buffer, pH 7.8: a solution containing 0.1 $M$ triethanolamine, and 20 m$M$ EDTA.

*Procedure.* To 0.94 ml of TEA-EDTA buffer in a quartz cuvette add 0.02 ml of fructose 6-phosphate, 0.01 ml of DPNH, 0.02 ml of erythrose 4-phosphate, 0.001 ml of the suspension of $\alpha$-glycerophosphate dehydrogenase-triosephosphate isomerase mixture, and 0.01 ml of transaldolase, diluted to give a change in absorbance of 0.005–0.060 per minute. Measure the change in absorbance at 340 nm at 25°. Run a control cuvette without transaldolase, and correct for the oxidation of DPNH in the absence of transaldolase.

*Definition of Unit and Specific Activity.* One unit is defined as the quantity of enzyme necessary to cleave 1 $\mu$mole of fructose 6-phosphate per minute under the conditions of the assay. Specific activity is expressed as units per milligram of protein. Determine protein by the turbidimetric method of Bücher.[13]

*Analytical Chromatography of Transaldolase.* Suspend DEAE-Sephadex A-50 (Pharmacia Fine Chemicals, Piscataway, New Jersey) in water and allow to swell overnight. Remove the fine particles by repeated decantation. Wash the exchanger successively with 0.5 $N$ HCl, water, 0.5 $N$ NaOH, and water, using a large suction flask to filter. Transfer the washed exchanger to a large beaker, suspend in water, and add 0.5 $M$ NaH$_2$PO$_4$ until the pH reaches 6.5. Then dilute the suspension until the phosphate concentration is 50 m$M$. Allow the exchanger to settle, decant the supernatant fluid, and resuspend it twice in fresh 50 m$M$ phosphate buffer, pH 6.5. Store it in the cold. To prepare a column, decant the supernatant and dilute the stock DEAE-Sephadex suspension at room temperature with 4 volumes of 50 m$M$ phosphate buffer, pH 6.5. Allow the suspension to settle for 20 min, decant the fine particles, and repeat the dilution and decanting procedure. Pack a column 0.8 × 26 cm with the freshly settled DEAE-Sephadex by a single addition of the slurry to the column which has previously been partially filled with buffer. Store the column in the cold overnight. To this column add 1 ml of transaldolase solution containing no more than 40 mg of protein which has previously been dialyzed against 50 m$M$ sodium phosphate buffer, pH 6.5. Elution is

---

[13] T. Bücher, *Biochim. Biophys. Acta* **1**, 292 (1947).

begun as soon as the enzyme is on the column, using a linear gradient prepared with 100 ml of 0.1 $M$ KCl and 100 ml of 0.2 $M$ KCl, both in 50 m$M$ phosphate buffer, pH 6.5. Collect fractions (2–4 ml) and analyze for transaldolase. Isoenzyme I can be expected to appear at 0.11 $M$ KCl, isoenzyme II at 0.13 $M$ KCl, and isoenzyme III at 0.15 $M$ KCl. When the analytical column is applied to extracts, these can first be concentrated with ammonium sulfate or by ultrafiltration to contain 20–30 units of transaldolase per milliliter before they are placed on the column.

### Purification Procedure for Transaldolase Isoenzyme I

*Step 1. Extraction.* Into each of several 4-liter beakers place 750 g of frozen yeast cake.[14] Allow to thaw at room temperature overnight. To each beaker add 1500 ml of cold deionized water and stir the mixture for 10 min. Adjust the pH to 4.8 from 5.6 with 87 ml of 0.5 $M$ $H_3PO_4$. Stir the mixture for 10 min more and then centrifuge at 9000 rpm (13,700 $g$) in a Sorvall RC2-B refrigerated centrifuge. Adjust the pH of the supernatant solution to 6.5 with 57 ml of 1 $N$ NaOH. Repeat the procedure with the contents of the second beaker and combine the supernatant solutions (Extract, 3750 ml, Table I).

*Step 2. Double-Bed Column Chromatography.* Carry out all operations in the cold. Pack a column, 5.0 × 40 cm, with DEAE-Sephadex to a height of 16 cm, and then add a bed of Cellex-D to a height of 10 cm.[15] Load the column with most of the extract, 3500 ml, 20,700 units, corresponding to 1400 g of yeast cake, over a period of 45 hr. Wash it with 350 ml of 50 m$M$ sodium phosphate pH 6.5, and then elute with a gradient of 50 m$M$ to 0.25 $M$ KCl in the same buffer, 2.0 liters per vessel. Start collecting fractions (21 ml per 15 min) when the gradient is connected, and assay activity and measure $A_{280}$ every third fraction. Pool fractions 42–88 (isoenzyme I, eluate, 970 ml, Table I), 89–134, (isoenzyme II, eluate, 990 ml, Table I) and 135–153 (isoenzyme III, eluate,

---

[14] *Candida utilis*, frozen, can be purchased from the Lake States Yeast and Chemical Division of the St. Regis Paper Co., Rhinelander, Wisconsin. The frozen cake can be stored at −20° for up to one year with no loss in enzymic activity.

[15] Preparation of double-bed column: Suspend Cellex-D (Bio-Rad Laboratories, Richmond, California), exchange capacity 0.78 mEq/g, 40 g, in 2 liters of 50m$M$ sodium phosphate, pH 6.5; allow to settle 15 min, remove the fines by suction, repeat the suspension, settling, and suction five times. Take a column, 5.0 × 40 cm, fitted with a Teflon metering valve (Kimax No. 41002-F, 2 mm) for easy reproducibility of flow rates, and pack it to 16 cm with DEAE-Sephadex in 50 m$M$ sodium phosphate, pH 6.5, prepared as described in the section on analytical chromatography. On top of the DEAE-Sephadex pack Cellex-D to 10 cm height. Wash the double-bed column with 50 m$M$ sodium phosphate, pH 6.5, overnight in the cold.

TABLE I

SEPARATION OF THE THREE ISOENZYMES OF TRANSALDOLASE

| Step | Volume (ml) | Total units | Specific activity, (units/mg) | Yield (%) |
|---|---|---|---|---|
| Extract from 1500 g of frozen yeast | 3750 | 22,200 | 1.15 | |
| Extract loaded on double-bed column | 3500 | 20,700 | 1.15 | 100 |
| Isoenzymes eluted: | | | | |
|   Iso I, eluate | 970 | 14,700 | 6.2 | 71 |
|     concentrate | 116 | 13,500 | 5.9 | |
|   Iso II, eluate | 990 | 2,420 | 0.71 | 12 |
|     concentrate | 132 | 2,130 | 0.63 | |
|   Iso III, eluate | 400 | 800 | 0.81 | 4 |
|     concentrate | 50 | 725 | 0.83 | |
| | | | | $\overline{87}$ |

400 ml, Table I). Concentrate each pooled solution 8-fold on an Amicon ultrafiltration cell, using a PM-30 Diaflo membrane. Add 0.2 ml toluene as preservative during concentration. Clarify the concentrates by centrifugation and store frozen at −16° (isoenzyme I concentrate, 116 ml; isoenzyme II concentrate, 132 ml; isoenzyme III concentrate, 50 ml, Table I).

*Step 3. Ammonium Sulfate Precipitation I.* To concentrated isoenzyme I, 116 ml, 13,500 units, add 33.8 g of ammonium sulfate to 50% saturation.[16] Centrifuge the suspension and bring the supernatant solution to pH 5.0 with 2.2 ml of 2 *N* acetic acid. Centrifuge off the precipitate, and to the supernatant add 21 g of ammonium sulfate to a saturation of 75%. Centrifuge and dissolve the precipitate in TEA-EDTA buffer II[17] (AS I, 28.2 ml, Table II).

*Step 4. CM-Sephadex Chromatography.* Prepare CM-Sephadex C-50, (Pharmacia Fine Chemicals, Piscataway, New Jersey) by suspending 40 g in water, allowing the exchanger to settle overnight, filtering on a Büchner funnel, and washing with 0.5 *N* NaOH, deionized water, 0.5 *N* HCl, and deionized water. Transfer the exchanger to a 4-liter cylinder, add water to the 1500-ml mark, and then add 1 *N* sodium acetate to

[16] For the ammonium sulfate fractionations use Fisher Granulated ASC Grade ammonium sulfate. Allow each precipitate to stand in ice for 10 min before centrifugation. Collect the precipitates by centrifuging at 0° for 10 min in a Sorvall centrifuge, Model RC2-B at 12,000 rpm.

[17] TEA-EDTA Buffer II: Triethanolamine–EDTA, 40–10 m*M*, pH 7.6.

TABLE II
PURIFICATION OF TRANSALDOLASE ISOENZYMES I AND III

| Step | Volume (ml) | Total units | Specific activity (units/mg) | Yield (%) |
|---|---|---|---|---|
| *Isoenzyme I* | | | | |
| Isoenzyme I concentrate | 116 | 13,500 | 5.9 | 100% |
| Ammonium sulfate fractionation (AS I) | 28.2 | 12,400 | 8.6 | 92% |
| Dialysis (AS I′) | 38.5 | 11,200 | 11.5 | 83% |
| CM-Sephadex chromatography (CM-Eluate) | 79.0 | 10,500 | 58 | 78% |
| Ammonium sulfate fractionation II (AS II) | 6.3 | 8,350 | 62 | 62% |
| Ammonium sulfate fractionation III (AS III) | 5.0 | 8,000 | 68 | 59% |
| Ammonium sulfate extract (55 S) | 40.0 | 3,640[a] | 60 | 75% |
| Crystals | 4.5 | 2,900 | 61 | 59% |
| *Isoenzyme III* | | | | |
| Isoenzyme III concentrate | 50.0 | 725 | 0.83 | 100% |
| Crystals | 5.8 | 387 | 44 | 54% |

[a] For the crystallization 4900 units of AS III were taken.

pH 5.4. Dilute with water to make the sodium acetate concentration equal to 80 m$M$. Allow to settle, decant, and add 80 m$M$ acetate buffer, pH 5.4. Repeat settling, decantation, and buffer addition one more time and then deaerate the suspension. Pack a column, $3.2 \times 142$ cm, with the exchanger, wash overnight in the cold with the same buffer. In the meantime dialyze the enzyme preparation, AS I, against 1 liter of 50 m$M$ sodium acetate, pH 5.2, overnight in the cold. Clarify the solution by centrifugation (AS I′, 38.5 ml, Table II) and load on the CM-Sephadex column. Elute with 80 m$M$ sodium acetate, pH 5.4, and collect fractions, 4.1 ml, every 25 min. Analyze each fraction for activity and absorbance at 280 nm. Pool fractions 145–163, containing the activity (CM-Eluate 79.0 ml, Table II).

*Step 5. Ammonium Sulfate Precipitation II.* Adjust the pH of CM-Eluate, 79 ml, from pH 5.35 to 4.5 with 6.8 ml of 1 $N$ acetic acid. Add ammonium sulfate, 28.0 g, to bring the saturation to 55% and centrifuge.[16] To the supernatant solution add 5.6 g of ammonium sulfate to bring the saturation to 65%. After stirring the suspension for 10 min collect the precipitate by centrifugation and dissolve in TEA-EDTA buffer II (AS II, 6.3 ml, Table II).

*Step 6. Ammonium Sulfate Precipitation III.* Add TEA-EDTA buffer

II to AS II to bring the protein concentration to 11 mg/ml (final volume: 13.5 ml). Add cold 100% saturated ammonium sulfate solution, pH 7.5, 23.2 ml, to the enzyme solution to increase the ammonium sulfate saturation from 5% to 65%. Stir the solution 10 min and then centrifuge. To the supernatant solution add 37.5 ml of 100% saturated cold ammonium sulfate, pH 7.5, to 80% saturation. Stir the suspension and centrifuge 10 min later. Dissolve the precipitate in TEA-EDTA buffer II, assay and store at −16° (AS III, 5.0 ml, Table II).

*Step 7. Crystallization.* Prepare ammonium sulfate solutions of 75, 65, 60, 55, and 50% saturation in 0.2 $M$ sodium acetate pH 4.5, by mixing appropriate amounts of cold saturated ammonium sulfate, 0.8 $M$ sodium acetate, pH 4.50, and water. To the solution of isoenzyme I, fraction AS III, 2.6 ml, 1890 units/ml, 30.7 mg/ml, ammonium sulfate saturation 30%, add 6.5 ml of cold saturated ammonium sulfate solution to 80% saturation. Keep at 0° for 15 min and then centrifuge. Suspend the precipitate in 8.0 ml of 65% ammonium sulfate in sodium acetate pH 4.5, centrifuge and suspend the residue in 8.0 ml of 60% ammonium sulfate in sodium acetate pH 4.5. Repeat the extraction of the residue with a sufficient amount of 55% ammonium sulfate in sodium acetate pH 4.5 to dissolve almost all of it (about 40 ml). Centrifuge the solution and keep the supernatant (55 S, 40 ml). Store the 55 S fraction at 2° for 2 weeks until crystals appear. Increase the ammonium sulfate saturation to 56% by adding 2 ml of 75% ammonium sulfate in sodium acetate, pH 4.5, and store in the cold. Centrifuge the crystals a few days later and suspend in 60% ammonium sulfate in sodium acetate pH 4.5. Dissolve an aliquot of the suspended crystals in two volumes of TEA-EDTA buffer II and assay for activity and protein (crystals, 4.5 ml, Table II).

## Purification Procedure for Transaldolase Isoenzyme III

*Steps 1 and 2.* Proceed as in steps 1 and 2 of the purification of isoenzyme I in the section above.

*Step 3. Crystallization.* To 50 ml of Iso III-concentrate (Table I), 725 units, specific activity 0.83 units/mg, add 11.3 g ammonium sulfate to bring the saturation to 40%. After stirring for 10 min, centrifuge the suspension, and add 2.9 g of ammonium sulfate to the supernatant solution to bring the saturation to 50%. Stir again the suspension for 10 min and centrifuge. Adjust the pH of the supernatant from 5.9 to 5.0 with 2 $N$ acetic acid. After centrifuging off the cloudiness, add 1.0 g of ammonium sulfate to 52.5% saturation and store the solution in the cold overnight. Place it then at room temperature for half an hour to complete the crystallization. Collect the crystals by centrifugation and wash them by suspending in 1.0 ml of cold water and centrifuging. Wash the crystals by suspending them and centrifuging four more times in 0.5, 0.3, 0.2, and 0.1 ml portions of cold water. After the last centrifugation suspend the crystals in 60% ammonium sulfate solution, assay after dissolving 0.1 ml with 0.1 ml of TEA-EDTA buffer II (crystals, 5.8 ml, Table II). The specific activity increases to about 55 units/mg on dialysis of the purified enzyme (Crystals) against 1 liter of TEA-EDTA buffer II in the cold overnight.

### Properties

Transaldolase isoenzymes I and III have identical $K_m$ values for their substrates. The $K_m$ for D-fructose 6-phosphate is 0.8 m$M$ and for erythrose 4-phosphate 20 $\mu M$. Other, more slowly reacting substrates have been listed elsewhere.[18] The optimum activity for both isoenzymes is at pH 7.8, falling off to 50% of maximum at pH 6.3 and pH 9.5.[18]

The purified isoenzyme I shows a single band in disc electrophoresis. Isoenzyme III, in addition to one major band, shows a faster traveling minor band with transaldolase activity, distinct from the other isoenzymes.[2] The isoenzymes are homogeneous by sedimentation velocity analysis. The molecular weights, estimated from sedimentation equilibrium experiments, are 68,000 for isoenzyme I and 65,200 for isoenzyme III. The amino acid analyses of the present preparations agree with those published earlier.[3] The absorbance of 1 mg of protein per milliliter in 0.1 $N$ NaOH at 280 nm for isoenzyme I is 0.850, and the corresponding value for isoenzyme III is 0.674. The ratio of absorbance at 290/280, which is used as an indication of the purity of the isoenzymes, is 1.24 for isoenzyme I and 1.31 for isoenzyme III.[3]

Both isoenzymes are stable in the cold in the crystalline form for periods of over one year. In solution, they are stable at neutral pH at room temperature for a few hours and in the cold for several days.

[18] O. Tsolas and B. L. Horecker, in "The Enzymes" (P. D. Boyer, ed.), 3rd ed., Vol. 7, p. 259. Academic Press, New York, 1972.

# Section III
# Dehydratases

## [48] D-Gluconate Dehydratase from *Alcaligenes*[1]

*By* K. KERSTERS and J. DE LEY

D-Gluconate → 2-keto-3-deoxy-D-gluconate + $H_2O$

## Assay Method

*Principle.* Periodate oxidation of 2-keto-3-deoxy-D-gluconate (KDG) yields formylpyruvate. The chromogen formed by the reaction between thiobarbituric and formylpyruvic acids is measured at 549 nm.

*Reagents*

Sodium-D-gluconate, 0.4 $M$
Tris·HCl buffer, 0.2 $M$, pH 8.5
$MgCl_2$, 10 m$M$
Trichloroacetic acid, 20%
Periodic acid, 0.025 $M$, in 0.125 $N$ $H_2SO_4$
Sodium arsenite, 2%, in 0.5 $N$ HCl
2-Thiobarbituric acid, 0.3%

*Procedure.* The incubation mixture (0.2 ml) contains 0.15 ml of Tris·HCl buffer, 0.02 ml of $MgCl_2$, 0.02 ml of sodium D-gluconate and about 0.004 unit of D-gluconate dehydratase preparation. Incubation is performed at 30° for 20 min and terminated by the addition of 0.1 ml of trichloroacetic acid. Distilled water (0.6 ml) is added, precipitated protein is removed by centrifugation, and a 0.2-ml sample of the supernatant is used for the periodate-thiobarbituric acid test.[2] Care should be taken that the amount of periodate is not limiting. Include one control mixture with heat-inactivated enzyme, and another one without enzyme.

KDG can also be measured as its semicarbazone according to the procedure of MacGee and Doudoroff,[3] provided that purified enzyme is used.

*Definition of Unit and Specific Activity.* One unit of enzyme activity is defined as the quantity that yields 1 $\mu$mole of KDG per minute under the above conditions. Specific activity is expressed as units per milligram of protein.

[1] K. Kersters, J. Khan-Matsubara, L. Nelen, and J. De Ley, *Antonie van Leeuwenhoek J. Microbiol. Serol.* **37**, 233 (1971).
[2] A. Weissbach and J. Hurwitz, *J. Biol. Chem.* **234**, 705 (1959).
[3] J. MacGee and M. Doudoroff, *J. Biol. Chem.* **210**, 617 (1954).

## Purification Procedure

*Alcaligenes* species M250[4] is grown for 40 hr at 30° on a solid medium in Roux flasks containing per liter of tap water: calcium gluconate, 10 g; yeast extract, 5 g; peptone, 5 g; NaCl, 3 g; $MgSO_4 \cdot 7H_2O$, 0.2 g, and agar 20 g; final pH 7.0. Cells are harvested by centrifugation and washed twice with 10 m$M$ phosphate buffer, pH 7.0.

*Step 1. Crude Extract.* Ten grams of cell paste are suspended in 20 ml of 10 m$M$ phosphate buffer, pH 7.0, containing 1 m$M$ $MgCl_2$, and disrupted in a 10 kHz, 250 W Raytheon sonic oscillator for 15 min at 4° in a hydrogen atmosphere. Cell debris is removed by centrifugation at 15,000 $g$ for 15 min. Unless otherwise indicated, all subsequent steps are performed at 0–4°, with buffers containing 1 m$M$ $MgCl_2$ and 1 m$M$ sodium D-gluconate.

*Step 2. Particle-Free Extract.* The crude cell-free extract is centrifuged for 2 hr at 78,000 $g$ (rotor 40, Spinco preparative ultracentrifuge). The precipitate is discarded, and the supernatant can be stored at −12° for at least 1 week.

*Step 3. Streptomycin Treatment.* Streptomycin sulfate (20% in 0.1 $M$ phosphate buffer, pH 7.8) is slowly added to the particle-free extract to make a final concentration of 1%. The mixture is kept in ice for 30 min and the precipitate is discarded by centrifugation.

*Step 4. Ammonium Sulfate I.* Solid ammonium sulfate is added to 30% of saturation (0°), and the precipitate is discarded. Ammonium sulfate concentration is raised to 40% of saturation, the precipitate collected by centrifugation, and dissolved in 20 m$M$ phosphate buffer, pH 7.8.

*Step 5. Ammonium Sulfate II.* Ammonium sulfate fraction I is again treated with solid ammonium sulfate. The 25–30% saturated fraction is dissolved in 20 m$M$ phosphate buffer, pH 7.8. This fraction is "dialyzed" on a Sephadex G-25 column, previously equilibrated with 10 m$M$ Tris·HCl buffer, pH 8.8.

*Step 6. Acrylamide Gel Electrophoresis.*[5] Acrylamide gel electrophoresis is carried out in gel cylinders (0.7 × 2.5 cm) according to the continuous buffer system.[6] One electrophoretic run with 16 cylinders allows the

[4] The culture was obtained from Dr. M. J. Pickett, Bacteriology Department, UCLA, Los Angeles, California, as *Achromobacter* M250. Synonymous strain numbers are ATCC 9220 and AB 61 from Dr. H. Lautrop, Copenhagen, Denmark. According to the nomenclature of the forthcoming Bergey's Manual of Determinative Bacteriology (8th edition), the strain would be called *Alcaligenes faecalis*. From extensive experiments (Kersters and De Ley, unpublished, 1973), we think that the strain should be best renamed *Alcaligenes denitrificans*.

[5] See also this series Vol. 22 [34] and [39].

[6] S. Hjertén, S. Jerstedt, and A. Tiselius, *Anal. Biochem.* 11, 219 (1965).

PURIFICATION OF D-GLUCONATE DEHYDRATASE FROM *Alcaligenes*

|  | Volume (ml) | Protein (mg/ml) | Total units | Specific activity | Yield (%) |
|---|---|---|---|---|---|
| Crude extract | 190 | 18.4 | 362 | 0.103 | 100 |
| Particle-free extract | 170 | 11.2 | 316 | 0.165 | 87 |
| Streptomycin treatment | 168 | 10.8 | 291 | 0.160 | 80 |
| Ammonium sulfate I | 20 | 8.4 | 240 | 1.43 | 66 |
| Ammonium sulfate II | 9.5 | 4.1 | 76 | 1.95 | 21 |
| Acrylamide gel electrophoresis | 11 | 0.34 | 39 | 10.2 | 11 |

purification of 6.6 mg of protein. The composition of the gel solution is as follows: 7% acrylamide, 0.183% bisacrylamide, 0.03% (v/v) TEMED and 0.07% ammonium persulfate in 64 m$M$ Tris·HCl buffer, pH 8.8. The same buffer, supplemented with 1 m$M$ MgCl$_2$ and 1 m$M$ gluconate, is used as electrode buffer. After preelectrophoresis (1 hr, 3 mA/gel), 100 $\mu$l of the dialyzed fraction from step 5 (to which is added 5% sucrose) is applied on top of each gel. Electrode buffer is renewed and electrophoresis is carried out for 10 min at 1.25 mA/gel, followed by 150 min at 4 mA/gel. During electrophoresis, a temperature of about 8° is maintained by circulating electrode buffer through an ice bath. D-Gluconate dehydratase is localized in one gel cylinder by incubation for 20 min at 30° in 5 ml of a mixture containing 0.1 $M$ Tris·HCl buffer (pH 8.5), 1 m$M$ MgCl$_2$ and 10 m$M$ sodium D-gluconate. The gel is then transferred for 10 min at 25° in a test tube containing 0.15% $o$-phenylenediamine in 10% trichloroacetic acid. Heating the gel at 105° for 10 min and inspection under UV light reveals the KDG formed as a yellow fluorescent band. The enzyme zone can now be excised from the other gels with a razor blade and electrophoretically eluted with an apparatus similar to the one described by Sulitzeanu et al.[5,7] The elution buffer (Tris·HCl, 30 m$M$, pH 8.2) contains 1 m$M$ MgCl$_2$ and 0.1 m$M$ sodium D-gluconate.

Analytical disc gel electrophoresis of the purified enzyme shows one band. The purification is summarized in the table.

### Properties

The enzyme is rather labile. Some degree of stabilization is obtained in 30 m$M$ Tris·HCl buffer, pH 8.0–8.8, containing 1 m$M$ MgCl$_2$ and 1 m$M$ sodium D-gluconate. When stored under these conditions at 0°, a 50% loss of activity occurs after 7 days. Maximal activity is obtained

[7] D. Sulitzeanu, M. Slavin, and E. Yecheskeli, *Anal. Biochem.* **21,** 57 (1967).

between pH 8.4 and 8.8 in Tris·HCl or glycine-NaOH buffer. The molecular weight, determined by thin-layer chromatography on Sephadex G-200 is 270,000 ± 25,000.

D-Gluconate dehydratase, either purified or in the particle-free extract (step 2), can be used for enzymic synthesis of KDG.[8,9]

*Substrate Specificity.* D-Gluconate dehydratase from *Alcaligenes* is specific for aldonic acids with 5 or 6 carbon atoms having hydroxyl groups at C-2 and C-3 in L-*threo* configuration. Relative activities (D-gluconate, 100) are as follows: D-xylonate, 89; D-fuconate, 15; D-galactonate, 14; L-arabonate, 2. No activity was observed with D-glucose, D-glucitol, 5-keto-D-gluconate, D-glucarate, D- and L-gulonate, L-galactonate, D-mannonate, D-mannarate, D-galactarate, D-talonate, α-D-glucoheptonate, α-D-galaheptonate, D-arabonate, D-ribonate, D-xylarate, D-, L-, and *meso*-tartaric acid, D-glycerate, L-rhamnitol, L-fucitol, 6-phospho-D-gluconate, and 6-phospho-D-galactonate. $K_m$ for D-gluconate is 20.8 m$M$.

*Effect of Metals and Inhibitors.* D-Gluconate dehydratase is strongly activated (more than 100%) by 1 m$M$ Mg$^{2+}$. No absolute requirement for Mg$^{2+}$ was detected.[1] Mn$^{2+}$ activates for 50% at 0.1 m$M$ and 30% at 1 m$M$. Ca$^{2+}$, Co$^{2+}$, Ni$^{2+}$, and Zn$^{2+}$ are inhibitory at 1 m$M$. EDTA, 0.1 m$M$, inhibits 70% of the enzyme activity. *p*-Chloromercuribenzoate is an effective inhibitor (62% inhibition at 1 m$M$ and 30% at 10 μ$M$), whereas thiol compounds do not activate.

*Distribution.* An enzyme with similar properties has been purified from *Agrobacterium tumefaciens* B6, grown on D-gluconate.[10] D-Gluconate dehydratase activity has been detected in the 64–70% GC subgroup of the *Achromobacter–Alcaligenes* bacteria,[9] in *Aspergillus niger*,[11] several clostridia,[12,13] and *Rhodopseudomonas spheroides*.[14] A D-gluconate dehydratase with different properties has been described in *Clostridium pasteurianum*.[15]

[8] See also, this series, Vol 9 [117] and Vol. 41 [22].
[9] J. De Ley, K. Kersters, J. Khan-Matsubara, and J. M. Shewan, *Antonie van Leeuwenhoek J. Microbiol. Serol.* **36**, 193 (1970).
[10] A. Reynaerts and K. Kersters, unpublished data, 1973.
[11] T. A. Elzainy, M. M. Hassan, and A. M. Allam, *J. Bacteriol.* **114**, 457 (1973).
[12] R. Bender, J. R. Andreesen, and G. Gottschalk, *J. Bacteriol.* **107**, 570 (1971).
[13] J. R. Andreesen and G. Gottschalk, *Arch. Mikrobiol.* **69**, 160 (1969).
[14] M. Szymona and M. Doudoroff, *J. Gen. Microbiol.* **22**, 167 (1960).
[15] R. Bender and G. Gottschalk, *Eur. J. Biochem.* **40**, 309 (1973).

# [49] D-Fuconate Dehydratase

By RICHARD L. ANDERSON and A. STEPHEN DAHMS

D-Fuconate → 2-keto-3-deoxy-D-fuconate + $H_2O$
L-Arabonate → 2-keto-3-deoxy-L-arabonate + $H_2O$

This enzyme functions in the metabolism of D-fucose and L-arabinose in pseudomonad MSU-1.[1]

## Assay Method[1]

*Principle.* The $\alpha$-keto acid formed during the reaction is measured as its semicarbazone by the method of MacGee and Duodoroff.[2]

*Reagents*

$N,N$-Bis(2-hydroxyethyl)glycine (Bicine buffer), 0.4 $M$, pH 7.2
2-Thioethanol, 15 m$M$
D-Fuconate,[3] 0.25 m$M$
$MgCl_2$, 0.25 $M$
Semicarbazide reagent: 1% semicarbazide hydrochloride and 1.5% sodium acetate trihydrate in water

*Procedure.* The reaction mixture consists of 0.05 ml of Bicine buffer, 0.01 ml of 2-thioethanol, 0.02 ml of $MgCl_2$, 0.02 ml of D-fuconate, D-fuconate dehydratase, and water to a volume of 0.15 ml. The reaction is initiated by the addition of the enzyme. The reaction mixture is incubated at 30° for 30 min and is then quenched by the addition of 1.0 ml of the semicarbazide reagent. After a 15-min incubation at 30°, the mixture is diluted to 5.0 ml with water. A sample is then poured into a quartz cuvette with a 1-cm light path. The absorbance is read in a spectrophotometer at 250 nm against a diluted reagent blank. Controls are necessary to correct for the absorbance of protein and other reaction components.

*Definition of Unit and Specific Activity.* One unit is defined as the amount of enzyme that catalyzes the formation of 1 $\mu$mole of $\alpha$-keto acid per minute, calculated on the basis of a molar absorption coefficient of 10,200.[2] Specific activity is in terms of units per milligram of protein. Protein is conveniently measured by the method of Warburg and Chris-

[1] A. S. Dahms and R. L. Anderson, *J. Biol. Chem.* **247,** 2233 (1972).
[2] J. MacGee and M. Doudoroff, *J. Biol. Chem.* **210,** 617 (1954).
[3] D-Fuconate can be prepared from D-fucose by the method of S. Moore and K. P. Link, *J. Biol. Chem.* **133,** 293 (1940).

tian,[4] but with crude extracts and fractions high in nucleic acid content, the biuret method[5] may be used.

*Qualitative Assay.* When a rapid qualitative assay for 2-keto-3-deoxyaldonic acid formation is desired (as in scanning fractions from chromatographic columns), the periodate-2-thiobarbituric acid procedure of Weissbach and Hurwitz[6] may be used. However, 2-thioethanol should not be used in the assay because it depresses chromogen formation.

## Purification Procedure[1]

The enzyme was purified from pseudomonad MSU-1.[7] The organism was grown as described in another article,[8] except that the carbon source was L-arabinose instead of D-glucose. The preparation of cell extracts was also as described,[8] except that the cells were suspended in 0.10 $M$ Bicine buffer, pH 7.4. The following procedures were performed at 0 to 4°. A summary of the purification procedure is given in the table.

*Protamine Sulfate Treatment.* To a cell extract containing 0.2 $M$ ammonium sulfate was added an amount of 2% (w/v) protamine sulfate solution (pH 7.0 in 0.10 $M$ Bicine buffer) to give a final concentration of 0.33%. After 30 min, the precipitate was removed by centrifugation and discarded.

*Ammonium Sulfate Fractionation.* The protein in the supernatant from the protamine step was fractionated by the addition of crystalline ammonium sulfate. The protein precipitating between 40 and 60% saturation was collected by centrifugation and was dissolved in 0.10 $M$ Bicine buffer (pH 7.4).

*Chromatography on Sephadex G-200.* The above fraction was chromatographed on a column (6 × 60 cm) of Sephadex G-200 equilibrated with 0.10 $M$ Bicine buffer, pH 7.4. Fractions (15 ml) were collected during elution with the same buffer, and those with the highest specific activity were combined.

*Chromatography on DEAE-Cellulose.* The combined Sephadex G-200 fractions were reduced in volume from 75 ml to 5 ml with a Diaflo ultrafiltration cell (Amicon Corp.) containing a type UM-10 membrane. This concentrate was applied to a column (2 × 6 cm) of DEAE-

[4] O. Warburg and W. Christian, *Biochem. Z.* **310**, 384 (1941).

[5] A. G. Gornall, C. J. Bardawill, and M. M. David, *J. Biol. Chem.* **177**, 751 (1941).

[6] A. Weissbach and J. Hurwitz, *J. Biol. Chem.* **234**, 705 (1959).

[7] This organism obtainable from the American Type Culture Collection as strain ATTC 27855.

[8] See this series, Vol. 41 [34].

PURIFICATION OF D-FUCONATE DEHYDRATASE

| Fraction | Volume (ml) | Total protein (mg) | Total activity (units) | Specific activity (units/mg protein) | $A_{280}:A_{260}$ |
|---|---|---|---|---|---|
| Cell extract | 380 | 8435 | 659 | 0.077 | 0.68 |
| Protamine sulfate supernatant | 455 | 6598 | 648 | 0.098 | 0.84 |
| (NH$_4$)$_2$SO$_4$ precipitate | 75 | 1575 | 491 | 0.312 | 1.22 |
| Sephadex G-200 | 79 | 869 | 117 | 1.35 | 1.48 |
| DEAE-Cellulose | 20 | 16 | 37.3 | 2.33 | 1.60 |

cellulose which was pretreated as recommended by Sober et al.[9] and which had been equilibrated with 10 m$M$ sodium phosphate buffer (pH 7.0). The protein was then eluted by the stepwise addition of 60-ml volumes of the same buffer containing 0, 0.05, 0.10, 0.20, and 0.30 $M$ NaCl. D-Fuconate dehydratase eluted with 0.30 $M$ NaCl. Prolonged contact of the enzyme with DEAE-cellulose resulted in its inactivation, so this step should be performed quickly.

The above procedures resulted in a 30-fold purification of D-fuconate dehydratase with an overall recovery of 6%.

### Properties[1]

*Substrate Specificity.* Of 19 aldonic acids tested, only D-fuconate ($K_m$ = 4.0 m$M$) and L-arabonate ($K_m$ = 4.3 m$M$) served as substrates for this dehydratase. Compounds that did not serve as substrates (<2% of the rate with D-fuconate) at 60 m$M$ concentrations were L-fuconate, D-arabonate, D-galactonate, 6-iodo-6-deoxy-D-galactonate, D-ribonate, D-xylonate, D-lyxonate, D-mannonate, L-rhamnonate, D-gluconate, L-galactonate, L-gluconate, D-galactarate, D-glucarate, D-glucuronate, D-galacturonate, and cellobionate.

*pH Optimum.* Dehydratase activity as a function of pH is maximal at about pH 7.1 to 7.4.

*Metal Requirement.* Dehydratase activity was nil in the presence of 0.33 $M$ ethylenediaminetetraacetate. The further addition of various salts (13.3 m$M$) resulted in the following relative rates (percentage of original activity): MgCl$_2$ or MgSO$_4$, 100; MnCl$_2$, 85; FeSO$_4$, 37; CuSO$_4$, 29;

[9] H. A. Sober, F. J. Gutter, M. Wyckoff, and E. A. Peterson, *J. Amer. Chem. Soc.* **78**, 756 (1956).

$CoCl_2$, 18; $NiCl_2$ 8; $CaCl_2$ or $ZnCl_2$, 0. The optimal $Mg^{2+}$ concentration was about 20 m$M$.

*Effect of Thiols.* Although an absolute requirement for a thiol could not be demonstrated, omission of 2-thioethanol from the assay mixture resulted in about a 30–45% reduction in activity.

*Stability.* The purified enzyme was stable unfrozen at 0° to 4° (<10% loss per week), but was about 50% inactivated when stored frozen overnight.

# [50] L-2-Keto-3-deoxyarabonate Dehydratase[1]

*By* ALLEN C. STOOLMILLER

$$
\begin{array}{c}
\text{COO}^- \\
|\\
\text{C}{=}\text{O} \\
|\\
\text{CH}_2 \\
|\\
\text{HO}-\text{C}-\text{H} \\
|\\
\text{H}-\text{C}-\text{OH} \\
|\\
\text{H}
\end{array}
\quad\longrightarrow\quad
\begin{array}{c}
\text{COO}^- \\
|\\
\text{C}{=}\text{O} \\
|\\
\text{CH}_2 + \text{H}_2\text{O} \\
|\\
\text{CH}_2 \\
|\\
\text{CHO}
\end{array}
\qquad (1)
$$

L-2-Keto-3-deoxyarabonate       α-Ketoglutarate semialdehyde

## Assay Method

*Principle.* The continuous spectrophotometric assay is based on the following sequence of reactions:

$$
\text{L-2-Keto-3-deoxy-arabonate} \xrightarrow{\substack{\text{L-2-keto-3-deoxy-}\\\text{arabonate dehydratase}}} \text{α-ketoglutarate semialdehyde} + \text{H}_2\text{O} \qquad (2)
$$

$$
\text{α-Ketoglutarate semialdehyde} + \text{NADP}^+ \xrightarrow{\substack{\text{aldehyde}\\\text{dehydrogenase}}} \text{α-ketoglutarate} + \text{NADPH} + \text{H}^+ \qquad (3)
$$

In the presence of excess aldehyde dehydrogenase the rate of α-ketoglutarate semialdehyde formation is equivalent to the rate of reduction of $NADP^+$, which is measured by the absorbance increase at 340 nm.

## Reagents

$K_2HPO_4$–$KH_2PO_4$, 0.2 $M$, pH 7.4, containing 20 m$M$ 2-mercaptoethanol

---

[1] A. C. Stoolmiller and R. H. Abeles, *Biochem. Biophys. Res. Commun.* **19**, 438 (1965); A. C. Stoolmiller and R. H. Abeles, *J. Biol. Chem.* **241**, 5764 (1966); A. C. Stoolmiller, Ph.D. Thesis, Univ. of Michigan, 1966.

Potassium DL-2-keto-3-deoxyarabonate, 20 m$M$. See Vol. 41 [23].
NADP⁺, 5 m$M$
Aldehyde dehydrogenase from *P. oleovorans*, 0.2 unit/ml. Preparation and some properties of this enzyme are given in the second part of this chapter.

*Procedure.* Place 0.4 ml of buffer, 0.1 ml of DL-2-keto-3-deoxyarabonate, 0.1 ml of NADP⁺, and 0.2 ml of aldehydye dehydrogenase in a 1-ml quartz cell with a light path of 1 cm. The reaction is initiated by adding 0.2 ml of diluted enzyme solution (replaced by buffer in blank) which produces an optical density change of 0.005–0.050 per minute at 340 nm.

*Definition of Unit and Specific Activity.* The unit of enzyme activity is defined as the amount required to produce 1 $\mu$mole of $\alpha$-ketoglutarate semialdehyde per minute at 30°. Specific activity is expressed as units per milligram of protein as determined by the method of Lowry *et al.*[2] or by the optical density at 280 nm after correcting for the absorption of nuelcic acid at 260 nm.[3]

## Purification Procedure for L-2-Keto-3-deoxyarabonate Dehydratase

### Enzyme Source

*Pseudomonas saccharophila* (ATCC 9114)[4] are grown and maintained on a minimal salt medium described by Doudoroff,[5] except that ferric ammonium sulfate is substituted for ferric ammonium citrate. Liquid cultures contain either 0.25% sucrose or 0.25% L-arabinose; agar plates (2% Difco Bacto-Agar) are prepared with 0.5% sucrose. Cultures are maintained by growing and plating the cells weekly in medium and on plates containing sucrose. The mass cultivation of *P. saccharophila* on L-arabinose (35 g per 150 ml of H$_2$O) and a trace-salts solution containtently yield high quality cells. The major problem seems to involve aeration control. However, cells of high quality can be obtained using the procedure described below.

Cultures are grown in 5-gallon bottles which contain 38 g of KH$_2$PO$_4$, 31 g of Na$_2$HPO$_4$, 15 g of NH$_4$Cl, and 7.5 g of MgCl$_2 \cdot$7H$_2$O in 19 liters

[2] O. H. Lowry, N. J. Rosebrough, A. L. Farr, and R. J. Randall, *J. Biol. Chem.* **193**, 265 (1951).

[3] O. Warburg and W. Christian, *Biochem. Z.* **310**, 384 (1941). See also this series, Vol. 3 [73].

[4] Other cultures of *P. saccharophila,* including ATCC 15946, would presumably contain the dehydratase when grown on L-arabinose.

[5] M. Doudoroff, *Enzymologia* **9**, 59 (1940).

of distilled water. Additionally, prepare two concentrated solutions of L-arabinose (35 g per 150 ml of $H_2O$) and a trace-salts solution containing 0.75 g of ferric ammonium sulfate and 0.075 g of $CaCl_2 \cdot 2H_2O$ separately. The bottles are equipped with stainless steel caps to permit sampling, aeration, and addition of fluids. Arabinose (35 g), trace salts, and a 5% inoculum of a culture grown on arabinose are added to each bottle of medium. After about 6 hr when the turbidity has reached 150 Klett units (No. 54 filter with water as reference), a second 35-g quantity of arabinose is added. Cultures are grown at 30° for about 18 hr with vigorous aeration (6 cubic feet of air per minute) and pH control (6.4 to 6.9 with 5 $M$ $NH_4OH$). Foam is controlled with Antifoam A (Dow Corning). When the turbidity of cultures reaches 375 to 400 Klett units, the cells are harvested by continuous-flow centrifugation, washed with cold 33 m$M$ $K_2HPO_4$–$KH_2PO_4$ buffer, pH 6.8. Each 5-gallon bottle yields approximately 110 g of cells. The cells can be used immediately or stored at −20° for over a year.

### Purification Procedure

All operations are carried out at 0–4° unless otherwise indicated. Potassium phosphate buffers, pH 7.5, are used extensively and will be referred to as "buffer" (exceptions are noted). Fractional precipitation of protein with $(NH_4)_2SO_4$ is carried out by adding the amounts of solid reagent specified to gently stirring solutions over a period of at least 10 min. Suspensions are allowed to stand for 30 min prior to centrifugation. Centrifugations are performed at 20,000 $g$ unless otherwise stated.

*Preparation of Sonic Extract.* Frozen cells (350 g) are suspended in 350 ml of 0.1 $M$ buffer and sonicated for 4 min with a MSE 500-W sonifier while maintaining the temperature between 5 and 15°. The sonic extract is centrifuged for 60 min and the residue is discarded.

*Streptomycin Sulfate Precipitation.* The supernatant fraction of the sonic extract is adjusted to pH 7.0 with HCl, and 0.1 volume of 20% streptomycin sulfate in 0.1 $M$ buffer, pH 7.0, is added with stirring. After 10 min the suspension is centrifuged for 60 min and the precipitate is discarded.

*Ammonium Sulfate Precipitation.* The streptomycin-treated extract is adjusted to pH 7.5 with KOH. Solid $(NH_4)_2SO_4$ is added (156 g per liter) while maintaining the pH at 7.5 with additions of KOH. The precipitate which forms is removed by centrifugation for 90 min. Additional $(NH_4)_2SO_4$ is added (133 g/liter) to the supernatant solution while continuing to maintain the same pH with KOH. The precipitated protein is collected by centrifugation for 90 min, dissolved in 50 ml of 50 m$M$

buffer and dialyzed for 15 hr against three 6-liter quantities of 12 m$M$ buffer.

*First DEAE-Cellulose Chromatography.* The dialyzed solution is applied to a column (5.6 × 50 cm) of DEAE-cellulose equilibrated with 12 m$M$ buffer. The stepwise elution is carried out with 1800 ml of 12 m$M$ buffer, 1000 ml of 40 m$M$ buffer, and 1600 ml of 70 m$M$ buffer. The enzyme activity is eluted with 70 m$M$ buffer. Fractions containing more than 0.2 unit of enzyme activity per milliliter are pooled and the protein is concentrated by precipitation with $(NH_4)_2SO_4$ (313 g/liter), The suspension is centrifuged for 60 min and the supernatant fraction is discarded. The precipitate is dissolved in 50 ml of 10 m$M$ buffer and dialyzed for 16 hr against three 6-liter quantities of 2 m$M$ buffer. Any precipitate which appears during dialysis is removed by centrifugation at 8000 $g$ for 15 min.

*Calcium Phosphate Gel-Cellulose Column Chromatography.* The clarified solution is applied to a column (3.6 × 15 cm) of calcium phosphate gel-cellulose. The column-bed is prepared from a mixture of cellulose powder (Whatman, standard grade)[6] and calcium phosphate gel[7] as described by Massey[8] and equilibrated with 2 m$M$ buffer. The column is eluted stepwise with 400 ml of 2 m$M$ buffer, 150 ml of 6 m$M$ buffer and 800 ml of 12 m$M$ buffer.

The dehydratase activity is eluted with 12 m$M$ buffer, fractions containing more than 0.1 unit per milliliter are pooled, and the enzyme is concentrated with $(NH_4)_2SO_4$ (313 g/liter). The precipitated protein is collected by centrifugation, the pellet is dissolved in 1–3 ml of 10 m$M$ buffer, and the solution is dialyzed for 6 hours against two 500-ml quantities of 12 m$M$ buffer.

*Second DEAE-Cellulose Column.* The dialyzed solution is applied to a column (0.7 × 22 cm) of DEAE-cellulose equilibrated with 12 m$M$ buffer. The column is eluted with a linear salt gradient (200 ml of 12 m$M$ buffer in one vessel and 200 ml of buffer containing 3.45 g $(NH_4)_2SO_4$ in the other) and 10-ml fractions collected. The fractions containing dehydratase activity are pooled according to specific activities as follows: fractions with specific activity greater than 25 units/mg (fraction 1); 15–25 units/mg (fraction 2); those with less than 15 units/mg were discarded. Fraction 2 may be saved and rechromatographed in subsequent

---

[6] The designation "standard grade" is obsolete, but presumably Whatman Cellulose Powder, CF 12 (Reeve Angel) would be suitable.

[7] Prepare the gel according to the procedure of S. M. Swingle and A. Tiselius, *Biochem. J.* **48,** 171 (1951) as modified by V. Massey. See this series Vol. 9 [88], footnote 9.

[8] V. Massey, *Biochim. Biophys. Acta* **37,** 310 (1960).

PURIFICATION OF L-2-KETO-3-DEOXYARABONATE DEHYDRATASE

| Fraction | Volume (ml) | Total units | Specific activity (units/mg protein) | Yield (%) |
|---|---|---|---|---|
| Sonic extract | 367 | 980 | 0.13 | 100 |
| Streptomycin sulfate | 407 | 855 | 0.07 | 87 |
| $(NH_4)_2SO_4$ precipitate | 130 | 885 | 0.18 | 90 |
| First DEAE-cellulose column | 92 | 765 | 3.4 | 78 |
| Calcium phosphate gel-cellulose column | 13.5 | 675 | 28 | 69 |
| Second DEAE-cellulose column | | | | |
| Fraction 1 | 19 | 340 | 36 | 35 |
| Fraction 2 | 36 | 120 | 19 | 12 |
| Sephadex G-200 column | 32 | 172 | 54 | 18 |
| Crystalline enzyme | — | — | 54 | — |

preparations. The protein in Fraction 1 is concentrated by precipitation with $(NH_4)_2SO_4$ (313 g/liter) as described previously and dissolved in 1 ml of 10 m$M$ buffer, pH 7.5.

*Sephadex G-200 Chromatography.* The highly purified dehydratase is passed through a column (2.2 × 100 cm) of Sephadex G-200 equilibrated with 10 m$M$ buffer and 5-ml fractions are collected. Six fractions (tubes 8–13) with peak specific activity of 52–54 units/mg are combined; yield 3–6 mg of protein. The enzyme is concentrated by precipitation with $(NH_4)_2SO_4$ (313 g per liter) and is stable at 4° as a 0.1% protein solution in 10 m$M$ buffer containing 20% $(NH_4)_2SO_4$ (110 g/liter).

*Crystallization.* Following the last precipitation with $(NH_4)_2SO_4$, the enzyme is dissolved in 1 ml of neutral 25% $(NH_4)_2SO_4$ (144 g/liter) and allowed to stand for several days at 4°. The dehydratase crystallizes as slender needles.

A typical enzyme preparation is summarized in the table.

## Properties

*Specificity.* The enzyme does not act on glycerol, 1,2-propanediol, or ethylene glycol, the only additional substrates tested.

*pH Optimum.* The enzyme has a broad pH activity profile (pH 6.9–7.5) with a maximum at pH 7.3. The enzyme exhibits 50% activity at pH 5.8 and 9.2, respectively.

*Kinetic Properties.* The $K_m$ for L-2-keto-3-deoxyarabonate is 70 $\mu M$. DL-2-Keto-3-hydroxyvalerate and pyruvate are competitive inhibitors with $K_i$'s of 0.16 m$M$ and 4.6 m$M$, respectively.[9]

[9] D. Portsmouth, A. C. Stoolmiller, and R. H. Abeles, *J. Biol. Chem.* **242,** 2751 (1967).

*Molecular Properties.* Purified L-2-keto-3-deoxyarabonate migrates as a single component upon electrophoresis in starch gel at either pH 7.0 or 5.5; the position of the stained protein and the dehydratase activity are coincident. Ultracentrifugation of a solution of the enzyme containing 3 mg of protein per milliliter shows a single asymmetric peak with an $s_{20,w}$ value of 6.3 S. The molecular weight of the enzyme estimated by gel filtration[10] is 85,000 ± 5000.

*Cofactors.* There are no known cofactors. The enzyme contains no significant amount of vitamin $B_{12}$ derivatives and charcoal treatment, dialysis, gel filtration, and exposure to sunlight have no effect on the activity.

*Inhibitors.* EDTA, 2,2′-bipyridine, *o*-phenanthroline, iodoacetate, and *N*-ethylmaleimide were not inhibitory at 1 m*M*. *p*-Hydroxymercuric benzoate was somewhat inhibitory; 0.5 m*M* gives 10% inhibition and 4 m*M* produces 50% inhibition.

The enzyme is inactivated by treatment with $NaBH_4$ in the presence of the substrate or other carbonyl compounds.

*Mechanism.* The enzyme conversion of L-2-keto-3-deoxyarabonate to α-ketoglutarate semialdehyde in HTO or $D_2O$ leads to the incorporation of isotopic hydrogen at C-3 and C-4. In $D_2O$, one deuterium atom is incorporated stereospecifically at each position. Treatment of the enzyme with $NaBH_4$ in the presence of substrate leads to inactivation of the enzyme and to covalent binding of 1 mole of substrate to 1 mole of enzyme. It is thought that the substrate and enzyme interact to form a Schiff base. With substrate of known specific radioactivity an apparent minimum molecular weight of 81,000[11] can be calculated after reduction with $NaBH_4$. A mechanism for the overall reaction has been proposed in which the enzyme-substrate Schiff base participates and which accounts for the observed isotope incorporation.[9]

## Partial Purification of an Aldehyde Dehydrogenase
### from *Pseudomonas oleovorans*

$$R\text{-}CHO + NADP^+ \rightarrow R\text{-}COOH + NADPH + H^+ \tag{4}$$

## Assay Method

*Principle.* The spectrophotometric assay is based on the rate of reduction of $NADP^+$ at 340 nm in the presence of aldehydic substrates. Glutaraldehyde and α-ketoglutarate semialdehyde are the best known substrates.

---

[10] P. Andrews, *Biochem. J.* **91**, 222 (1964).

[11] This value is in good agreement with a molecular weight of 85,000 determined by gel filtration.

*Reagents*

$K_2HPO_4–KH_2PO_4$, 0.2 $M$, pH 7.4, containing 10 m$M$ 2-mercapto-
ethanol
NADP$^+$, 5 m$M$
Glutaraldehyde, 50 m$M$
Enzyme, 0.2 unit/ml

*Procedure.* Place 0.2 ml of water, 0.4 ml of buffer, 0.1 ml of glutaraldehyde, and 0.2 ml of diluted enzyme solution in a 1-ml quartz cell with a
light path of 1 cm. The reaction is initiated by addition of 0.1 ml of
NADP$^+$ (replaced by water in blank). Record the optical density at
1-min intervals for 10–30 min against the blank.

*Definition of Unit and Specific Activity.* The unit of enzyme activity
is defined as the amount which causes the reduction of 1 $\mu$mole of NADP$^+$
per minute at 30°. Specific activity is expressed as units per milligram
of protein as determined by the method of Lowry *et al.*[2]

*Application of Assay Method to Fresh Cell Extracts.*[12] When assaying
enzyme preparations from freshly harvested cells, there is an appreciable
interference due to the nonspecific oxidation of NADPH by contaminating enzymes, and a correction for this must be made when calculating
enzyme activity.

## Purification Procedure

*Growth of Pseudomonas oleovorans.*[13] *Pseudomonas oleovorans* is
cultured at room temperature in 2800-ml Fernbach flasks which contain
800 ml of medium containing 0.1% $KH_2PO_4$, 0.1% $Na_2HPO_4$, 0.1%
$(NH_4)_2SO_4$, 0.02% $MgSO_4 \cdot 7H_2O$, 0.002% $CaCl_2$, and 0.001% $FeCl_3$.[14]
Solutions of the phosphate, sulfate, and chloride salts are sterilized in
separate containers. The carbon source, 20 ml of hexane, is placed in the
center well (a 50-ml beaker weighted with sand). After inoculation, seal
the flask to prevent loss of hexane vapor and grow the cultures for 3–5
days. Growth is monitored by withdrawing aliquots and measuring the
turbidity with a Klett photometer (No. 54 filter with water as reference).
When the cultures reach 150 Klett units, the cells are harvested by con-

---

[12] Cells (or cell extracts) stored at $-20°$ for 6–12 months are more suitable for
preparation of aldehyde dehydrogenase because aged cells contain little or no
NADPH-oxidase activity compared to freshly harvested cells.
[13] This particular strain of *P. oleovorans* is available from Dr. Minor J. Coon,
Department of Biological Chemistry, University of Michigan, Ann Arbor, Michigan
48104 and is not presently held by the ATCC.
[14] J. N. Baptist, R. K. Gholson, and M. J. Coon, *Biochim. Biophys. Acta* **69**, 40
(1963).

tinuous-flow centrifugation. The packed cells are suspended in 33 m$M$ $K_2HPO_4$–$KH_2PO_4$, pH 6.8, containing 10 m$M$ 2-mercaptoethanol, and re-centrifuged; the packed cell paste is stored at $-20°$. Cells may be stored 6–12 months with no significant loss of aldehyde dehydrogenase activity. About 1 g of cells (wet weight) are obtained per liter of medium.

*Preparation of Cell-Free Extracts.* Perform all operations at $0°$ un-less otherwise indicated. Twenty grams of frozen cells (6–12 months of age[12]) are suspended in 40 ml of 0.1 $M$ $K_2HPO_4$–$KH_2PO_4$, pH 7.5, con-taining 10 m$M$ 2-mercaptoethanol and sonicated for 3 min with a 500-W MSE sonifier while maintaining the temperature between $5°$ and $15°$. The suspension is centrifuged at 35,000 $g$ for 45 min.

*Streptomycin Fraction.* To the cell extract is added 0.1 volume of 20% streptomycin sulfate (adjusted to pH 7 with HCl). After stirring for 10 min, the suspension is centrifuged at 35,000 $g$ for 45 min and the pre-cipitate discarded.

*DEAE Chromatography.* The supernatant solution is adjusted to pH 7.5 with KOH and applied to a column (2.2 $\times$ 25 cm) of DEAE-cellulose which is equilibrated with 0.1 $M$ $K_2HPO_4$–$KH_2PO_4$, pH 7.5, containing 10 m$M$ 2-mercaptoethanol. The enzyme is not absorbed to this column and emerges in the first protein fraction. Fractions containing enzyme activity are pooled, divided into small aliquots, and stored at $-20°$. The enzyme remains active for several months; the specific activity of the aldehyde dehydrogenase is approximately 0.02 unit per milligram of protein.

## Properties

Gluteraldehyde and $\alpha$-ketoglutarate semialdehyde are oxidized at sim-ilar rates in the presence of $NADP^+$; caproaldehyde reacts at one-fifth of this rate. Formaldehyde, glycolaldehyde, acetaldehyde, lactaldehyde, benzaldehyde, and capraldehyde were not oxidized at significant rates. $NAD^+$ does not substitute for $NADP^+$ in the reactions catalyzed by this enzyme. The enzyme is inactivated by sulfhydryl reagents.

## [51] Glycerol Dehydratase from *Aerobacter aerogenes*

*By* B. Connor Johnson, A. Stroinski, and Z. Schneider

$$CH_2OH\text{—}CHOH\text{—}CH_2OH \xrightarrow[\text{coenzyme B}_{12}]{\text{glycerol dehydratase}} CH_2OH\text{—}CH_2\text{—}CHO + H_2O$$

Glycerol dehydratase from *Aerobacter aerogenes*, an enzyme inducible by glycerol, is made up of two different protein subunits and coenzyme

$B_{12}$. The dissociation and association of the apoenzyme into subunits is influenced by pH, monovalent cations, and glycerol.

## Assay Method

### Principle

Glycerol dehydratase activity is determined by assay of the $\beta$-hydroxy-propionaldehyde formed in the enzymic reaction. Two methods have been used: (1) The tryptophan method of Smiley and Sobolov[1] is based on the conversion of $\beta$-hydroxypropionaldehyde to acrolein, which reacts with tryptophan to give a red color assayed colorimetrically. (2) The MBTH method of Paz et al.[2] is based on the ability of $\beta$-hydroxypropion-aldehyde to react with 3-methyl-2-benzothiazolinone hydrazone (MBTH) Eastman Kodak Co. to form an azine, which is measured spectrophotometrically.

### Reagents

For both methods:
  Glycine-KOH buffer, 0.2 $M$, pH 8.6
  Potassium chloride, 1 $M$
  Potassium sulfite, 1 $M$
  Glycerol, 0.2 $M$
  Coenzyme $B_{12}$ (5'-deoxyadenosylcobalamin), 2 $\mu M$, protected from light at all times
  Glycerol dehydratase apoenzyme preparation in 0.1 $M$ glycine-KOH buffer, pH 8.6
For the tryptophan method only:
  Tryptophan, 15 m$M$ in 0.1 $M$ HCl
  Concentrated hydrochloric acid, sp. d. 1.19
  Acrolein, freshly distilled
For the MBTH method only:
  Potassium citrate buffer, 0.1 $M$, pH 3.6
  3-Methyl-2-benzothiazolinone hydrazone, 0.1% solution in water.

Procedure. THE TRYPTOPHAN METHOD. The reaction mixture contains 0.5 ml of glycine-KOH buffer, pH 8.6, 0.1 ml KCl solution, 0.05 ml $K_2SO_3$ solution, 0.01 ml glycerol solution, and 1–3 units of glycerol dehydratase apoenzyme. Total volume is made up to 1 ml with water. The mixture is preincubated for 10 min at 30°. The reaction is started by the introduc-

[1] K. L. Smiley and M. Sobolov, Arch. Biochem. Biophys. 97, 538 (1966).
[2] M. Z. Paz, O. O. Blumenfeld, M. Rojkind, E. Henson, C. Furfine, and P. M. Gallop, Arch. Biochem. Biophys. 109, 548 (1965).

tion, in the dark, of 0.1 ml of coenzyme $B_{12}$ solution and is continued in the dark for 5 min at 30°. The reaction is stopped by the addition of 0.5 ml of tryptophan reagent and 3 ml of concentrated HCl. The samples are heated at 60° for 5 min and cooled in an ice bath; the optical density is determined colorimetrically at 550 nm against a blank, which contains all components except the coenzyme $B_{12}$. The amount of $\beta$-hydroxypropionaldehyde is calculated from a standard curve, prepared for every series of determinations with freshly distilled acrolein. The optical density is proportional to the concentration of acrolein over the range of 0.5–3.0 $\mu$moles.

THE MBTH METHOD. The reaction mixture contains 0.5 ml of glycine-KOH buffer, pH 8.6, 0.1 ml of KCl solution, 0.05 ml of $K_2SO_3$ solution, 0.01 ml of glycerol solution and 0.2–0.6 unit of glycerol dehydratase apoenzyme. Total volume is made up to 1 ml with water. The mixture is preincubated for 10 min at 30°. The reaction is started by the introduction, in the dark, of 0.1 ml of coenzyme $B_{12}$ solution and is continued, in the dark, for 2 min at 30°. The reaction is stopped by the addition of 0.8 ml of potassium citrate buffer, pH 3.6, followed by the introduction of 0.2 ml of MBTH reagent. The samples are heated at 100° for 10 min, cooled in an ice bath and, after the addition of 4 ml of water to each sample, the optical density is determined spectrophotometrically at 330 nm against a blank, which contains all components except the coenzyme $B_{12}$. The assay range for $\beta$-hydroxypropionaldehyde is from 0.05 to 0.25 $\mu$mole.

*Definition of Unit.* One unit of enzyme activity is defined as the amount of enzyme which produces 1 $\mu$mole of $\beta$-hydroxypropionaldehyde per minute under the assay conditions.

## Purification

Purification procedures for glycerol dehydratase have been based on either separate isolation[3] of the two apoenzyme subunits or on isolation of a stable but inactive complex[4] of apoenzyme and hydroxocobalamin. Recently, a more satisfactory method for the preparation of glycerol dehydratase has been described, based on purification of the intact[5] apoenzyme. This method is given below.

*Growth of Cells.* *Aerobacter aerogenes*, strain 572 PZH,[6] available

[3] Z. Schneider, K. Pech, and J. Pawelkiewicz, *Bull. Acad. Polon. Sci.* **14**, 7 (1966).
[4] Z. Schneider, E. G. Larsen, G. Jacobson, B. C. Johnson, and J. Pawelkiewicz, *J. Biol. Chem.* **245**, 3388 (1970).
[5] A. Stroinski, J. Pawelkiewicz, and B. C. Jonhson, *Arch. Biochem Biophys.* **162**, 321 (1974).
[6] National Institute of Hygiene, Warsaw, Poland.

PURIFICATION OF GLYCEROL DEHYDRATASE APOENZYME FROM *Aerobacter aerogenes*[a]

| Purification step | Protein content (mg) | Total activity (units) | Specific activity (units/mg) | Yield of activity (%) |
|---|---|---|---|---|
| Crude enzyme | 1124 | 2300 | 2.1 | 100 |
| 1 | 690 | 2070 | 3.0 | 90 |
| 2 | 34.2 | 1265 | 37.0 | 55 |
| 3 | 6.8 | 851 | 125 | 37 |
| 4 | 2.4 | 368 | 153 | 16 |
| 5 | 2.5 | 570 | 228 | 25 |

[a] Purification steps as described in text. Holoenzyme activity was determined by the tryptophan method. Protein concentration was estimated by ultraviolet absorption.

also as *Enterobacter aerogenes*, ATCC 25955, is grown[7] on a medium consisting of: Difco yeast extract 10 g, casein hydrolysate 20 g, $K_2HPO_4$ 14 g, $KH_2PO_4$ 6 g $MgSO_4$ 5 g, $MnSO_4$ 0.5 g, $K_2SO_4$ 10 g, glycerol 200 ml, made up to a final volume of 8.5 liters with distilled water and adjusted to pH 7.0 with KOH. The organism is grown at 30° with vigorous aeration[8] for approximately 18 hr. The pH is adjusted to 8.6–9.2 (pH paper) with 2 N KOH and the cells are harvested in a Sharples centrifuge.

*Extraction of the Enzyme.* The harvested packed wet cells are ruptured by grinding with $Al_2O_3$ (Bacteriological Grade Alcoa A-305)—or alternatively by sonication[7]—and the enzyme is extracted 3 times, at room temperature, with 0.1 M potassium phosphate buffer, pH 8.6, in 0.1 M glycerol, using for each extraction 500 ml of this solution per 30 g wet weight of cells. Each extraction is followed by centrifugation for 15 min at 2000 rpm in the Sorvall refrigerated centrifuge. The combined supernatants, which contain about 1 mg of protein per milliliter, are dialyzed overnight, in the cold, against 0.1 M glycine-KOH buffer, pH 8.6, containing 0.1 M glycerol.

*Step 1.* The dialyzate is used for purification. Nucleic acids are precipitated at 5° by the addition of 1 ml of 1 M $MnCl_2$ per 100 ml of enzyme solution and are removed by centrifugation. The supernatant is dialyzed against 0.1 M potassium phosphate buffer, pH 8.6, containing 0.1 M glycerol. Any manganese phosphate precipitate, which forms dur-

[7] J. Pawelkiewicz and B. Zagalak, *Acta Biochim. Polon.* **12**, 207 (1965).

[8] G. Jacobson, Z. Schneider, and B. C. Johnson: No decrease in enzyme content is obtained when the organism is grown aerobically and there is a much greater yield of cells than when it is grown anaerobically (unpublished data).

ing dialysis, is removed by centrifugation. The solution is then concentrated 10-fold by the Flodin[9] method against Sephadex G-50 to give a protein concentration of about 10 mg/ml.

*Step 2.* This concentrated enzyme solution is dialyzed against 20 m$M$ potassium phosphate, pH 8.6, containing 0.2 $M$ glycerol and is applied to a Sephadex G-200 column (15 $\times$ 95 cm), which has been equilibrated with this dialysis solution. The fractions with the highest specific activity (>30) are combined and concentrated, using dry Sephadex G-25,[9] to a protein concentration of 5 mg/ml.

*Step 3.* This concentrated enzyme solution is dialyzed against 0.1 $M$ glycine buffer, pH 8.6, containing 0.1 $M$ glycerol and applied to a Sephadex G-200 column (5 $\times$ 95 cm), which has been equilibrated with the same dialysis solution. Fractions with a specific activity above 100 are combined and concentrated, using dry Sephadex G-25[9], to a protein concentration of approximately 2 mg/ml.

*Step 4.* This concentrated enzyme solution is dialyzed again against 0.1 $M$ glycine buffer, pH 8.6, containing 0.1 $M$ glycerol, and applied to a Sepharose 6B column (2.5 $\times$ 90 cm), equilibrated with the same dialysis solution. Fractions with specific activity above 150 are combined and concentrated as before.

*Step 5.* Since it has been found that at the end of step 4 more of subunit B (the larger subunit) is present than of subunit A, owing to dissociation, it is necessary, in order to obtain a maximum yield of apoenzyme (AB complex), to make up such a deficiency by the addition of an adequate amount of subunit A.

*Preparation of Subunit A.*[10] Crude extract of cells, obtained as described above, is dialyzed overnight against water. Potassium phosphate, pH 8.0 is added to a final concentration of 0.2 $M$. The solution is placed in 15 mm i.d. test tubes and heated in a water bath at 70° for 6 min. The tubes are cooled, and the solution is centrifuged at 20,000 $g$ for 20 min. The supernatant solution contains subunit B. The precipitate contains subunit A in an inactive, polymerized form, which is further purified and reactivated. The pellet is suspended in 5 volumes of cold water, followed by immediate centrifugation at 30,000 $g$ for 20 min. This procedure is repeated twice to wash out the phosphate. The washed precipitate is suspended in three times the initial volume of cold water and dilute HCl (1:1) is added with constant stirring—to a pH of 2.0 $\pm$ 0.5. The suspension is immediately centrifuged for 6 min at 30,000 g and the supernatant solution, which contains the active subunit A, adjusted to pH 8.6

[9] P. Flodin, in Dextran gels and their application in gel filtration. Pharmacia Uppsala, 1962.

[10] Z. Schneider and B. C. Johnson, unpublished data.

with KOH. This subunit A is about 5% pure. Approximately 30% purity can be obtained by repeating the cycle of heating and acidic reactivation.

As can be seen from the table, 100-fold purification of glycerol dehydratase apoenzyme has been obtained, to yield—in the presence of coenzyme $B_{12}$—a holoenzyme with a specific activity of approximately 230 units per milligram of apoenzyme protein. (This can be compared to 3 units per milligram of protein obtained by reactivation of purified hydroxocobalamin complex.) The yield of apoenzyme was 25%. The high purity of the isolated apoenzyme has been demonstrated by polyacrylamide disc gel electrophoresis, which showed only traces of three other, more rapidly migrating, proteins.

## Properties

*Composition and Molecular Characterization.* All three components of the enzyme, subunit A, subunit B, and coenzyme $B_{12}$, are necessary for enzymic activity. The separated subunits A and B are inactive, and neither of them separately binds coenzyme $B_{12}$. Coenzyme $B_{12}$ binding requires AB complex indicating that only in the association of A and B is a coenzyme $B_{12}$ binding site formed.

Holoenzyme can be obtained by the addition of coenzyme $B_{12}$ to A and B, or by replacement of hydroxocobalamin from its stable complex with AB, by coenzyme $B_{12}$ in the presence of $Mg^{2+}$ and sulfite. Certain of the chemical characteristics of the enzyme are best studied using the more stable $AB \cdot B_{12}OH$ complex, others can be determined using purified A and B or AB.

The molecular weight of the $AB \cdot B_{12}OH$ complex determined by sucrose gradient centrifugation using alcohol dehydrogenase as standard has been reported[4] as 188,000. The complex is orange and contains 1 mole of hydroxocobalamin per mole of AB protein. Its amino acid composition has been published.[4]

The molecular weights of each of the subunits have been estimated. Subunit B gave a molecular weight of $189,000 \pm 22,000$ by gel filtration on Sephadex G-200 in 50 m$M$ ethanolamine-HCl buffer, pH 8.6.[5] When this buffer contained in addition 0.1 $M$ KCl, subunit B dissociated into sub-subunits of $90,000 \pm 25,000$ molecular weight.[5]

The molecular weight of subunit A was estimated to be $22,000$[11] by gel filtration on Sephadex G-100. At low ionic strength, e.g., 20 m$M$ potassium phosphate, and at more acid pH's e.g., 6.7, subunit A dissociated into subsubunits of approximately 11,000 molecular weight. These data

[11] Z. Schneider and J. Pawelkiewicz, *Acta Biochim. Polon.* **13**, 311 (1966).

indicate a molecular weight of approximately 211,000 for the apoenzyme AB.

*Effect of Monovalent Ions.* Potassium ions are essential for enzymic activity, and their effect on the enzyme can be totally blocked by sodium ions. This highly specific antagonistic effect of sodium versus potassium is characteristic of glycerol dehydratase. In the absence of other monovalent ions optimum potassium concentration is between 0.08 and 0.15 $M$. While this requirement has been interpreted as indicating a role of potassium ions for AB subunit association, recent work with the purified apoenzyme indicates that K⁺ reduces the forces binding A and B subunits.[5] In this report the highest degree of association of AB in apoenzyme was obtained in glycine-KOH buffer, pH 8.6 (0.1 $M$ glycine, 5 m$M$ K⁺) in the presence of glycerol. The presence of potassium ions at 0.1 $M$ (potassium chloride or phosphate) brought about dissociation of the apoenzyme. This appears to be substantiated by the finding that both potassium and sodium ions at increasing concentrations increasingly promote the spontaneous inactivation of the holoenzyme. This similar action of potassium and sodium contrasts sharply with their antagonistic roles with respect to the activity of the enzyme.

*Specificity.* Glycerol dehydratase catalyzes the conversion of glycerol to $\beta$-hydroxypropionaldehyde. The enzyme has also been reported[7] to catalyze the conversion of ethylene glycol to acetaldehyde and of propane-1,2-diol to propionaldehyde.

*pH Optimum.* The pH optimum for the reaction glycerol to $\beta$-hydroxypropionaldehyde has been shown to be 8.0 to 9.0 with its peak at 8.6.[11]

*Activators and Inhibitors.* In addition to activation by potassium ions the enzyme also requires a reducing environment (e.g., $K_2SO_3$) to prevent destruction by oxidation. The enzyme contains an active thiol which is present in subunit B. Hence, the enzymic reaction is blocked by 1.0 m$M$ $N$-ethylmaleimide and by 1.0 m$M$ $p$-chloromercuribenzoate.[7]

EDTA (and salicylic acid), but not other chelating agents studied, cause inactivation by dissociation of the apoenzyme into its subunits.[4] (It does not cause dissociation of AB·B₁₂OH.) The inhibition by EDTA is reversed by dialysis, by $Mg^{2+}$, by glycerol or by addition of excess of either subunit. This latter suggests a reversible association of EDTA with the subunits.

*Stability of the Holoenzyme.* The holoenzyme is highly unstable and undergoes rapid, irreversible inactivation in the presence or absence of substrate. Oxygen participates in the inactivation (as in the case of the spontaneous inactivateion of propandiol dehydratase[12]) and the inacti-

[12] O. W. Wagner, H. Lee, P. A. Frey, and R. H. Abeles, *J. Biol. Chem.* **241**, 1751 (1966).

vated enzyme contains hydroxocobalamin and 5'-deoxyadenosine. The holoenzyme can be protected from oxygen by the addition of $K_2SO_3$ to the solution (50 m$M$).

*Factors Affecting Association and Dissociation of the Apoenzyme.* In addition to monovalent ions and EDTA discussed above, gel filtration studies of subunit interaction have shown[5] that the addition of substrate (glycerol) has a stabilizing effect on the apoenzyme AB, perhaps through the formation of an AB-glycerol complex. In studies of association-dissociation in a number of different 0.1 $M$ pH 8.6 buffer solutions the highest degree of association was found in glycine-KOH buffer (0.1 $M$ glycine, 5 m$M$ K$^+$) in the presence of 0.1 $M$ glycerol.[5] Potassium ions in the form of 0.1 $M$ KCl or 0.1 $M$ potassium phosphate brought about high dissociation of the apoenzyme.

*Effect of Heating on the Apoenzyme and Its Subunits and on* $AB \cdot B_{12}$-*OH.* The response of glycerol dehydratase to heat treatment is one of the most interesting aspects of this enzyme. The apoenzyme has been found to be very unstable to heat. Four minutes at 46° resulted in 50% destruction of apoenzyme. In contrast $AB \cdot B_{12}OH$ was found to be much more stable, no destruction occurring with 4 min heating at 70° and 50% destruction only after 4 min at 82°.

This heating of $AB \cdot B_{12}OH$ over 70° produces an active subunit B, an inactivated subunit A, and a $B_{12}$ derivative. Li$^+$ and Na$^+$ ions promote this thermal dissociation, while K$^+$ ions inhibit it.[13]

The thermal lability of the apoenzyme is due to the thermal lability of subunit A. Subunit B in contrast is stable up to 70° above which it becomes highly thermolabile.

Heated separately subunit A is very heat labile but is stabilized by high (above 2 $M$) concentrations of NaCl, RbCl, KCl, or CsCl, but not at all by LiCl. The inactivation of subunit A at 70° is reversible[10] and appears to be due to the formation of a polymer from subunit monomers. The inactive subunit A can be reactivated by heating to over 90°, by sonication, by urea treatment, or by exposure to either high or very low pH. The reversibly inactive form of subunit A is stabilized by the same monovalent ions which stabilize subunit A. These ions inhibit the irreversible inactivation of both active subunit A and reversibly inactive A.

*Allosteric Nature of the Enzyme.* The enzyme exhibits homotropic effects with regard to glycerol binding sites which can be extended from positive cooperativity (in the presence of glycine buffer) to negative cooperativity (in the presence of ethanolamine buffer).[5] The high cooperativity between glycerol binding sites on the enzyme in glycine buffer, can be abolished by the addition of phosphate. By decreasing

[13] Z. Schneider and B. C. Johnson, *Arch. Biochem. Biophys.* **144**, 491 (1971).

the value of the Hill coefficient and increasing the $V_{max}$ of the enzymic reaction, phosphate appears to act as an allosteric activator.

*Kinetic Data.* In 0.1 $M$ potassium phosphate buffer the enzyme exhibits linear Michaelis-Menten kinetics with a $K_m$ of 1.25 m$M$ and a $V_{max}$ of the order of 0.18 $\mu M/2$ min. In 0.1 $M$ glycine–KOH buffer the $V_{max}$ is of the order of 0.12 $\mu M/2$ min.

## [52] Enolase from *Escherichia coli*

*By* THOMAS G. SPRING and FINN WOLD

D-Glycerate 2-phosphate $\rightleftarrows$ enolpyruvate phosphate $+$ H$_2$O

### Assay Method

The details of substrate preparation and assay method for enolase have been described in detail by Westhead in a previous volume[1] and in this volume.[2] For *E. coli* enolase, the following standard assay medium is used:

D-Glyceric acid 2-phosphate, 1 m$M$
Magnesium sulfate, 1 m$M$
Potassium chloride, 100 m$M$
Tris·HCl buffer, final pH 8.1, 50 m$M$ in Tris
EDTA, 10 $\mu M$

The complete assay medium can be stored frozen without significant deterioration and should be thawed immediately prior to use. The thawed assay medium can also be stored at 4° for several days without apparent deterioration. The above assay conditions differ significantly from those used for yeast or vertebrate enolases, especially in the more alkaline pH optimum for the *E. coli* enzyme (8.1). The magnesium and potassium ion concentrations are approximately those required for optimal activity. Substrate concentration is not quite saturating ($K_m$ is 0.1 m$M$) but is adequate to give reasonably linear assays. The EDTA is included as a heavy metal chelator to prevent inhibition of activity, which sometimes occurs.

As discussed elsewhere,[2] it is convenient routinely to use the arbitrary unit of enolase activity defined as the amount of enzyme that causes a change in $A_{240}$ of 0.1 per minute in the standard assay at 25°. The molar absorbancy of enolpyruvate phosphate under the conditions of the

[1] E. W. Westhead, see this series, Vol. 9, p. 667.
[2] F. Wold, see this volume, [53].

$E.$ $coli$ enolase assay is about 1500 $M^3$, and the arbitrary unit thus, is converted to the international enzyme unit (micromoles of product per minute in 3 ml reaction volume) by multiplication with the factor $0.1 \times 3/1.5 = 0.2$. Protein concentrations are most easily' determined from the absorbance at 280 nm. An $A_{280}$ unit of protein is arbitrarily defined as the amount of protein in 1 ml of a solution with $A_{280} = 1$. Especially in the early steps, however, the contribution of the nucleic acids to the ultraviolet absorbance is substantial and direct protein determination is recommended. A 1 mg/ml solution of the purified enzyme has an absorbance of 0.57 at 280 nm.

## Preparation of Enzyme

A convenient amount of starting material is 500 g of frozen $E.$ $coli$ B cells[4] from which about 50–100 mg of pure enzyme can be recovered. The following preparation is scaled to 500 g of starting material.[5]

*Step 1. Acetone Powder Extraction.* The frozen cells are thawed partially at room temperature and then thoroughly dispersed in a Waring Blendor for 30 sec. The thick cell suspension is poured slowly into 8 volumes of cold acetone (precooled to $-20°$) and stirred vigorously. If complete dispersal is a problem, more vigorous agitation in a Waring Blendor is suggested. The insoluble material is collected by suction filtration, resuspended in 4 volumes of cold acetone, filtered again, and washed finally with 2 volumes of absolute ether (precooled to $-20°$). The filter cake is dried on a large surface area of absorbent paper and stored as a dry powder.[6] The yield of acetone powder is about 20% of the wet weight of the cells. Extraction of enolase from the acetone powder is accomplished by adding the powder slowly to a cold (4°) solution of TME buffer (10 mM Tris, 1 mM $MgSO_4$, 10 mM EDTA, adjusted to pH 8.0 with HCl) with vigorous stirring. The volume of buffer used is 18 ml per gram of acetone powder. Stirring is continued until the solution is homogeneous, usually for several hours. It was found advantageous to perform the extraction overnight, so that steps 2–5 could be carried out the next day without interruption. The extract is centrifuged at 10,000 $g$ for 15 min to remove insoluble material. Reextraction of the pellet does not appreci-

---

[3] F. Wold and C. E. Ballou, $J.$ $Biol.$ $Chem.$ **227**, 301 (1957).

[4] Purchased from General Biochemicals, Catalog No. 150040, grown on Kornberg medium, and harvested in late log phase.

[5] T. G. Spring and F. Wold, $J.$ $Biol.$ $Chem.$ **246**, 6797 (1971).

[6] The acetone powder should be stable indefinitely if kept dry in a freezer, but we have seldom stored the powder more than a few days before extracting it. We have no data on enolase activity of acetone powders stored for long periods.

ably improve the yield of enolase, which is 3500–7000 units per gram of acetone powder (depending on the batch of acetone powder and the length of the extraction period).

*Step 2. Heat Treatment.* The aqueous extract (about 1200 ml) is made 1% w/v in $MgSO_4 \cdot 7H_2O$ by addition of $MgSO_4 \cdot 7H_2O$ dissolved in a minimum volume of TME buffer. The heat step is conveniently performed on about 500 ml of extract at one time. The temperature of the extract is raised rapidly to 55° by immersion in a 70° water bath and maintained at 55° for 2 min. The heated extract is then cooled on ice to 10° and centrifuged at 10,000 $g$ for 15 min. The insoluble material is inactive and can be discarded. Occasionally the heat step results in a small (10%) increase in total activity.

*Step 3. Acetone Fractionation.* Reagent grade acetone, 43 ml/100 ml of solution from step 2 is cooled to −20° and added to the stirred protein solution in a fine stream over a period of 2–3 min. The temperature of the solution is kept below 10° at all times. In the initial stages, a gelatinous precipitate forms, which coalesces as more acetone is added. The precipitate from this 30% acetone solution is removed by centrifugation at 10,000 $g$ for 15 min and discarded. An additional 57 ml of cold acetone per 100 ml of original solution from step 2 is added to the 30% acetone supernatant as above. The precipitate, collected by centrifugation as above, is suspended in a volume of TME buffer equal to 10% of the original volume from step 2. It is essential for both optimal yield and purification to use the prescribed volume of buffer here, and not to try to dissolve the large amount of insoluble material. After stirring the suspension for several minutes at 4°, it is centrifuged as above. The clear supernatant should contain all the enolase activity in approximately 90–100% yield at a concentration of about 2700 units/ml and an $A_{280}$ of ∼80. The pH and conductivity of the solution are approximately the same as TME buffer.

*Step 4. TEAE-Cellulose Titration.* This step is incorporated prior to the column step to remove a large amount of material which adsorbs strongly to TEAE cellulose. This procedure prevents overloading the TEAE cellulose column and improves resolution. TEAE-cellulose (Brown Company, coarse grade, 0.81 meq/g) is prewashed with 1 $N$ NaOH and 1 $N$ HCl to remove colored impurities and then equilibrated with TME buffer. Increments of moist TEAE-cellulose are added to the fraction from step 3 to adsorb protein and nucleic acid impurities. The $A_{280}$ and enolase activity are monitored after each addition. The titration is stopped when the enolase activity begins to decrease, which generally occurs after the addition of about 6 mg of moist TEAE-cellulose per

$A_{280}$ unit of protein present in the solution. TME buffer is added, when necessary, to keep the cellulose in a thick slurry. The TEAE-cellulose is removed by gentle suction filtration and washed once with about 1 ml of TME buffer per gram of TEAE-cellulose to remove residual unbound protein. At this point, the specific activity is about 70 units/$A_{280}$ unit and the $A_{280}/A_{260}$ ratio about 1.0. The combined filtrate and wash are applied directly to the TEAE-cellulose column.

*Step 5. TEAE-Cellulose Chromatography.* The protein solution from step 4 is applied to a $4 \times 50$ cm TEAE-cellulose column equilibrated with TME buffer. The column is then washed with 300 ml of TME buffer and finally eluted with a 0 to 0.25 $M$ linear NaCl gradient in TME buffer (total elution volume 2 liters). The first 500 ml of effluent can be discarded, after which 3-min fractions are collected (flow rate 7.5 ml/min). Enolase elutes at a buffer conductivity of 5–10 mmho and is reproducibly found to elute between an orange and a yellow colored peak. Fractions containing enolase with specific activity of 200 or greater are pooled and the protein is precipitated by adding 53 g of solid ammonium sulfate per 100 ml of eluate. The precipitated protein is centrifuged and resuspended in a minimum volume of TME buffer (about 3 ml/100 mg of protein), and clarified by centrifugation if necessary.

This particular column step was chosen as a compromise between resolution and speed. The flow rate is fast enough to allow for completion of the column step in 5 hr, with a 3-fold purification over step 4 and an overall yield of 40 to 50% of the original activity. (Rechromatography of side fractions can improve the yield, but is time-consuming.) Resolution in the column is not optimal since protein often occupies 50–70% of the total column capacity, and the enolase peak is usually skewed.

*Step 6. Gel Filtration on Sephadex G-100.* Protein from step 5 (30–40 mg/ml in TME buffer) is applied to a $3 \times 85$ cm Sephadex G-100 column equilibrated with TME buffer and eluted in downward flow with TME buffer at a flow rate of up to 30 ml/hr. Up to 500 mg of protein can be put on the column at once. Fractions of 5 ml are collected and those fractions containing enolase with specific activity of 540 or more are combined and precipitated with ammonium sulfate as in step 5. The yield at this stage is 20–30% of the original activity, and specific activity is about 700. (Again, the yield can be improved by rechromatography of side fractions.)

*Step 7. Rechromatography on Sephadex G-100.* The material from step 6 is rechromatographed as described in step 6 and, by proper selection of peak fractions, nearly pure enolase can be obtained with a specific activity of about 1200 and an overall yield of 20–30%. Pure enolase is not always obtained at this stage, and a third Sephadex chromatography and/or crystallization is sometimes required. Crystallization is recommended for long-term storage of the enzyme.

*Step 8. Crystallization.* To a concentrated enolase solution (10–20 mg/ml) of at least 70–80% pure enolase is added 0.44 g solid ammonium sulfate per milliliter of solution. The pH is adjusted to 6.0 with dilute acetic acid if necessary. Increments of a saturated ammonium sulfate solution (pH 6) are added until turbidity is evident. Crystallization occurs within a few hours. At early stages, amorphous precipitates sometimes occur, especially when contaminating proteins are present. The precipitates are usually inactive and can be discarded. The crystalline enzyme is not always 100% pure, but crystallization can still be used as a purification step, especially when the enzyme is nearly pure. Suspensions of the crystalline enzyme in ammonium sulfate at pH 6 have been stored for more than a year at 4° without activity loss. The only critical factor observed in the crystallization is the pH, which must be 6.0 or lower. In fact, the crystals can be dissolved merely by raising the pH of the ammonium sulfate suspension from 6.0 to 8.0. The purest preparation obtained by this method is estimated to be at least 95% pure by disc gel electrophoresis. Electrophoretically pure enolase has a specific activity of 780 units per milligram of protein. A summary of the purification is presented in the accompanying table.

## Properties of the Bacterial Enzyme[5]

The purified enzyme has a native molecular weight of approximately 90,000 and is composed of two subunits of identical molecular weight. The native molecule has also been shown to contain two active sites, presumably one per subunit.[7] The catalytic properties ($Mg^{2+}$ activation, fluoride inhibition, $K_m$ and $V_m$) are all very similar to those for yeast enolase, although *E. coli* enolase has a slightly higher pH optimum (8.1) than yeast enolase (7.8). In addition, *E. coli* enolase can be crystallized, a property which has not yet been reported for the yeast enzyme, except as the inactive $Hg^{2+}$ salt.[8] Unlike yeast enolase, which consists of at least three chromatographic forms,[9] *E. coli* enolase is a single chromatographic and electrophoretic species. The amino acid composition is similar to that of yeast enolase in the relatively low cysteine content (3.4 moles per 90,000 g compared with 2 per 88,000 g for yeast enolase[10]). Recently,

---

[7] T. G. Spring and F. Wold, *Biochemistry* **10**, 4655 (1971).

[8] O. Warburg and W. Christian, *Biochem. Z.* **210**, 384 (1942).

[9] V. Shanbag, G. Blomquist, G. Johansson, and A. Hartman, *FEBS Lett.* **22**, 105 (1972).

[10] J. M. Brewer, T. Fairwell, J. Travis, and R. E. Lovins, *Biochemistry* **9**, 1011 (1970).

PURIFICATION PROCEDURE FOR ENOLASE FROM *Escherichia coli*[a]

| Step | Volume (ml) | Activity (units)[b] | Protein (g) | Specific activity | | Yield (%) | $A_{280}/A_{260}$ |
|---|---|---|---|---|---|---|---|
| | | | | Units[a]/ $A_{280}$ | Units[b,c]/ mg protein | | |
| 1. Extraction of acetone powder | 1200 | 252,000 | 12.6 | 3.3 | 20.4 (4.1) | 100 | 0.64 |
| 2. Heat treatment | 1180 | 265,000 | 8.0 | 4.4 | 33 (6.6) | 105 | 0.58 |
| 3. Acetone fractionation | 100 | 299,000 | 3.7 | 37 | 82 (16.4) | 119 | 0.84 |
| 4. TEAE-cellulose titration | 210 | 272,000 | 2.6 | 75 | 104 (20.8) | 108 | 1.03 |
| 5. TEAE-cellulose chromatography | 275 | 139,000 | 0.47 | 275 | 296 (59.2) | 55 | 0.93 |
| 6. Sephadex G-100 gel filtration I | 30 | 82,000 | 0.194 | 741 | 421 (84) | 32 | 1.85 |
| 7. Sephadex G-100 gel filtration II | 30 | 68,000 | 0.098 | 1210 | 694 (139) | 29 | — |
| 8. Crystallization | 3 | 54,400 | 0.074 | 1292 | 734 (147) | 22 | 1.75 |

[a] From 83 g of acetone powder obtained from 500 g of *E. coli* cells.

[b] Arbitrary enzyme units ($10 \times \Delta A_{240}$/min at 25°).

[c] Protein was determined directly. Values in parentheses are international enzyme units (micromole of product formec per minute at 25° in 3 ml assay volume).

enolases from two thermophilic bacteria have been purified.[11,12] These enolases are similar in catalytic parameters to the *E. coli* enzyme, but differ significantly in their extreme thermal stability (70–90°) and in their octameric rather than dimeric subunit structure. The individual subunits for the thermophilic enolases are, however, in the molecular weight range observed for other enolases (40,000–50,000). Both the thermophilic enolases lack cysteine, in agreement with the observed low cysteine content of *E. coli* and yeast enolases compared to the mammalian enolases.

[11] E. Stellwagen, M. C. Cronlund, and L. D. Barnes, *Biochemistry* **12**, 1552 (1973).
[12] L. D. Barnes and E. Stellwagen, *Biochemistry* **12**, 1559 (1973).

# [53] Enolase from Fish Muscle

*By* Finn Wold

2-Phosphoglycerate $\rightleftarrows$ enolpyruvate phosphate $+ H_2O$

The preparation of enolase from yeast and rabbit muscle has been described in Vol. 9 of this series.[1] Since that time several preparations of enolase from a variety of sources have been reported; these have been summarized in a review,[2] and more recently the preparations of enolase from monkey muscle,[3] swine muscle,[4] and thermophilic bacteria[5] have also been published.

The white muscle of the Salmonidae is one of the richer sources of enolase. Based on activity and electrophoretic screening of extracts of 8 different species from the major genera of the salmonoids, *Salmo, Salvelinus,* and *Oncorhynchus,* the enolase level has been found to be 5–8 mg per gram of wet muscle, whereas most other sources seem to fall in the range of 0.5–4 mg per gram weight of starting cells or tissue. Enolase has been crystallized from three species of Salmonidae, rainbow trout (*Salmo gairdnerii gairdnerii*),[6] coho salmon (*Oncorhynchus kisutch*[7]), and chum salmon (*Oncorhynchus keta*).[7] The isolation procedures developed for the three fish enolases differ only in minor details, and in the following

[1] E. W. Westhead, see this series, Vol. 9, p. 669.
[2] F. Wold, in "The Enzymes" (P. D. Boyer, ed.), 3rd ed., Vol. 5, p. 499. Academic Press, New York, 1971.
[3] J. A. Winstead, *Biochemistry* **11**, 1046 (1972).
[4] E. Wolna, M. Wolny, and T. Baranowski, *Acta Biochem. Polon.* **18**, 87 (1971).
[5] E. Stellwagen, M. C. Cronlund, and L. D. Barnes, *Biochemistry* **12**, 1552 (1973).
[6] R. P. Cory and F. Wold, *Biochemistry* **5**, 3131 (1966).
[7] R. C. Ruth, D. M. Soja, and F. Wold, *Arch. Biochem. Biophys.* **140**, 1 (1970).

discussion all three procedures will be presented together, identifying the individual variations in each step.

## Assay Methods

Enzyme activity was determined by the direct spectrophotometric method of Warburg and Christian,[8] monitoring enolpyruvate phosphate production at 240 nm according to standard procedure. The substrate (D-glycerate 2-phosphate, barium salt) was converted to and used as the water-soluble tricyclohexylammonium salt.[1] The assay medium contained 1 m$M$ substrate, 1 m$M$ MgSO$_4$, 0.2 $M$ KCl, and 50 m$M$ imidazole buffer at a final pH of 7.0. Protein determinations were based on the 280 nm absorbance and eventually related to accurate dry weight of the pure enzymes. For the crude samples the absorbance of a 1 mg/ml solution of protein was assumed to be 1; the actual value determined for each of the pure enzymes is 0.74 for trout and coho enolase and 0.87 for chum enolase.[6,7] An arbitrary activity unit, defined as the amount of enzyme that will give an absorbance change at 240 nm of 0.1 per minute at 25° is convenient to use in day-to-day operation and will also be reported here. To convert this arbitrary unit to international enzyme units, one only needs to know the molar extinction coefficient of enolpyruvate phosphate under the conditions (pH and metal ion concentration) used in the assay. This value is available in the literature[9] and has been found to be about 1300 $M^{-1}$ for the assay conditions used. Thus, the arbitrary absorbancy units can be converted to international enzyme units ($\mu$moles of product formed per minute in 3-ml reaction volume) by multiplication with the factor 0.1 $\times$ 3/1.3 = 0.23.

## Preparation of Enzymes

Starting materials are generally commercially available through fish farms or hatcheries. For some, the process of obtaining the starting materials by more direct, individual involvement brings an added challenge and excitement to the enzyme preparation. If the fish is frozen reasonably soon after killing, it can be stored at $-10$–$20°$ for several months without any change in either quality or quantity of enolase. It is convenient to start with 600–800 g of fish muscle, and it is consequently recommended to freeze the fish muscle in individual packages of that appropriate size.

*Step 1. Extraction.* Appropriate amounts of frozen muscle tissue (600–800 g) are allowed to thaw at 4°. When fully thawed, the meat

[8] O. Warburg and W. Christian, *Biochem. Z.* **310**, 384 (1942).
[9] F. Wold and C. E. Ballou, *J. Biol. Chem.* **227**, 301 (1957).

is ground in a meat grinder, then homogenized in a Waring Blendor for 5 min with two parts (by weight) of cold EDTA solution (0.05% of EDTA). The homogenate is centrifuged immediately at 0° for 30–60 min at 9000 $g$. The supernatant is decanted and filterd through glass wool to remove lipid material. From this point there are some differences in the details of the isolation procedures for the three enzymes. To clearly distinguish between the methods, the procedure for the coho enzyme will be described in detail, and at the end of each step, the modification used for the chum and trout enzymes will be stated very briefly.

*Step 2. Acetone Fractionation.* Coho enzyme: The acetone (reagent grade) is precooled to —10°. An appropriate amount is added to the extract to bring the final concentration to 33% (492 ml acetone per liter of extract, assuming the volumes to be additive). The acetone should be added quickly, with stirring to ensure complete mixing, and the solution is centrifuged immediately at 0° for 30 min at 9000 $g$. The supernatant is decanted, brought to 45% acetone (326 ml of acetone per liter of original extract) and centrifuged immediately at 0° for 30 min at 9,000 g. The supernatant is decanted and the sides of the tubes are wiped with tissue to remove as much acetone solution as possible. The 33–45% precipitate is dissolved in cold imidazole buffer (50 m$M$ imidazole, pH adjusted to 7.8 with HCl). The volume of buffer should be approximately one-tenth that of the original extract. Chum enzyme: The protein fraction precipitating between 30% (430 ml of acetone per liter of extract) and 50% (570 ml of acetone per liter of extract) acetone is collected and dissolved in 50 m$M$ imidazole buffer, pH 7.0. Trout enzyme: the protein fraction precipitating between 35% (540 ml of acetone per liter of extract) and 46% (313 ml of acetone per liter of extract) acetone is collected and dissolved in 50 m$M$ imidazole buffer (pH 7.9), containing 1 m$M$ MgSO$_4$.

*Step 3. Heat Treatment.* (Except for the difference in pH, this step is identical for the three enzymes.) The acetone fraction in imidazole buffer is placed in a shallow dish in the cold room. To this is added enough solid MgSO$_4$·7H$_2$O to bring the solution to 0.5% (w/v) MgSO$_4$·7H$_2$O, and the solution is stirred in the dish at 4° overnight in order to remove as much acetone as possible. The solution is then brought, with constant stirring, to 55° by immersion in a 70° water bath. It is held at 55° for 3 min, then quickly cooled in an ice bath to approximately 10° and centrifuged for 30 min at 9000 $g$ and 0°. The supernatant can be filtered through glass wool to remove light insoluble material and is then ready for fractionation with ammonium sulfate.

*Step 4. Ammonium Sulfate Fractionation.* All ammonium sulfate steps should be carried out at 4°. The heat supernatant is decanted directly into a 250-ml centrifuge bottle. Nitrogen is passed over the top of the solution, and all subsequent steps involving stirring the fish enolases in ammonium sulfate solutions should be carried out under an anaerobic atmosphere, as the probability of denaturing the enzyme in ammonium sulfate solutions has been found to increase in the presence of air. Coho enzyme: Solid ammonium sulfate (312 g per liter of heat supernatant) is added to the supernatant from the heat step and the solution is stirred for 15 min and then centrifuged at 9000 $g$ for 1 hr. The supernatant (which contains the active enzyme) is decanted into another centrifuge bottle, and a further addition of ammonium sulfate (162 g per liter of heat supernatant) is made. This solution is also stirred for 15 min, then centrifuged at 9000 $g$ for 1 hr. The precipitate obtained in the range 312–474 g of ammonium sulfate per liter is suspended in 30 ml/100 ml of original heat supernatant of a "60% saturated" ammonium sulfate solution prepared by mixing 3 parts of 4°-saturated ammonium sulfate[10] and 2 parts of 50 m$M$ imidazole buffer, pH 7.8 containing 1 m$M$ MgSO$_4$. The suspension is stirred gently for 8–16 hr. It is then centrifuged at 9000 $g$ for 1 hr; the supernatant is decanted and saved for attempts at crystallization. The precipitate should be reextracted with the same volume of the "60% saturated" ammonium sulfate solution. After this second extraction, little or no activity remains in the precipitate, which can be discarded. Chum enzyme: The fraction precipitating between 353 and 473 g of solid ammonium sulfate per liter of heat supernatant is collected and extracted twice with 30 ml of the "60% saturated" ammonium sulfate solution. Trout enzyme: The fraction precipitating between 350 and 519 g of solid ammonium sulfate per liter of heat supernatant is collected and extracted as for the other enzymes.

Step 5. Crystallization. (Identical for all enzymes.) Saturated ammonium sulfate solution[10] is added to the above extracts in sufficient quantity to give slight turbidity, and the solutions are left at 4°. Crystal formation should generally start within the first few hours, and an optimal yield is obtained after 2–3 days. The very fine crystals can be collected by centrifugation at 9000 $g$ for 1 hr.

Step 6. Recrystallization. (Identical for all enzymes.) Recrystallization can be achieved by dissolving the crystals in minimal amounts of the appropriate imidazole buffer and adding saturated ammonium sulfate solution until slight turbidity is obtained. In our hands, the first one or two recrystallizations have improved the specific activity, but further

---

[10] We have found it convenient to prepare this saturated ammonium sulfate solution by dissolving 760 g of ammonium sulfate and 100 mg of MgSO$_4$ in 1 liter of water, and after addition of conc. NH$_4$OH to pH 7.9, allowing the excess of ammonium sulfate to crystallize out at 4° over a period of at least 2 days.

PURIFICATION OF ENOLASE FROM FISH MUSCLE

| Step | Coho salmon (820 g) | | | Chum salmon (630 g) | | | Rainbow trout (685 g) | | |
| --- | --- | --- | --- | --- | --- | --- | --- | --- | --- |
| | Activity (units[a]) | Protein (g) | Specific activity (units/mg[a,b]) | Activity (units[a]) | Protein (g) | Specific activity (units/mg[a,b]) | Activity (units[a]) | Protein (g) | Specific activity (units/mg[a,b]) |
| 1. Extraction | 1960 | 51.6 | 38 (8.7) | 1320 | 40.3 | 33 (7.6) | 1510 | 61 | 25 (5.8) |
| 2. Acetone fractionation | 1440 | 13.4 | 107 (24.6) | 1093 | 17.9 | 61 (14) | 1230 | 18.2 | 68 (15.6) |
| 3. Heat treatment | 1320 | 4.3 | 307 (71) | 810 | 3.7 | 219 (50) | 1010 | 3.5 | 289 (66) |
| 4. Ammonium sulfate fractionation | 1266 | 3.9 | 325 (75) | 675 | 2.4 | 281 (65) | 630 | 1.85 | 341 (78) |
| 5. Crystallization | 666 | 1.9 | 351 (81) | 335 | 0.9 | 372 (86) | 367 | 1.04 | 353 (81) |
| 6. Recrystallization | 249 | 0.69 | 361 (83) | 200 | 0.5 | 400 (92) | 220 | 0.58 | 379 (87) |
| 7. Sephadex G-100 gel filtration | | | — | | | — | | | 410 (94) |

[a] Arbitrary units ($10 \times \Delta A_{240}/\text{min}$) at 25°.
[b] Numbers in parentheses are international enzyme units (micromoles of product formed per minute at 25° in 3-ml assay volume).

recrystallizations generally have not brought about any further purification.

*Step 7. Gel Filtration.* A slight improvement in the specific activity of the trout enzyme has been observed after gel filtration on a 120 × 2.1 cm column of Sephadex G-100, equilibrated at 4° with 25 m$M$ Tris phosphate buffer, pH 8.6, containing 1 m$M$ MgSO$_4$. A concentrated solution of the recrystallized protein (200 mg dissolved in 2.5 ml of equilibration buffer) is dialyzed against two changes of the same buffer at 4° and is then subjected to gel filtration at a flow rate of 30 ml/hr.

A summary of yield and purification in each step is presented in the table. The final recrystallized enzymes have been found to be at least 95% pure by electrophoresis on starch and acrylamide gels. Whereas solutions of the enzymes in concentrated ammonium sulfate solutions appear to be quite labile, the crystal suspensions in ammonium sulfate solutions are stable. The fish enolases have been stored for several years in this form in well-sealed containers in the cold room without any apparent loss of activity.

### Properties of the Fish Enzymes

The molecular weight of the three fish enolases is about 95,000. Although solid physical and chemical data are available only for the two salmon enzymes, it is reasonable to conclude that all three enzymes are dimers, made up of two very similar subunits[7] and containing two active sites. This is consistent with the established properties of the enolases from yeast, rabbit muscle, and *E. coli.*[2] The amino-terminal groups of all three fish enzymes appear to be blocked; this property is shared by the rabbit muscle enzyme which contains $N$-acetylalanine at the amino-terminal position. One interesting feature of the fish enolases is their occurrence in multiple molecular forms. Indeed, all the 8 species of *Salmonidae* investigated to date have been found to contain 3 electrophoretically distinct forms of enolase.[11] This isozyme pattern is not significantly altered during the purification, and the crystalline enzyme preparations appear to contain the same characteristic ratio of the three forms as does the original extract. The electrophoretic mobility and the ratio of isozymes are unique for each species investigated, however, and appear to be independent of both the age and the sex of the fish.[11]

The catalytic properties of the three fish enolases are quite similar, with specific activities from 83 to 94 international units per milligram, $K_m$ at 40 $\mu M$ and pH optima between 6.9 and 7.1. The magnesium activation and fluoride inhibition are also quantitatively identical for the three enzymes.

[11] H. Tsuyuki and F. Wold, *Science* **146**, 535 (1964).

## [54] Enolase from Human Muscle

By T. Baranowski and E. Wolna

$$\text{D-CH}_2\text{OH—CHOPO}_3\text{H}_2\text{—COOH} \rightleftarrows \text{CH}_2\text{=COPO}_3\text{H}_2\text{—COOH} + \text{H}_2\text{O}$$

## Assay Method

The standard assay is run at 25° in a solution containing:
Sodium 2-phospho-D-glycerate, 1.0 m$M$
Magnesium sulfate, 3.0 m$M$
Potassium chloride, 0.4 $M$
Imidazole·HCl buffer, 50 m$M$, pH 6.8
Aliquots (3 ml) are used in cuvettes of 1 cm optical path. Upon addition of enolase, the linear absorbancy increase of 0.100 at 240 nm is determined corresponding to the conversion of 0.226 $\mu$mole of the substrate. The specific activity is expressed in enzyme units (EU) per milligram of protein.[1]

The protein in purified enzyme preparations is determined by the method of Warburg and Christian.[2] In crude extracts the protein content is estimated by the turbidimetric method of Mejbaum-Katzenellenbogen.[3] Enzyme concentration is calculated using an absorbancy coefficient of 0.9 at 280 nm for 1 mg of enolase per 1 ml of solution as found for rabbit muscle enolase.[4]

## Preparation of the Enzyme[5]

All operations are carried out in cold (0–4°) unless otherwise stated.

*Step 1. Extraction.* Human skeletal muscles are passed through a commercial meat grinder. The mince is homogenized in a Waring Blendor with 2 parts (by weight) of cold water and stirred for 30 min. The homogenate is centrifuged at 2000 $g$ for 30 min and the supernatant is decanted.

*Step 2. Heat Fractionation.* To each liter of extract 50 g of solid MgSO$_4$·H$_2$O is added and the solution is heated for 3 min in a water bath at 53° with continuous stirring. The solution is quickly cooled in an ice bath and left overnight in cold. The resulting precipitate of denatured protein is removed by centrifugation of filtration.

[1] "Report of the Commission on Enzymes of the International Union of Biochemistry," p. 45. Pergamon, Oxford, 1961.
[2] O. Warburg and W. Christian, *Biochem. Z.* **310**, 384 (1941).
[3] W. Mejbaum-Katzenellenbogen, *Acta Biochim. Pol.* **2**, 139 (1955).
[4] A. Holt and F. Wold, *J. Biol. Chem.* **236**, 3227 (1961).
[5] T. Baranowski, E. Wolny, and A. Morawiecki, *Eur. J. Biochem.* **5**, 119 (1968).

*Step 3. First Ammonium Sulfate Fractionation.* Solid ammonium sulfate is added to the clear solution to give a final concentration of 2.4 *M*. After the preparation has stood for several hours, the precipitate is removed by centrifugation and the supernatant is brought to 2.75 *M* concentration with respect to ammonium sulfate. The resulting protein precipitate is allowed to equilibrate overnight and is collected by centrifugation.

*Step 4. Gel Filtration on Sephadex G-75.* The precipitate is dissolved in a minimal amount of water and clarified by centrifugation. The protein solution is applied to a column (60 × 2.5 cm) of Sephadex G-75 equilibrated with water. The maximal load is 15 ml of the solution containing 600–800 mg of protein. Protein is eluted with distilled water and assayed in 1-ml fractions by measurement of absorbancy at 280 nm. The absorbance at 410 nm is measured in parallel in order to examine the effectiveness of the separation of impurities absorbing at this wavelength. The enzyme is eluted in the second protein peak after the first one which contains lower molecular weight impurities strongly absorbing at 410 nm. The absorbance at 410 nm should be small in the enzyme peak otherwise the crystallization may fail.

*Step 5. Second Ammonium Sulfate Fractionation.* To the combined eluates containing the enzyme, solid ammonium sulfate is added to give a final concentration 2.3 *M*. The resulting protein precipitate is allowed to equilibrate for several hours and is clarified by centrifugation. The supernatant solution is brought to 2.75 *M* with respect to ammonium sulfate. After storing overnight in a cold room the precipitate is centrifuged down and the supernatant is discarded.

*Step 6. Crystallization.* The precipitate is dissolved in a minimal volume of 50 m*M* imidazole·HCl buffer, pH 7.8, and to the clear solution powdered ammonium sulfate is added with cautious stirring to avoid foaming. As soon as a slight turbidity appears (sometimes schlieren is observed indicating crystal formation), the solution is allowed to stand at 0° in a refrigerator. The crystal growth is enhanced by slowly warming the solution to room temperature and cooling back to 0° repeated several times a day. After a few days, crystallization is complete. The final concentration of ammonium sulfate is about 2.3 *M* and the final pH about 6.2. Enolase crystallizes in form of long rectangular prisms or plates.

Attempts to crystallize the human muscle enolase from buffer systems other than imidazole were unsuccessful.

*Step 7. Recrystallization.* The crystals are harvested by centrifugation at 26,500 *g*, washed with 2.3 *M* ammonium sulfate solution, and dissolved in one volume of imidazole buffer, pH 7.8. If turbid, the solution is centrifuged and the clear supernatant is treated with powdered ammo-

TABLE I
PREPARATION OF ENOLASE FROM HUMAN MUSCLE (1 kg)

| Fraction | Volume (ml) | Total protein (g) | Total units (EU) | Specific activity (EU/mg) | Yield (%) |
|---|---|---|---|---|---|
| Crude extract | 1600 | 45 | 50,000 | 1.13 | 100 |
| Supernatant after heat treatment | 1500 | 24 | 49,500 | 2.06 | 99 |
| First ammonium sulfate fractionation | 49 | 3.5 | 42,500 | 11.8 | 83 |
| Gel filtration on Sephadex G-75 | 100 | 0.89 | 30,000 | 34 | 60 |
| Second ammonium sulfate fractionation | 24 | 0.60 | 25,800 | 43 | 51.5 |
| Crystallization | 24 | 0.18 | 15,000 | 84 | 30 |

nium sulfate (or saturated solution) until a slight turbidity appears. Recrystallization proceeds easily in the cold. The repeated recrystallizations give no further increase in specific activity of the enzyme.

A summary of the procedure giving 74-fold purification is presented in Table I.

The enolase crystals can grow to considerable size (several millimeters) after prolonged storage in mother liquor in the cold.

## Properties

*Catalytic Activity.*[5,6] The highest specific activity of crystalline enzyme was found to be 84 EU per milligram of protein. The crystals are devoid of the following activities: aldolase, glyceraldehyde-3-phosphate dehydrogenase, pyruvate kinase, and phosphoglyceromutase (sample 100 $\mu$g).

In imidazole·HCl buffer, pH 6.8 the $K_m$ for 2-phosphoglycerate is 300 $\mu M$ and or $Mg^{2+}$ 6 m$M$.

In imidazole and Tris buffers (50 m$M$) the highest activity is found at pH 6.8 and in phosphate buffer (50 m$M$) at pH 7.1.

The influence of metal ions on the activity of enolase is shown in Table II.

Ninety thousand grams of protein contain 14 moles of SH groups assayed by titration with *p*-chloromercuribenzoate in 0.33 $M$ acetate buffer, pH 7.1, in presence of $MgCl_2$ ($I = 0.10$).

[6] E. Wolna, *Arch. Immunol. Ther. Exp.* **15**, 389 (1967).

## TABLE II
### INFLUENCE OF METAL IONS ON ACTIVITY OF ENOLASE

| Ion | Optimal concentration $(M)$ | Relative activity $(\%)$ |
|---|---|---|
| Mg | $3 \times 10^{-3}$ | 100 |
| Mn | $3 \times 10^{-4}$ | 50 |
| Zn | $3 \times 10^{-4}$ | 36 |
| Ni | $1 \times 10^{-5}$ to $1 \times 10^{-2}$ | Trace |
| Ca | $1 \times 10^{-5}$ to $1 \times 10^{-2}$ | None |

*Molecular Properties.*[5,7] The crystalline enzyme is homogeneous by chromatographic, electrophoretic, and ultracentrifugal criteria. The enzyme sediments in imidazole buffer, pH 7.1 ($MgCl_2$, $I = 0.10$) as a single symmetrical peak. The sedimentation constant $s_{20,w}$ is $5.55 \times 10^{-13}$ sec. The intrinsic viscosity of protein in the same buffer is 4.10 ml/g, and the diffusion constant is $5.6 \times 10^{-7}$ cm$^2$ sec.

The molecular weight calculated from sedimentation and diffusion data and independently from sedimentation and viscosity measurements is found to be $90,000 \pm 3000$. Frictional coefficient $f/f_0$ is found to be 1.29.

Gel filtration on Sephadex G-50 after treatment of protein with sodium dodecyl sulfate reveals a single symmetrical peak corresponding to a molecular weight of 45,000. This is an indication that the enolase molecule is composed of two subunits of the same molecular weight.

[7] E. Wolna and A. Morawiecki, unpublished observations.

Section **IV**

Phosphatases

# [55] Sucrose-phosphate Phosphohydrolase (Sucrose Phosphatase) from Plants

*By* J. S. HAWKER and M. D. HATCH

## Assay Method

*Principle.* Hydrolysis of [fructosyl-$^{14}$C]sucrose phosphate is measured as the appearance of radioactive sucrose following chromatography of reaction samples on paper.[1,2] Enzyme relatively free of nonspecific phosphatases can be assayed with higher concentrations of unlabeled sucrose phosphate by measuring the appearance of orthophosphate. We used the former procedure largely because it provides a higher degree of specificity with crude preparations (see below).

## Reagents

Tris-maleate buffer 50 m$M$, pH 6.7, containing 10 mg of bovine serum albumin per milliliter

MgCl$_2$, 0.1 $M$

Enzyme, samples diluted in 2 m$M$ Tris·HCl buffer, pH 7.0, containing 5 mg/ml of bovine serum albumin so that 20 $\mu$l contains sufficient activity to hydrolyze between 0.02 and 1 nmole of sucrose phosphate per minute

[fructosyl-$^{14}$C]Sucrose phosphate, 0.25 m$M$ (2.1 × 10$^6$ dpm/ml). To prepare radioactive sucrose phosphate, incubate about 9 $\mu$moles of [U-$^{14}$C]fructose 6-phosphate (70 × 10$^6$ dpm/ml) with 25 $\mu$moles of UDP-glucose, 25 $\mu$moles of Tris·HCl, pH 7.1, and sucrose phosphate synthetase (prepared by the procedure of Mendicino[3]) in a total volume of 2 ml. The reaction is stopped by heating at 100°, and any precipitated protein is removed. The time required for maximum conversion of fructose-6-P to sucrose phosphate should be checked in small-scale trials. Stop trial reaction samples at intervals, treat with commercially available calf intestine alkaline phosphatase, and chromatograph to detect free radioactive sucrose and fructose as described below for the assay of sucrose phosphatase (Procedure section).

[1] M. D. Hatch, *Biochem. J.* **93**, 521 (1964).
[2] J. S. Hawker and M. D. Hatch, *Biochem. J.* **99**, 102 (1966).
[3] J. Mendicino, *J. Biol. Chem.* **235**, 3347 (1960).

*Purification of Sucrose Phosphate.* The following methods have been used to purify the sucrose phosphate:

a. Residual UDP-glucose, fructose 6-phosphate, and any free sugars are removed by treating the reaction mixture on a column of Dowex 1 essentially as described by Leloir and Cardini.[4] The reaction mixture is applied to a 20-ml column (2 cm$^2$ × 10 cm, chloride form) and eluted initially with 50 ml of 1 m$M$ NH$_4$OH and then a linear gradient developed with 200 ml of 20 m$M$ NH$_4$Cl and 10 m$M$ sodium borate in the mixing chamber and 200 ml of 30 m$M$ NH$_4$Cl in the reservoir. Sucrose phosphate emerges after the application of about 300 ml of this gradient and fructose 6-phosphate is retained on the column. Fractions containing radioactive sucrose phosphate are pooled and evaporated to dryness under vacuum. Borate is removed by twice adding 25 ml of methanol and evaporating to dryness under vacuum. Most of the remaining salt is removed in an electrolytic desalting apparatus (until the current drops to 0.4 A), and the preparation is chromatographed as a band on Whatman No. 3 MM paper for 30 hr using propanol–ammonia (sp. gr. 0.88)–water (6:3:1) as the developing solvent. The radioactive sucrose phosphate band is eluted from this chromatogram with water. More than 99% of the radioactivity is in sucrose phosphate.

b. If all the fructose 6-phosphate is consumed in the original reaction, it is adequate for most purposes to simply chromatograph the original reaction mixture as a band on Whatman 3 MM paper using the solvent described above.

c. If only minor proportions of radioactive fructose 6-phosphate remain then a major part of the sucrose phosphate can be freed of this substrate by concentrating the reaction to 0.5 ml, applying this to a column of Sephadex G-10 (100 × 1.5 cm), and eluting with water at a rate of 8 ml/hr. Collect 2-ml fractions and check small samples of the radioactive fractions (usually fractions 25–35) for purity by hydrolyzing with alkaline phosphatase and then chromatographing free sugars as described in the Procedure section. At least half of the sucrose phosphate peak is free of fructose 6-phosphate. Remaining UDP and UDP-glucose are removed by passing the fractions from Sephadex G-10 through a 1 ml column of activated charcoal previously washed with 5 m$M$ HCl. The eluate should be quickly neutralized and then concentrated to the desired volume.

A chemical procedure for the synthesis of sucrose phosphate has recently been described.[5]

[4] L. F. Leloir and C. E. Cardini, *J. Biol. Chem.* **214**, 157 (1955).
[5] J. G. Buchanan, D. A. Cummerson, and D. M. Turner, *Carbohyd. Res.* **21**, 283 (1972).

*Procedure.* To small test tubes add 50 μl of the buffer–bovine serum albumin mixture, 10 μl of MgCl$_2$, and 20 μl of [fructosyl-U-$^{14}$C]sucrose phospate. Start the reaction by adding 20 μl of enzyme to duplicate tubes. Incubate the tubes at 30° and stop individual reactions at 5 and 10 min by heating in a boiling water bath. To determine the proportion of sucrose phosphate hydrolyzed, 30 μl samples of reaction mixtures are spotted 3.5 cm apart on sheets of Whatman No. 1 paper [underspotted with 10 μl of a mixture of 1% (w/v) each of sucrose, glucose and fructose in 70% ethanol] and then chromatographed for 16 hr with ethyl acetate–pyridine–water (8:2:1 by volume) as the developing solvent. Sucrose phosphate remains at the origin of the chromatograms. The position of sugars is determined by dipping the sheets in a solution containing 1% (v/v) aniline, 1% (w/v) diphenylamine, and 4% (w/v) phosphoric acid in acetone. The sheets are dried in air and then heated for a few minutes at about 95° in an oven. Radioactivity remaining in sucrose phosphate and that appearing in sucrose is determined by counting the areas with a 2-inch diameter Geiger-Muller tube after masking adjacent radioactive spots with an aluminium foil shield. Radioactivity in fructose should also be checked if enzyme extracts contain invertase. Alternatively, areas of radioactivity can be located with a chromatogram scanner and then excised and counted by a liquid scintillation procedure.

With the procedure described here, the concentration of the substrate is initially about one-third of its Michaelis constant. Relative rates can be determined if the proportion of substrate hydrolyzed is restricted to about 15% or less. However, the reaction under these conditions follows first-order kinetics so that initial rates can readily be calculated with little error for reactions in which as much as 70% of the substrate has been hydrolyzed. For a particular enzyme, maximum velocities can be calculated from this data if the $K_m$ for sucrose phosphate is separately determined.

*Application to Crude Enzyme Extracts.* This assay procedure can be used with crude enzyme extracts prepared and assayed as follows. Chilled plant tissue (3 g) is ground in a mortar with 7 ml of 0.35 $M$ Tris·HCl buffer, pH 8.5, containing 20 m$M$ EDTA, 10 m$M$ sodium diethyldithiocarbamate, 15 m$M$ cysteine·HCl, and 6% Carbowax 4000 at 0–4°. The homogenate is filtered through cheesecloth and centrifuged at 30,000 $g$ for 15 min. The supernatant is diluted up to 100-fold with the Tris·maleate–bovine serum albumin mixture and assayed as described above. This procedure has been found to be effective with many plant tissues and serves to minimize denaturation of the enzyme by tannins, phenols, low pH, and dilution. With the assay pH of 6.7, and the low concentrations of plant extract and sucrose phosphate in the assay, there is little interference by nonspecific phosphatases.[6] Specificity can be checked for by adding either EDTA (20 m$M$), sucrose (100 m$M$), or maltose (100 m$M$) to the assay system. Sucrose phosphatase is inhibited by EDTA and these sugars whereas nonspecific phosphatases are only slightly inhibited by EDTA and not at all by the sugars.[6-8] In tissues containing low activities of sucrose phosphatase and high activities of invertase, counts can appear in fructose during assays.

## Purification Procedures

The enzyme from immature stem tissue of sugarcane was most extensively purified. However, since carrot tissue is more readily available details of a partial purification from this source is also included. Purification data are provided in the table.

### Carrot Root[2]

*Step 1. Isolation of Mitochondria.* Peeled carrot roots are sliced and cooled to 2°; 500 g are blended with 500 ml of 100 m$M$ Tris·HCl buffer (pH 7.6) containing 300 m$M$ mannitol, 10 m$M$ EDTA, 10 m$M$ MgCl$_2$, 10 m$M$ sodium diethyldithiocarbamate, 15 m$M$ of cysteine-HCl, and 500 mg of bovine serum albumin. This and subsequent operation are conducted at 2°. The filtrate obtained by pressing this extract through muslin is centrifuged for 5 min at 2000 $g$, and the resulting supernatant is centrifuged at 10,000 $g$ for 15 min. This pellet is washed by suspending in 50 ml of 1 m$M$ Tris·HCl (pH 7.2) containing 300 m$M$ mannitol and centrifuging at 10,000 $g$ and then resuspended in 5 ml of the same medium and frozen and thawed twice. Sucrose phosphatase solubilized by this procedure is separated from particulate material by centrifuging at 100,000 $g$ for 30 min.

*Step 2. Sephadex G-100.* To a 175 ml (2.5 × 35 cm) column of Sephadex G-100, previously washed with 2 m$M$ Tris·HCl (pH 7.0), add 4 ml of the 100,000 $g$ supernatant. Adjust the flow-rate to 0.5 ml/min and elute the column with 2 m$M$ Tris·HCl (pH 7.0). Collect 5-ml fractions and immediately add 25 mg of bovine serum albumin to each. Sucrose phosphatase begins to emerge about 15 ml after the void volume and is collected in the next 15 ml.

### Sugarcane Stem[6]

*Step 1. Extraction.* Forty grams of the soft, rapidly expanding internodal tissue from the top of sugarcane stems is cooled to 2° and then

[6] J. S. Hawker, *Phytochemistry* **5**, 1191 (1966).
[7] J. S. Hawker, *Phytochemistry* **10**, 2313 (1971).
[8] J. S. Hawker, *Biochem. J.* **102**, 401 (1967).

PURIFICATION OF SUCROSE PHOSPHATASE FROM PLANTS

| Step and fraction | Units[a] | Specific activity ($\mu$moles/hr/ mg protein) | Ratio of maximum velocities, sucrose phosphate:fructose 6-phosphate |
|---|---|---|---|
| | Carrot root | | |
| 1. Crude extract | 51 | — | 7 |
| Washed mitochondria | 0.94 | 3.4 | 30 |
| 100,000 $g$ supernatant | 0.40 | 10.8 | 44 |
| 2. Sephadex G-100 | 0.08 | 86 | 64 |
| | Sugarcane stem | | |
| 1. Original filtrate | 9.9 | — | 1.5 |
| 2. 33–42% satd. $(NH_4)_2SO_4$ | 2.9 | — | 2.5 |
| 3. Sephadex G-100 | 3.8 | — | 22 |
| 4. Hypatite C | 1.9 | — | 48 |

[a] Micromoles per hour per gram of tissue extracted (maximum velocities were calculated as described in the Procedure section).

sliced and ground in a large mortar with 20 ml of 400 m$M$ Tris·HCl buffer (pH 8.0) containing 60 m$M$ cysteine, 60 m$M$ sodium diethyldithiocarbamate, and 10 m$M$ EDTA. This homogenate is filtered through muslin. All subsequent operations are conducted at 2° and completed in one day.

*Step 2. Ammonium Sulfate.* Add to the muslin-filtrate solid $(NH_4)_2SO_4$ to give 33% saturation. After standing for 20 min, centrifuge at 10,000 $g$ for 15 min, discard the precipitate, and add additional $(NH_4)_2SO_4$ to give 42% saturation. The protein precipitated after standing for 20 min is recovered by centrifuging and dissolved in 2.5 ml of 2 m$M$ Tris·HCl, pH 7.0.

*Step 3. Sephadex G-100.* The dissolved 33–42% saturated $(NH_4)_2SO_4$ fraction is applied to a 175 ml (2.5 cm × 35 cm) column of Sephadex G-100 previously equilibrated with 2 m$M$ Tris·HCl buffer, pH 7.0. The column is eluted with the same buffer at a flow rate of 0.5 ml/min and 2-ml fractions are collected after the void volume (10 mg of dextran blue can be added with the enzyme). Most of the activity appears in fractions 8–14. This is checked in trial runs so that the enzyme can be immediately applied to calcium phosphate gel.

*Step 4. Hypatite C.* The pooled active fractions from the previous step are applied to a 3 ml column of calcium phosphate gel (Hypatite C., Clarkson Chemical Co. Williamsport, Pennsylvania) previously washed with 5 m$M$ potassium phosphate buffer, pH 6.7. The column is then washed with 9 ml of 20 m$M$ potassium phosphate (pH 6.7) and sucrose phosphatase is eluted by the application of 6 ml of 80 m$M$ potassium phosphate, pH 6.7. This fraction of approximately 6 ml is supplemented with 30 mg of bovine serum albumin. To remove phosphate the fraction is then applied to a 25 ml column of Sephadex G-25 previously equilibrated with 2 m$M$ Tris·HCl, pH 7.0. After the emergence of the void volume the next 8 ml are collected and stored at $-15°$.

## Properties

*Sources and Intracellular Location.* Specificity of phosphatase activity toward sucrose phosphate in crude tissue extracts has been judged by the fact the specific sucrose phosphatase requires $Mg^{2+}$ and is inhibited by EDTA and sugars.[6-8] On the basis of these criteria an enzyme with a high degree of specificity toward sucrose phosphate is present in root, stem, leaf, and seed tissue from a variety of species.[2,6,7] Using procedures for extracting intact organelles, a proportion of the total activity always remains associated with washed pellets containing mitochondria.[2,6] This activity can be solubilized by freezing and thawing. For sugarcane stem tissue, the properties of the enzymes from soluble and particulate fractions appear identical.[6]

*Specificity and Affinity for Substrates.* Partially purified preparations that hydrolyze sucrose phosphate at least 50 times faster than a variety of other phosphorylated compounds have been obtained from sugarcane and carrot tissue.[2,6] For these enzymes the $K_m$ for sucrose phosphate was approximately 0.15 m$M$. The activity of both enzymes was increased manyfold by 5 m$M$ $Mg^{2+}$, and this requirement was partially replaced by $Mn^{2+}$ but not $Ca^{2+}$ ions.[2] The $K_m$ for $Mg^{2+}$ is about 0.3 m$M$.

*Activators and Inhibitors.* With the partially purified enzyme from carrot[2] and sugarcane[6] and crude extracts from various other sources[6] the $Mg^{2+}$-stimulated activity is inhibited by an excess of EDTA. EDTA also completely inhibits the small activity seen without added $Mg^{2+}$. Inhibition observed with 30 m$M$ inorganic phosphate and pyrophosphate[2] was probably due to these compounds binding with $Mg^{2+}$. In contrast, the enzyme hydrolyzing fructose 6-phosphate in crude extracts does not require $Mg^{2+}$ and is not inhibited by EDTA. Sucrose inhibits sucrose phosphatase from all sources tested.[7,8] For the sugarcane enzyme inhibition was partially competitive with a $K_i$ for sucrose of about 10 m$M$. Melezitose and 6-kestose (trisaccharides containing the sucrose moiety) and maltose were also inhibitory, but several other mono- and disaccharides had no effect on sucrose phosphatase activity.

*Stability.* Concentrated crude extracts retained activity when stored at $-15°$.[2,6] The partially purified enzyme from both sugarcane and carrot lost about 50% of the original activity in 7–8 days at $-15°$ but further loss of activity was slow. The enzyme is unstable at pH values below 6.0 and in dilute solution. Bovine serum albumin reduces the loss of activity in dilute solutions.

## [56] Fructose-1,6-diphosphatase and Sedoheptulose-1,7-diphosphatase from *Candida utilis*

*By* Sandro Pontremoli and Serena Traniello

### Assay Method for Fructose-1,6-diphosphatase

$$\text{Fructose 1,6-diphosphate} + H_2O \xrightarrow{\text{FDPase}} \text{fructose 6-phosphate} + P_i$$

*Principle.* The formation of F-6-P is measured spectrophotometrically by following the reduction of NADP at 340 nm in the presence of excess of glucose-6-phosphate dehydrogenase and glucose-6-phosphate isomerase. The amount of NADPH formed is proportional to the concentration of FDPase. For analytical purpose, the enzymic activity can be tested between pH 7.5 and 9.2; pH 9.2 is the value generally used in the specific test system.

*Reagents*

    Fructose 1,6-diphosphate, sodium salt, 10 m$M$, pH 7.5

    Glycine buffer 0.2 $M$, pH 9.4 (final pH in the cell 9.1–9.2), or triethanolamine buffer 0.2 $M$, pH 7.5

    $MgCl_2$, 0.1 $M$

    EDTA, 10 m$M$

    NADP, 10 m$M$

    Glucose-6-phosphate isomerase, 1 mg/ml

    Glucose-6-phosphate dehydrogenase, 1 mg/ml; both enzymes purchased from Boehringer, Mannheim, Germany, as crystalline suspensions in ammonium sulfate.

    FDPase: Dilute the enzyme solution to be tested to a concentration of 1–5 units/ml.

*Procedure.* Place into a cell having a 1-cm light path, in a final volume of 1 ml: 0.2 ml of buffer, 0.01 ml of $MgCl_2$, 0.01 ml of EDTA, 0.005 ml each of glucose-6-P isomerase and glucose-6-P dehydrogenase, 0.01 ml of NADP, and the amount of enzyme to be assayed. Record the absorbance at 340 nm until no further increase occurs. Add 0.01 ml of F-1,6-P and read at 1-min intervals for 6–8 min.

*Definition of Unit and Specific Activity.* One unit of enzyme is defined as the amount that causes the formation of 1 $\mu$mole of F-6-P per minute, at room temperature, under the above conditions. Specific activity is expressed as units per milligram of protein. Protein is determined by the methods of Bücher[1] or Lowry et al.[2]

## Assay Method for Sedoheptulose-1,7-diphosphatase

$$\text{Sedoheptulose 1,7-diphosphate} + H_2O \xrightarrow{\text{SDPase}} \text{Sedoheptulose 7-phosphate} + P_i$$

*Principle.* The formation of S-7-P is measured by following the release of inorganic phosphate detected by the procedure of Fiske and SubbaRow.[3]

The activity of SDPase is tested at pH 6.0.

*Reagents*

Sedoheptulose 1,7-diphosphate, sodium salt, 25 m$M$ (see this series, Vol. 3 [30])

Maleate buffer, 1 $M$, pH 6.0

EDTA, 50 m$M$

SDPase: Add the enzyme in an amount sufficient to release 0.05–0.1 $\mu$mole of $P_i$ in 5 min at 37°.

*Procedure.* Place in a centrifuge tube in a final volume of 1 ml: 0.1 ml of buffer, 0.1 ml of S 1,7-P and 0.01 ml of EDTA. The assay is started by the addition of the enzyme and incubated at 37°. The reaction is stopped by the addition of 0.1 ml of 85% trichloroacetic acid, and the protein precipitate is removed by centrifugation. The inorganic phosphate released is determined on an aliquot of the supernatant solution.

*Definition of Unit and Specific Activity.* One unit of enzyme is defined as the amount which hydrolyzes 1 $\mu$mole of $P_i$ per minute at 37°, under the above conditions.

## Purification Procedure (FDPase)

Dried *Candida utilis* is obtained from the Lake States Yeast Corporation, Rhinelander, Wisconsin.

Unless otherwise stated all operations are carried out in the cold and all centrifugations are performed in a refrigerated centrifuge at 0° at 20,000 $g$.

*Step 1. Autolyzate.* Suspend 200 g of dried cells in 800 ml of distilled

[1] T. Bücher, *Biochim. Biophys. Acta* **1**, 192 (1947).
[2] O. H. Lowry, N. J. Rosebrough, A. L. Farr, and R. J. Randall, *J. Biol. Chem.* **193**, 265 (1951).
[3] C. H. Fiske and Y. SubbaRow, *J. Biol. Chem.* **66**, 375 (1925).

water containing 1 m$M$ EDTA and allow to autolyze for 1 hr at room temperature with constant stirring. Centrifuge the suspension and collect the supernatant. Add to the latter a 2% protamine (obtained from Sigma Chemical Co., St. Louis, Missouri) solution in the amount of 10 ml for each 100 ml of enzyme solution. The precipitate is discarded by centrifugation.

*Step 2. Heat Precipitation.* The clear supernatant is brought to pH 5.8 by small addition of 5 $N$ acetic acid and heated in a water bath at 50° for 10 min. The mixture is immediately chilled and centrifuged to remove the precipitated proteins.

*Step 3. Phosphocellulose Column Fractionation I.* Before use, the powder (Whatman phosphocellulose P-11 obtained from C. Erba, Milan) was washed alternatively with alkali (0.5 $N$ NaOH) and acid (0.5 $N$ HCl) and then with distilled water until neutral. Filter the slurry and add the moist powder to the enzyme solution as described below. Dilute the heated fraction with an equal volume of distilled water and adjust the pH to 6.3 with 2 $N$ NaOH. Then add phosphocellulose with continuous stirring until 10% of the total enzymic activity is adsorbed. Maintain the pH at 6.3 during the addition of cellulose by small volumes of 2 $N$ NaOH. Remove the phosphocellulose by filtration on a Büchner filter, and adjust the pH of the solution to 5.7 with a few drops of 5 $N$ acetic acid. Add again phosphocellulose, keeping the pH constant and stirring, until approximately 90% of the enzymic activity is adsorbed. Pour the slurry into a glass column (4.2 cm in diameter by 30 cm in length) and wash with 0.2 $M$ acetate buffer, pH 7.5, at the rate of 4 ml per minute, until the absorbance of the effluent at 280 nm is less than 0.02. Decrease the column flow rate to 2 ml/min and elute the enzyme with a solution of 2 m$M$ F-1,6-P in acetate buffer, pH 6.3. Collect 1-ml fractions and pool those with specific activity between 25 and 30. The enzyme emerges as a sharp peak between fraction 20 and 23. Dialyze the enzyme solution for 4 hr against distilled water containing 0.1 m$M$ EDTA.

*Step 4. Phosphocellulose Column Fractionation II.* Dilute the dialyzed enzyme solution with 9 volumes of distilled water and readjust the pH to 5.7 with 1 $N$ acetic acid. Prepare a column of phosphocellulose (1.7 $\times$ 13 cm) and apply the diluted solution of FDPase. All the enzymic activity is retained. Wash the column successively with 0.15 $M$ acetate buffer, pH 5.7, and 0.25 $M$ acetate buffer, pH 5.8, until the effluent is free of protein: then elute the enzyme with a solution containing 2 m$M$ F-1,6-P in 0.25 $M$ acetate buffer, pH 5.8. The flow rate of the column is 0.25 ml/min. Collect 0.5 ml fractions, and combine those containing the enzyme with a specific activity around 75. The peak containing the FDPase generally emerges between fraction 5 and 8. The overall purifica-

TABLE I
PURIFICATION OF FRUCTOSE-1,6-DIPHOSPHATASE FROM *Candida utilis*

| Step and fraction | Total units | Specific activity (units/mg) | Recovery (%) |
|---|---|---|---|
| 1. Autolyzate | 435 | 0.02 | 100 |
| 2. Heated fraction | 325 | 0.13 | 75 |
| 3. Phosphocellulose eluate I | 107 | 27.5 | 25 |
| 4. Phosphocellulose eluate II | 19 | 73 | 4.5 |

tion is 3400-fold with a yield of 4–5% (Table I). The enzyme solution can be frozen for several months without loss of activity.

## Purification Procedure (SDPase)

Unless otherwise stated all operations are carried out in the cold and all centrifugations are performed in a refrigerated centrifuge at 0° at 20,000 $g$.

*Step 1. Autolyzate.* Suspend 150 g of dried *C. utilis* in 450 ml of 0.1 $M$ sodium bicarbonate and allow to autolyze for 1 hr, at room temperature and with constant stirring. The precipitate is discarded by centrifugation.

*Step 2. Ammonium Sulfate Fractionation I.* The autolyzate is adjusted to pH 4.8 with 2 $N$ acetic acid. Discard the precipitate by centrifugation. Bring to pH 5.3 with 0.1 $N$ KOH the clear supernatant and add 0.243 g of ammonium sulfate per milliliter. Centrifuge and collect the supernatant. Add 0.132 g of ammonium sulfate per milliliter. The precipitate is collected by centrifugation and suspended in water.

*Step 3. Acetone Fractionation.* Dilute the ammonium sulfate fraction (540 ml) with an equal volume of 0.1 $M$ acetate buffer, pH 5.0. Cool the enzyme solution and slowly add 0.43 volume of acetone previously cooled at −40°. Centrifuge immediately in a refrigerated centrifuge at −15° and discard the precipitate. Treat the supernatant as above with 0.35 volume of acetone and collect again the supernatant by centrifugation. The further addition of 0.37 volume of cold acetone yields a precipitate, which is collected by centrifugation, containing the enzymic activity. Dissolve the precipitate in 0.2 $M$ malonate buffer containing 1 m$M$ EDTA, pH 6.3, to achieve a protein concentration of 10 mg/ml. Treat this solution with 1 volume of saturated (0°) ammonium sulfate solution at pH 8.6 and discard the precipitate formed. The enzyme is then precipitated with 1.25 volume of the same ammonium sulfate solution and centrifuged, the precipitate is dissolved in malonate buffer as before. Dialyze

TABLE II

Purification of Sedoheptulose-1,7-diphosphatase from *Candida utilis*

| Step and fraction | Total units | Specific activity (units/mg) | Recovery (%) |
|---|---|---|---|
| 1. Autolyzate | 176 | 0.0055 | |
| 2. Ammonium sulfate fraction | 203 | 0.035 | 100 |
| 3. Acetone fraction | 92 | 0.295 | 45 |
| 4. Phosphocellulose eluate I | 18.9 | 4 | 9 |
| 5. Phosphocellulose eluate II | 11.4 | 11.5 | 5 |

the enzyme solution for 4 hr against 20 m$M$ malonate buffer pH 6.3, containing 1 m$M$ EDTA.

*Step 4. Phosphocellulose Column Fractionation I.* Prepare a phosphocellulose[4] column (1.6 × 24 cm) by suspending the resin in a 20 m$M$ malonate buffer, pH 6.3, containing 1 m$M$ EDTA and allow to equilibrate. Apply the enzyme solution to the column and wash extensively with the equilibrating buffer, until no significant absorbance at 280 nm is measured in the effluent. Elute with a linear gradient of malonate buffer, pH 6.3, from 0.02 to 0.2 $M$, containing 1 m$M$ EDTA. Collect 6.8-ml fractions and pool those containing SDPase with specific activity around 4. The enzyme usually emerges between fractions 16 and 18. Add 0.55 g of ammonium sulfate per milliliter of enzyme solution. Centrifuge and dissolve the precipitate in a small volume of 0.2 $M$ malonate buffer, pH 6.3, containing 1 m$M$ EDTA.

*Step 5. Phosphocellulose Column Fractionation II.* Dialyze the eluate for 4 hr against 50 m$M$ malonate buffer, pH 6.3, containing 1 m$M$ EDTA, and apply the dialyzate solution into a phosphocellulose column (0.7 × 7 cm) previously equilibrated with the same buffer. Elute the column with a linear gradient from 50 m$M$ to 0.2 $M$ malonate buffer, pH 6.3. Collect fractions of 1.5 ml and combine those containing SDPase with the highest specific activity (approximately 11–12). The enzyme emerges as a single peak at a malonate concentration of 0.15–0.18 $M$. The overall purification is 2100-fold with the yield of approximately 5–6% (Table II). The enzymic solution can be frozen for several months without loss of activity.

## Properties (FDPase)

*Specificity. Candida* FDPase specifically catalyzes the hydrolysis of fructose 1,6-diphosphate. No other mono- or diphosphate sugar, including

---

[4] The phosphocellulose powder is washed as described under the purification procedure for FDPase.

sedoheptulose 1,7-diphosphate, is hydrolyzed to any appreciable extent.

*Effect of pH and EDTA.* The purified enzyme shows maximum activity between pH 8.6 and 8.8, when the pH curve is carried out in 1:1 mixture of triethanolamine buffer and glycine buffer in the entire pH range.[5] Addition of 0.1 m$M$ EDTA enhances the activity, without significant change in the pH profile. FDPase activity in the autolyzate, assayed in the same buffer medium, shows maximum activity between pH 8.3 and 8.6 in the absence of a chelating agent. The addition of 0.1 m$M$ EDTA shifts the maximum to pH 7.8 and increases the activity in the neutral pH range by 3–4-fold. The difference in the pH curve when the enzyme activity is assayed in the autolyzate or in the purified preparation may be ascribed to partial proteolytic digestion of a native enzyme form. This has been shown to occur during the preparation of FDPase from rabbit liver.[6]

*Effect of Substrate Concentration.* Purified FDPase assayed at pH 7.5 and 9.2, in the presence of 0.1 m$M$ EDTA, exhibits maximum activity at F-1,6-P concentrations of approximately 0.1 m$M$. Higher concentrations of substrate are inhibitory. The $K_m$ calculated from Lineweaver-Burk plots are found to be about 8 $\mu M$ at either pH.

*Metal Requirement.* FDPase activity is strictly dependent upon the addition of $Mg^{2+}$ (1 m$M$ or $Mn^{2+}$ (0.1 m$M$), the latter being less effective as activating agent.

*Activators and Inhibitors.* Oxidized CoA and cystamine are activators of *C. utilis* FDPase, as observed for the enzyme from mammalian sources.[7] Both compounds yield a 2–2.5-fold activation which can be detected at pH 7.5 and 9.2 when the enzyme is assayed with $Mg^{2+}$ as activating cation. Other disulfides tested (glutathione, cysteine, and pantetheine) fail to activate the enzyme. AMP at concentrations of 20 $\mu M$ to 0.5 m$M$ inhibits up to 50% the FDPase activity at pH 7.5.[6]

*Homogeneity and Molecular Weight.* The purified FDPase is homogeneous in disc gel electrophoresis and shows a sharp peak in sucrose density gradient. The molecular weight is approximately 130,000, as calculated by sedimentation analysis in sucrose density gradient.

*Subunit Structure.* Treatment of the purified enzyme with SDS, followed by analysis in SDS polyacrylamide gel electrophoresis, reveals the presence of a single component of molecular weight of approximately

[5] S. Traniello, M. Calcagno and S. Pontremoli, *Arch. Biochem. Biophys.* **146**, 603 (1971).

[6] S. Traniello, E. Melloni, S. Pontremoli, C. L. Sia, and B. L. Horecker, *Arch. Biochem. Biophys.* **149**, 222 (1972).

[7] K. Nakashima, S. Pontremoli, and B. L. Horecker, *Proc. Nat. Acad. Sci. U.S.* **64**, 947 (1969).

35,000. Since the molecular weight of the native enzyme is 130,000, *C. utilis* FDPase is assumed to be composed of four identical polypeptide chains.

## Properties (SDPase)

*Specificity.* *Candida utilis* SDPase specifically hydrolyzes sedoheptulose 1,7-diphosphate; no other mono- or diphosphate sugar, including fructose 1,6-diphosphate, is hydrolyzed to any appreciable extent.[5]

*Effect of pH and EDTA.* Both purified SDPase and autolyzate extract show maximum activity at about pH 6.0. EDTA has little effect on the enzyme activity in crude extract, but increases the activity of purified enzyme at pH 6.0 by almost 2-fold.

*Effect of Substrate Concentration.* No substrate inhibition is observed even at concentration of S-1,7-P as high as 10 m$M$. The $K_m$ value calculated from Lineweaver-Burk plots is found to be 1 m$M$ at pH 6.0.

*Metal Requirement.* *Candida utilis* SDPase does not require a divalent cation for its catalytic activity.

*Activators and Inhibitors.* Treatment of SDPase with oxidized CoA, cystamine, or glutathione does not modify the catalytic activity. AMP at concentration as high as 3 m$M$ does not inhibit the purified enzyme.

*Homogeneity and Molecular Weight.* The enzyme preparation yields a single band in disc gel electrophoresis and a sharp peak in sucrose density gradient. The molecular weight as determined by sucrose gradient method, has been estimated to be approximately 75,000.

*Subunit Structure.* Treatment of purified SDPase with SDS allows dissociation into a component of approximately 35,000 MW, as revealed by the acrylamide gel pattern.[5] SDPase from *C. utilis* appears therefore to be composed of two subunits of equal molecular weight.

*Comments.* In *C. utilis* two specific enzymes are present to catalyze the hydrolysis of fructose 1,6-diphosphate and sedoheptulose 1,7-diphosphate. In mammalian tissues (liver,[8] muscle,[9] and kidney[10]), a single enzyme catalyzes the hydrolysis of both substrates.

[8] S. Pontremoli, S. Traniello, B. Luppis, and W. A. Wood, *J. Biol. Chem.* **240**, 3459 (1965).

[9] J. Fernando, M. Enser, S. Pontremoli, and B. L. Horecker, *Arch. Biochem. Biophys* **126**, 599 (1968).

[10] S. Traniello, E. Melloni, and S. Pontremoli, unpublished observations, 1970.

## [57] Fructose-1,6-diphosphatase from Rabbit Liver

*By* Sandro Pontremoli and Edon Melloni

Fructose-1,6-diphosphatase $+ H_2O \rightarrow$ fructose 6-phosphate $+ P_i$

### Assay Method

*NADP Reduction Method*

The formation of F-6-P is measured spectrophotometrically following NADP reduction at 340 nm in the presence of excess glucose-6-phosphate isomerase and glucose-6-phosphate dehydrogenase. The amount of NADPH formed is proportional to the concentration of FDPase.

The activity of fructose-1,6-diphosphatase (FDPase) can be tested at pH 7.5 or 9.2 (the former is the actual pH in the specified test system).

*Reagents*

Buffer, diethanolamine 0.1 $M$–triethanolamine 0.1 $M$, pH 7.5 or pH 9.2

$MgCl_2$, 0.2 $M$

EDTA, 10 m$M$

NADP, 10 mM

Glucose-6-phosphate isomerase, 1 mg/ml

Glucose-6-phosphate dehydrogenase, 1 mg/ml. Both enzymes have been purchased from Boehringer Mannheim, Germany.

Enzyme. Dilute the solution to be tested to a concentration of 1–5 units per milliliter with water (see definition below).

*Procedure.* Place in a cell having a 1-cm light path in a final volume of 1 ml: 0.2 ml of buffer, 0.01 ml of $MgCl_2$, 0.01 ml of EDTA, 0.005 ml each of glucose-6-phosphate isomerase and glucose-6-phosphate dehydrogenase, 0.01 ml of NADP, and the amount of enzyme to be assayed. The reaction is started by the addition of 0.01 ml of F-1,6-P with readings at 340 nm at 1-min intervals for 8–10 min.

*Definition of Unit and Specific Activity.* One unit of enzyme activity is defined as the amount which causes the formation of 1 $\mu$mole of F-6-P per minute at pH 7.5 at 23° under the above conditions. Specific activity is expressed as units per milligram of protein. Protein concentration is measured by the method of Bücher[1] or Lowry *et al.*[2] or calculated from

[1] T. Bücher, *Biochim. Biophys. Acta* **1**, 292 (1947).
[2] O. H. Lowry, N. J. Rosebrough, A. L. Farr, and R. J. Randall, *J. Biol. Chem.* **193**, 265 (1951).

the absorbance at 280 nm. The absorbance of a dialyzed solution containing 1 mg of purified FDPase (dry weight) per milliliter in a 1.0-cm light path is 0.73 at 280 nm and 0.37 at 260 nm.

## Phosphate Liberation Method

The hydrolysis of F-1,6-P or sedoheptulose 1,7-diphosphate (S-1,7-P) is measured by estimation of the rate of appearance of inorganic phospate, using the method of Fiske and SubbaRow.[3]

### Reagents

F-1,6-P, 10 m$M$ tetrasodium salt
S-1,7-P, 10 m$M$ sodium salt
Buffer, diethanolamine 0.1 $M$–triethanolamine 0.1 $M$, pH 7.5 or pH 9.2
$MgCl_2$, 0.2 $M$
EDTA, 10 m$M$
Enzyme. Add the enzyme solution in an amount sufficient to release 0.05–0.1 $\mu$moles of $P_i$ in 10 min at 23°.

*Procedure.* Place 0.04 ml of F-1,6-P or 0.04 ml of S-1,7-P, 0.1 ml of buffer, 0.01 ml of $MgCl_2$, 0.01 ml of EDTA, and 0.34 ml of distilled water in a centrifuge tube. The assay is started by the addition of the enzyme solution. The reaction is stopped by the addition of 0.1 ml of 5 $N$ sulfuric acid. The protein precipitate is removed by centrifugation, and an aliquot of the supernatant solution is analyzed for the inorganic phosphate.

*Definition of Unit and Specific Activity.* One unit enzyme is defined as the amount which liberates 1 $\mu$mole of $P_i$ per minute under the above conditions. Specific activity is expressed as units per milligram of protein.

## Purification Procedure

All operations are performed at 2–4°.

*Step 1. Extraction.* One kilogram of freshly collected rabbit livers is minced in a hand meat mincer. The paste is then extracted with vigorous stirring for 1 min with 2.2 liters of 0.25 $M$ sucrose solution at pH 7.0. The suspension is centrifuged at 17,000 $g$ for 30 min (crude extract, 2300 ml).

*Step 2. Phosphocellulose Treatment.* The crude extract is adjusted

---

[3] C. H. Fiske and Y. SubbaRow, *J. Biol. Chem.* **66**, 375 (1925).

PURIFICATION OF FRUCTOSE-1,6-DIPHOSPHATASE

| Step and fraction | Total units[a] | Specific activity[b] (units/mg) | Recovery (%) | Ratio pH 9.2: pH 7.5 |
|---|---|---|---|---|
| 1. Crude extract | 2250 | 0.039 | 100 | 0.4 |
| 2. Phosphocellulose treated extract | 2020 | 0.092 | 90 | 0.4 |
| 3. Phosphocellulose eluate | 970 | 14.6 | 43 | 0.4 |

[a] Assayed at pH 7.5 with 2 m$M$ Mg$^{2+}$.
[b] Calculated from activities measured at pH 7.5 with Mg$^{2+}$.

to pH 6.5 with 2 $N$ acetic acid, and phosphocellulose P-11 suspension, corresponding to 30 g of the original dry powder, is added with constant stirring. The pH is maintained constant by addition of 2 $N$ NaOH. The phosphocellulose is removed by filtration in a Büchner funnel. At the clear filtrate is added an equal amount of phosphocellulose keeping the pH at 6.1. The resin is again discarded by filtration as above (phosphocellulose-treated extract, 2300 ml).

*Step 3. Absorption and Chromatography.* The phosphocellulose treated extract is diluted with an equal volume of distilled water and EDTA is added to a final concentration of 0.1 m$M$. The pH of the extract is adjusted to 5.6 by addition of 2 $N$ acetic acid. Phosphocellulose is again added to the extract in an amount sufficient to absorb almost all the enzymic activity. The suspension is filtered and the clear solution is discarded. The resin, collected on the Büchner funnel, is washed with 2 liters of 0.12 $M$ acetate buffer pH 5.6 containing 0.1 m$M$ EDTA. The phosphocellulose is transferred to a glass column (4.3 cm in diameter and 40 cm high) and washed extensively with 5 liters of 0.18 $M$ acetate buffer pH 5.75 containing 0.1 m$M$ EDTA until the absorbance of the effluent at 280 nm is less than 0.02. FDPase is eluted from the column with a solution of 2 m$M$ F-1,6-P in 0.23 $M$ acetate buffer pH 6.3 containing 0.1 m$M$ EDTA. FDPase activity emerges as a sharp peak and the fractions with specific activity approaching 14 units/mg are combined (phosphocellulose eluate, 30–35 ml).

The enzyme is precipitated by the addition of solid ammonium sulfate to a final concentration of 80% of saturation. The precipitate is collected by centrifugation and suspended in a small volume of 80% saturated ammonium sulfate. The overall purification is 380 with a yield of 43%. The enzyme suspension can be stored for several months at 2–4° without significant loss of activity.

## Properties

*pH Optimum and Effect of Cations.* The pH optimum of FDPase is very much dependent on the concentration of the activating cation when the enzyme activity is tested in the absence of a chelating agent. In these conditions at $Mg^{2+}$ concentration lower than 1 m$M$ maximum activity is observed in the alkaline region (8.2–8.5); at higher concentrations of this cation the pH optimum is shifted to values around 7.6–7.8.

If EDTA is added in a concentration of 0.1 m$M$, maximum activity is observed around pH 7.2–7.3 for any concentrations of $MgCl_2$ ranging from 1 m$M$ to 5 m$M$.

The chelating agent also increases the catalytic activity by 2-fold in the neutral pH region, but little or no effect of chelator is observed in the alkaline pH region.

The EDTA effect has been explained, on the basis of results obtained with rabbit kidney,[4] as an indication of the existence in the enzyme protein of a binding site for the free metal distinct from a specific binding site for the metal-EDTA chelate.

*Catalytic Specificity.* Rabbit liver FDPase catalyzes the hydrolysis of the inorganic phosphate from the C1 position of fructose 1,6-diphosphate and sedoheptulose 1,7-diphosphate.

S-1,7-P is dephosphorylated at 70% the rate of F-1,6-P. The relative activities with F-1,6-P or S-1,7-P do not change significantly throughout the purification and even in the purified homogeneous enzyme form. On the basis of this evidence, as well as on the results of competition experiments,[5] it has been concluded that liver contains a single enzyme protein capable of hydrolyzing both substrates.

A number of monophosphate and diphosphate esters such as glucose 6-phosphate, glucose 1-phosphate, fructose 1-phosphate, fructose 6-phosphate, ribose 5-phosphate, ribulose 5-phosphate, xylulose 5-phosphate, ribulose 1,5-diphosphate, sedoheptulose 1-phosphate, and sedoeptulose 7-phosphate have been tested as substrates for the purified FDPase. None of these compounds is cleaved at a significant rate even at high concentrations.

*Molecular Structure.* The purified FDPase is homogeneous by electrophoretic and ultracentrifugal criteria. The molecular weight has been estimated to be 144,000. FDPase is presumably made up of four identical subunits of MW 36,000 as indicated by analysis in SDS disc-gel electro-

[4] J. S. Rosenberg, Y. Tashima, B. L. Horecker, and S. Pontremoli, *Arch. Biochem. Biophys.* **154**, 283 (1973).

[5] A. Bonsignore. G. Mangiarotti, M. Mangiarotti, A. De Flora, and S. Pontremoli, *J. Biol. Chem.* **238**, 3151 (1963).

phoresis and by the release of 4 alanine residues, per mole of enzyme, at the COOH terminal end.[6]

*Catalytically Active Sites.* There are at least four distinct binding sites for each ligand (one per subunit?).[7] This has been shown for F-1,6-P, $Mn^{2+}$, and AMP, which is the most potent allosteric inhibitor.

At pH 7.5 the association constant for F-1,6-P has been calculated to range between $0.77 \times 10^7$ $M^{-1}$ and $5.8 \times 10^7$ $M^{-1}$. These increasing values are the result of cooperative binding of F-1,6-P to the enzyme protein. The value of the association constant for AMP is $0.5 \times 10^5$ $M^{-1}$; while that for $Mn^{2+}$ corresponds to $0.5 \times 10^5$ $M^{-1}$.

Binding of $Mn^{2+}$ to the catalytic sites occurs only in the presence of the substrate F-1,6-P, thus indicating the following compulsory mechanism in the enzyme reaction:

$$\text{FDPase} + \text{F-1,6-P} \rightarrow \text{FDPase F-1,6-P}$$
$$\text{FDPase} - \text{F-1,6-P} + Mn^{2+} \rightarrow \text{FDPase} + \text{F-6-P} + P_i + Mn^{2+}$$

Experimental evidence indicate that each class of four binding sites for F-1,6-P and AMP is correlated, both at the binding and at the catalytic level, to the presence of two sets of four tyrosine residues each.

Thus both catalytic and allosteric properties of FDPase each seem to involve a specific amino acid residue per polypeptide chain.

Also 1 cysteine residue per subunit seems to participate in the modulation of the catalytic activity through a disulfide exchange mechanism between FDPase and specific biological compounds. Finally in the case of $Mn^{2+}$ histidyl residues and carboxylate ions seem to be involved in the binding of the activating cation to the FDPase molecule.

*Activators and Inhibitors.* FDPase undergoes a significant increase in catalytic activity when 4 cysteine residues per mole of enzyme have reacted with either cystamine, CoA, acyl-carrier protein or homocysteine to form the E-S-S-R mixed disulfides. The activated enzyme has been shown to undergo a conformational transition from a native to an "active" form.[8] Oleic acid[9] and certain phospholipids[10] (phosphatidic acid and diphosphoinositide) have been shown to be potent activators of FDPase.

As previously discussed several chelating agents (EDTA, histidine, etc.) increase the FDPase activity in the neutral pH region. The role

[6] S. Traniello, E. Melloni, S. Pontremoli, C. L. Sia, and B. L. Horecker, *Arch. Biochem. Biophys.* **149**, 222 (1972).

[7] S. Pontremoli and B. L. Horecker, *in* "The Enzymes" (P. D. Boyer, ed.), 3rd ed., Vol. 4, p. 611. Academic Press, New York, 1971.

[8] S. Pontremoli, E. Melloni, F. Balestrero, A. De Flora, and B. L. Horecker, *Arch. Biochem. Biophys.* **156**, 255 (1973).

[9] R. C. Baxter, C. W. Carlson, and B. M. Pogell, *J. Biol. Chem.* **247**, 2969 (1972).

[10] M. B. Allen and J. Mc D. Blair, *Biochem. J.* **130**, 1167 (1972).

of these chelators in the regulation of FDPase and specifically in their metal-chelate form may turn out to be of great physiological significance.

AMP is the most powerful inhibitor of FDPase and has been considered the physiological allosteric regulator of the enzyme activity. All known FDPases from different sources are inhibited to a different degree by this adenylic nucleotide.[11-14]

*Interconversion of Neutral to Alkaline FDPase.* Proteolytic digestion[15] converts the neutral FDPase form, AMP sensitive, to an alkaline form, AMP insensitive, and with a pH optimum in the alkaline region (9.0–9.2). In a model system subtilisin catalyzes such conversion through the removal from each of the four subunits of a peptide, of molecular weight of about 6000, containing the only tryptophan residue in the molecule. The peptide is removed from the amino-terminal end while the carboxy-terminal region does not appear to be affected by proteolysis.[15]

*Distribution.* In mammals specific FDPase has been described and purified to homogeneity, in kidney and muscle.[7]

Several properties such as specificity, metal requirement, AMP inhibition, number of binding sites for the ligands are similar in liver, kidney, and muscle FDPases. The three enzymes differ, however, in other catalytic properties and in amino acid composition. The kidney enzyme reacts with antibodies prepared against purified rabbit liver FDPase; the muscle enzyme does not react.[16] The question of identity or nonidentity of liver and kidney FDPases from the same species remains, however, to be resolved.

[11] K. Taketa and B. M. Pogell, *Biochem. Biophys. Res. Commun.* **12**, 229 (1963).
[12] S. Traniello, S. Pontremoli, Y. Tashima, and B. L. Horecker, *Arch. Biochem. Biophys.* **146**, 161 (1971).
[13] C. Gancedo, M. L. Salas, A. Giner, and A. Sols. *Biochem. Biophys. Res. Commun.* **20**, 15 (1965).
[14] J. Mendicino and F. Vasarhely, *J. Biol. Chem.* **238**, 3528 (1963).
[15] S. Pontremoli, E. Melloni, A. De Flora, and B. L. Horecker, *Proc. Nat. Acad. Sci. U.S.* **70**, 661 (1973).
[16] M. Enser, S. Shapiro, and B. L. Horecker, *Arch. Biochem. Biophys.* **129**, 377 (1969).

## [58] Purification of Fructose-1,6-diphosphatase from *Polysphondylium pallidum*

### By ORA M. ROSEN

Fructose 1,6-diphosphate + $H_2O$ → fructose 6-phosphate + $P_i$

## Assay Method

*Principle.* Fructose-1,6-diphosphatase (FDPase) catalyzes the hydrolysis of D-fructose-1,6-di-P to D-fructose-6-P and $P_i$. Enzymic activity can be followed spectrophotometrically by converting fructose-6-P to glucose-6-P and coupling the oxidation of glucose-6-P to the reduction of NADP. Alternatively, the inorganic phosphate formed by hydrolysis of fructose-di-P can be measured by the method of Fiske and SubbaRow.[1]

*Reagents for the Spectrophotometric Assay*

Glycine buffer, (0.2 $M$), pH 9.5
NADP, 50 m$M$, sodium salt
D-Fructose-di-P, 20 mM sodium salt
$MgCl_2$, 0.1 $M$
Glucose-6-P dehydrogenase, 1 mg/ml, crystalline, Boehringer & Son
Phosphohexose isomerase, 1 mg/ml crystalline, Boehringer & Son
EDTA, 50 m$M$, neutralized to pH 7.5 with 1 $N$ NaOH

*Procedure.* The incubation mixture (1.0 ml) contains 20 m$M$ glycine buffer, pH 9.5, 0.5 m$M$ NADP. 0.2 mM fructose-di-P, 1.0 m$M$ $MgCl_2$, glucose-6-P dehydrogenase (5.0 $\mu$g), and phosphohexose isomerase (5.0 $\mu$g). The reaction is initiated by the addition of an aliquot of FDPase. NADP reduction is followed spectrophotometrically at 340 nm with a Gilford recording spectrophotometer. The amount of NADP reduced is calculated using the extinction coefficient of $6.22 \times 10^6$ cm$^2$ mole$^{-1}$.[2] When the reaction is performed at pH 7.5, 20 m$M$ triethanolamine buffer, pH 7.5, is used in place of glycine buffer. Where indicated, EDTA (0.2 m$M$) is added. All reactions are measured in silica cells of 1-cm light path and 1 ml capacity at room temperature (22–25°).

*Definition of Enzyme Unit.* A unit of enzyme is the amount required to catalyze the conversion of 1 $\mu$mole of fructose-di-P to fructose-6-P in 1 min under the conditions of the assay. Specific activity is expressed

[1] C. H. Fiske, and Y. SubbaRow, *J. Biol. Chem.* **66**, 375 (1925).
[2] B. L. Horecker and A. Kornberg, *J. Biol. Chem.* **175**, 385 (1948).

as units per milligram of protein. Protein is determined by the method of Lowry *et al.*[3] using bovine serum albumin as standard.

## Purification

### *Reagents*

Frozen *P. pallidum* myxamebae. Myxamebae are grown axenically in liquid medium,[4] harvested by centrifugation at 20,000 $g$ for 15 min, washed once in growth medium and stored frozen at $-20°$.

Tris·HCl buffer, 50 m$M$ pH 7.9

Protamine sulfate, 1% Eli Lilly & Co.

Ammonium sulfate, reagent grade

Sodium malonate buffer, 5 m$M$ and 10 m$M$, pH 6.0.

Sephadex G-25, Pharmacia Chemicals (equilibrated with 5 m$M$ malonate buffer pH 6.0)

Fructose-di-P, sodium salt, 2.0 m$M$ in 15 m$M$ sodium malonate

CM-Sephadex C-50, coarse; Pharmacia Chemicals. The resin is prepared by suspending it in 0.2 $M$ malonic acid, adjusting the slurry to pH 5.0 with 10 $N$ NaOH, and pouring it into a column (0.5 $\times$ 10 cm). The column is then washed exhaustively with 5 m$M$ malonate buffer, pH 6.0, until the effluent reaches pH 6.0.

*Procedure.* All operations are performed at 4°. Centrifugation is carried out at 10,000 rpm in the SS No. 34 rotor of a refrigerated Sorvall centrifuge (RC-2). The table provides a summary of the purification procedure.[5]

*Step 1. Cell-Free Extract.* Ten grams of frozen myxamebae are ground for 5 min in a chilled mortar with 20 g of alumina. The enzyme is extracted into 50 m$M$ Tris·HCl buffer, pH 7.9, the suspension centrifuged and the residue discarded.

*Step 2. Protamine Sulfate Precipitation.* One-eighth volume of 1% protamine sulfate is added to the supernatant fluid with stirring. The precipitate which forms is removed by centrifugation.

*Step 3. Ammonium Sulfate Fractionation.* Crystalline ammonium sulfate (24.3 mg/10 ml of solution) is added with stirring, and the resultant precipitate is removed by centrifugation. Solid ammonium sulfate (14.3 mg/10 ml) is again added to the supernatant fluid, and the precipitate

[3] O. H. Lowry, N. J. Rosebrough, A. L. Farr, and R. J. Randall, *J. Biol. Chem.* **193**, 265 (1951).

[4] M. Sussman, *Science* **139**, 338 (1963).

[5] O. M. Rosen, *Arch. Biochem. Biophys.* **114**, 31 (1966).

PURIFICATION OF FDPase FROM *Polysphondylium pallidum*[a]

PURIFICATION OF FDPase FROM *Polysphondylium pallidum*[a]

| Fraction | Total volume (ml) | Total activity (units) | Specific activity (units/mg) | Overall recovery (%) |
|---|---|---|---|---|
| 1. Cell-free extract | 14.8 | 2.68 | 0.013 | — |
| 2. Protamine sulfate fraction | 18.5 | 2.68 | 0.025 | 100 |
| 3. Ammonium sulfate fraction (40–60% saturation) | 20.0 | 2.17 | 0.045 | 81 |
| 4. Sephadex G-25 eluate | 40.0 | 2.14 | 0.045 | 80 |
| 5. CM Sephadex eluate[b] | 10.0 | 0.66 | 5.50 | 25 |

[a] O. M. Rosen, *Arch. Biochem. Biophys.* **114,** 31 (1966).

[b] If CM cellulose is used instead of CM Sephadex, FDPase with a specific activity of 10.0–12.0 units/mg may be obtained. The overall recovery, however, is only one-half that found when CM-Sephadex is used.

is collected by centrifugation and dissolved in 20 ml of 5 m$M$ malonate buffer, pH 6.0.

*Step 4. Sephadex G-25 Eluate.* The enzyme solution is passed through a 150-ml column of Sephadex G-25 which had previously been equilibrated with 5 m$M$ malonate buffer, pH 6.0. Fractions containing enzymic activity and free of ammonium sulfate (as determined by conductivity measurements) are pooled.

*Step 5. CM-Sephadex Chromatography.* The solution containing enzyme is applied at a rate of 1.0 ml per minute to a 0.5 × 10 cm column of CM-Sephadex equilibrated with 5 m$M$ malonate buffer pH 6.0. The column is then washed successively with 5 m$M$ and 10 m$M$ malonate buffer, pH 6.0. In each instance a protein peak is eluted and the washing is continued until the $A_{280}$ of the effluent is less than 0.01. FDPase is then selectively eluted with 2.0 m$M$ fructose-di-P in 15 m$M$ malonate buffer, pH 6.0. Fractions (2 ml) containing activities equal to or greater than 4.5 units of protein per milligram are pooled.

This preparation of FDPase can be stored at —20° for several weeks without appreciable loss of activity.

## Properties

*Stoichiometry and Substrate Specificity.* The purified enzyme catalyzes the hydrolysis of the phosphate group on C-1 of fructose-di-P and sedoheptulose-di-P to give fructose-6-P and sedoheptulose-7-P, respectively. Sedoheptulose-di-P is hydrolyzed at about 80% the rate of fructose-di-P. No hydrolysis of 1 m$M$ glucose-6-P, fructose-6-P, fructose-1-P, or glucose-1-P occurs when incubations are carried out for 30 min and assayed for release of inorganic phosphate.

*Effect of pH, EDTA, and Metals.* Maximal activity takes place at pH 9–9.5. In the presence of EDTA (0.1–0.2 m$M$), activity at pH 9 is doubled and an equal amount of activity appears at pH 7.5. High concentrations of EDTA (1 m$M$) inhibit all activity. FDPase has an absolute requirement for a divalent cation best satisfied by 1 m$M$ Mg$^{2+}$.

*Regulation.* Unlike most FDPases,[6] the enzyme from *P. pallidum* is not significantly inhibited by 5'-AMP. Activity, measured at pH 7.5 and pH 9.0, is the same in cell-free extracts derived from four different stages of the developmental cycle of *P. pallidum* (myxamebae, multicellular aggregate, preculmination stage, and differentiated fruiting body).

[6] B. L. Horecker, S. Pontremoli, O. M. Rosen, and S. Rosen, *Fed. Proc., Fed. Amer. Soc. Exp. Biol.* **25**, 1521 (1966).

# [59] Fructose-1,6-diphosphatase from Bovine Liver

*By* Arthur M. Geller and William L. Byrne

Fructose 1,6-diphosphate + H$_2$O → fructose 6-phosphate + P$_i$

Fructose-1,6-diphosphatase is a required enzyme in the gluconeogenic pathway. The purified enzyme from bovine liver requires a metal cofactor for activity, and is maximally active at neutral pH.[1-3] Activity is inhibited by high concentrations of the metal cofactor, by high concentrations of the substrate, fructose-1,6-P$_2$, and by 5'-AMP.[3,4]

## Assay Method

*Principle.* Fructose-1,6-diphosphatase activity is assayed either by measuring the formation of inorganic phosphate using a discontinuous assay[1] or by a continuous spectrophotometric assay which measures the reduction of NADP$^+$ when fructose 6-phosphate formation is coupled with phosphoglucose isomerase and glucose-6-phosphate dehydrogenase.[3]

*Method I. Phosphate Release Assay.* The assay mixture contains 25

[1] W. L. Byrne, G. T. Rajagopolan, L. D. Griffin, E. H. Ellis, T. M. Harris, P. Hochachka, L. Reid, and A. M. Geller, *Arch. Biochem. Biophys.* **146**, 118 (1971).
[2] A. M. Geller and W. L. Byrne, *Arch. Biochem. Biophys.* **153**, 526 (1972).
[3] C. J. Marcus, A. M. Geller, and W. L. Byrne, *J. Biol. Chem.* **248**, 8567 (1973).
[4] H. J. Cohen, T. M. Harris, A. M. Geller, and W. L. Byrne, *Arch. Biochem. Biophys.* **146**, 144 (1971).

m$M$ Tris–25 m$M$ histidine, pH 6.5, 5 m$M$ MgSO$_4$, 0.1 m$M$ EDTA, 5.0 m$M$ fructose 1,6-diphosphate, and enzyme in a final volume of 2.0 ml at 37°. The Tris-histidine buffer can be used between pH 6.0 and 9.5. At 0-, 15-, and 30-min intervals, 0.5-ml aliquots are withdrawn, pipetted into 1.0 ml of 10% trichloroacetic acid, and centrifuged if necessary. The deproteinized samples are analyzed for inorganic phosphate by the method of Fiske and SubbaRow.[5]

*Method II. Spectrophotometric Assay*

$$\text{Fructose 1,6-diphosphate} + H_2O \rightarrow \text{fructose 6-phosphate} + P_i \quad (1)$$
$$\text{Fructose 6-phosphate} \rightleftharpoons \text{glucose 6-phosphate} \quad (2)$$
$$\text{Glucose 6-phosphate} + NADP^+ \rightleftharpoons \text{6-phospho-δ-gluconolactone}$$
$$+ NADPH + H^+ \quad (3)$$

In this assay fructose-1,6-diphosphatase (reaction 1) is coupled with phosphoglucose isomerase (reaction 2) and glucose-6-phosphate dehydrogenase (reaction 3). The reaction is followed by measuring NADPH formation at 340 nm. The standard assay mixture contains 8.4 $\mu M$ Fru-1,6-P$_2$, and phosphoglucose isomerase and glucose-6-phosphate dehydrogenase in excess, in a final volume of 1.0 ml. Solutions containing Tris, histidine, and EDTA are prepared and adjusted to the desired pH at 37°; the temperature at which the assay is carried out. Since NH$_4^+$ has been shown to stimulate fructose-1,6-diphosphatase, phosphoglucose isomerase and glucose-6-phosphate dehydrogenase are diluted in 10 m$M$ Tris·HCl, pH 7.5, and dialyzed at 4° for 15 hr against 500 volumes of the same buffer to remove (NH$_4$)$_2$SO$_4$.

Reactions are always initiated by the addition of enzyme, because it has been observed that there is a difference in the initial rate when the assay is started with substrate or magnesium. In order to use the coupled assay at micromolar substrate concentrations, very dilute enzyme solutions are required. A typical assay of the purified enzyme uses about 27 ng of protein. The diluted enzyme is unstable, so it is stored at a concentration of 2.7 $\mu$g/ml in bovine serum albumin (10 mg/ml) and 5.0 m$M$ MgSO$_4$. The dilution into the final assay is generally 100-fold.

*Units.* One unit of fructose-1,6-diphosphatase activity is defined as that amount of enzyme which causes the liberation of 1.0 $\mu$mole of inorganic phosphate per minute at 37°. Specific activity is expressed as units per milligram of protein.

## Purification Procedure

All purification steps are carried out at 6°, unless otherwise stated, however, the pH for all solutions is adjusted at room temperature. The

[5] C. H. Fiske and Y. SubbaRow, *J. Biol. Chem.* **66**, 375 (1925).

PURIFICATION OF FRUCTOSE-1,6-DIPHOSPHATASE FROM BOVINE LIVER

| Fraction | Protein (mg/ml) | Activity[a] (units/ml) | Total activity (units) | Specific activity (units/mg) |
|---|---|---|---|---|
| 1. Homogenate[b] | 12 | 0.1 | 35,000 | 0.083 |
| 2. Methanol[b] | 31.5 | 11.5 | 15,167 | 0.37 |
| 3. Ammonium sulfate | 42 | 21.8 | 5,300 | 0.52 |
| 4. DEAE-Sephadex | 1.5 | 6.8 | 490 | 4.5 |
| 5. Cellex-P | 3.2 | 57.7 | 138 | 18.0 |

[a] Activity was measured by measuring the release of inorganic phosphate at pH 6.5 at 37°.

[b] These two steps are carried out at the New England Enzyme Center, Boston, Massachusetts.

livers used generally weight about 15 pounds. Fresh or frozen liver (cut up into small pieces) give equivalent results. Protein determinations are carried out by the method of Lowry et al.,[6] and bovine serum albumin is used as the color standard. Bovine fructose-1,6-diphosphatase, when assayed as described in Method I, is normally maximally active at pH 6.5, with less activity at pH 9.0. Generally the ratio of activity at 6.5 to the activity at pH 9.0 is approximately 2.0. Under some circumstances this ratio decreases during purification. This may be due to proteolysis, since most hepatic fructose-1,6-diphosphatases are very sensitive to the action of proteolytic enzymes.[7,8] During the purification procedure, each step is monitored by determining the pH 6.5 to pH 9.0 activity ratio and by measuring the percentage inhibition by 0.1 m$M$ 5'-AMP.[9] Typical results are summarized in the table.

*Methanol Fractionation.* Two hundred grams of liver are homogenized in 1 liter of 50 m$M$ sodium acetate, pH 5.0, 3 m$M$ MgSO$_4$, and 1.0 m$M$ EDTA. The homogenate is then centrifuged for 10 min at 20,000 $g$ at 0°, and the precipitate is discarded. Methanol (precooled to −18°) is added to the supernatant until the mixture is 35% methanol (v/v). This solution is stirred for 10 min at −18°, and then centrifuged at 20,000 $g$ at −18° for 10 min. The precipitate is discarded. The supernatant is again placed in a bath at −18°, and sufficient methanol is added to bring the methanol concentration to 60% (v/v). The solution is recentri-

[6] O. H. Lowry, N. J. Rosebrough, A. L. Farr, and R. J. Randall, *J. Biol. Chem.* **193**, 265 (1951).

[7] A. M. Geller, G. T. Rajagopolan, E. H. Ellis, and W. L. Byrne, *Arch. Biochem. Biophys.* **146**, 134 (1971).

[8] K. Nakashima and B. L. Horecker, *Arch. Biochem. Biophys.* **146**, 153 (1960).

[9] R. A. Arneson, Dissertation, Duke University, Durham, North Carolina, 1971.

fuged at 20,000 $g$ for 10 min at $-18°$. The precipitate is taken up in 10 m$M$ Tris, pH 7.9, 3.0 m$M$ MgSO$_4$, and 1.0 m$M$ EDTA, and then dialyzed against the same buffer.

The methanol fractionation step described above is for laboratory scale purification. This step has been scaled up at the New England Enzyme Center, Tufts University, Boston, Massachusetts, so that the homogentate of an entire liver can be processed at one time. The yield for this step is low (between 30 and 45%), but the overall savings in effort compensates for the poor yield.

*Ammonium Sulfate Fractionation.* The dialyzed methanol preparation is diluted to 7.0 units/ml with 10 m$M$ Tris, pH 7.9, 3.0 m$M$ MgSO$_4$ and 1.0 m$M$ EDTA. Ammonium sulfate is added slowly until 0.50 saturation (add 31.0 g/100 ml) is reached. The solution is stirred for 20 min, centrifuged at 10,000 $g$ for 15 min, and the precipitate is discarded. The supernatant is brought to 0.65 saturation (add 9.7 g/100 ml) with ammonium sulfate, stirred for 20 min, and centrifuged again at 10,000 $g$ for 15 min. The precipitate is dissolved in 50 m$M$ Tris, pH 8.0. (The buffer volume should be about 10% of the volume of the dialyzed methanol preparation.) This solution is then dialyzed against 10 m$M$ Tris, pH 7.8, 0.5 m$M$ EDTA, 1.0 mM Fru-1,6-P$_2$ (calcium salt), and 80 m$M$ NaCl.

*DEAE-Sephadex Step.* Before use DEAE-Sephadex was washed with 0.5 $N$ NaOH, 0.5 $N$ HCl, and distilled water. The resin was poured at room temperature into a 4.5 × 67 column, and then allowed to cool at 6°. The resin was then washed with 10 m$M$ Tris, pH 7.8, 0.5 m$M$ EDTA, 1 mM Fru-1,6-P$_2$ (calcium salt), and 80 m$M$ NaCl until the pH of the effluent was 7.8.

Before being placed on the column the dialyzed enzyme was incubated at 37° for 60 min and then centrifuged at 10,000 $g$ for 10 min. This step is necessary in order to keep materials from precipitating during the time the enzyme was on the column.

The sample is placed on the column, and washed into the column with 150 ml of the buffer with which the column is equilibrated. When the buffer level reaches the top of the column bed, an additional 200 ml of buffer is added. The sample is then eluted with 4 liters of a 80 m$M$ to 0.13 $M$ NaCl linear gradient made up in 10 m$M$ Tris, pH 7.8, 0.5 m$M$ EDTA, and 1 m$M$ Fru-1,6-P$_2$ (Ca$^{2+}$ salt). The height of the 4-liter reservoir is adjusted to give the smallest hydrostatic pressure that will allow flow of liquid through the column, since higher hydrostatic pressure decreased the flow rate due to compression of the DEAE-Sephadex. The eluent is collected in 20-ml samples. Fructose-1,6-diphosphatase is eluted from the column at concentrations of NaCl between 0.11 $M$ and 0.12 $M$. About 45% of the protein put on the column comes off before the fructose-1,6-diphosphatase peak. Approximately 4% to 5% of the total protein is present in those tubes which contain fructose-1,6-diphosphatase activity. The tubes having a specific activity of about 3.3 units/ml or higher are combined and dialyzed against 10 m$M$ Tris, pH 7.3, 0.1 m$M$ EDTA, 70 m$M$ NaCl, and 1 m$M$ MgSO$_4$.

*Cellex-P Step.* Cellex-P (phosphocellulose from Bio-Rad Laboratories) was washed with 0.5 $M$ NaOH and distilled water, poured into a 4.5 × 90 cm column, and equilibrated with 10 m$M$ Tris, pH 7.3, 0.1 m$M$ EDTA, and 70 m$M$ NaCl. For optimal results only 0.45 mg of protein may be placed on the column per milliliter of column bed.

Just prior to placing the enzyme on the column, enough 0.1 $M$ EDTA is added to the dialyzed sample to make the final concentration 2 m$M$, thus making the Mg$^{2+}$ unavailable to the enzyme. The presence of 1.0 m$M$ Mg$^{2+}$ during dialysis is required to ensure that none of the Fru-1,6-P$_2$ used in the DEAE-Sephadex procedure remains. It is important to eliminate the Fru-1,6-P$_2$ since it influences the binding of the enzyme to the column, and, as indicated below, Fru-1,6-P$_2$ is used as a specific eluting agent.

After the enzyme is placed on the column, the column is washed with 2500 ml of 10 m$M$ Tris, pH 7.3, 0.1 m$M$ EDTA, and 70 m$M$ NaCl. About 14% of the protein and none of the fructose-1,6-diphosphatase comes off in this wash solution. Then the column is connected to a reservoir containing 2000 ml of 10 mM Tris, pH 7.3, 0.1 m$M$ EDTA, 70 m$M$ NaCl, 1 m$M$ Fru-1,6-P$_2$ (Ca$^{2+}$ salt), and 1 m$M$ AMP. Fructose-1,6-diphosphatase is eluted from the column in a sharp symmetrical peak just after the void volume; this peak contains about 16% of the protein put onto the column. The tubes containing fructose-1,6-diphosphatase activity are combined, lyophilized, and then taken up in 10 mM Tris, pH 8.0, 3 m$M$ MgSO$_4$, and 1 m$M$ EDTA. The sample is then dialyzed against this same buffer and stored at $-15°$.

## Properties

*Molecular Weight.* The molecular weight of bovine liver fructose 1,6-diphosphatase as determined by sedimentation equilibrium and sedimentation velocity is approximately 130,000.[1] The sedimentation coefficient is concentration-dependent, giving an extrapolated value of 7.61 S for $s_{20,w}$.

The carboxy-terminal amino acid has been found to be lysine. No amino terminal residues have been detected. Purified enzyme gives a single band on gel electrophoresis.[8]

*Kinetics.* Bovine fructose-1,6-diphosphatase has a $K_m$ for Fru-1,6-P$_2$ of about 1.7 $\mu M$ (at 5 m$M$ MgSO$_4$).[3] High concentrations of Fru-1,6-P$_2$

are inhibitory. Maximal activity is observed at about 10 $\mu M$, and higher concentrations inhibit. A maximum of 50% inhibition is observed. The enzyme also requires a metal cofactor, either magnesium or manganese. The concentration of metal ion required for half-maximal activity at pH 6.5 is 0.26 m$M$ for magnesium and 30 $\mu M$ for manganese. High concentrations of metal cofactor are inhibitory.

$\beta$-Glycerophosphate is a substrate for bovine fructose-1,6-diphosphatase at pH 9.0, with a $K_m$ of 10 m$M$, but it is not a substrate at pH 6.5.[4]

5'-AMP is a specific allosteric inhibitor of fructose-1,6-diphosphatase, with a $K_i$ of 60 $\mu M$. The $K_i$ is temperature dependent, decreasing with decreasing temperature.

*pH Optimum.* The pH optimum of bovine hepatic fructose-1,6-diphosphatase is between pH 6.5 and 7.5. The shape of the pH-activity curve is dependent upon the concentration of both the substrate, Fru-1,6-P$_2$, and the metal cofactor.[3]

*Activators and Inhibitors.* Proteolytic enzymes such as Nagarse, papain, and Pronase stimulate activity 2- to 3-fold at pH 9.0, with a simultaneous loss of activity at pH 6.5.[7] Reagents such as $N$-ethylmaleimide, 1-fluoro-2,4-dinitrobenzene, 5,5'-dithiobis-(2-nitrobenzoic acid), and $p$-mercuribenzoate have similar, although not identical, effects on activity. After incubation with these latter reagents, fructose-1,6-diphosphatase activity is inhibited up to 90% at neutral pH (pH 6.5–7.0) and slightly increased at alkaline pH (pH 9.0), when magnesium is the metal cofactor in the assay. However, when magnanese is used as the metal cofactor, activity is increased up to 3-fold at alkaline pH, with less activation at neutral pH. Iodoacetate inhibits at both neutral and alkaline pH.[2] Glutaraldehyde treatment inhibits activity at both pH 6.5 and 9.0, shifts the pH optimum to about pH 9.0, and desensitizes the enzyme to inhibition by 5'-AMP and Fru-1,6-P$_2$.[10]

*Stability.* Concentrated solutions of the enzyme stored at pH 7.5–8.0 in 10 m$M$ Tris, 1.0 m$M$ EDTA, 3.0 MgSO$_4$, or 20 m$M$ $\beta$-glycerophosphate–1.0 m$M$ Fru-1,6-P$_2$ may be stored for many months at −20°. Bovine fructose 1,6-diphosphatase is unstable after incubation at 50°, losing about 50% of its activity within 15 min. Fru-1,6-P$_2$, Mg$^{2+}$, and Mn$^{2+}$ protect against the loss of activity, while 5'-AMP has no protective effect.

[10] V. J. Aloyo, A. M. Geller, C. J. Marcus, and W. L. Byrne, *Biochem. Biophys. Acta* **289**, 242 (1972).

## [60] Fructose-1,6-diphosphatase from Rabbit Liver

By Edgar H. Ulm, Burton M. Pogell, Margaret M. deMaine,
Carol B. Libby, and Stephen J. Benkovic

Fructose 1,6-diphosphate + H$_2$O → D-fructose 6-phosphate + P$_i$

Other methods of purification of rabbit liver fructose-1,6-diphosphatase (FDPase) have yielded products which have been subjected to various degrees of proteolysis.[1,2] This modification[3] of purification by substrate elution from CM-cellulose[4] yields enzyme from frozen liver with properties of unproteolyzed enzyme, i.e., high specific activity in the presence of EDTA at neutral pH compared with that at pH 9.3.

### Assay Method

*Principle.* The production of fructose 6-phosphate is coupled to the reduction of NADP by glucose-6-P isomerase and glucose-6-P dehydrogenase.[5] The increase in absorbance at 340 nm in this assay can be used at all stages of purification. Measurement of P$_i$ release can be used as an alternative assay.[5,6]

*Reagents.* All enzyme reagents were adjusted to neutrality (or pH indicated) with NaOH when necessary.

Tris·HCl, 0.1 $M$, pH 7.5
> *or*

Glycine, 0.2 $M$, pH 9.6
MgSO$_4$, 0.1 $M$
KCl, 3.0 $M$
EDTA, 10 m$M$
NADP, 15 m$M$
Fructose-1,6-P$_2$, 10 m$M$
Glucose-6-P isomerase, 700 units/ml, and glucose-6-P dehydrogenase, 140 units/ml, dialyzed twice against 100 volumes of 0.3 $M$ ammonium sulfate (pH 7) (auxiliary enzymes)

[1] S. Traniello, S. Pontremoli, Y. Tashima, and B. L. Horecker, *Arch. Biochem. Biophys.* **146**, 161 (1971).
[2] M. G. Sarngadharan and B. M. Pogell, *Biochem. Biophys. Res. Commun.* **46**, 1247 (1972).
[3] C. W. Carlson, R. C. Baxter, E. H. Ulm, and B. M. Pogell, *J. Biol. Chem.* **248**, 5555 (1973).
[4] B. M. Pogell, and M. G. Sarngadharan, this series, Vol. 22 [32].
[5] K. Taketa and B. M. Pogell, *J. Biol. Chem.* **240**, 651 (1965).
[6] B. N. Ames, this series, Vol. 8 [10]

*Procedure.* The assay is performed at 30° in a thermostated spectrophotometer. In a 0.5-ml cuvette, place 0.25 ml of buffer, 0.025 ml of $MgSO_4$, 0.025 ml of KCl, 0.005 ml of EDTA, 0.005 ml of NADP, 0.005 ml of auxiliary enzymes, FDPase, and water to a total volume of 0.495 ml. After 5 min incubation minus substrate, start the reaction by adding 0.005 ml FDP.

*Definition of Unit and Specific Activity.* One unit of FDPase activity is defined as that quantity of enzyme which results in the reduction of 1 μmole of NADP per minute at 30° in the coupled assay. Specific activity is defined as units per milligram of protein. In crude fractions, protein is determined by the Lowry method[7] using bovine serum albumin as standard. Pure protein is determined from the absorbance at 280 nm, assuming an absorbance of 0.890 for a 0.1% solution.[8]

## Purification Procedure

All procedures are performed at 0–4° unless otherwise noted.

*Homogenization.* Individually wrapped and frozen livers can be obtained from Pel Freez Inc., Rogers, Arkansas (type I, from young, 20–24 hr fasted rabbits, average weight, 52 g). The frozen livers (400 g) are broken into small pieces and added, still frozen, to a Waring Blendor (1-gallon capacity) containing 1400 ml of fresh 5 m$M$ $NaHCO_3$, pH 8.1. Homogenization is carried out at high speed (19,000 rpm) in two 1-min periods, separated by 0.5 min. The homogenate is then strained through 4 layers of gauze.

*Heat Treatment.* The homogenate is heated as rapidly as possible to 65° and held at that temperature for 10 min. By pumping at 500 ml/min (Masterflex Pump, Cole-Parmer) through a stainless steel coil (outside diameter 0.0375 inch, wall 0.028 inch, length 215 inch) immersed in a 70° water bath (Forma), we can heat the extract from 5° to 65° in about 35 sec. The coil is then rinsed with 100 ml of 5 m$M$ $NaHCO_3$. After 10 min at 65°, the mixture can be cooled by pumping it through the coil immersed in an ice-water bath. The coil is again rinsed with 100 ml $NaHCO_3$ solution. Overall the ratio (w/v) of liver to total buffer is 1:4 at the end of this step.

The mixture is centrifuged at 13,000 $g$ for 25 min. The supernatant has a pH of 7 and can be stored frozen at this stage.

*Dialysis and pH Adjustment.* The heated supernatant is dialyzed against deionized distilled water until the tissue $Cl^-$ concentration in a perchloric acid supernatant of the retentate is less than 0.1 m$M$ (mea-

[7] E. Layne, this series, Vol. 3 [73].
[8] S. Pontremoli, this series, Vol. 9 [112a].

sured with $AgNO_3$). We have employed countercurrent dialysis for this step (Omnivector continuous countercurrent dialyzer, Model CD-16, distributed by Bio-Rad Laboratories). The pH of the retentate is adjusted to 6.0 by the addition of glacial acetic acid and the preparation is allowed to stand with occasional stirring for 30 min. Any precipitate is removed by centrifugation at 13,000 $g$ for 40 min.

*Adsorption to CM-Cellulose.* Two hundred grams, wet weight, of CM-cellulose (preswollen, CM-52, Whatman catalog No. 22522) is suspended in 1 liter of water, and the pH is adjusted to 6.0 by addition of 1 $M$ malonic acid. After settling, the fines are decanted and the CM-52 is resuspended in 5 m$M$ sodium malonate, pH 6.0. The settling and defining are repeated, and the cellulose is collected in a coarse glass Büchner funnel. After equilibration with the malonate buffer by five additional washes in the funnel, the CM-52 is suspended to a volume of 1 liter of buffer. Settled gel volume is approximately 370 ml/liter slurry.

The dialyzed liver supernatant is added to a continuously stirred slurry of CM-cellulose [ca. 10 ml of supernatant (15 units of enzyme) per milliliter of settled gel = 0.6 g original wet weight of gel]. We use a peristaltic pump for this addition, which takes several hours, and always stir the gel by means of a motor-driven blade. Supernatant is added until ca. 5% of the total activity remains unbound.

*Washing and Elution of the CM-Cellulose.* The gel with enzyme bound is washed batchwise 3 or 4 times with 5 m$M$ malonate, pH 6.0 (wash volume = 3–4 times the settled gel volume). Most of the wash supernatant is decanted through a coarse glass Büchner funnel, and the remainder is removed by collecting the cake on the Büchner under vacuum. The gel is resuspended as above in malonate and the pH is raised to 6.8 by the addition of 2 $M$ NaOH. After the gel is stirred overnight, it is washed extensively with 5 m$M$ malonate, pH 6.8, until the absorbance at 280 nm is less than 0.02 and the pH of the wash is identical to the wash buffer.

The gel is suspended in pH 6.8 buffer and poured into a column (8 cm, i.d.). The bed is consolidated by passage of several column volumes of buffer after which FDPase is eluted in a single peak of protein by inclusion of a fresh dilution of 0.06 m$M$ FDP in the 5 m$M$ sodium malonate, pH 6.8 (Fig. 1). Fractions with $A_{280}$ greater than 0.1 and specific activities greater than 20 are pooled. Usually a long trail of activity at low protein concentration elutes after the main peak of activity. Pooled fractions 13–21 had a specific activity of 28.

The pooled fractions are adjusted to pH 6.0 with 1.0 $M$ malonic acid. The enzyme is concentrated either by lyophilization and subsequent reconstitution with water, by ultrafiltration (Amicon Diaflo XM-50), or

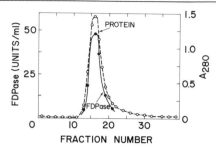

Fig. 1. Substrate elution of FDPase from CM-cellulose. See text for details. Fraction volumes were 5 ml. ●——●, Enzyme activity per milliliter; ○---○, $A_{280}$.

by precipitation with saturated ammonium sulfate. Before use, FDP is removed by incubation of the concentrated enzyme with 10 m$M$ MgSO$_4$ at room temperature for 30 min, followed by dialysis against 5 m$M$ malonate, pH 6.0.[9] Concentrated stocks (5–12 mg/ml) are stored frozen at —65°.

Typical results obtained during a purification are summarized in the table.

*Comments.* Dilutions of enzyme from concentrated stock solutions in 0.5 $M$ KCl (containing 50 m$M$ Tris·HCl (pH 7.2) and either 0.1 m$M$ MnCl$_2$ or 5 m$M$ MgSO$_4$) and assays in the presence of 0.15 $M$ KCl have been found to give the most reproducible results, increasing the specific activity as much as 2-fold.[3,10] Incubation of diluted enzyme for at least 5 min in the final assay mixture before addition of FDP is also necessary for maximal initial rates of activity. Some interference has been noted in attempts to measure the low activities of FDPase found in the supernatant after adsorption with CM-cellulose. The rates start very low and then slowly increase.

To obtain enzyme with minimum evidence of proteolysis, it is absolutely essential to use frozen rabbit livers from *young* animals. The initial heated extracts from mature frozen rabbit liver had much lower pH 7.3 to pH 9.3 activity ratios, presumably because of more labile lysosomes in the older animals. A similar variation in the properties of isolated liver FDPase because of seasonal changes in liver lysosome stability has been discussed in detail.[11]

[9] K. Taketa, A. Watanabe, M. G. Sarngadharan, and B. M. Pogell, *Biochemistry* **10**, 565 (1971).

[10] E. H. Ulm, R. C. Baxter, C. W. Carlson, and B. M. Pogell, unpublished observations.

[11] S. Pontremoli, E. Melloni, F. Balestrero, A. T. Franzi, A. de Flora, and B. L. Horecker, *Proc. Nat. Acad. Sci. U.S.* **70**, 303 (1973).

PURIFICATION OF FRUCTOSE-1,6-DIPHOSPHATASE FROM RABBIT LIVER

| Step | Volume (ml) | Total protein (mg) | Total activity (units) | Yield (%) | Specific activity (units/mg protein) |
|---|---|---|---|---|---|
| Crude homogenate | 1450 | — | — | — | 0.06 |
| Heated supernatant, thawed | 1000 | 5500 | 1930 | 100 | 0.4 |
| pH 6.0 supernatant | 1225 | 4050 | 1520 | 79 | 0.4 |
| FDP eluted peak[a] | 46 | 36 | 1019 | 53 | 28. |

[a] Yield from the CM-cellulose peak can be increased by pooling fractions eluting after the major peak.

Precipitation of a large portion of liver protein during the heating step further increases the amount of FDPase which is bound per unit weight of CM-cellulose, and elution with substrate at pH 6.8 now gives a single peak of pure enzyme.

We had some success in preparations when dialysis of the heated supernatant was omitted,[3] but more reproducible elution and better recovery of FDPase from CM-cellulose is found with dialysis included. Part of the problem in preparations without dialysis was traced to hydrolysis of FDP during the CM-cellulose elution procedure, apparently because enough $Mg^{2+}$ and $Mn^{2+}$ was bound to the cellulose to permit FDPase activity.

Higher overall yields (up to 80%) have been attained by inserting the pH adjustment step of the heated supernatant before dialysis. No advantage in enzyme purification was found with dialyzed, heated liver supernatant if a layer of unsaturated CM-cellulose was included below the FDPase-saturated CM-cellulose.[3]

## Properties

Purification of FDPase from frozen rabbit livers without any apparent proteolysis is possible by first homogenizing and heating rapidly in bicarbonate buffer. There was no evidence for any enzymic activity which modified the enzyme after the heating step.[3] FDPase prepared by this method has a ratio of activity measured at pH 7.3 to activity at pH 9.3 in excess of 3.5 with a specific activity at pH 7.3 and 30° ranging from 28 to 35. It elutes as a single symmetrical peak from Sephadex G-200 and yields one principal polypeptide band on SDS gel electrophoresis[12]

[12] K. Weber and M. Osborn, J. Biol. Chem. 244, 4406 (1969).

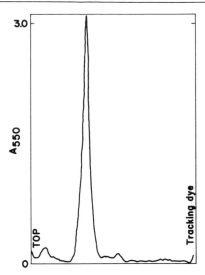

Fig. 2. Spectrophotometer scan of SDS gel electrophoresis. FDPase (after G-200 gel chromatography) was dissociated with SDS; ca. 20 $\mu$g of treated protein was electrophoresed on SDS containing, 10% acrylamide gels [K. Weber and M. Osborn, *J. Biol. Chem.* **244**, 4406 (1969)]; and the gel was fixed and stained with Coomassie blue [G. Fairbanks, T. L. Steck, and D. F. H. Wallach, *Biochemistry* **10**, 2606 (1971)]. The tracing of the spectrophotometer scan at 550 nm indicates the protein was 90% in the form of a single polypeptide.

after staining with Coomassie blue[13] (Fig. 2). The enzyme has a neutral pH optimum and is much more sensitive to inhibition by AMP [$K_i$ (cooperative) of 15–20 $\mu M$ at 30° and pH 7.0 in the presence of EDTA].[3] Fatty acids, preferably oleate,[3] and an uncharacterized protein factor[10] can replace EDTA as an essential activator at neutral pH for homogeneous enzyme. At neutral pH, the kinetic $K_{FDP}$, 3–9 $\mu M$, remains the same as found with partially proteolyzed enzyme[14] and is unaltered in the presence of effectors such as AMP and oleate.[10]

## Acknowledgments

We wish to thank Robert C. Baxter, Charles W. Carlson, Joseph J. Villafranca, and William A. Frey for their advice and contributions during the development of this procedure.

[13] G. Fairbanks, T. L. Steck, and D. F. H. Wallach, *Biochemistry* **10**, 2606 (1971).
[14] H. Aoe, M. G. Sarngadharan, and B. M. Pogell, *J. Biol. Chem.* **245**, 6383 (1970).

# [61] Fructose-1,6-diphosphatase, Phosphofructokinase, Glycogen Synthetase, Phosphorylase, and Protein Kinase from Swine Kidney

*By* Joseph Mendicino, Hussein Abou-Issa, Rudolf Medicus, and Nancy Kratowich

Enzymes that catalyze physiologically unidirectional reactions in the glycolytic and gluconeogenic multienzyme systems in kidney were isolated from the same homogenate.[1-5] The reactions catalyzed by these enzymes are shown in Eq. (1)–(5).

$$\text{Glycogen}_{(n)} + \text{UDP-D-glucose} \rightarrow \text{glycogen}_{(n+1)} + \text{UDP} \qquad (1)$$

$$\text{D-Fructose-6-P} + \text{ATP} \rightarrow \text{D-fructose-1,6-P}_2 + \text{ADP} \qquad (2)$$

$$\text{D-Fructose-1-6-P}_2 + \text{H}_2\text{O} \rightarrow \text{fructose-6-P} + \text{P}_i \qquad (3)$$

$$\text{Glycogen}_{(n)} + \text{P}_i \rightarrow \text{glycogen}_{(n-1)} + \text{D-glucose-1-P} \qquad (4)$$

$$
\begin{array}{ccc}
\text{Dephosphoglycogen synthetase} & & \text{phosphoglycogen synthetase} \\
or & + \text{ATP} \rightarrow & or & + \text{ADP} \\
\text{dephosphohistone} & & \text{phosphohistone}
\end{array}
$$

$$(5)$$

The methods used for the sequential separation of these five enzymes from the same swine kidney homogenate are based on a comprehensive procedure developed to obtain all the enzymes with the best possible yields, in each case. The method was developed for the large-scale preparation of homogeneous enzymes in order to use these preparations as substrates to study the mechanisms involved in the regulation of their activities in kidney. Unless otherwise stated, all the procedures are carried out at 3° and centrifugations are for 15 min at 27,000 $g$.

## I. Isolation and Properties of Glycogen Synthetase

### Assay Method

The activity of the enzyme in crude extracts and partially purified preparations is measured by determining the rate of transfer of [14C]glu-

[1] J. Mendicino and N. Kratowich, *J. Biol. Chem.* **247**, 6643 (1972).
[2] H. Abou-Issa, and J. Mendicino, *J. Biol. Chem.* **248**, 685 (1973).
[3] H. Abou-Issa, N. Kratowich, and J. Mendicino, *Eur. J. Biochem.* **42**, 461 (1974).
[4] R. Medicus and J. Mendicino, *Eur. J. Biochem.,* **40**, 63 (1973).
[5] T. Shih, and J. Mendicino, manuscript in preparation.

cose from UDP-[$^{14}$C]glucose to glycogen.[6,7] With purified preparations a spectrophotometric assay based on the determination of UDP formed in the reaction was also used.

*Reagents*

Tris·HCl, 1.0 $M$, pII 7.5
Cysteine, 0.1 $M$, pH 7.5
UDP-[$^{14}$C]glucose, 50 m$M$ (50,000 cpm/$\mu$mole)
Glucose-6-P, 0.1 $M$
Rabbit liver glycogen (100 mg/ml)
UDP-glucose, 50 m$M$
NADH, 10 m$M$
KCl, 1.0 $V$
P-enolpyruvate, 0.1 $M$
MgCl$_2$, 0.2 $M$
Pyruvic kinase, 133 units/ml
Lactic dehydrogenase, 80 unis/ml
Standard buffer routinely used to dissolve, dilute, dialyze, and store glycogen synthetase contain 50 m$M$ Tris·HCl, pH 7.5–0.3 $M$ sucrose–20 m$M$ 2-mercaptoethanol, and 5 m$M$ EDTA

*Procedure.* One-tenth milliliter of cysteine, 0.04 ml of Tris·HCl, 0.03 ml of glycogen and 0.17 ml of enzyme plus water are added to a small test tube. The incubation mixture is placed in a water bath at 37° for 1 min, and the reaction is initiated by the addition of 0.01 ml of UDP-[$^{14}$C]glucose; the mixture is incubated for 10 min at 37°. The reaction is terminated by the addition of 1 ml of 10% trichloroacetic acid and the precipitated protein is removed by centrifugation. Glycogen is precipitated from the supernatant solution by the addition of 4 ml of 95% ethanol. The turbid solution is centrifuged and the pellet is dissolved in 1 ml of water. The glycogen is washed two more times by this procedure, and the final pellet is dissolved in 1 ml of water, transferred to a scintillation vial, and counted.[8] The amount of glucose transferred is determined by dividing the total counts incorporated by the specific activity of the UDP[$^{14}$C]-glucose. When the total glycogen synthetase activity is being measured, 0.04 ml of glucose 6-phosphate is added to the reaction mixture.

[6] L. F. Leloir, J. M. Olavarria, S. H. Goldemberg, and H. Carminatti, *Arch. Biochem. Biophys.* **81,** 508 (1959).
[7] J. Mendicino and M. Pinjani, *Biochim. Biophys. Acta* **89,** 242 (1964).
[8] M. D. Patterson, and R. C. Green, *Anal. Chem.* **37,** 854 (1965).

In the spectrophotometric assay mixture, 0.05 ml of Tris·HCl, 0.3 ml of cysteine, 0.1 ml of glycogen, 0.01 ml of NADH, 0.05 ml of $MgCl_2$, 0.08 ml of KCl, 0.02 ml of P-enolpyruvate, 0.02 ml of pyruvic kinse, 0.02 ml of lactic dehydrogenase, and 0.32 ml of enzyme plus water are added to a cuvette at room temperature. The reaction is started by the addition of 0.03 ml of UDP-glucose and the rate of oxidation of NADH is measured at 340 nm.

The same activities are obtained by either assay, when a correction is made for the difference in temperature. One unit of activity is defined as the amount of enzyme which catalyzes the formation of 1 $\mu$mole of UDP or the transfer of 1 $\mu$mole of glucose from UDP-glucose to glycogen per minute, and specific activity is expressed as units per milligram of protein at 37°. The concentration of protein is measured by the method of Lowry et al.[9] and by the biuret procedure[10] with crystalline bovine serum albumin as a standard.


## Purification Procedure

*Step 1.* Kidneys are removed from animals soon after slaughter, and they are cleaned and immediately frozen at −20° until needed. A crude extract is prepared by homogenizing 1000 g of partially thawed kidneys for 2 min in a Waring Blendor at medium speed with 1.5 liters of 50 m$M$ Tris·HCl, pH 8.0–50 m$M$ 2-mercaptoethanol–0.1 m$M$ EDTA. The disruption of the tissue is carried out in the presence of 1 m$M$ of the proteolytic inhibitor, phenylmethylsulfonyl fluoride (Sigma Chemical Company, St. Louis, Missouri), in order to avoid the possibility of a partial hydrolysis of some of the enzymes. The thick homogenate is centrifuged, and the precipitate is collected and homogenized again with 1 liter of the same buffer. This suspension is centrifuged and the two supernatant solutions are combined and filtered through cheesecloth.

*Step 2.* The crude extract is stirred at 3° and 150 ml of alumina Cγ[11] (0.5 mg dry weight of gel per milligram of protein) is added. The thick suspension is centrifuged, and the precipitate, which is practically devoid of activity, is discarded. This step does not yield a large purification; however, it greatly reduces the amount of calcium phosphate gel which is required to absorb all the glycogen synthetase in the subsequent step.

[9] O. H. Lowry, N. J. Rosebrough, A. L. Farr, and R. J. Randall, *J. Biol. Chem.* **193**, 265 (1951).

[10] H. W. Robison and C. G. Hogden, *J. Biol. Chem.* **135**, 707 (1950).

[11] Amphojel (aluminum hydroxide gel) was obtained from Wyeth Laboratories Inc., Philadelphia, Pennsylvania.

Furthermore, it also improves the reproducibility of the calcium phosphate gel elution procedure.

*Step 3.* The supernatant solution from the previous step which usually contains about 30–35 mg of protein per milliliter is stirred at 3°, and a slurry containing calcium phosphate gel (50 mg dry weight per milliliter) is slowly added. The gel is prepared by the method of Keilin and Hartree.[12] About 200 ml of the suspension of aged gel is usually added to obtain a final ratio of about 0.2 mg of gel per milligram of protein. Since the adsorptive capacity of each batch of gel varies somewhat, the dry weight and age of the gel are not used as a basis for determining the amount of gel to be added. Instead, the amount of gel required to absorb about 80% of the glycogen synthetase is routinely added. If more than this amount of gel is used, a large amount of phosphofructokinase is also absorbed to the gel. The calcium phosphate gel is collected by centrifugation, and this precipitate is washed by resuspending it in 600 ml of ice-cold 50 m$M$ Tris·HCl, pH 7.5, and centrifuging the resulting slurry. The supernatant solution which is devoid of glycogen synthetase activity is discarded. The enzyme is then eluted from the washed gel by two successive extractions with 300 ml of 20 m$M$ potassium citrate. The gel is removed by centrifugation after each extraction, and the supernatant solutions are combined and 29.1 g/100 ml of solid ammonium sulfate is added at 3°. The precipitate is collected by centrifugation, and it is dissolved in 250 ml of the standard buffer. The solution is dialyzed against several changes of 2 liters of the same buffer for 3 hr.

*Step 4.* The pH of the dialyzed solution is adjusted to 5.8 by the careful addition of 1 $N$ acetic acid. The resulting turbid solution is centrifuged, and almost all the phosphorylase kinase present in crude kidney extracts is removed in this precipitate. The pH of the supernatant solution was then adjusted to 4.9 by the addition of more 1 $N$ acetic acid. This precipitate which contains the glycogen synthetase, is collected by centrifugation and it is dissolved in 95 ml of the standard buffer.

*Step 5.* The ratio of the activity of the enzyme in the presence of glucose 6-P to the activity in the absence of this activator is about 2.8 to 3.2 in all the preparations up to this stage of purification, which indicates that the two interconvertible forms of glycogen synthetase are probably both present in the crude extracts. The glycogen synthetase in this partially purified preparation can be converted to a form which is completely dependent on the presence of glucose-6-P for activity by incubating the enzyme with a kidney particulate protein kinase preparation.[2,3] When this form of the enzyme is needed, the dialyzed enzyme solution from the previous step is incubated for 15 min at 37° with 10 mg of washed

[12] D. Keilin and E. F. Hartree, *Proc. Roy. Soc. Ser. B* **124**, 397 (1938).

kidney particulates,[2] 10 mM ATP, 10 mM $MgCl_2$, 250 mM sucrose, and 10 mM cysteine. The particles are removed by centrifugation at 34,000 $g$ for 10 min after incubation. As shown in Table I, this procedure changes the ratio of the activity in the presence of glucose-6-P to the activity in the absence of this cofactor from about 3 to greater than 95. Essentially all the glycogen synthetase activity is usually recovered in this step. When this treatment is omitted from the purification procedure the ratio of the final homogeneous enzyme is 3, and the same yield and final specific activity in the presence of glucose-6-P is obtained.

*Step 6.* Four preparations taken through the previous step are combined (200 units, 5200 mg of protein) and the solution is diluted to 1000 ml with the standard buffer. The solution is cooled to 0° and 175 ml of 95% ethanol at −20° is added with stirring to obtain a final ethanol concentration of 13%. The resulting precipitate which is removed by centrifugation at −10° is discarded. Glycogen is added to the supernatant solution to a final concentration of 7.5 mg/ml. More ethanol, 170 ml, is then added with constant stirring to a final concentration of 30%. The fine precipitate which forms is collected by centrifugation at −10° and it is dissolved in 500 ml of the standard buffer.

*Step 7.* This solution is diluted to 1000 ml with buffer, and 250 ml of 95% ethanol at −20° is slowly added with stirring. The precipitate is again collected by centrifugation at −10° and it is dissolved in 320 ml of buffer and dialyzed against three changes of 2 liters of the standard buffer for 3 hr. The basis for the purification in steps 6 and 7 is to dilute the protein concentration to a level where it will not precipitate with ethanol. Then only those proteins which bind to glycogen would be removed from solution upon addition of ethanol, since the polysaccharide will still be very insoluble under these conditions.

*Step 8.* The enzyme is adsorbed to a DEAE-cellulose column (2.2 × 5 cm) which was previously equilibrated with the standard buffer. The column is then washed with about 300 ml of a solution containing 0.15 M Tris·HCl, pH 7.5–4 mM EDTA–0.3 M sucrose–20 mM 2-mercaptoethanol until the absorption at 280 nm reaches the same value as that observed with the solvent alone. The enzyme is then eluted with a solution of Tris·HCl, pH 7.5 formed as a linear gradient from 100 ml of the above buffer in the mixing chamber, and 100 ml of the buffer with 0.3 M Tris·HCl, pH 7.5, in the reservoir. The glycogen synthetase activity is found in a single protein peak which emerges from the column after about 40 ml of the eluting solution had been collected. The fractions containing significant activity are combined and placed in dialysis tubing. The solution is concentrated by blowing air at room temperature over the dialysis tubing. The enzyme is then dialyzed against four changes of 100-ml volumes of the standard buffer. The values in Table I after step 5 are divided by 4 because four preparations from previous steps were combined in this procedure. The same specific activity is obtained when the preparation from step 8 is subjected to chromatography on Bio-Gel A1.5 or Sepharose 6B columns.

*Step 9.* The enzyme from the previous step is adsorbed to a hydroxyapatite column (2.2 × 4 cm) which was previously equilibrated against the standard buffer and the column is washed with this buffer until the absorption at 280 nm reaches the same value as that obtained with the buffer alone. The enzyme is then eluted with 200 ml of 50 m$M$ potassium phosphate, pH 7.5 and this solution is dialyzed against saturated ammonium sulfate made up in the standard buffer at 3°. The fine precipitate which forms is dissolved in 5 ml of the standard buffer and dialyzed against 100 ml of the same buffer for 3 hr. The enzyme was essentially pure at this stage and a final specific activity of 15 $\mu$moles per minute per milligram of protein is obtained with a yield of 15–30% in different preparations.

A typical purification is summarized in Table I.

## Properties

The specific activities of the final purified preparations varied from 9 to 15 units/mg; however, only a single band was observed on electrophoresis in 1% sodium dodecyl sulfate with these preparations. The purified enzyme was stable for at least 2 months when frozen at −20° in a buffer containing 5 m$M$ Tris·HCl, pH 7.5–0.3 $M$ sucrose–20 m$M$ 2-mercaptoethanol, and 5 m$M$ EDTA. Maximum stability is observed in the presence of both 0.3 $M$ sucrose and 20 m$M$ 2-mercaptoethanol, and it is necessary to keep the enzyme dissolved in a solution containing these compounds at all stages of purification.

The molecular weight of both the phospho and dephospho forms of kidney glycogen synthetase measured by several different physical methods is 370,000 ± 10,000. Both forms of the enzyme contains 4 identical subunits with molecular weights of 92,000 ± 3000. The enzymes have a Stokes radius of 63 Å and diffusion constants of 3.4 × 10$^{-7}$ cm$^2$/sec.

The $K_m$ of the form of the enzyme that is dependent on glucose-6-P for activity is 0.11 m$M$ for UDP-glucose at 1 m$M$ glucose-6-P and 67 $\mu M$ at 11 m$M$ glucose-6-P. The $K_m$ of the form of the enzyme which is not dependent on glucose-6-P activity is 0.87 m$M$ for UDP-glucose. The two forms of the enzyme can be interconverted by protein kinases and phosphoprotein phosphatases which have been isolated from swine kidney extracts.[2,3]

TABLE I

PURIFICATION PROCEDURE FOR GLYCOGEN SYNTHETASE

| Step | Volume (ml) | Protein (mg/ml) | Total activity (units) | Specific activity (units/mg) | Recovery (%) | Ratio, +glucose-6-P: −glucose-6-P |
|---|---|---|---|---|---|---|
| 1. Crude extract | 2500 | 52 | 106 | 0.0008 | 100 | 3.2 |
| 2. Treatment with alumina Cγ gel | 2700 | | 95 | 0.0010 | 90 | 3.1 |
| 3. Calcium phosphate gel eluate | 600 | 14 | 66 | 0.0078 | 62 | 3.2 |
| 4. Acid precipitation, pH 5.8 and pH 4.9 | 100 | 13.8 | 51 | 0.037 | 48 | 2.8 |
| 5. Treatment with particulate protein kinase | 100 | 13 | 51 | 0.04 | 48 | >95 |
| 6. First alcohol precipitation | 125 | 0.84 | 46 | 0.43 | 44 | >95 |
| 7. Second alcohol precipitation | 80 | 0.12 | 42 | 4.4 | 39 | >95 |
| 8. Chromatography on DEAE-cellulose | 5 | 0.55 | 25 | 9.1 | 23 | >95 |
| 9. Elution from hydroxyapatite | 5 | 0.46 | 25 | 15 | 23 | >95 |

## II. Isolation and Properties of Phosphofructokinase

### Assay Methods

The activity of the enzyme is determined by two spectrophotometric assays in a series of coupled reactions which measure the rate of formation of either fructose-1,6-$P_2$ or ADP. The standard assay procedure for following the course of purification of the enzyme was based on the determination of fructose-1,6-$P_2$ with aldolase, triosephosphate isomerase, and glycerophosphate dehydrogenase. In kinetic studies where it is necessary to use small amounts of substrates, another assay, which contains a fructose-6-P and ATP regenerating system is used. In this system the rate of formation of ADP is measured with pyruvate kinase and lactic dehydrogenase, and fructose-1,6-$P_2$ is continuously hydrolyzed to fructose-6-P by fructose-1,6-diphosphatase.[1]

*Reagents*

Tris·HCl, 10 m$M$ pH 8.0
ATP, 0.1 $M$
MgCl$_2$, 0.1 $M$
2-Mercaptoethanol, 0.1 $M$, pH 8.0
NADH, 10 m$M$
Fructose-6-P, 0.1 $M$
KCl, 1.54 $M$
Aldolase, 100 units/ml
Triosephosphate isomerase, 100 units/ml
Glycerophosphate dehydrogenase, 100 units/ml
Standard buffer used to dissolve, dialyze, and dilute the enzyme contained 20 m$M$ Tris·HCl, pH 8.0, 0.1 m$M$ ATP, 10 m$M$ fructose 1,6-$P_2$, 1 m$M$ EDTA, and 50 m$M$ 2-mercaptoethanol

*Procedure.* For routine assay 0.7 ml of Tris·HCl, pH 7.5, 0.02 ml of ATP, 0.01 ml of MgCl$_2$, 0.01 ml of 2-mercaptoethanol, 0.02 ml of NADH, 0.1 ml of KCl, 0.02 ml of aldolase, 0.02 ml of triosephosphate isomerase and glycerophosphate dehydrogenase, and 0.08 ml of phosphofructokinase plus water are added to a 1-ml cuvette at room temperature. Phosphofructokinase is maximally activated by preliminary incubation in 10 m$M$ Tris·HCl, pH 8.0–50 m$M$ 2-mercaptoethanol for 5 min at 30° before it is added to the reaction mixture, and it is diluted in the same solution to give a rate of 0.04 to 0.1 absorbance unit at 340 nm/min. The reaction is initiated by the addition of 0.03 ml of fructose-6-P and the rate of oxidation of NADH at 340 nm is measured.

One unit of activity represents the amount of enzyme which catalyzes the formation of 1 $\mu$mole of fructose-1,6-$P_2$ per minute, and specific activity is expressed as units per milligram of protein.[9,10]

## Purification Procedure

*Steps 1, 2, and 3.* These procedures are the same as those described previously for the isolation of glycogen synthetase.

*Step 4.* After removal of the calcium phosphate gel by centrifugaton the turbid supernatant solution from the previous step was stirred at 3°, and a solution containing 10 g of protamine sulfate (Sigma, Grade II from salmon) in 400 ml of distilled water was slowly added. The thick suspension was stirred for 10 min after the addition of protamine and it was centrifuged. The precipitate was dispersed in 1 liter of 0.2 $M$ potassium phosphate, pH 8.0, containing 5 m$M$ EDTA and 1 m$M$ phenylmethylsulfonyl fluoride. The suspension was stirred while the flask was heated to 60° in a boiling water bath. After the temperature reached 60°, about 5 min, the solution was immediately cooled to 3° in an ice bath. The precipitate was removed by centrifugation and discarded. Phosphofructokinase was precipitated by the addition of 43.6 g/100 ml of solid ammonium sulfate and it was stored at 3°.

*Step 5.* The precipitate from the previous step was collected by centrifugation and it was dissolved in 300 ml of 20 m$M$ Tris·HCl, pH 8.0–1 m$M$ EDTA–0.1 m$M$ ATP–0.01 m$M$ fructose-1,6-$P_2$ 50 m$M$ 2-mercaptoethanol. The solution was passed through a modified Blue Dextran-Sepharose 4B column (16 × 5 cm) which was previously equilibrated against the same buffer. Blue Dextran[13] was covalently attached to Sepharose 4B[13] by the procedure of Cuatrecasas.[14] The column was set up in a Büchner funnel so that a slight vacuum could be applied to regulate the flow rate. The enzyme passed through the column under these conditions, but a large amount of tightly bound protein was removed. The column was washed with 600 ml of the same buffer. The application solution and wash were combined and diluted to 3 liters with distilled water. This solution was passed into a second Blue Dextran-Sepharose 4B column

---

[13] Blue Dextran and Sepharose 4B may be purchased from Pharmacia Chemicals AB, Uppsala, Sweden. Blue Dextran is prepared by attaching a dye, Procion Blue H-B, to dextran which has a molecular weight of about 2 million. The dye in Blue Dextran contains an aromatic amino group which reacts with Sepharose 4B after it has been activated with CNBr.[14] The authors would like to thank Dr. James Travis and Mr. Ralph Pannell of the Department of Biochemistry, University of Georgia for suggesting the use of this procedure and for making large amounts of Blue Dextran–Sepharose 4B available to us.

[14] P. Cuatrecasas, *J. Biol. Chem.* **245**, 3059 (1970).

(16 ×12 cm), and afterward the column was washed with 500 ml of buffer. Phosphofructokinase was then eluted with 2 liters of the buffer solution containing 0.7 $M$ (NH$_4$)$_2$SO$_4$. The eluent was placed in dialysis tubing and it was concentrated by blowing air over the tubing at room temperature. The volume was reduced to 500 ml, and this solution was dialyzed against a saturated solution of ammonium sulfate at 3°. The precipitate was collected by centrifugation, and it was dissolved in 500 ml of buffer and applied to a small Blue Dextran–Sepharose 4B column (5 × 20 cm). The column was washed with 100 ml of buffer and the enzyme was eluted with 700 ml of buffer containing 1 m$M$ ATP, 1 m$M$ fructose 1,6-P$_2$, and 0.3 $M$ (NH$_4$)$_2$SO$_4$. The solution was concentrated and the enzyme was precipitated with ammonium sulfate as described previously.

*Step 6.* The precipitate from the last step was collected by centrifugation, and it was dissolved in 100 ml of the standard buffer. It was dialyzed against 500 ml of this same solution for 30 min, and the solution was then applied to a DEAE-cellulose column (2.5 × 7 cm). The column was washed with 100 ml of buffer, and the enzyme was eluted with 200 ml of the standard buffer containing 0.5 $M$ (NH$_4$)$_2$SO$_4$. The solution was concentrated 2-fold by evaporation, and the enzyme was precipitated by the addition of 43.6 g/100 ml of ammonium sulfate as described previously.

*Step 7.* The precipitated enzyme was collected by centrifugation and it was dissolved in 10 ml of the standard buffer; the solution was applied

TABLE II

PURIFICATION OF PHOSPHOFRUCTOKINASE FROM SWINE KIDNEY

| Step | Total volume (ml) | Protein (mg/ml) | Total activity (units) | Specific activity (units/mg) | Yield (%) |
|---|---|---|---|---|---|
| 1. Crude extract | 4300 | 34 | 2500 | 0.017 | 100 |
| 2. Treatment with alumina C$_\gamma$ gel | 4700 | 20 | 2450 | 0.026 | 98 |
| 3. Treatment with calcium phosphate gel | 5300 | 8 | 2200 | 0.03 | 88 |
| 4. Protamine sulfate precipitation and heat treatment | 1050 | 16 | 1200 | 0.26 | 48 |
| 5. Chromatography on Blue Dextran–Sepharose 4B | 700 | 0.1 | 1070 | 15.0 | 43 |
| 6. Chromatography on DEAE-cellulose | 260 | 0.08 | 950 | 45.0 | 38 |
| 7. Chromatography on Sepharose 4B | 10 | 0.69 | 900 | 130.0 | 36 |

to a Sepharose 6B column $(60 \times 2.3$ cm). The column was previously equilibrated with a solution containing the standard buffer, 0.5 $M$ $(NH_4)_2SO_4$ and 2 m$M$ EDTA. The same solution was used to elute the enzyme which emerged from the column after 80 ml of eluent had been collected. Fractions between 80 and 100 ml which contained a single protein peak were pooled and the solution was concentrated by evaporation and stored at 3° after precipitation with 43.6 g/100 ml of $(NH_4)_2SO_4$.

A typical purification is summarized in Table II.

## Properties

The purified enzyme was homogeneous and stable for 2 months when it was stored in the standard buffer as a precipitate in saturated ammonium sulfate.

The molecular weight of the active species of kidney phosphofructokinase was 380,000. The enzyme readily dimerizes to a species which has a molecular weight of 720,000. Aggregates with much higher molecular weights which are very difficult to dissociate are also formed in dilute salt solutions. The enzyme contains 4 very similar subunits with molecular weights of about 95,000. The $K_m$ of the enzyme for fructose-6-P is 20 $\mu M$ and the $K_m$ for ATP is 8.3 $\mu M$. High concentrations of ATP, 1 m$M$, inhibit the enzyme, and the extent of inhibition is dependent as the concentration of fructose-6-P. The enzyme has a pH optimum at 7.8 under the standard assay conditions.

## III. Isolation and Properties of Fructose-1,6-diphosphatase

### Assay Methods

Three different assays have been used to measure the activity of fructose-1,6-diphosphatase.[15] The formation of fructose-6-P can be determined spectrophotometrically in the presence of NADP and excess phosphohexose isomerase and glucose-6-P dehydrogenase.[15] This is the most reliable assay for fructose-1,6-diphosphatase in crude tissue extracts. Another spectrophotometric assay which contains a fructose-1,6-$P_2$-regenerating system can also be used to avoid the potent inhibition of this enzyme by its substrate.[15] In this procedure the hydrolysis of fructose-1,6-$P_2$ in the presence of excess phosphofructokinase, pyruvic kinase, and lactic dehydrogenase is followed by measuring the rate of oxidation of NADH.[15] The standard assay procedure for following the course of purification of

[15] J. Mendicino and F. Vasarhely, *J. Biol. Chem.* **238**, 3528 (1963).

the enzyme is based on the formation of $P_i$ from fructose-1,6-$P_2$. The activity of the enzyme in crude tissue extracts is measured by applying a small aliquot of the solution to a cellulose phosphate column ($0.9 \times 2$ cm). The enzyme can be quantitatively removed from the column with 25 ml of 0.5 $N$ NaCl–50 m$M$ Tris·HCl, pH 8.0 and assay for activity.

*Reagents*

Tris·HCl, 1.0 $M$, pH 7.5
MgCl$_2$, 0.5 $M$
Cysteine, 0.1 $M$, pH 8.0
Fructose-1,6-$P_2$, 10 m$M$
KCl, 1.54 $M$
Solution containing H$_2$SO$_4$, 5 $N$ and ammonium molybdate, 2.5%

*Procedure.* For routine assay, 0.3 ml of Tris·HCl, 0.3 ml of MgCl$_2$, 0.4 ml of cysteine, 0.1 ml of KCl, and 7.72 ml of fructose-1,6-diphosphatase plus water are added to a test tube, which is placed in a water bath at 30°. The reaction is initiated by the addition of 0.07 ml of fructose-1,6-$P_2$, and the reaction mixture is incubated for 5 min. The reaction is stopped by the addition of 1 ml of the H$_2$SO$_4$–ammonium molybdate mixture. Then 0.2 ml of reducer is added and $P_i$ is determined by a modification of the method of Fiske and SubbaRow.[16,17] One unit of activity in this assay represents the formation of 1 $\mu$mole of $P_i$ per minute at 30° and specific activity is expressed as units per milligram of protein. In concentrated solutions protein is determined by a modification of the method of Lowry *et al.*[9,18] and in dilute solutions it is measured by the spectrophotometric procedure.[19]

## Purification Procedure

*Steps 1, 2, 3, and 4.* These procedures are the same as those described previously for the isolation of phosphofructokinase.

*Step 5.* The supernatant solution, 2900 ml, obtained after removing the protamine precipitate in step 4 of the previous procedure, is diluted with an equal volume of distilled water, the pH is adjusted to 7.0 with 1 $N$ HCl, and then 2.0 ml of 0.25 $M$ EDTA are added. The enzyme is applied to two large cellulose-phosphate columns ($14 \times 12.5$ cm) in 3 batches. The columns were washed with 400 ml of 10 m$M$ Tris·HCl,

---

[16] C. H. Fiske and Y. SubbaRow, *J. Biol. Chem.* **66**, 375 (1925).
[17] K. Lohmann and L. Jenrassik, *Biochem. J.* **178**, 419 (1926).
[18] E. F. Hartree, *Anal. Biochem.* **48**, 422 (1972).
[19] E. Layne, this series, Vol. 3, p. 447.

pH 7.0–1 m$M$ EDTA after each application. Batchwise application with intermittent washing removed loosely bound proteins and permitted greater loading of the columns with fructose-1,6-diphosphatase. After the last wash, 300 ml of 50 m$M$ Tris·HCl, pH 8.0–0.5 $M$ NaCl–0.2 m$M$ EDTA is passed into each column. The columns can be stored in the cold room overnight at 3° at this stage.

Fructose-1,6-diphosphatase is eluted from each column with 2 liters of 50 m$M$ Tris·HCl, pH 8.0–0.5 $M$ NaCl–1 m$M$ EDTA. The eluates are combined and diluted with 5 volumes of cold buffer containing 0.2 m$M$ EDTA–5 m$M$ 2-mercaptoethanolamine, and the diluted solution is passed into two smaller columns of cellulose phosphate (5.8–13.0 cm) overnight. The columns are washed with 150 ml of 0.5 $M$ NaCl–50 m$M$ Tris·HCl, pH 8.0–1.0 m$M$ EDTA, and fructose-1,6-diphosphatase is then eluted with 400 ml of the same solution. The combined eluates are dialyzed against two changes of 4 liters of 10 m$M$ Tris·HCl, pH 8.0–1 m$M$ EDTA–5 m$M$ 2-mercaptoethanolamine for 1 hr. The enzyme is diluted 2.5-fold with 50 m$M$ Tris·HCl, pH 8.0, and it is applied to eight small cellulose-phosphate columns (2.0 × 6.0 cm) at 3°. Each of the columns is washed with 5 ml of 0.5 $M$ NaCl–0.05 $M$ Tris–HCl, pH 8.0–1.0 m$M$ EDTA, and then the enzyme is eluted with 30 ml of the same solution.

*Step 6.* The eluates are combined and dialyzed at 3° against 500 ml of saturated ammonium sulfate containing 1.0 m$M$ EDTA and 5.0 m$M$ 2-mercaptoethanolamine, and solid ammonium sulfate is added until the solution inside the dialysis tubing is saturated with ammonium sulfate. About 6 hr of dialysis is required to attain equilibrium. This procedure reduces the volume of the solution, thereby increasing the yield of enzyme. The precipitate is collected by centrifugation and dissolved in 20 ml of 50 m$M$ Tris·HCl, pH 8.0. This concentrated enzyme solution is dialyzed for 3 hr against 300 ml of a solution containing 29.1 g/100 ml of ammonium sulfate containing 1.0 m$M$ EDTA and 5 m$M$ mercaptoethanolamine. The precipitate which forms is removed by centrifugation, and the supernatant solution is dialyzed at 3° against saturated ammonium sulfate containing EDTA and mercaptoethanolamine as described previously until no more precipitate forms. The precipitate is collected by centrifugation and dissolved in 20 ml of 50 m$M$ Tris·HCl, pH 8.0.

*Step 7.* The solution is diluted to 60 ml with 50 m$M$ Tris·HCl, pH 8.0–30 m$M$ cysteine and incubated for 15 min at room temperature to completely activate the enzyme. It is then diluted to 600 ml with 50 m$M$ Tris·HCl, pH 8.0–0.2 m$M$ EDTA–5 m$M$ mercaptoethanolamine, and this solution was applied to a cellulose-phosphate column (4.6 × 20 cm). The column is washed with 50 ml of 50 m$M$ Tris·HCl, pH 8.0, and the enzyme is then eluted with 400 ml of 50 m$M$ Tris·HCl, pH 8.0–1 m$M$ AMP–1

TABLE III
PURIFICATION PROCEDURE FOR FRUCTOSE-1,6-DIPHOSPHATASE

| Fraction | Volume (ml) | Protein (mg/ml) | Total activity (units) | Specific activity (unit/mg) | Recovery (%) |
|---|---|---|---|---|---|
| 1. Crude extract | 2500 | 52 | 8900 | 0.07 | 100 |
| 2. Treatment with alumina $C_\gamma$ | 2700 | 35 | 8810 | 0.09 | 99 |
| 3. Supernatant from calcium phosphate gel adsorption | 2800 | 24 | 8740 | 0.13 | 98 |
| 4. Supernatant from protamine precipitation | 2900 | 13.5 | 8620 | 0.22 | 97 |
| 5. Chromatography on cellulose phosphate | 240 | 1.7 | 7640 | 18.7 | 86 |
| 6. Precipitation with ammonium sulfate | 21.5 | 9.1 | 6660 | 34.0 | 75 |
| 7. Elution from cellulose phosphate with AMP and ATP | 29.6 | 3.5 | 5190 | 50.1 | 58 |
| 8. Chromatography on Sephadex G-25 | 29.6 | 3.5 | 5040 | 48.6 | 57 |

m$M$ ATP. The eluate is dialyzed against saturated ammonium sulfate containing 1.0 m$M$ EDTA and 5 m$M$ mercaptoethanolamine as described previously. The precipitate is collected by centrifugation, and it is dissolved in 30 ml of 50 m$M$ Tris·HCl, pH 8.0.

*Step 8.* Aliquots of 3 ml are passed through Sephadex G-25 columns (2 × 31 cm) to remove the last traces of AMP and ATP. The solution used to develop these columns contain 0.15 $M$ NaCl–15 m$M$ Tris·HCl, pH 8.0 and fructose-1,6-diphosphatase is eluted immediately after the void volume, whereas AMP and ATP appear in later fractions. The enzyme is precipitated by dialysis against saturated ammonium sulfate and the precipitate is dissolved in 30 ml of 50 m$M$ Tris·HCl, pH 8.0.

A typical purification is summarized in Table III.

## Properties

The final preparations is homogeneous and it has a specific activity of about 50 units per milligram of protein. Enzyme preparations can be stored at −20° for months. Some preparations begin to lose activity after several months; however, all the original activity can be restored by incubating the enzyme at 30° for 15 min in the presence of 30 m$M$ cysteine.

The enzyme has a molecular weight of 135,000, and it contains 4 identical subunits with molecular weights of 34,000.[1] The COOH-terminal amino acid of the polypeptide chain is alanine and the $NH_2$-terminal amino acid is serine.

Cysteine, potassium chloride, and magnesium ion are required for maximum activity.

The $K_m$ for fructose-1,6-$P_2$ is 5 $\mu M$, and the enzyme requires magnesium ion for activity. It is inhibited by AMP and high concentrations of fructose-1,6-$P_2$.[13] The purified enzyme binds about 4 equivalents each of AMP, ATP, and fructose-1,6-$P_2$.[5] The pH optimum of the kidney enzyme is about 7.0 to 7.5 at low concentrations of fructose-1,6-$P_2$.

## IV. Purification and Properties of Phosphorylase

### Assay Method

The activity of the enzyme is determined in the direction of glycogenolysis in a coupled enzyme system which measured the rate of formation of D-glucose-1-P with phosphoglucomutase and glucose-6-P dehydrogenase according to the procedure of Helmreich and Cori.[20]

#### Reagents

Buffer solution containing: imidazole, pH 7.0, 50 m$M$; MgCl$_2$, 10 m$M$ EDTA, 2 m$M$; bovine serum albumin, 0.01%
Glycogen (Sigma, Type V), 10%
Potassium phosphate, pH 7.5, 1.0 $M$
NADP, 30 m$M$
Protamine sulfate, 1%
AMP, 0.1 $M$
$\alpha$-D-Glucose-1,6-$P_2$, 1 m$M$ (see this series Vol. 41 [17])
Glucose-6-P dehydrogenase, 100 units/ml
Phosphoglucomutase, 100 units/ml

*Procedure.* For routine assay 0.72 ml of buffer solution, 0.11 ml of glycogen, 0.07 ml of potassium phosphate, 0.01 ml of NADP, 0.04 ml of protamine sulfate, 0.01 ml of $\alpha$-D-glucose 1,6-$P_2$, 0.01 ml of AMP, 0.01 ml of glucose-6-P dehydrogenase, and 0.01 ml of phosphoglucomutase are added to a 1-ml cuvette at room temperature. The reaction is initiated by the addition of 0.01 ml of diluted enzyme to yield a rate of 0.02–0.06 absorbance units at 340 nm/min. The phosphorylase is routinely diluted in the standard buffer solution containing 1% glycogen, and it is maximally activated by incubation for at least 5 min at room temperature

[20] E. Helmreich and C. F. Cori, *Proc. Nat. Acad. Sci. U.S.* **51**, 131 (1964).

before it is added to the assay mixture. One unit of activity in this assay is defined as the amount of enzyme that will catalyze the formation of 1 μmole of glucose-1-P per minute, and specific activity is expressed as units per milligram of protein.[9,10]

### Purification Procedure

*Steps 1, 2, 3, and 4.* These procedures are the same as those described previously for the isolation of fructose-1,6-diphosphatase.

*Step 5.* This step is also described in the preparation of fructose-1,6-diphosphatase in the previous procedure. Fructose-1,6-diphosphatase is adsorbed to the cellulose phosphate column, whereas essentially all the phosphorylase activity is found in the solutions that pass through the column. All of the application solutions and the washes are combined for the isolation of phosphorylase in the preceding step.

*Step 6.* Two grams of glycogen (shellfish, type II, Sigma) are dissolved in 75 ml of distilled water and added to the supernatant solution obtained from the preceding step. The solution, 6700 ml, is continuously stirred, and it is cooled to 3° in a dry-ice acetone bath while 3800 ml of acetone at —60° is slowly added. The temperature is maintained between 0° and —5° during the addition of acetone over a period of 3–5 min. The resulting suspension is centrifuged at —10° for 5 min. The centrifuge tubes containing the precipitate are carefully rinsed with distilled water to remove excess acetone without disturbing the tightly packed pellet. The precipitate is then suspended in 200 ml of 30 m$M$ Tris pH 7.5. The preparation can be stored at this stage by lyophilizing it to dryness to remove all traces of acetone. The dry powder can be kept under vacuum at —20° for at least 2 months with no loss of activity.

*Step 7.* The dry powder, 36 g, from 3 preparations taken through the previous step is suspended in 750 ml of 30 m$M$ Tris, pH 7.5, and a large amount of insoluble protein is removed by centrifugation and discarded. About 75 ml of 0.1 $M$ sodium pyrophosphate, pH 8.0, is added to the supernatant solution to remove residual protamine. The suspension is centrifuged and 1-ml aliquots of 0.1 $M$ sodium pyrophosphate, pH 8.0, are added to the supernatant solution until no more precipitate forms. Glycogen is added to the clear solution to a final concentration of 0.2%, and it is cooled to 0°. Then 277 ml of 95% ethanol, previously cooled to —60°, is slowly added to 840 ml of the solution containing the enzyme. The temperature was maintained at —5° during the addition of ethanol. Afterward, the suspension is centrifuged at 48,000 $g$ for 10 mim at —10°. The supernatant solution is kept at —10° for 5 min, and if no further precipitate forms it is discarded. Occasionally, a second centrifugation is required to remove all the precipitate at this stage. The pellets are dissolved in 90 ml of 30 m$M$ Tris·HCl, pH 7.5, and 11 ml of 95% ethanol is added as described previously; the resulting turbid solution is centrifuged. About 19 ml of ethanol is added to the supernatant solution to bring the ethanol concentration to 33%, and the resulting precipitate is collected by centrifugation and dissolved in 90 ml of 30 m$M$ Tris·HCl, pH 7.5. Some insoluble protein is removed by centrifugation. All the values in Table IV after step 6 are divided by 3 because three preparations were combined at this stage.

*Step 8.* Solid ammonium sulfate, 26 g, is added to 90 ml of the solution from the previous step and the precipitate, which contains about 50 units of activity, is removed by centrifugation. Then 43 g/100 ml of ammonium sulfate is added to the solution at 3° and the resulting suspension is stirred at 3° for 5 min and centrifuged. The supernatant solution is adjusted to 0.75 saturation by the addition of 43 g of solid ammonium sulfate. The suspension is stirred at 3° for 5 min and centrifuged. The precipitate is dissolved in 25 ml of 30 m$M$ Tris·HCl, pH 7.5, and dialyzed overnight against 4 liters of the same buffer.

The solution is then diluted with an equal volume of distilled water at 3°, and it is quickly applied to a DEAE column (4.4 × 15 cm) which was previously equilibrated with 30 m$M$ Tris·HCl, pH 7.5. The column is washed with 100 ml of 30 m$M$ Tris·HCl, pH 7.5, and then with 100 ml of 50 m$M$ Tris·HCl, pH 7.5. The enzyme is eluted with 210 ml of 0.2 $M$ Tris·HCl, pH 7.5.

*Step 9.* Glycogen is added to the solution from the previous step to a final concentration of 0.4%, and 84 ml of acetone at −60° is slowly added while the temperature is maintained at −5°. The solution is centrifuged at −10°, and the precipitate is dissolved in 30 ml of 30 m$M$ Tris·HCl, pH 7.5, and dialyzed against the same buffer.

*Step 10.* The enzyme isolated through this stage of purification is essentially completely in the dephosphorylated form. The dephosphophosphorylase in this partially purified preparation can be partially converted to the phosphorylated or "active" form of the enzyme by incubating it with a partially purified kidney protein kinase[16,21] which is isolated from

---

[21] A protein kinase which activates phosphorylase can be partially purified from the precipitate obtained after acid precipitation at pH 5.7 (step 4) in the preparation of glycogen synthetase. This precipitate is dissolved in 90 ml of 25 m$M$ glycerol-P, pH 7.5 containing 0.15 $M$ sucrose, and the solution is applied to a DEAE-column (4.4 × 10 cm) which was previously equilibrated with the same buffer. After washing with 200 ml of buffer containing 0.1 $M$ KCl, the protein kinase was eluted with 100 ml buffer containing 0.3 $M$ KCl. The eluate is dialyzed against a solution of saturated ammonium sulfate, and the resulting precipitate is collected by centrifugation and dissolved in 50 m$M$ glycerol-P–30 m$M$ sucrose pH 7.5. The

TABLE IV
PURIFICATION OF GLYCOGEN PHOSPHORYLASES FROM SWINE KIDNEY

| Step | Volume (ml) | Protein (mg/ml) | Total activity (units) | Specific activity (units/mg) | Re-covery (%) |
|---|---|---|---|---|---|
| 1. Crude extract | 2500 | 52 | 250 | 0.0019 | — |
| 2. Treatment with alumina $C_\gamma$ gel | 2700 | 35 | 275 | 0.003 | — |
| 3. Supernatant from calcium phosphate gel adsorption | 2800 | 24 | 270 | 0.004 | — |
| 4. Supernatant from protamine precipitation | 2900 | 13.5 | 318 | 0.008 | 100 |
| 5. Solution which passes through cellulose phosphate | 6700 | 5.1 | 310 | 0.009 | 97 |
| 6. Precipitation with acetone | 280 | 16 | 225 | 0.050 | 70 |
| 7. Precipitation with ethanol | 90 | 11 | 200 | 0.2 | 63 |
| 8. Chromatography on DEAE | 70 | 0.70 | 130 | 2.6 | 41 |
| 9. Second precipitation with acetone | 10 | 0.90 | 100 | 10.0 | 31 |
| 10. Treatment with protein kinase | | | | | |
| Dephospho form | 20 | 0.20 | 80 | 20 | 25 |
| Phospho form | 19 | 0.21 | 190 | 47 | 25 |
| 11. Chromatography on hydroxyapatite | | | | | |
| Dephospho form | 5 | 0.42 | 64 | 30 | 20 |
| Phospho form | 5 | 0.48 | 167 | 70 | 22 |

the same swine kidney extracts. When this form of the enzyme is required, the preparation from the previous step is incubated for 90 min at 30° with 6 m$M$ ATP, 12 m$M$ MgCl$_2$, and 100 μg of the partially purified protein kinase. Afterward, the solution is dialyzed against 30 m$M$ Tris·HCl, pH 7.5, at 3° and it is passed into a DEAE column (2.2 × 5 cm); the enzyme is eluted as described previously (step 8), and the active fractions are combined and concentrated at −5° by the addition of 1.5% glycogen and cold acetone to a final concentration of 28%. The precipitate is collected by centrifugation and dissolved in 60 ml of 30 m$M$ Tris·HCl, pH 7.5, and dialyzed overnight against this same buffer. Each of the phosphorylase preparations are treated in an identical manner in subsequent purification steps. The activities of both the treated and un-

enzyme is purified about 150-fold at this stage and can be used to prepare the phosphorylated form of kidney phosphorylase. This preparation is stable for several weeks when aliquots are stored frozen at −20°. High concentrations of salt, 0.15 $M$, completely inhibit the enzyme, but full activity is observed after dialysis against 50 m$M$ glycerol-P–0.3 $M$ sucrose. The enzyme has a broad pH optimum with maximum activity at about pH 7.5.

treated samples taken through this step are shown in Table IV. The specific activity of the untreated sample doubled after chromatography on DEAE and reprecipitation with acetone. The specific activity of the sample treated with protein kinase increased from 10 to 47 units per milligram in this step.

*Step 11.* The treated or untreated preparation was applied to a hydroxyapatite column (1 × 4 cm) previously equilibrated with 30 m$M$ Tris·HCl, pH 7.5. The column was washed with 20 ml of 30 m$M$ Tris·HCl, pH 7.5, and then with 20 ml of 50 m$M$ potassium phosphate, pH 7.5. The enzyme was eluted with 30 ml of 80 m$M$ potassium phosphate, pH 7.5. This solution was diluted with an equal volume of 0.1 $M$ Tris·HCl, pH 7.5 and the concentration of glycogen was adjusted to 0.2%. The enzyme is precipitated at −10° by the addition of 24 ml of acetone as described previously. The precipitate is dissolved in 5 ml of 30 m$M$ Tris·HCl, pH 7.5, and it is dialyzed against the same buffer for 3 hr. The data derived in a typical preparation are summarized in Table IV. The final purification from the crude extract was about 15,000-fold with a yield of 20%.

## Properties

The phospho and dephospho forms of the enzyme are stable for about 2 weeks when they were stored at 3°. The preparations can be lyophilized and the dry powders have been stored under vacuum at −20° for 4 months with a loss of about 30% of the original activity. The final preparation of phosphorylase were completely free of amylases, phosphofructokinase, glycogen synthetase, protein kinases, and phosphoprotein phosphatases, when assayed with [32]P-labeled phosphorylase and glycogen synthetase as substrates. Phosphofructokinase was removed at the protamine sulfate precipitation step; glycogen synthetase and protein kinases, which phosphorylated phosphorylase, were removed at the calcium phosphate adsorption step. In the presence of saturating concentrations of AMP and protamine, the specific activity of the dephospo form of the homogeneous enzyme was only 30 units per milligram whereas the specific activity of the phosphorylated form was 70 units per milligram under the same conditions.

The molecular weights of the phospho and dephospho forms of swine kidney phosphorylase calculated from data obtained with several different methods is 190,000 ± 5000. The Stokes radius of the enzymes is 48 Å, and they both had a diffusion constant of about 4 × 10⁻⁷ cm²/sec. Both enzymes contain two identical subunits with molecular weights of 95,000 ± 5000. The $K_m$ of the phospho form of the enzyme for $P_i$ was 2.8 m$M$

compared to 100 m$M$ for the dephospho form in the presence of 0.2 mg of glycogen and 2 m$M$ AMP. The corresponding values for glycogen at about 4 m$M$ P$_i$ were 1 4 m$M$ and 20 m$M$ for the phospho and dephospho forms, respectively.

## V. Isolation and Properties of Protein Kinases

### Assay Methods

An enzymic and isotopic procedure are used to measure the activity of these enzymes. The enzymic assay is based on the ability of protein kinase to catalyze conversion of glycogen synthetase to a form which is dependent on the presence of glucose-6-P for activity.[22] The rate of decrease in the activity of glycogen synthetase is measured in the absence of glucose-6-P. Two incubations are required in this assay. In the first reaction mixture the protein kinase is allowed to phosphorylate glycogen synthetase. This reaction is terminated by the addition of EDTA which binds magnesium ion and completely inhibits the protein kinase. Then aliquots are incubated in a second reaction mixture in the absence of magnesium to determine the glycogen synthetase activity remaining.

In order to examine the activity with proteins which have no enzymic activity an assay method based on the rate of phosphorylation of the protein by [$\alpha$-$^{32}$P]ATP is used. The reaction mixture is incubated under the same conditions used in the enzymic assay so that the results obtained in the two assays would be comparable.

### Reagents

Sucrose, 1 $M$
Tris·HCl, pH 7.0, 1 $M$
Cysteine, pH 7.0, 0.1 $M$
MgCl$_2$, 0.1 $M$
[$\alpha$-$^{32}$P]ATP, 20 m$M$ (5 × 10$^7$ cpm/$\mu$mole)
Cyclic AMP, 0.1 m$M$
ATP, 20 m$M$
Glycogen synthetase,[23] 1 mg/ml
Bovine serum albumin, 1%
Trichloroacetic acid, 10%

---

[22] D. L. Friedman and J. Larner, *Biochemistry* **2**, 669 (1963).

[23] Kidney glycogen synthetase isolated by the procedure described in this article was used as the substrate in this reaction mixture.

EDTA, 0.25 $M$
HCl, 10 $N$
Glycogen, 10%
UDP [U-$^{14}$C]glucose, 50 m$M$ (specific activity, 50,000 cpm/$\mu$mole)
Histone (calf thymus), 20 mg/ml

*Procedure.* In a small test tube, mix 0.02 ml of Tris·HCl, 0.03 ml of sucrose, 0.25 ml of cysteine, 0.1 ml of MgCl$_2$, 0.1 ml of cyclic AMP, 0.1 ml of glycogen synthetase, and 0.18 ml of protein kinase plus water. The reaction mixture is incubated at 30° for 3 min, and the reaction is initiated by the addition of 0.02 ml of ATP. Aliquots of 0.2 ml are removed at 0, 3, and 6 min and they are added to test tubes containing 0.03 ml of EDTA, 0.03 ml glycogen, 0.03 ml Tris·HCl, and 0.04 ml of distilled water to terminate the protein kinase reaction. Then 0.02 ml of UDP [U-$^{14}$C]glucose are added to these tubes to initiate the glycogen synthetase reaction. These tubes are incubated at 37° for 10 min and then 1 ml of trichloroacetic acid is added to stop the reaction and the resulting suspension is centrifuged to remove protein. Glycogen is isolated from the supernatant solution by repeated precipitation with ethanol as described in the assay for glycogen synthetase.[2] A unit of activity in this assay is defined as the amount of protein kinase required to convert 1 $\mu$mole of glycogen synthetase to a form dependent on glucose 6-P for activity per minute. Specific activity is expressed as units per milligram of protein. This assay was used to obtain the data in Table V. One micromole of kidney glycogen synthetase will catalyze the transfer of about 5550 $\mu$moles (15 $\mu$moles/min per milligram $\times$ 370 mg/$\mu$mole) of glucose from UDP-glucose to glycogen per minute.

Histone is used as the protein substrate in the isotopic assay. In a small test tube mix 0.02 ml of Tris·HCl, 0.03 ml of sucrose, 0.25 ml of cysteine, 0.1 ml of MgCl$_2$, 0.1 ml of cyclic AMP, 0.1 ml of histone, and 0.18 ml of protein kinase plus water. The mixture is incubated for 3 min at 30° and then the reaction is initiated by the addition of 0.02 ml of [$\alpha$-$^{32}$P]ATP. Aliquots of 0.2 ml are removed at 0, 3, and 6 min, and they are added to small test tubes containing 3 ml of trichloroacetic acid and 1 mg of bovine serum albumin as carrier protein. The solution is mixed and the precipitate is collected by centrifugation. The pellet is dissolved in 0.5 ml of 1 $N$ NaOH. After the precipitate dissolves 0.5 ml water is added and the protein is reprecipitated by the addition of 3 ml of trichloroacetic acid. The precipitate is collected and washed once more by this procedure. The final pellet is dissolved in 0.5 ml of 1 NaOH  and 0.5 ml of water. The sample is transferred to a vial containing 10 ml of scintillation solvent and 0.06 ml of HCl and it is mixed and counted. Only about 80% of the protein substrate is routinely recovered in this isolation procedure and a correction can be made for this loss. A unit of activity in this assay is defined as the amount of enzyme required to transfer 1 $\mu$mole of $^{32}$P from [$\alpha$-$^{32}$P]ATP to the protein substrate per minute at 30°.

## Purification Procedure

*Steps 1, 2, and 3.* The procedure used in these steps are exactly the same as those described for the isolation of swine kidney glycogen synthetase.

*Step 4.* The dialyzed calcium phosphate gel eluate from step 3 was stirred at 3° and the pH was adjusted to 4.9 by the addition of 1 $N$ acetic acid. The resulting turbid suspension was centrifuged and the pH of the supernatant solution was readjusted to 7.0 by the addition of 1 $M$ Tris base.

*Step 5.* This solution is then passed into a DEAE-cellulose column (2.2 × 14 cm) which was previously equilibrated with 10 m$M$ Tris·HCl, pH 7.0–0.001 $M$ EDTA. The enzyme is eluted from this column with a linear gradient formed with 200 ml of 10 m$M$ Tris·HCl, pH 7.0–1 m$M$ EDTA in the mixing chamber and 200 ml of 0.5 $M$ Tris·HCl, pH 7.0–1

TABLE V

PURIFICATION OF SOLUBLE PROTEIN KINASE FROM SWINE KIDNEY

| Step | Volume (ml) | Protein (mg/ml) | Total activity (units) | Specific activity ($\rho$moles/ min/mg) | Yield (%) |
|---|---|---|---|---|---|
| 1. Crude extract | 2500 | 52 | 0.139 | 1.0 | 100 |
| 2. Treatment with alumina C$_\gamma$ | 2700 | 35 | 0.105 | 1.1 | 76 |
| 3. Calcium phosphate gel eluate | 600 | 14 | 0.096 | 11.4 | 69 |
| 4. Supernatant from precipitation at pH 4.9 | 210 | 3.0 | 0.062 | 98 | 44 |
| 5. DEAE-cellulose | | | | | |
|    Peak I | 15 | 2.0 | 0.011 | 364 | 8 |
|    Peak II | 20 | 3.5 | 0.020 | 285 | 14 |
| 6. Chromatography on Sephadex G-200 | | | | | |
|    Peak I | 7 | 2.0 | 0.010 | 713 | 7 |
|    Peak II | 7 | 4.0 | 0.019 | 680 | 13 |
| 7. Chromatography on Blue Dextran–Sepharose 4B | | | | | |
|    Peak I | 5 | 0.6 | 0.009 | 3000 | 6 |
|    Peak II | 5 | 0.9 | 0.015 | 3000 | 11 |

m$M$ EDTA in the reservoir. Two peaks containing protein kinase activity are eluted from the column.

*Step 6.* The active fractions from each peak are combined, and the enzyme is precipitated by the addition of 32.6 g/100 ml of ammonium sulfate. After centrifugation, the pellets are dissolved in 20 ml of 50 m$M$ Tris·HCl, pH 7.5 and the enzyme from each peak is dialyzed against the same buffer. The two preparations are then further purified by chromatography on two Sephadex G-200 columns (2.2 × 33 cm). Each enzyme eluted in a single well defined peak with a buffer containing 20 m$M$ Tris·HCl pH 7.5–1 m$M$ EDTA. The fractions with activity were collected, concentrated and dialyzed against the same buffer.

*Step 7.* The preparations obtained from the preceding step were passed into Blue Dextran-Sepharose 4B columns (4.0 × 10 cm) and the columns were then washed with 200 ml of 50 m$M$ Tris·HCl, pH 7.5. The enzyme was eluted with a solution containing 50 m$M$ Tris·HCl, pH 7.5, and 1 m$M$ ATP. A protein peak containing kinase activity emerged from the column after about 30 ml of this solution passed through the column. The active fractions were combined, concentrated by evaporation as described previously and the solution was dialyzed against 20 m$M$ Tris·HCl, pH 7.0–1 m$M$.

A typical purification is summarized in Table V.

## Properties

The two enzyme preparations are stable for at least 2 months when aliquots are frozen at −20° in 20 m$M$–Tris·HCl, pH 7.0–1 m$M$ EDTA.

The pH optimum of both preparations is 6.5, and they require magnesium ion for optimum activity. The apparent $K_m$'s for ATP are 7 $\mu M$ and 8 $\mu M$ for the first peak and the second peak, respectively. Cyclic AMP did not significantly alter the apparent $K_m$ for ATP. The activity of both preparations was stimulated about 50% in the presence of 50 $\mu M$ cyclic AMP. The molecular weight of both preparations was about 180,000.

## [62] Ferredoxin-Activated Fructose-1,6-diphosphatase System of Spinach Chloroplasts

*By* Bob B. Buchanan

Fructose diphosphatase (FDPase), which hydrolyzes P$_i$ from the C-1 position of fructose 1,6-diphosphate, plays a key regulatory role in carbohydrate synthesis in both gluconeogenesis[1] and photosynthesis.[2,3] In photosynthesis, FDPase appears to function as part of an enzyme system that consists of: (a) the FDPase component (D-fructose-1,6-biphosphate 1-phosphohydrolase, EC 3.1.3.11), which is insensitive to inhibition by AMP or fructose diphosphate; (b) a protein factor component; (c) ferredoxin; and (d) $Mg^{2+}$. All these components are located in chloroplasts, the subcellular organelles that photosynthetically convert $CO_2$ to carbohydrate.

The FDPase component, in the presence of the protein factor, is activated by photoreduced ferredoxin. We have proposed that the FDPase reaction is the initial regulatory reaction in a sequence that involves, in turn, ribulose-1,5-diphosphate carboxylase (which is activated by fructose 6-phosphate and deactivated by fructose diphosphate[4]) and ADP-glucose pyrophosphorylase (which is activated by 3-phosphoglycerate[5]). This regulatory system uses light to initiate the regulation of the photosynthetic conversion of $CO_2$ to starch in chloroplasts.

The role of the protein factor and reduced ferredoxin in the activation of the FDPase component is not yet established; current evidence indicates that the FDPase is activated by a reduction of —S—S— to SH groups in accordance with Eq. (1). The activated enzyme catalyzes the conversion of fructose diphosphate to fructose 6-phosphate in accordance with Eq. (2).

$$\text{FDPase}_{\text{inactive}} \xrightarrow[\text{protein factor}]{\text{reduced ferredoxin}} \text{FDPase}_{\text{active}} \qquad (1)$$
$$(\text{—S—S—}) \qquad\qquad\qquad (\text{—SH—HS—})$$

$$\text{Fructose 1,6-diphosphate} + H_2O \xrightarrow[\text{Mg}^{2+}]{\text{active FDPase}} \text{fructose 6-phosphate} + P_i \qquad (2)$$

The ferredoxin, FDPase, and protein factor components of the FDPase system can be isolated from the aqueous extract of isolated spinach chloroplasts. For preparation on a large scale, procedures have been devised for purification of each of these components from spinach leaves. Procedures are given below for purification of the FDPase and protein factor components; a method for the purification of the ferredoxin component has been given in a previous volume of this series.[6] The FDPase component was extensively purified earlier from spinach leaves

[1] H. A. Krebs, *Proc. Roy. Soc. Ser. B* **159**, 545 (1964).
[2] J. A. Bassham, *Science* **172**, 526 (1971).
[3] B. B. Buchanan, P. Schürmann, and P. P. Kalberer, *J. Biol. Chem.* **246**, 5952 (1971).
[4] B. B. Buchanan and P. Schürmann, *J. Biol. Chem.* **248**, 4956 (1973).
[5] H. P. Ghosh and J. Preiss, *J. Biol. Chem.* **241**, 4491 (1966).
[6] B. B. Buchanan and D. I. Arnon, this series, Vol. 23, p. 413.

by Racker and Schroeder[7] and by Preiss et al.[8] Its crystallization has been reported by El-Badry and Bassham.[9]

## Preparation of FDPase Component of Chloroplast FDPase System

### Assay Method

*Principle.* FDPase activity is determined by following the formation of $P_i$ from fructose diphosphate in the presence of $Mg^{2+}$. The amount of $P_i$ released is determined colorimetrically by a modified Fiske-SubbaRow procedure.[10] Activity of the enzyme can also be determined by continuous measurement of fructose 6-phosphate formation in the presence of excess phosphohexose isomerase, glucose-6-phosphate dehydrogenase, and NADP.[8]

*Reagents for Assay*

Tris·HCl buffer, 1 $M$, pH 8.5
$MgCl_2$, 0.16 $M$
Ethylenediaminetetraacetate (EDTA), disodium salt, 1 m$M$
Fructose 1,6-diphosphate, sodium salt, 60 m$M$
FDPase component (0.2–0.6 units)

*Assay Procedure.* The reaction is carried out in a 1.5 × 12 cm test tube containing 0.5 ml of $H_2O$, 0.1 ml of Tris buffer, and 0.1 ml each of $MgCl_2$, EDTA, and FDPase component. The reaction is initiated by adding 0.1 ml of fructose diphosphate solution and is continued for 10 min at room temperature. The reaction is stopped by adding 0.5 ml of 10% trichloroacetic acid. If turbid, the precipitate is centrifuged off and 0.5 ml of the clear supernatant solution is analyzed for $P_i$. The method is satisfactory for assay of the FDPase component in crude extracts of spinach leaves.

*Reagents for $P_i$ Analysis*

$H_2SO_4$, 9 $N$
Ammonium molybdate [$(NH_4)_6 Mo_7O_{24}·4H_2O$], 6.6 g/100 ml

[7] E. Racker and E. A. R. Schroeder, *Arch. Biochem. Biophys.* **74**, 326 (1958).
[8] J. Preiss, M. L. Biggs, and E. Greenberg, this series, Vol. 23, p. 691.
[9] M. El-Badry and J. A. Bassham, *Abstr. Amer. Chem. Soc. Chicago, Sept. 1970*, No. 138.
[10] J. B. Sumner and G. F. Somers, "Laboratory Experiments in Biological Chemistry," p. 71. Academic Press, New York, 1949.

$FeSO_4 \cdot 7H_2O$, 2.0 g/25 ml of a solution containing 24.5 ml of $H_2O$ and 0.5 ml of 0.9 $N$ $H_2SO_4$.

*Procedure for $P_i$ Analysis.* The reaction is carried out in a 1.5 $\times$ 12 cm test tube containing 3.0 ml of $H_2O$ and 0.5 ml each of $H_2SO_4$, the sample (containing up to 0.5 $\mu$mole of $P_i$), molybdate, and $FeSO_4$ solutions. The preparation is allowed to stand for 5 min, then absorbance at 660 nm is measured in a cuvette (3-ml capacity, 1-cm light path); 0.5 $\mu$mole of $P_i$ gives an absorbance of about 0.3 under these conditions.

*Definition of Unit of Specific Activity.* One unit of enzyme is defined as that amount which catalyzes the release of 1 $\mu$mole of $P_i$ per minute under the assay conditions. The specific activity is expressed as units of enzyme activity per milligram of protein. Protein concentration is determined by a modified phenol reagent procedure[11] with crystalline bovine serum albumin as a standard.

## Purification Procedure

Unless indicated otherwise, all purification steps of the FDPase component given below are carried out at 4°.

*Step 1. Preparation of Leaf Extract.* Fresh market spinach leaves, 5 kg, are washed and frozen overnight at −20°. The frozen leaves (in 1-kg batches) are blended in 1400 ml of water for 3 min in a Waring Blendor (gallon size, Model CB-5). Ten milliliters of 1 $M$ $K_2HPO_4$ (sufficient to give a final pH of 7) are added prior to blending. The homogenate is filtered through four layers of cheesecloth, and the residue is discarded. The remaining particulate material is sedimented by centrifugation at 13,000 $g$ for 5 min and is also discarded.

*Step 2. pH 4.5 Precipitation.* The green supernatant fraction (10,700 ml) is adjusted to pH 4.5 with 1 $N$ formic acid and centrifuged (5 min, 12,000 $g$). The yellow supernatant fraction obtained in this step contains the bulk of the protein factor needed for activation of the FDPase component by reduced ferredoxin; the fraction is neutralized with 1 $M$ Tris buffer (pH unadjusted) and fractionated as described below.

The green precipitate containing the FDPase component is suspended in a HEPES–EDTA solution (25 m$M$ HEPES–NaOH buffer, pH 7.6, containing 2.5 $\times$ 10$^{-4}$ $M$ EDTA-Na) and then brought to a volume of 1000 ml. The pH is adjusted to 6.5 with 0.1 $N$ $NH_4OH$. The suspension is centrifuged for 10 min at 35,000 $g$ and the residue is discarded. Final volume of the supernatant fraction containing the FDPase is 700 ml.

[11] W. Lovenberg, B. B. Buchanan, and J. C. Rabinowitz, *J. Biol. Chem.* **238**, 3899 (1963).

*Step 3. Ammonium Sulfate Fractionation.* Solid ammonium sulfate is added to the redissolved pH 4.5 precipitate fraction to give 50% saturation. The solution is centrifuged (10 min, 13,000 g) and the precipitate is discarded. Ammonium sulfate is added to the supernatant fraction to 90% saturation. The solution is centrifuged as above, the supernatant fraction is discarded, and the precipitate (containing the bulk of the FDPase activity) is dissolved in 30 mM Tricine buffer, pH 8.0, to a final volume of 80 ml. This slightly turbid solution is centrifuged (15 min, 35,000 g) to clarify.

*Step 4. Sephadex G-100 Chromatography.* The 50–90% ammonium sulfate fraction is applied to a Sephadex G-100 column (5 × 90 cm) equilibrated beforehand and developed with 30 mM Tricine buffer, pH 8.0. Fractions of 10 ml are collected with a fraction collector. FDPase chromatographs just behind the excluded volume and is eluted before the bulk of the applied protein. The slightly yellow fractions containing FDPase activity are pooled, and HEPES–NaOH buffer, pH 7.6, and NaCl are added to give a final concentration of 50 mM HEPES buffer, pH 7.6, and 0.2 M NaCl.

*Step 5. DEAE-Cellulose Chromatography.* The combined FDPase fractions are applied to a DEAE-cellulose column (5 × 45 cm) equilibrated beforehand with a solution containing 50 mM HEPES–NaOH buffer, pH 7.6, and 0.15 M NaCl. The column is washed with a solution containing 50 mM HEPES–NaOH buffer, pH 7.6, and 0.3 M NaCl until the eluate is free of protein (determined by absorbance at 280 nm), and the column is attached to a fraction collector and eluted in 10-ml fractions with a solution containing 50 mM HEPES–NaOH buffer, pH 7.6, and 0.5 M NaCl. FDPase is eluted just behind the solvent front; the active fractions are pooled and concentrated by vacuum dialysis.

The table shows a summary of the purification and yield of the alkaline FDPase component in each of the purification steps.

PURIFICATION OF CHLOROPLAST ALKALINE FRUCTOSE-1,6-DIPHOSPHATASE
COMPONENT FROM SPINACH LEAVES

| Fraction | Total protein (mg) | Total activity (units) | Yield (%) | Specific activity (units/mg) |
|---|---|---|---|---|
| Leaf extract | 48,100 | 2900 | 100 | 0.06 |
| pH 4.5 precipitate | 8,400 | 3200 | 110 | 0.38 |
| 50–90% (NH₄)₂SO₄ | 1,760 | 2720 | 94 | 1.54 |
| Sephadex G-100 | 384 | 2160 | 75 | 5.62 |
| DEAE-cellulose | 84 | 2050 | 71 | 24.4 |

The FDPase component purified by this procedure is relatively stable and can be stored in HEPES–EDTA buffer at −20° for several weeks with little loss of activity. Repeated freezing and thawing cause a slow loss of activity.

## Some Properties of the FDPase Enzyme Component

The freshly purified FDPase component appears to be homogeneous.[3] The purified protein shows a single peak in the ultracentrifuge and in Sephadex G-200 chromatography. The FDPase component travels in polyacrylamide gel electrophoresis as one main band trailed by diffuse protein material. This slow-moving material, present in all preparations, is probably due to a change in the homogeneous enzyme during electrophoresis—an interpretation that is supported by the finding that under certain conditions (particularly after the enzyme is concentrated by vacuum or pressure dialysis) the FDPase component dissociates into subunits.

The molecular weight of the pure FDPase component determined by sedimentation equilibrium and sedimentation velocity ultracentrifugation, by gel filtration, and by amino acid composition is 145,000. The subunits show a molecular weight of 73,000, thus indicating that the active enzyme is a dimer comprised of two identical subunits.

Like the FDPases from rabbit muscle and liver,[12] the FDPase component of spinach chloroplasts is characterized by a high content of glutamate, aspartate, and glycine, a low content of histidine, and the absence of tryptophan. The chloroplast enzyme also appears to contain an unusually large number of cystine (—S—S—) residues.[3]

The chloroplast FDPase component is characterized spectrally by a single peak at 278 nm and shows no absorption in the visible region. An absorbance at 278 nm (1-cm light path) is equal to 1.3 mg protein per milliliter. The enzyme contains about 2% carbohydrate.

In contrast to its counterpart from liver cells[13-15], the chloroplast FDPase component is not inhibited by AMP or fructose diphosphate[3,15,16] However, analogous to the liver enzyme,[17] the chloroplast component is activated by such —S—S— reagents as cystamine and DTNB. The pro-

[12] J. Fernando, S. Pontremoli, and B. L. Horecker, *Arch. Biochem. Biophys.* **129**, 370 (1969).

[13] K. Taketa and B. M. Pogell, *Biochem. Biophys. Res. Commun.* **12**, 229 (1963).

[14] E. A. Newsholme, *Biochem. J.* **89**, 38P (1963).

[15] J. Mendicino and F. Vasarhely, *J. Biol. Chem.* **238**, 3528 (1963).

[16] J. Preiss, M. L. Biggs, and E. Greenberg, *J. Biol. Chem.* **242**, 2292 (1967).

[17] S. Pontremoli, S. Traniello, M. Enser, S. Shapiro, and B. L. Horecker, *Proc. Nat. Acad. Sci. U.S.* **58**, 286 (1967).

tein factor has no effect on the FDPase component under these conditions. The physiological significance of this type of activation is an open question.

## Preparation of Protein Factor Component of Chloroplast FDPase System

### Assay Method

*Principle.* The activity of the protein factor is determined by the increase of $P_i$ released from fructose diphosphate in the presence of photoreduced ferredoxin and homogeneous FDPase component.[3]

*Reagents for Assay*

Tris·HCl buffer, 1 $M$, pH 8.0
Ethylenediaminetetraacetate (EDTA), disodium salt, 1 m$M$
MgCl$_2$, 10 m$M$
2,6-Dichlorophenolindophenol (DPIP), 1 m$M$
Sodium ascorbate, 0.1 $M$
Reduced glutathione, neutralized with KOH, 50 m$M$
FDPase component (purified as above), 0.4 mg/ml
Spinach ferredoxin, 1.2 mg/ml
Protein factor, 1.5 mg/ml
Spinach chloroplast fragments ($P_1S_1$ prepared as given by Buchanan and Arnon[6]), 1 mg chlorophyll/ml
Fructose 1,6-diphosphate, sodium salt, 60 m$M$

*Assay Procedure.* The reaction is carried out at 20° in Warburg flasks. The main compartment contains 0.4 ml of $H_2O$ and 0.1 ml each of Tris buffer, EDTA, MgCl$_2$, DPIP, sodium ascorbate, reduced glutathione, FDPase component, spinach ferredoxin, and protein factor. The sidearm contains 0.1 ml each of fructose 1,6-diphosphate and chloroplast fragments. The flask is equilibrated with argon (or $N_2$) for 6 min, and the reaction is started by adding fructose diphosphate and chloroplast fragments from the sidearm. The reaction is continued for 30 min under illumination (10,000 lux) and is stopped by adding 0.5 ml of 10% trichloroacetic acid. The precipitate is centrifuged off ·and 0.5 ml of the clear supernatant solution is analyzed for $P_i$ as described above.

The assay is an excellent indicator of the presence of protein factor; however, it has not been developed to the extent that it will give a precise quantitative measure of protein factor activity. Hence no unit of enzyme activity has been defined for the protein factor.

## Purification Procedure

*Step 1. Acetone Fractionation.* The neutralized pH 4.5 supernatant fraction from step 2 of the FDPase procedure is used as a source of the protein factor. Acetone, cooled to $-20°$, is added with constant stirring to a final concentration of 75%; the solution is left 1 hr at $-20°$ to allow the precipitate to settle. The supernatant fraction is decanted and discarded; the precipitate (containing the protein factor) is collected by centrifugation (3 min, 1000 $g$) and suspended in 30 m$M$ Tris·HCl buffer, pH 8.0. The turbid solution is then dialyzed 24 hr against 30 m$M$ Tris·HCl buffer, pH 8.0. Denatured protein is centrifuged off (15 min, 13,000 $g$).

*Step 2. Ammonium Sulfate Fractionation.* Solid ammonium sulfate is added to the dialyzed acetone fraction to 50% saturation. The heavy precipitate is sedimented by centrifugation (10 min, 13,000 $g$) and discarded. The ammonium sulfate concentration in the supernatant is increased to 90% saturation and the precipitate (containing the protein factor) is collected by centrifugation as above and dissolved in 100 ml of 20 m$M$ Tris·HCl buffer, pH 8.0. The slightly turbid solution is centrifuged (5 min, 35,000 $g$) to clarify.

*Step 3. Sephadex G-100 Chromatography.* The 50–90% ammonium sulfate fraction is applied to a Sephadex G-100 column (5 × 90 cm) equilibrated beforehand and developed with 30 m$M$ Tris·HCl buffer, pH 8.0. Fractions (13 ml) are collected, and those containing protein factor activity are pooled and stored at $-20°$.

The protein factor is stable for several months at $-20°$.

## Properties of the Protein Factor Component

The above procedure yields a protein factor preparation that shows several bands on polyacrylamide gel electrophoresis. The partially purified protein factor is heat sensitive (5 min, 80°) and, based on Sephadex gel filtration, has a molecular weight of 40,000. Other properties are unknown.

## Some Properties of the Ferredoxin-Activated FDPase System

Ferredoxin, photoreduced by chloroplasts (or reduced in the dark with hydrogen gas in the presence of hydrogenase[18]), enhances the release of $P_i$ from fructose diphosphate up to 40-fold. The release of $P_i$ required,

[18] B. B. Buchanan, P. P. Kalberer, and D. I. Arnon, *Biochem. Biophys. Res. Commun.* **29,** 74 (1967).

in addition to reduced ferredoxin, $MgCl_2$, fructose diphosphate, the alkaline FDPase component, and the protein factor. The reaction is stimulated by an —SH reagent (reduced glutathione or 2-mercaptoethanol) supplied in addition to reduced ferredoxin. Native spinach ferredoxin can be partially replaced by the nonphysical dye methyl viologen, but not by reduced NAD or NADP.

The homogeneous spinach leaf FDPase preparation of Preiss et al.[8] behaves like the above preparation in the ferredoxin-dependent release of $P_i$ from fructose diphosphate. A homogeneous preparation of the enzyme from rabbit liver, however, showed no response to reduced ferredoxin and protein factor.

*Effect of $Mg^{2+}$ Concentration on Ferredoxin-Activated FDPase.* The effect of reduced ferredoxin on FDPase activity is most pronounced at 0.67 to 2 m$M$ $MgCl_2$, but a consistent stimulation is observed at $MgCl_2$ concentrations at least as great as 4 m$M$.

*Activation of Chloroplast Alkaline FDPase Component by Dithiothreitol.* The nonphysiological —SH reagent dithiothreitol can replace reduced ferredoxin in activation of the chloroplast FDPase component. $P_i$ release from fructose diphosphate by the dithiothreitol-activated FDPase component occurs in the dark without chloroplasts and requires $MgCl_2$ and the protein factor. Cysteine, reduced glutathione, 2-mercaptoethanol, and sodium dithionite do not replace dithiothreitol in activation of the FDPase component.

*Effect of Some Inhibitors on Ferredoxin-Activated FDPase from Chloroplasts.* Ferredoxin-activated FDPase is not affected by KCN or the copper-chelating agents diethyldithiocarbamate or cuprizone (at 1 m$M$) but is 97% inhibited by 0.5 m$M$ EDTA. In certain preparations, a low concentration of EDTA (70 $\mu M$) increased 2- to 4-fold the release of $P_i$ from fructose diphosphate.

# [63] 3-Phosphoglycerate Phosphatase

*By* D. D. RANDALL and N. E. TOLBERT

$$\begin{array}{ccc}
\text{COOH} & & \text{COOH} \\
| & & | \\
\text{CHOH} + H_2O \rightarrow & \text{CHOH} + H_3PO_4 \\
| & & | \\
\text{CH}_2\text{O}\cdot\text{PO}_3\text{H}_2 & & \text{CH}_2\text{OH}
\end{array}$$

## Assay Method

*Principle.* Orthophosphate ($P_i$) is determined colorimetrically following enzymic hydrolysis of 3-phosphoglycerate. Cacodylate buffer is used because it does not inhibit the enzyme nor interfere with the $P_i$ determination.

### Reagents

Sodium 3-phosphoglycerate, 10 m$M$, pH 5.9
Cacodylate buffer, 200 m$M$, pH 5.9
Trichloroacetic acid, 12%
Reagents for determination of $P_i$[1-3]

*Procedure.* The reaction mixture, containing 160 $\mu$moles (0.8 ml) cacodylate buffer at pH 5.9, and 6 $\mu$moles (0.6 ml) 3-phosphoglycerate is equilibrated at 30° for 10 min. The reaction is initiated by the addition of 1.0 ml of enzyme solution which has been held in an ice bath. After 10 min the reaction is terminated by the addition of 0.5 ml of 12% trichloroacetic acid, denatured proteins are removed by a brief centrifugation, and the $P_i$ is determined on a 1.0-ml aliquot. The appropriate controls, one without enzyme and one without substrate, are used to determine endogenous $P_i$. With the purified enzyme the volumes of reactants can be reduced to half, and $P_i$ can be determined on the acidified reaction mixture without centrifugation. Protein determinations can be by the method of Lowry *et al.*[4]

*Units and Specific Activity.* One unit is the amount of enzyme needed for the production of 1 $\mu$mole of $P_i$ per minute. Specific activity is expressed as units per milligram of protein.

### Purification Procedure

This procedure was developed using the leaves of sugarcane, *Saccharum*, grown in the greenhouse from stem nodes.[5] Preparations of the enzyme from spinach leaves did not give as high a specific activity. The enzyme has also been prepared from bean leaves. A summary of the purification procedure is shown in the table.

*Initial Extract.* Freshly harvested leaves are washed, deribbed, and diced into small pieces at room temperature, but the remaining steps in the procedure are all done at 0–4°. The tissue is immersed for 5–15 min in 5 volumes of a medium containing 20 m$M$ cacodylate buffer at pH 6.3, 1 m$M$ EDTA, 2 m$M$ ascorbate or isoascorbate, and 2% (w/v) Poly-

---

[1] C. H. Fiske and Y. SubbaRow, *J. Biol. Chem.* **66**, 375 (1925).
[2] P. S. Chen., Jr., T. Y. Toribara, and H. Warner, *Anal. Chem.* **28**, 1756 (1956).
[3] I. Berenblum and E. Chain, *Biochem. J.* **32**, 295 (1938).
[4] O. H. Lowry, N. J. Rosebrough, A. L. Farr, and R. J. Randall, *J. Biol. Chem.* **193**, 265 (1951).
[5] D. D. Randall and N. E. Tolbert, *J. Biol. Chem.* **246**, 5510 (1971).

PURIFICATION OF 3-PHOSPHOGLYCERATE PHOSPHATASE FROM SUGARCANE LEAVES

| Fraction | Unit ($\mu$moles/min) | Specific activity | Yield (%) | Total protein (mg) |
|---|---|---|---|---|
| Crude extract | 5048 | 0.2 | 100 | 24,038 |
| pH fractionation | 4854 | 0.5 | 93 | 10,112 |
| First acetone fractionation | 2864 | 9.0 | 67 | 309 |
| (NH$_4$)$_2$SO$_4$ fractionation | 2223 | 23.1 | 52 | 96 |
| Second acetone fractionation | 2006 | 50.6 | 46 | 40 |
| G-200 Sephadex chromatography | 1190 | 100 | 30 | 39 |
| Ion exchange chromatography and concentrated | 498 | 740 | 4.4 | 0.6 |

clar AT (polyvinylpolypyrrolidone, Sigma). The tissue is then homogenized for 2 min in the same medium with a Waring Blendor running at high speed, and the homogenate is pressed through 6 layers of cheesecloth. The filtrate is immediately adjusted to pH 4.5 with 3 $N$ HCl, clarified by centrifugation, and readjusted to pH 6.3 with 3 $N$ NaOH. The enzyme preparation is now stable. The extraction and pH fractionation should be completed as rapidly as possible to reduce detrimental action of phenolic compounds. For this reason large preparations should consist of several smaller batches so that each can be processed quickly. All centrifugations in the preparations are done at 14,000 $g$ for 20 min.

*First Acetone Fractionation.* A volume of reagent grade acetone maintained at —5° and equal to 35% of the volume of the enzyme is added dropwise through polyethylene tubes (1 mm i.d.) at about 5 ml/min. The enzyme solution is continuously stirred during all fractionations. After allowing the system to equilibrate for 15–30 min, the precipitate is removed by centrifugation and discarded. The phosphatase is precipitated by the addition of another volume of acetone equal to 35% of the starting volume of enzyme. The precipitate is resuspended in a buffer of 20 m$M$ cacodylate at pH 6.3 and 1 m$M$ EDTA equal in volume to 20% of the original extract. The resuspended material is centrifuged to remove insoluble material and designated the first acetone fraction.

*(NH$_4$)$_2$SO$_4$ Fractionation.* An initial (NH$_4$)$_2$SO$_4$ fraction between 0 and 33 g (NH$_4$)$_2$SO$_4$ per 100 ml of solution is discarded. The second fraction between 33 and 55 g of (NH$_4$)$_2$SO$_4$ per 100 ml original enzyme solution precipitates the phosphatase. This precipitate is dissolved in 20 m$M$ cacodylate at pH 6.3 and 1 m$M$ EDTA buffer equal in volume to 10% of the first acetone fraction.

*Second Acetone Fractionation.* This step is performed similarly to the first acetone fraction. This time the phosphatase is precipitated between additions of acetone equal to 63% to 110% of the volume of the ammonium sulfate fraction. The precipitate is resuspended in a few ml of 20 m$M$ cacodylate at pH 6.3 and 1 m$M$ EDTA buffer, and the insolubles are removed by centrifugation. This should be a highly concentrated solution of protein for the gel chromatography procedure.

*Gel Filtration and Ion Exchange Chromatography.* The enzyme solution is made to 10% (w/v) with sucrose and chromatographed on a Sephadex G-200 column (2.5 × 46 cm) equilibrated and eluted with 20 m$M$ cacodylate at pH 6.3 and 1 m$M$ EDTA buffer. The active fractions are pooled and dialyzed for 36 hr against 40 volumes of 50 m$M$ acetate at pH 4.5 and 1 m$M$ EDTA buffer. The enzyme in this solution is then adsorbed on a phosphocellulose column (1.4 × 10 cm) that has been equilibrated with the 50 m$M$ acetate buffer at pH 4.5. The column is washed with 240 ml of this acetate buffer and eluted with a 500 ml gradient of 0 to 0.5 $M$ NaCl in the acetate buffer. The enzyme elutes at aboute 0.25 $M$ NaCl. The active fractions are pooled and dialyzed for 18 hr against 40 volumes of buffer of 50 m$M$ acetate at pH 5.0, and 1 m$M$ EDTA. The enzyme is then absorbed on a carboxymethyl-Sephadex (C-50 medium) column (1.4 × 10 cm). The column is washed with 200 ml 50 m$M$ acetate buffer at pH 5.0 and a 500-ml linear gradient of 0 to 0.5 $M$ NaCl is used for elution. The phosphatase elutes at about 0.25 $M$ NaCl. The enzyme is dialyzed overnight against 40 volumes of 20 m$M$ cacodylate at pH 5.9 and then concentrated by vacuum dialysis.

## Properties

3-Phosphoglycerate phosphatase[5] is stable between pH 4.5 and 7.5, but it is inactivated rapidly outside of this range. The enzyme is stable to freezing and prolonged, sterile storage at 4°. The pH optimum is about 5.7–6.3 and the isoelectric point is about 6.6. The enzyme does not require any cofactors for activity. Heavy metal cations such as $Pb^{2+}$, $Cu^{2+}$, $Hg^{2+}$, and $Zn^{2+}$ inhibit the enzyme. The $K_m$ is 0.28 m$M$. The phosphatase is not completely specific for 3-phosphoglycerate but will hydrolyze phosphocnol-pyruvate and $p$-nitrophenyl phosphate at two-thirds the rate of 3-phosphoglycerate; of 42 other phosphate esters examined, only 7 are hydrolyzed at rates greater than 25% of 3-phosphoglycerate. Phosphoglycerate phosphatase has little activity toward phosphoglycolate hydrolysis and phosphoglycolate phosphatase does not hydrolyze phosphoglycerate.[6]

The 3-phosphoglycerate phosphatase is not found in roots, stems, or

[6] D. E. Anderson and N. E. Tolbert, see Vol. 9, this series, p. 646.

etiolated plant tissue. In surveys this phosphatase activity has been found in all green leaf tissue including 38 terrestrial plants and fresh water algae[7] and in 14 marine plants and algae.[8] It is particularly active in plants with the $C_4$-dicarboxylic acid pathway of photosynthesis, but the $C_3$ bean plants also have a very high level of the phosphatase. In sugarcane leaves 3-phosphoglycerate phosphatase is in the cytoplasm, but in spinach and sugarbeet leaves part of the activity may be in chloroplasts.[9] Another form of 3-phosphoglycerate phosphatase is sometimes found associated with starch particles in spinach leaves.[10]

[7] D. D. Randall, N. E. Tolbert, and D. Gremel, *Plant Physiol.* **48**, 480 (1971).
[8] D. D. Randall, *J. Marine Biol.* in press.
[9] M. J. Abbate and N. E. Tolbert, unpublished observations.
[10] D. D. Randall and N. E. Tolbert, *Plant Physiol.* **48**, 488 (1971).

# [64] 1,3-Diphosphoglycerate Phosphatase

*By* GIAMPIETRO RAMPONI

1,3-Diphosphoglycerate $+ H_2O \rightarrow$ 3-phosphoglycerate $+ P_i$

1,3-Diphosphoglycerate phosphatase splits off the labile phosphate of the 1-position of 1,3-diphosphoglycerate; it hydrolyzes also a large variety of compounds with carboxyl-phosphate bond, and for this reason it is usually called acyl phosphatase (EC 3.6.1.7).

The enzyme, described for the first time by Lipmann[1] as acetyl phosphatase, has been purified from various sources, such as horse muscle[2] and liver[3] bovine brain,[4] heart muscle,[5] human erythrocytes,[6] and partially also from chicken breast muscle[7] and yeast.[8] As regards the enzyme purified and crystallized from rabbit muscle,[9] the hydrolytic activity on 1,3-diphosphoglycerate was not reported.

[1] F. Lipmann, *Advan. Enzymol.* **6**, 231 (1946).
[2] G. Ramponi, A. Guerritore, C. Treves, P. Nassi, and V. Baccari, *Arch. Biochem. Biophys.* **130**, 362 (1969).
[3] G. Ramponi, P. Nassi, G. Cappugi, C. Treves, and G. Manao, *Biochim. Biophys. Acta* **284**, 485 (1972).
[4] D. A. Diederich and S. Grisolia, *J. Biol. Chem.* **244**, 2412 (1969).
[5] D. A. Diederich and S. Grisolia, *Biochim. Biophys. Acta* **227**, 192 (1971).
[6] E. T. Rakitzis and G. C. Mills, *Arch. Biochem. Biophys.* **134**, 372 (1969).
[7] J. F. Pechere, *Bull. Soc. Chim. Biol.* **49**, 897 (1967).
[8] A. M. Firenzuoli, S. Bartoli, and A. Guerritore, *Boll. Soc. Ital. Biol. Sper.* **36**, 1990 (1960).
[9] H. Shiokawa and L. Noda, *J. Biol. Chem.* **245**, 669 (1970).

## Assay Methods

### NADH Oxidation Method

*Principle.* The assay procedure is based on spectrophotometric determination at 340 nm of the residual 1,3-diphosphoglycerate after the enzymic hydrolysis. 1,3-Diphosphoglycerate is converted to glyceraldehyde 3-phosphate by NADH oxidation in the presence of glyceraldehyde-3-phosphate dehydrogenase.

### Reagents

Acetate buffer, 0.1 $M$, pH 5.3

Triethanolamine-HCl buffer (TRA), 0.1 $M$, pH 7.9, containing disodium ethylenediaminetetraacetate (EDTA), 10 m$M$

MgCl$_2$, 0.5 $M$

Glutathione, reduced form, 0.120 $M$, neutralized with NaOH before the use (GSH)

(NH$_4$)$_2$SO$_4$, 2.5 $M$

NADH solution, containing 10 mg/ml

1,3-Diphosphoglycerate, 10 m$M$, prepared according to Ramponi *et al.*[10]

Crystalline glyceraldehyde-3-phosphate dehydrogenase, suspension in 3.2 $M$ ammonium sulfate solution, containing 2 mg/ml (GAPDH)

*Procedure.* Place in a tube 2 $\mu$moles of 1,3-diphosphoglycerate and 50 $\mu$moles of acetate buffer in a total volume of 1 ml; the reaction is started by the addition of 10 $\mu$l of acyl phosphatase appropriately diluted before addition. The acyl phosphatase reaction, carried out at 25°, is stopped, after suitable periods of time, by cooling in ice and adding ammonium sulfate (final concentration 0.4 $M$), which inhibits the enzyme. The residual substrate is measured immediately by a spectrophotometric optical test at 340 nm in a cuvette of 1-cm light path containing in a total volume of 2 ml: TRA-EDTA, 1 ml; NADH, 30 $\mu$l; MgCl$_2$, 40 $\mu$l; GSH, 20 $\mu$l; GAPDH, 10 $\mu$l; the reaction is started by adding an aliquot of the enzymic reaction mixture containing 1,3-diphosphoglycerate.

### Benzoyl Phosphate Method

*Principle.* The assay, reported by Ramponi *et al.*,[11] is based on the formation of benzoate in the hydrolysis of benzoyl phosphate by acyl

---

[10] G. Ramponi, C. Treves, and A. Guerritore, *Experientia* **23**, 1019 (1967).
[11] G. Ramponi, C. Treves, and A. Guerritore, *Experientia* **22**, 705 (1966).

phosphatase. In the region 220–300 nm, benzoyl phosphate has a higher extinction coefficient than benzoate. The region chosen for the measurement of the enzyme activity is in the range 260–300 nm, where the maximum $\Delta E$ is at 283 nm; also lower wavelengths are suitable for measurement with low substrate concentrations.[12]

### Reagents

Acetate buffer, 0.2 $M$, pH 5.3

Lithium benzoyl phosphate, 15 m$M$, was synthesized according to Ramponi et al.[12]; the preparation had a 98% content of hydroxylamine reactive carboxyl-phosphate bonds, and a benzoic acid:P mole ratio of 1:1

*Procedure.* Put in the reference cell (0.5 cm light path): acetate buffer, 0.750 ml; lithium benzoyl phosphate, 0.300 ml; $H_2O$ to a final volume of 1.5 ml. The text mixture contains in a quartz cell (0.5 cm light path): acetate buffer, 0.750 ml; lithium benzoyl phosphate, 0.500 ml; enzyme solution 0.010 ml or more; $H_2O$ to a final volume of 1.5 ml. The reaction was started by the addition of the enzyme. Since the extinction change at 283 nm is 0.630 m$M^{-1}$ cm$^{-1}$, the micromoles of substrate hydrolyzed per minute are obtained by multiplying the $\Delta E$/min by 4.8.

## Acetyl Phosphate Method

*Principle.* This method, previously described by Lipmann and Tuttle,[13] is based on the reaction of acetyl phosphate with hydroxylamine to form hydroxamic acid; this, in the presence of ferric salts, gives a complex, colored red to violet, that may be determined quantitatively by colorimetric analysis.

### Reagents

Acetate buffer, 0.1 $M$, pH 5.3

Acetyl phosphate, 60 m$M$

Solution A: neutralized hydroxylamine·HCl solution. This is prepared just before use by mixing equal volumes of 4 $M$ $NH_2OH·HCl$ and 3.5 $M$ NaOH

Solution B: this is prepared by mixing equal volumes of 5% $FeCl_3$ in 0.1 $N$ HCl, 12% trichloroacetic acid, 3 $N$ HCl

*Procedure.* Place in a tube 0.1 ml of acetyl phosphate solution, 0.5 ml of the acetate buffer solution and $H_2O$ to a final volume of 1 ml (minus

[12] G. Ramponi, C. Treves, and A. Guerritore, *Arch. Biochem. Biophys.* **115**, 129 (1966).

[13] F. Lipmann and L. C. Tuttle, *J. Biol. Chem.* **159**, 21 (1945).

the amount of the enzyme). The enzyme reaction is started by adding
to the mixture, incubated at 25°, a small amount of acyl phosphatase
solution. After a suitable time, the reaction is stopped by adding 1 ml
of Solution A. After the preparation has stood for 5 min at room temperature, 1 ml of $H_2O$ and then 3 ml of Solution B are added and shaken
vigorously. A control tube without enzyme is also prepared and incubated
in the same way. The colored solution is read at 530 nm.

  *Units.* The units of activity are determined as follows: for horse muscle and liver acyl phosphatase, a unit of activity is defined as the amount
of enzyme which splits 1 $\mu$mole of benzoyl phosphate per minute at 25°
and pH 5.3; for brain and pork heart acyl phosphatase a unit is the
amount of enzyme that hydrolyzes 1 $\mu$mole of acetyl phosphate in 20
min at 27° and pH 6.0. For the erythrocyte acyl phosphatase, a unit
is defined as the amount of enzyme that hydrolyzes 1 $\mu$mole of acetyl
phosphate per hour at 37° and pH 5.4. Specific activity is defined as
units per milligram of protein.

### Purification Procedure

*Preparation of Horse Muscle Enzyme*

  The procedure given is that of Ramponi *et al.*[2] Horse muscle, 110
kg, is cleaned, cut in small pieces, and minced. The tissue is homogenized
in 550 liters of 0.116 $M$ HCl; after the homogenate had stood overnight,
it is centrifuged at 2000 $g$ for 30 min in a refrigerated centrifuge. The
above supernatant is then brought to pH 4.8 by addition of 10 $N$ NaOH
with efficient stirring. A copious precipitate is discarded by centrifuging
at 2000 $g$; to remove the fat, the supernatant fraction is filtered through
fiber glass; 75 ml of 6 $M$ $ZnCl_2$ are added slowly for each liter of the
above supernatant. The mixture, adjusted to pH 5.8 with 10 $N$ NaOH
is allowed to settle. The clear supernatant is siphoned off, and the residue
is centrifuged at 2000 $g$ for 20 min; the supernatant is discarded.

  To the collected precipitate, mixed with crushed ice, concentrated HCl
is added with efficient stirring until the pH of the mixture is 1.5. The
suspension obtained is dialyzed against 10 volumes of 0.05 $N$ HCl for
4 days. It is important to replace every 24 hr the HCl solution and to
resuspend the precipitate within the dialysis membrane. During this procedure, by the removal of Zn ions, there is a solubilization of the enzyme.
The dialyzed solution, adjusted to pH 5.1 with 10 $N$ NaOH, is centrifuged
for 60 min at 2000 $g$. The supernatant fraction is collected and brought
to pH 8.1. The precipitate is discarded by centrifugation for 60 min at

2000 g. The supernatant so obtained (19 liters) is diluted to 30 liters with distilled water in order to lower the ionic strength and then adjusted to pH 5.8 with concentrated HCl.

This solution is introduced into a column of CM-Sephadex C-25, 15 × 62 cm, equilibrated with 50 mM acetate buffer, pH 5.8. The CM-Sephadex bed is washed with 20 liters of equilibrating buffer and the proteins are eluted with a linear salt concentration gradient. The column effluent is fractionated into 45-ml portions at the rate of 5–6 fractions per hour. The protein content of each fraction is estimated spectrophotometrically at 280 nm, and the chloride according to the method of Volhard. Most of the acyl phosphatase activity is located in a peak of protein, at 0.17–0.22 M NaCl. A second peak of activity, much lower than the first one, located at 0.25–0.30 M NaCl, is discarded. Fractions 415–465 inclusive are pooled (corresponding to 2260 ml of eluate, containing 791 mg of protein).

The above solution is diluted 4 times with water and introduced into a column of CM-Sephadex C-25 (2 × 20 cm), equilibrated with 50 mM acetate buffer, pH 5.8. The gel bed is washed with 300 ml of the same buffer, and the enzyme is eluted with a linear salt concentration gradient. The effluent is collected in 4.5-ml fractions at the rate of 10 ml/hr. A single active peak is obtained, corresponding to the highest protein peak. Eluate fractions 160–172 inclusive, are pooled (57.5 ml containing 124 mg of protein). This solution has a specific activity of 3000 expressed as micromoles of benzoyl phosphate hydrolyzed per minute per milligram of protein, at 25° and pH 5.3. This indicates a 4225-fold purification. A summary of the procedure is given in Table I.

## Preparation of Horse Liver Enzyme

The procedure reported is that of Ramponi et al.[3] Tissue, 25 kg, is homogenized in 100 liters of 1.5 M acetic acid. The supernatant, collected by centrifugation for 60 min at 2000 g is brought to pH 4.9; the precipitate obtained by centrifugation is discarded and the supernatant is dialyzed against tap water for 40 hr. It is then chromatographed on a CM-Sephadex C-25 column (15 × 80 cm) equilibrated with 50 mM acetate buffer, pH 5.8. The column is eluted by a salt concentration gradient from 0 to 1 M NaCl. The elution profile gives four components with acyl phosphatase activity. Fractions 280–320 (45 ml each), corresponding to the major component with acyl phosphatase activity, are pooled.

The above fractions are rechromatographed on a CM-Sephadex C-25 column (5 × 55 cm) equilibrated with 50 mM acetate buffer, pH 4.8.

TABLE I

PURIFICATION OF HORSE MUSCLE ENZYME[a]

| Step | Fraction | Volume (ml) | Protein (mg) | Specific activity (units/mg) | Total activity (units) | Yield (%) | Purification (fold) |
|---|---|---|---|---|---|---|---|
| 1 | Acid extract | 382,000 | 8,861,000 | 0.71 | 6,382,000 | 100 | |
| 2 | Foreign-protein precipitation | 292,200 | 664,000 | 6.70 | 4,276,000 | 67.0 | 9.4 |
| 3 | ZnCl$_2$ precipitation | 19,000 | 44,270 | 30 | 1,330,000 | 20.8 | 42.2 |
| 4 | First CM-Sephadex chromatography | 2,260 | 791 | 897 | 709,000 | 11.1 | 1263 |
| 5 | Second CM-Sephadex chromatography | 57.5 | 124 | 3000 | 372,000 | 5.8 | 4225 |

[a] From G. Ramponi, A. Guerritore, C. Treves, P. Nassi, and V. Baccari, *Arch. Biochem. Biophys.* **130**, 362 (1969).

The elution is performed with a salt concentration gradient from 0 to 1 $M$ NaCl, and tubes 410–438 (30 ml each) are pooled. One-fifth of the above fraction is again rechromatographed on a CM-Sephadex C-25 column (1 × 20 cm) equilibrated with 50 m$M$ phosphate buffer, pH 5.8. Phosphate is a well known inhibitor of acyl phosphatase activity, and it is therefore chosen to attempt a specific elution of the enzyme. The column is eluted by a salt concentration gradient from 0 to 0.3 $M$ NaCl dissolved in the above phosphate buffer. Fractions 100–110 (2.4 ml each) are pooled. This procedure is repeated for all the preparation. The enzyme solution obtained has a specific activity of 2800, which indicates about 16,000-fold purification. In Table II the steps of the purification procedure are summarized.

## Preparation of Bovine Brain Enzyme

The procedure described is that of Diederich and Grisolia.[4] Batches of bovine brain, 25 kg, are processed daily. For preparation reported here, 500 kg were used. Portions of tissue (500 g) are homogenized for 60–90 sec in a Waring Blendor with 2 liters of 0.1 $M$ sodium acetate, pH 4.0; the homogenate is poured into a polyethylene container equipped with a spigot and allowed to settle for 8–12 hr. Beneath a layer of thick foamy protein and fat, a reddish clear layer is formed; this is drained off and filtered through coarse filter paper on 15-liter Büchner funnels. To the filtrate (fraction 1) 2 volumes of acetone are added and the mixture is allowed to settle overnight. The bulk of the supernatant fluid is siphoned off, the residue is centrifuged, and the supernatant is discarded. The precipitate is suspended in 250 ml of $H_2O$ per kilogram of starting tissue, stirred, and then centrifuged. The supernatant so obtained is defined as fraction 2. This, when frozen, is stable for at least 6 months.

Fraction 2 is diluted with water to a protein concentration of 7.5 mg/ml and then 0.14 volume of 0.15 $M$ sodium sulfosalicylate, pH 2.0, is added, with stirring. To the supernatant obtained by centrifugation at 10,000 $g$ for 10 min, two volumes of acetone are added and the pecipitate is allowed to settle. The bulk of the supernatant is siphoned off, and the residue is centrifuged at 3000 $g$ for 10 min. The precipitate, finely suspended in 20 m$M$ Tris·HCl, pH 6.75 (20 ml per kilogram of starting tissue), is centrifuged at 10,000 $g$ for 10 min. The supernatant (fraction 3) is dialyzed 4–6 hr against 20 m$M$ Tris·HCl, pH 6.7, and stored at −20°C.

Fraction 3, containing 8.0–8.5 g of protein, is introduced into a column of Bio-Rex 70 (5 × 100 cm), 200–400 mesh, and equilibrated with 50 m$M$ Tris·HCl, pH 6.9. Ten such columns are used at the same time.

## TABLE II
### PURIFICATION OF HORSE LIVER ENZYME[a]

| Step | Fraction | Volume (ml) | Protein (mg) | Specific activity (units/mg) | Total activity (units) | Yield (%) | Purification (fold) |
|---|---|---|---|---|---|---|---|
| 1 | Acetic acid extract | 93,000 | 1,848,780 | 0.17 | 319,700 | 100 | — |
| 2 | Foreign-proteins precipitation | 98,000 | 56,840 | 2.76 | 157,000 | 49.1 | 16 |
| 3 | First CM-Sephacex chromatography | 1,900 | 1,805 | 36.6 | 66,000 | 20.6 | 215 |
| 4 | Second CM-Sephadex chromatography | 800 | 116 | 386 | 44,800 | 14.0 | 2,270 |
| 5 | CM-Sephadex chromatography (specific elution) | 102.5 | 7.5 | 2800 | 21,500 | 6.7 | 16,470 |

[a] From G. Ramponi, P. Nassi, G. Cappugi, C. Treves, and G. Manao, *Biochim. Biophys. Acta* **284**, 485 (1972).

These are washed with 50 m$M$ Tris·HCl, pH 6.9, until the protein concentration of the effluent is less than 0.1 mg per ml (8–10 liters). The enzyme is eluted by changing to 0.162 $M$ Tris·HCl, pH 6.9 (about 3000 ml). The residual protein is removed from the resin by flowing through the column 2000 ml of 0.5 $M$ Tris·HCl, pH 6.9. The contents of all tubes with a specific activity above 1500 are pooled; 1000–1200 ml was so obtained. This fraction contained over 90% of the enzyme present in the column effluent with a specific activity of 4000–5000. In order to remove the bulk of buffer, the sample is concentrated to about 20 ml, diluted to 120 ml, and then reconcentrated to about 20 ml; the Model 400 Diaflo cell with the UM-2 membrane is used. This fraction, named fraction 4, is stable at —20° for at least 6 months.

Five such fractions (containing 1200–1500 mg of protein) are pooled and concentrated to a protein concentration of 25–30 mg/ml. This fraction is applied to a column of Bio-Rex 70 (2.5 × 100 cm), 200–400 mesh, with a flow rate of 8–10 ml/hr, and equilibrated with 50 m$M$ Tris·HCl buffer, pH 6.9. The column was washed with 50 m$M$ Tris·HCl, pH 6.9, until the effluent had less than 0.05 mg of protein per milliliter. The enzyme is eluted with 0.162 $M$ Tris·HCl buffer, pH 6.9. The tubes containing a constant specific activity of 25,750 ± 250 are pooled to give about 70 ml. This is then concentrated (with a Model 50 Diaflo cell) to about 5 ml (fraction 5). This fraction (300,000 units/ml) is stable at —20° for at least 1 year. A summary of the purification procedure is shown in Table III.

TABLE III
PURIFICATION OF BOVINE BRAIN ENZYME[a]

| Fraction | Volume (ml) | Total activity (units × 10⁴) | Total protein (mg) | Specific activity (units/mg protein) | Yield (%) |
|---|---|---|---|---|---|
| Homogenate | 125,000 | 7.5 | 6,000,000 | 1.5 | 100 |
| 1 | 88,000 | 5.1 | 176,000 | 30 | 70 |
| 2 | 5,000 | 4.5 | 41,000 | 100 | 61 |
| 3[b] | 5,000 | 17.35 | 69,000 | 250 | 46 |
| 4 | 100 | 6.94 | 1,500 | 4,500 | 19 |
| 5 | 25 | 2.25 | 90 | 25,000 | 5 |

[a] From D. A. Diederich and S. Grisolia, *J. Biol. Chem.* **244**, 2412 (1969).

[b] The equivalent of five batches was combined at this point and then processed in batches of 8.5 g which in turn were combined before processing to step 5. Thus, the data are given for the purification of a 25-kg batch to fraction 2 and for 125 kg for the remaining fractions.

*Preparation of Heart Muscle Enzyme*

The procedure is that of Diederich and Grisolia.[5] Batches (12 kg) of pork heart are processed daily; 500-g portions of ground muscle are homogenized in a Waring Blendor for 3 min with 2 liters of water. The homogenate is acidified to pH 1.7 with 1 $M$ HCl and then brought to 70° with continuous stirring in stainless steel containers immersed in a steam bath. This step should be carried on for no more than 7–8 min, in order to avoid a loss of enzyme activity. The material is then rapidly cooled to 10° by stirring in an ice-acetone bath. The pH is adjusted to 5.0 with 1.0 $M$ NaOH; and the precipitate which forms is removed by filtration through cloth towels. The filtrate (fraction 1) is mixed with 2 volumes of cold acetone and allowed to settle overnight. The supernatant fluid is siphoned off, and the residue is centrifuged. The precipitate, after the bulk of the acetone is evaporated by use of a hair dryer, is suspended in 450 ml of water per kilogram of starting tissue and stirred until finely dispersed. This fraction is centrifuged; the supernatant is fraction 2.

To fraction 2, 0.1 volume of 0.1 $M$ sodium acetate, pH 4.0, and 0.6 volume of cold acetone are added in this order; after centrifugation, 2.5 volumes of acetone are added to the supernatant fluid. The precipitate is allowed to settle and, after the bulk of the supernatant is decanted, the residue is centrifuged. The precipitate is resuspended in 50 m$M$ Tris·HCl, pH 7.0 (30 ml/kg of starting tissue); after vigorous stirring, the suspension is centrifuged and the precipitate is discarded. The supernatant (fraction 3) can be stored at —20° up to 6 months without loss of enzymic activity.

Fraction 3 obtained from two pooled 12-kg batches of pork heart are layered on a 4 × 100 cm column of Bio-Rex 70, 200–400 mesh, equilibrated with 50 m$M$ Tris·HCl, pH 6.9, pumped at 40 ml/hr. The 50 m$M$ Tris·HCl wash is continued until the protein concentration of the eluate is below 0.1 mg/ml. The enzyme elution is carried out with 2000 ml of 0.165 $M$ Tris·HCl buffer, pH 6.9. Three peaks of acyl phosphatase activity are eluted (A, B, and C). The tubes containing the bulk of the acyl phosphatase activity in peaks A (Fraction 4A) and B + C combined (Fraction 4B + 4C) are pooled separately and concentrated to about 20 ml each with a Diaflo ultrafiltration assembly using a UM-2 membrane. After concentration, fractions 4A and 4B + 4C are stored at —20°.

Fractions 4B + 4C from five such columns were pooled, diluted with water (2.5 volumes) to obtain a Tris·HCl concentration below to 50 m$M$ and then concentrated in a Model 50 Diaflo assembly to a protein concentration of about 25 mg/ml. This fraction was then put on 2.5 × 100 cm column of Bio-Rex 70, <400 mesh, equilibrated as for the preparative

TABLE IV
PURIFICATION OF PORK HEART ENZYME[a]

| Fraction | Volume (ml) | Total activity (units × 10³) | Total protein (mg) | Specific activity (units/ mg protein) | Yield (%) |
|---|---|---|---|---|---|
| Homogenate | 60,000 | 2,100 | 2,100,000 | 1.0 | 100 |
| Fraction 1 | 60,000 | 1,080 | 120,000 | 10.0 | 50 |
| Fraction 2 | 5,400 | 920 | 37,000 | 35.0 | 43 |
| Fraction 3 | 360 | 720 | 7,200 | 100.0 | 34 |
| Fraction 4: | | | | | |
| A | 380 | 108 | 520 | 208 | 5.1 |
| B + C | 800 | 410 | 716 | 570 | 19.5 |
| D | 310 | 84 | 370 | 226 | 4.0 |
| Fraction 5: | | | | | |
| B | 48 | 4 | 49.5 | 1,080 | 2.5 |
| C | 60 | 86 | 7.1 | 12,000 | 4.1 |

[a] From D. Diederich and S. Grisolia, *Biochim. Biophys. Acta* **227**, 192 (1971).

column. The column was washed at a flow rate of 10 ml/hr with 50 mM Tris·HCl buffer, pH 6.8, until the protein concentration is less than 0.05 mg/ml. The column buffer is changed to 0.16 M Tris·HCl, pH 7.0, and two well separated peaks of acyl phosphatase activity are eluted (fractions 5B and 5C). Tubes corresponding to the second peak with a specific activity above 10,000 are pooled (fraction 5C). Fraction 5C diluted with water to a Tris·HCl concentration of 50 mM, is then concentrated with a Diaflo assembly to a protein concentration of about 20 mg/ml. This fraction was stored at −20° and used for subsequent studies. Fraction 5C from separate columns had a specific activity of 12,000 ± 800. A summary of the purification procedure is reported in Table IV.

## Preparation of Human Erythrocyte Enzyme

The purification procedure is that of Rakitzis and Mills.[6] All steps are carried out at 4°. Outdated human bank blood is used routinely, although sometimes freshly drawn blood is used. The blood is centrifuged at 140 g for 10 min. The cells, after removal of the plasma and buffy coat by suction, are washed twice with 15 volumes of 0.15 M NaCl, containing 0.1% glucose. Packed cells (50 ml) are then lysed with 200 ml of water and stirred for 30 min. The pH is brought to 7.50 with 3.5 N NaOH, and the hemolysate is introduced into a CM-cellulose column (3 × 28 cm) previously equilibrated with 5 mM phosphate buffer, pH

TABLE V
PURIFICATION OF ACYL PHOSPHATASE FROM HUMAN RED CELLS[a]

| Fraction | Total volume (ml) | Total protein (mg) | Total activity[b] | Specific activity[c] | Yield (%) |
|---|---|---|---|---|---|
| Crude hemolysate | 250 | 30,600 | 23,000 | 0.75 | 100 |
| CM-cellulose eluate | 70 | 80.5 | 8,960 | 111 | 39 |
| DEAE-cellulose eluate | 10.3 | 7.1 | 5,120 | 721 | 22 |

[a] From E. T. Rakitzis and G. C. Mills, *Arch. Biochem. Biophys.* **134**, 372 (1969).
[b] In micromoles per hour at pH 5.4.
[c] In micromoles per hour per milligram of protein.

7.50. After the hemolysate has passed into the column, hemoglobin is removed by running through the column 300 ml of a 5 m$M$ sodium phosphate buffer, pH 7.50. A 0.5 $M$ NaCl solution is then applied to the top of the column, and 40-ml fractions are collected.

Fractions 7–10, containing acyl phosphatase activity, are pooled and placed inside a dialysis bag surrounded with Carbowax flakes. The enzyme preparation is concentrated to approximately 70 ml and then dialyzed overnight, with stirring, against 5 liters of water; after the dialysis, one-ninth volume of 0.1 $M$ Na$_2$HPO$_4$ solution is added in order to obtain a final phosphate concentration of 10 m$M$. The enzyme solution (pH 9.2–9.4) is passed through a DEAE-cellulose column (1 $\times$ 37 cm), previously equilibrated with a freshly prepared 10 m$M$ Na$_2$HPO$_4$ solution. A 10 m$M$ phosphate buffer, pH 6.0, is applied to the column, and 10-ml fractions are collected. Fractions 4–10, containing the enzyme activity, are pooled, concentrated with Carbowax, and dialyzed for 5 hr against 2 liters of water. The enzyme preparation (10.3 ml) contains 22% of the original hemolysate activity, with an overall 920-fold purification. A summary of the steps of the preparation procedure is given in Table V.

## Properties

*Hydrolytic Activity on 1,3-Diphosphoglycerate.* 1,3-Diphosphoglycerate phosphatase prepared from various sources, as reported previously, hydrolyzes 1,3-diphosphoglycerate, and relative rates of enzymic hydrolysis for this substrate and acetyl phosphate are reported in Table VI.

*Effect of 1,3-Diphosphoglycerate Concentration.* The kinetic properties, expressed as $K_m$, of acyl phosphatase prepared from various sources are reported comparatively in Table VI.

*Effect of pH (Optimum pH).* The optimal pH for horse muscle and

TABLE VI

SUBSTRATE SPECIFICITY AND $K_m$ FOR 1,3-DIPHOSPHOGLYCERATE OF
1,3-DIPHOSPHOGLYCERATE PHOSPHATASE FROM SEVERAL TISSUES

| Source | 1,3-Diphos-phoglycerate acetyl phosphate[a] | Carbamyl phosphate acetyl phosphate[a] | Benzoyl phosphate acetyl phosphate[a] | $K_m$ (m$M$) |
|---|---|---|---|---|
| Horse muscle | 1.20 | 0.09 | 6 | 1.6 |
| Chicken breast muscle | 0.12 | — | — | — |
| Heart pig | 0.09 | 0.09 | — | — |
| Beef brain | 0.11 | 0.10 | — | — |
| Horse liver | 0.70 | 0.10 | 7.50 | — |
| Human erythrocytes | 0.15 | — | — | 0.12 |

[a] The rate of hydrolysis of acetylphosphate is assumed as unit.

horse liver acyl phosphatase activity on 1,3-diphosphoglycerate is about 5.3.[3,10] The optimal pH for bovine brain acyl phosphatase determined on acetyl phosphate was 7.4–7.6.[4] The maximal activity at approximately pH 5 was found for the human erythrocyte enzyme, using acetyl phosphate as substrate.[6] The optimum pH range for the hydrolysis of acetyl phosphate by heart acyl phosphatase was 5.4–5.6.[5]

*Activity on Other Compounds.* Horse muscle acyl phosphatase hydrolyzes *p*-nitrobenzoyl phosphate, benzoyl phosphate, and carbamyl phosphate; it shows very low hydrolytic activity toward phosphocreatine, adenosine 5′-triphosphate, pyrophosphate, and *p*-nitrophenyl phosphate at pH 5.3; it does not split *p*-nitrophenyl phosphate at pH 10.4, phosphoenolpyruvate, and phosvitin.[2] The specificity of horse liver acyl phosphatase toward various compounds is very similar to that found for muscle acyl phosphatase: besides 1,3-diphosphoglycerate and acetyl phosphate, it hydrolyzes *p*-nitrobenzoyl phosphate, benzoyl phosphate, and carbamyl phosphate; it does not split the other compounds studied.[3] Acyl phosphatase prepared from bovine brain and pork heart hydrolyzes, besides 1,3-diphosphoglycerate and acetyl phosphate, also carbamyl phosphate.[4,5] Acyl phosphatase prepared from human erythrocytes splits 1,3-diphosphoglycerate and acetyl phosphate but does not show hydrolytic activity on carbamyl phosphate, 3-phosphoglycerate, phosphocreatine, DL-3-glycerophosphate, fructose 1,6-diphosphate, glucose 6-phosphate, fructose 6-phosphate, phosphoenolpyruvate, ATP, ADP, AMP, cyclic-AMP, 6-phosphogluconate phosphoserine, inorganic pyrophosphate, and *p*-nitrophenyl phosphate.[6]

*Molecular Weights.* The molecular weight of horse muscle acyl phosphatase, determined by the Archibald method, is 9400, the average value obtained with two enzyme concentrations. This value, used for the calculation of the amino composition, is also in good agreement with the values obtained by sedimentation and Sephadex G-75 gel filtration procedures.[2,14] Horse liver acyl phosphatase shows an apparent molecular weight of 8300 when determined by the Sephadex G-75 gel filtration method. Using the sodium dodecyl sulfate-polyacrylamide gel electrophoresis procedure,[15] the mobility of the liver enzyme overlaps that of muscle acyl phosphatase.[3] For bovine brain acyl phosphatase a molecular weight of 12,000 was found using a Sephadex G-75 column, and 13,800 using Bio-Gel column.[4] The molecular weight of fraction 5C of the pork heart acyl phosphatase estimated by the gel filtration method was 11,095,[5] and the value for the enzyme prepared from human erythrocytes has not been reported.

*Isoelectric Point.* An isoelectric point of 11.4 was obtained for the horse muscle enzyme.[14] The isoelectric point of fraction 5C of pork heart acyl phosphatase, determined by means of the electrofocusing technique, was 7.25–7.3.[5] Data for the enzyme prepared by other sources were not reported, although the retention of human erythrocyte acyl phosphatase on a carboxymethyl-cellulose column at pH 7.5 and on a diethylamino-ethyl-cellulose column at pH 9.4 indicates a basic protein.

*Amino Acid Composition.* Comparison of amino acid composition of acyl phosphatase obtained from various sources (horse muscle, horse liver, bovine brain) is reported in Table VII. Also the amino acid composition of horse muscle acyl phosphatase obtained by complete enzymic hydrolysis is reported in Table VII.

*End Groups.* For horse muscle acyl phosphatase, after tryptic digestion of carboxymethylated protein, the $NH_2$-terminal peptide was isolated, and it was found that the $NH_2$-terminal amino acid is blocked.[16] This peptide was sequenced by an enzymic method using carboxypeptidase B and A. The kinetics of amino acids release indicate this sequence: X-HN-Ser-Thr-Ala-Arg. For the determination of COOH-terminal amino acids, hydrazinolysis, selective tritiation, and digestion with carboxypeptidase A gave tyrosine as the COOH-terminal amino acid of CM-acyl phosphatase.[17] The native enzyme, has another $NH_2$-terminal amino acid: the glutamic acid, and another COOH-terminal amino acid,

[14] G. Ramponi, C. Treves, and A. Guerritore, *Arch. Biochem. Biophys.* **120**, 666 (1967).

[15] A. K. Dunker and R. R. Rueckert, *J. Biol. Chem.* **244**, 5074 (1969).

[16] G. Cappugi, P. Chellini, P. Nassi, and G. Ramponi, *Proc. Int. Cong. Biochem. 9th, Stockholm,* July 1973, n. 2p 10.

[17] G. Ramponi, G. Cappugi, C. Treves, and P. Nassi, *Biochemistry* **10**, 2082 (1971).

## TABLE VII
COMPARISON OF AMINO ACID COMPOSITION[a] OF 1,3-DIPHOSPHOGLYCERATE
PHOSPHATASE FROM VARIOUS SOURCES

| Amino acid | Native[b] (acid hydrolysis) | Horse muscle CM-enzyme[c] (acid hydrolysis) | CM-enzyme[d] (enzymic hydrolysis) | Horse liver[e] (acid hydrolysis) | Bovine brain[f] (acid hydrolysis) |
|---|---|---|---|---|---|
| Tryptophan | — | — | 1 | — | 2[g] |
| Lysine | 8 | 8 | 8 | 8 | 7 |
| Histidine | 1 | 1 | (1) | 1 | 2 |
| Arginine | 5 | 5 | 3 | 5 | 3 |
| CM-cysteine | — | 1 | 1 | — | — |
| Cystine (half) | 2 | — | — | 1 | — |
| Aspartic acid | 6 | 6 | 3 | 7 | 7 |
| Threonine | 5 | 5 | 4 | 5 | 5 |
| Serine + amides | — | — | 16 | — | — |
| Serine | 10 | 10 | — | 13 | 4 |
| Glutamic acid | 9 | 8 | 6 | 10 | 11 |
| Proline | 3 | 3 | 2 | 3 | 3 |
| Glycine | 7 | 6 | 6 | 9 | 7 |
| Alanine | 3 | 3 | 3 | 4 | 4 |
| Valine | 8 | 8 | 8 | 8 | 7 |
| Methionine | 2 | 2 | 1 | 1 | 1 |
| Isoleucine | 2 | 2 | 3 | 3 | 3 |
| Leucine | 3 | 3 | 3 | 3 | 5 |
| Tyrosine | 3 | 3 | 3 | 2 | 2 |
| Phenylalanine | 3 | 3 | 3 | 3 | 4 |

[a] Each amino acid is given as the number of residues per molecule of protein (nearest integral number).
[b] G. Ramponi, A. Guerritore, C. Treves, P. Nassi, and V. Baccari, *Arch. Biochem. Biophys.* **130**, 362 (1969).
[c] G. Ramponi, G. Cappugi, C. Treves, and P. Nassi, *Biochemistry* **10**, 2082 (1971).
[d] G. Ramponi, G. Cappugi, C. Treves, and P. Nassi, *Life Sci.* **10**, 983 (1971).
[e] G. Ramponi, P. Nassi, G. Cappugi, C. Treves, and G. Manao, *Biochim. Biophys. Acta* **284**, 485 (1972).
[f] D. A. Diederich and S. Grisolia, *J. Biol. Chem.* **244**, 2412 (1969).
[g] Determined spectrophotometrically.

glycine. The differences between native horse muscle acyl phosphatase and the carboxymethylated one are attributable to the fact that the native enzyme consists of two chains joined by an S—S bond. One chain corresponds to the CM-acyl phosphatase, and the other to glutathione.[17]

The inability to detect any NH$_2$-terminal amino acid derivative both

ENZYME—ACYLPHOSPHATE     ENZYME—ALKYLPHOSPHATE     ENZYME—INORGANIC PHOSPHATE

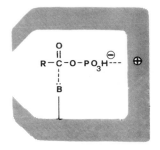 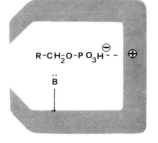 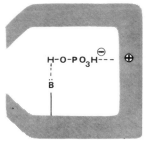

FIG. 1.

for the horse liver[3] and the bovine brain[4] enzyme suggests that also in this case the NH$_2$-terminal residue is substituted.

The COOH-terminal amino acid of horse liver acyl phosphatase[3] obtained by selective tritiation experiments is tyrosine. Lysine is the COOH-terminal amino acid residue of bovine brain acyl phosphatase.[4]

*Effect of Some Chemicals and Cations.* Some metabolic compounds which have been shown[18] to inhibit a less pure preparation of horse muscle acyl phosphatase were tested also with the pure enzyme. The inhibition of the enzyme activity by many phosphorylated compounds indicates the presence of a site on the enzyme for the phosphate group. Simple phosphate esters, such as methyl phosphate and benzyl phosphate, show moderate competitive inhibition ($K_i$ = 3.5 and 11 m$M$, respectively).[19] The high inhibition produced by inorganic phosphate could arise through hydrogen-bonding of the free OH-group of the phosphate to a base (Fig. 1). Adenosine triphosphate and fructose 1,6-diphosphate show noncompetitive inhibition ($K_i$ = 0.57 m$M$ and 3.2 m$M$, respectively), whereas adenosine monophosphate, phosphoethanolamine, and $\alpha$-glycerophosphate are not inhibitors. Pyridoxal 5′-phosphate produces a pH-dependent inhibition of horse muscle acyl phosphatase activity by its interaction with $\epsilon$-amino groups of lysine residues.[20] This inhibition is purely competitive with respect to substrate binding ($K_i$ = 0.32 m$M$). Horse liver acyl phosphatase is inhibited competitively by inorganic phosphate and noncompetitively by orotic acid.[21] As regards

[18] A. Guerritore, G. Ramponi, and A. Zanobini, *Eur. Symp. Med. Enzymol.* Milan p. 507 (1960).

[19] G. F. White, G. Manao, and G. Ramponi, *Acta Vitaminol. Enzymol.* **26** (5/6), 200 (1972).

[20] G. Ramponi, G. Manao, and G. F. White, in press.

[21] P. Nassi, G. Cappugi, A. Niccoli, and G. Ramponi, *Physiol. Chem. Phys.* **5**, 109 (1973).

bovine brain acyl phosphatase, $HgCl_2$ and iodoacetate at 40 $\mu M$ and 4 m$M$, respectively, are not inhibitors; on the other hand, 8 $M$ urea under certain conditions produces an inhibition of the enzyme activity of 100%. Studies have demonstrated enhancement of enzyme activity by 15 m$M$ KCl and 5 m$M$ $MgCl_2$ when the Tris salt of acetyl phosphate is used as substrate. Adenosine triphosphate inhibits competitively human erythrocyte acyl phosphatase with a $K_i$ of 4.4 m$M$. Also inorganic phosphate is a competitive inhibitor of this enzyme, with a $K_i$ of 3.4 m$M$. Carbamyl phosphate was found to inhibit competitively, with a $K_i$ of 6.9 m$M$, when acetyl phosphate was used as substrate. Various other metabolic intermediates show an inhibitory effect on the enzyme activity. $P_i$ and $S_i$ are competitive inhibitors of the purified preparation of pork heart acyl phosphatase, when acetyl phosphate was used as substrate. $HgCl_2$ in concentrations up to 2.0 m$M$ is without effect on the enzyme activity of the final preparation. Iodacetate (10 m$M$) and $p$-chloromercuribenzoate (1 m$M$) pretreatment of the enzyme shows only very low (<5%) inhibition of the acyl phosphatase activity. The effect of $P_i$, adenosinetriphosphate, 2,3-diphosphoglycerate and 3-phosphoglycerate upon the hydrolysis of 1,3-diphosphoglycerate by fraction 5C from heart was also studied: 2 m$M$ 2,3-diphosphoglycerate produced 12% inhibition; 5 m$M$ concentration of adenosine triphosphate, 3-phosphoglycerate, and $P_i$ resulted in 11%, 8%, and 37% inhibition, respectively.

*Stability.* Horse muscle acyl phosphatase, as previously reported, appears to be one of the smallest enzymes on record. Such a low molecular weight could explain the stability of the enzyme to various denaturating agents, its solubility in trichloracetic acid solutions, the difficulty in its precipitation by salts and its thermostability. The purification procedure for horse liver acyl phosphatase involves the acid, heat-stable enzyme, since another acyl phosphatase activity, heat labile, is present in the liver.[22] Acyl phosphatase purified from bovine brain shows an unusual stability to prolonged storage, lyophilization, acid and heat treatment. Heating the purified preparation of human erythrocyte acyl phosphatase (a solution of the enzyme in water) at 70° for 5 min, destroyed 40% of its activity, using acetyl phosphate as substrate. The thermal lability of erythrocyte acyl phosphatase contrasts with the stability at high temperatures of the muscle and brain enzymes. The effect of heat and pH on acyl phosphatase activity from pork heart can be summarized as follows: pH 1.5–2.0, 100% of activity; pH 9.0, 92% of residual activity; 85°, pH 2.0, 5 min, 95% of residual activity.[5]

*Comments.* 1,3-Diphosphoglycerate phosphatase commonly named

[22] A. Guerritore, A. Zanobini, and G. Ramponi, *Boll. Soc. Ital. Biol. Sper.* **35**, 2163 (1959).

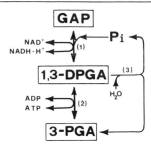

Scheme 1. GAP: Glyceraldehyde 3-phosphate; 1,3-DPGA: 1,3-diphosphoglycerate; 3-PGA: 3-phosphoglycerate; (1): glyceraldehyde-3-phosphate dehydrogenase (EC 1.2.1.12); (2): 3-phosphoglycerate kinase (EC 2.7.2.3); (3): 1,3-diphosphoglycerate phosphatase (EC 3.6.1.7).

acyl phosphatase, has been purified from various sources. It always appears to be an enzyme with a low molecular weight and with the characteristics of a basic protein. It is a widespread hydrolase, which acts specifically by splitting off the carboxyl phosphate bond. It is interesting to note that human erythrocyte acyl phosphatase does not split carbamyl phosphate, which inhibits the enzyme competitively. All preparations of acyl phosphatase here described attack 1,3-diphosphoglycerate. This enzyme would appear to act as a "safety valve," preventing the intracellular accumulation of 1,3-diphosphoglycerate by catalyzing the hydrolysis of this substrate to 3-phosphoglycerate. In the presence of 1,3-diphosphoglycerate phosphatase, an increase of the rate of yeast fermentation[23] and retina glycolysis[24] have been observed. Under the above conditions the decrease of the control exerted by ATP:ADP ratio and inorganic phosphate concentration on the transformation of glyceraldehyde 3-phosphate to 3-phosphoglycerate can occur according to Scheme 1. Furthermore recent studies[25] have demonstrated that 1,3-diphosphoglycerate can acylate histones, particularly the lysine-rich ones, and that 1,3-diphosphoglycerate phosphatase can prevent this acylation.

[23] I. Harary, *Biochim. Biophys. Acta* **26**, 434 (1957).
[24] V. Baccari, A. Guerritore, G. Ramponi, and M. P. Sabatelli, *Boll. Soc. Ital. Biol. Sper.* **36**, 360 (1960).
[25] G. Ramponi and S. Grisolia, *Biochem. Biophys. Res. Commun.* **38**, 1056 (1970).

Section V

Mutases

# [65] Phosphoglycerate Mutase from Wheat Germ
## (2,3-PGA-Independent)

*By* S. Grisolia and J. Carreras

The 2,3-DPGA-independent PGA mutase[1] catalyzes the reaction:

$$3\text{-PGA} \rightleftharpoons 2\text{-PGA} \tag{1}$$

In contrast with the 2,3-DPGA-dependent PGA mutases,[2] which require 2,3-DPGA as cofactor, the 2,3-DPGA-independent PGA mutases do not.[3-6] The direct intramolecular transfer of the phosphoryl between the two hydroxyl groups of the substrate is the only possible mechanism.[7] The acid-catalyzed interconversion of 2-PGA and 3-PGA[8] probably proceeds via the intermediate formation of a cyclic 2,3-phosphoglycerate, but it is unlikely that such a cyclic ester is involved in the PGA mutase mechanism.[7] It has been suggested[7] that the mutase reaction may proceed via the intermediate formation of a phosphoenzyme and free glyceric acid, which remains tightly bound to the enzyme until it is rephosphorylated.

## Assay Method

Although other methods may be used, the enolase-mutase-coupled assay[2] is very convenient for determining the activity of the DPGA-independent mutase. However, the assay is carried out at pH 8.7 since by this method, the optimal pH for the DPGA-independent PGA mutase is about 9.[5] Most preparations of 3-PGA contain 2,3-DPGA as contaminant ($\sim 0.2\%$), but, since the DPGA-independent PGA mutase is not markedly inhibited by DPGA, there is no need to use 3-PGA free of DPGA in the routine assay.

---

[1] Abbreviations used are 2,3-DPGA, D-2,3-diphosphoglyceric acid; PGA, phosphoglyceric acid; 3-PGA, D-3-phosphoglyceric acid; 2-PGA, D-2-phosphoglyceric acid.
[2] See this volume [66].
[3] S. Grisolia and B. K. Joyce, *J. Biol. Chem.* **234**, 1335 (1959).
[4] S. Grisolia, *in* "Homologous Enzymes and Biochemical Evolution" (Thoai and Roche, eds.), p. 167. Gordon & Breach, New York, 1968.
[5] N. Ito and S. Grisolia, *J. Biol. Chem.* **234**, 242 (1959).
[6] M. Fernandez and S. Grisolia, *J. Biol. Chem.* **235**, 2188 (1960).
[7] H. G. Britton, J. Carreras, and S. Grisolia, *Biochemistry* **10**, 4522 (1971).
[8] C. E. Ballou and H. O. L. Fisher, *J. Amer. Chem. Soc.* **76**, 3188 (1954).

*Reagents*

Tris·HCl buffer, 1.0 $M$, pH 8.7
3-PGA, 1.0 $M$
Enolase (mutase free) ∼100 units/ml
$MgSO_4$, 0.1 $M$

*Procedure.* The following components (in a quartz cell having 1-cm light path) are mixed in 3 ml: potassium 3-PGA, 50 μmoles; $MgSO_4$, 10 μmoles; Tris, pH 8.7, 100 μmoles; and 25 units of enolase. After temperature equilibration (30°) up to 0.5 unit of mutase (in 0.1 ml or less) is added. The rate of increase in optical density at 240 nm and at 30° is then followed. One unit of enzyme activity causes an increase in optical density of 0.1 per minute. The enzyme units may be converted[2] to micromoles of 3-PGA turnover by dividing by 3.9.

*Preparation of 3-PGA Free of 2,3-DPGA.* For certain studies it may be necessary to use 3-PGA free of 2,3-DPGA.[9] A simplified procedure[10] is as follows: 28.6 g of the acid barium salt of 3-PGA are slurried into 40 ml of $H_2O$; 16.0 ml of 10 $N$ $H_2SO_4$ is added and thoroughly stirred, and the mixture is centrifuged at 5000 $g$ for 10 min. The supernatant solution is retained, and the $BaSO_4$ residue is washed twice with 50-ml portions of $H_2O$. All supernatant portions are combined and diluted to 400 ml (pH should be 1.86 ± 0.02) and percolated slowly through a 1 × 8 cm Dowex column (Dowex AGI-X8, 200–400 mesh, chloride form Bio-Rad) with about 2 feet of pressure. The effluent is then neutralized with 20 g of Tris. This yields about 0.2 $M$ solution at pH 7. It may be kept frozen and is stable for years. For preparation of 2,3-DPGA, see this series, Vol. 6 [71].

*Preparation of Enolase Free of PGA Mutase.* Unless indicated otherwise, all centrifugations are carried out for 10 min at 5000 $g$. All steps are carried out at 0–5°. All volumes refer to the original volume for the particular step. The acetone and ethanol are measured at 0° and then cooled to about —50° in a dry-ice acetone bath.

Fresh baker's yeast cakes are broken by hand, passed through a wire gauze, and then dried slowly by spreading on filter paper. The yeast is occasionally mixed and is well dried after about a week at room temperature. It is then passed through a wire sieve. It can be stored for over a year.

Mix 300 g of the dry yeast with 900 ml of distilled water. Incubate at 38° and stir occasionally for the first 2 hr until the yeast is well suspended. Let it stand for 15 ± 1 hr at 38°. Centrifuge at 5000 $g$ for 20

[9] J. C. Towne, V. W. Rodwell, and S. Grisolia, *J. Biol. Chem.* **226**, 777 (1957).
[10] S. Grisolia, K. Moore, J. Luque, and H. Grady, *Anal. Biochem.* **31**, 235 (1969).

min, and discard the precipitate. The supernatant fluid is the "crude fraction." To each 100 ml of the crude fraction add 54 ml of acetone. The addition should be made rapidly, the temperature being maintained below 5°. Centrifuge and discard the precipitate. Add acetone to the supernatant fluid (38 ml per each 100 ml of the original crude fraction). Centrifuge, discard the supernatant fluid. Take the precipitate in water to about half the volume of the crude fraction. Centrifuge off and discard any insoluble material. Mix in rapidly sufficient 1 $M$ acetic acid to bring the pH of the preparation to 4.75 (about 7.5 ml) and then 0.1 volume of 1.2 $M$ KCl. Add 105 ml of ethanol to each 100 ml, centrifuge, and discard the precipitate. It is imperative that the ethanol fractionation be carried out rapidly. (It seems that the presence of K+ during the alcohol fractionation are responsible for the complete inactivation of PGA mutase. However, on protracted contact they inactivate the phosphopyruvate hydratase; more than 90% is destroyed when the preparation is allowed to stand for several hours at 0° after the ethanol addition.) Add ethanol to the supernatant fluid (65 ml per each 100 ml being fractionated) and centrifuge. This precipitate is the "ethanol fraction." This fraction is taken into about 0.1 volume of the crude fraction and centrifuged.

The specific activity of the ethanol fraction is comparable to that of the crystalline muscle enzyme and the fraction has about 80% of the activity of the crystalline yeast enzyme; under the conditions reported by Winstead and Wold,[11] it will catalyze the transfer of approximately 100 $\mu$moles/min per milligram of protein at 30° or 640 units/mg.

## Enzyme Purification

The DPGA-independent PGA mutase has been detected in some 50 tissues[3,4] and has been purified from wheat germ[5] and rice germ.[6]

### From Wheat Germ

The wheat germ enzyme was first purified by Ito and Grisolia.[5,12] The procedure yielded ~200-fold purification (specific activity of 400) and ~25% recovery of the starting activity. The method was very reproducible for a particular batch of wheat germ (variation <20%). However, some variability with different batches was observed.

Grisolia et al.[13] modified the original method to yield specific activity

[11] J. A. Winstead and F. Wold, in "Biochemical Preparations" (A. C. Maehly, ed.), Vol. XI, p. 31. Wiley, New York, 1966.
[12] This series, Vol. 5, [26].
[13] S. Grisolia, B. K. Joyce, and M. Fernandez, Biochim. Biophys. Acta 50, 81 (1961).

of 1200, and later further modified it.[7] The procedure[7] is very reliable (reproduced over 25 times); although the specific activity is not higher than previously reported,[13] the yield is substantially greater. The procedure is as follows: Unless indicated otherwise, all operations are carried out at 0–5° and centrifugations are for 10 min at 7000 $g$. Ammonium sulfate solutions are saturated at pH 5.5. All volumes refer to the original volume for the particular step.

*Extraction and Ammonium Sulfate Fractionation.* Wheat germ is extracted with 4 volumes of water for 30 min and then centrifuged. To each 100 ml of the supernatant fluid (fraction I) 60 ml of $(NH_4)_2SO_4$ are added. After 10 min, the solution is centrifuged and 66 ml of $(NH_4)_2SO_4$ are added to the supernatant. The solution is allowed to stand 10 min and centrifuged; the precipitate is dissolved in water (0.75 volume of fraction I) to give fraction II (protein concentration $14 \pm 2$ mg/ml).

*Calcium Phosphate Gel Adsorption.* About 1.4 volume of calcium phosphate gel (35 mg/ml) is added to fraction II. The mixture is stirred occasionally for 10 min and centrifuged; the precipitate is discarded. The supernatant is fraction III.

Calcium phosphate gel is prepared as follows: 170 g of dibasic anhydrous sodium phosphate and 260 g of calcium chloride dihydrate are mixed in 20 liters of tap water. The pH is adjusted to 7.4 with concentrated NaOH, and the mixture is diluted to about 40 liters in a tall, cylindrical jar. The supernatant is siphoned off, and the precipitate is washed six times with tap water and finally concentrated to about 6 liters (35 mg dry weight per milliliter). [The exact proportions of calcium phosphate and bentonite (see below) should be checked with each batch.]

*Bentonite Adsorption.* The supernatant (fraction III) is stirred with ~0.07 volume of bentonite (50 mg/ml) for 10 min and centrifuged.

Bentonite is washed by suspending (using a paint shaker) 1 part of bentonite with 20 parts of deionized water. The suspension is centrifuged at 200 $g$ for 30 sec, and the coarse precipitate is discarded. The supernatant fluid is blenderized in a Waring Blendor for 40 sec at low speed, and the mixture is centrifuged at 1000 $g$ for 15 min. The supernatant is discarded and the precipitate is suspended in deionized water to a concentration of about 50 mg/ml.

*Ammonium Sulfate Precipitation.* The enzyme is precipitated from the supernatant fluid (fraction IV) by the addition of 0.516 g/ml of solid ammonium sulfate. After standing for 30–60 min the preparation is centrifuged for 20 min. The loosely packed precipitate is transferred to a 40-ml centrifuge tube and centrifuged at 15,000 $g$ for 30 min. The pellet is taken up in a minimum of water (20–30 ml) and centrifuged at 15,000 $g$ to remove insoluble material. The supernatant is fraction V.

*Sephadex G-100 Gel Filtration.* This step is carried out in several batches. Fraction V (18 ml; 221 mg of protein, specific activity 94) is applied to the column (2.8 × 120 cm), equilibrated with 10 m$M$ sodium phosphate buffer (pH 7.2) and eluted with the same buffer at a flow rate of 12 ml/hr. Four-milliliter fractions are collected and assayed for protein (optical density at 280 nm) and for mutase. Fractions eluted from 260 to 320 ml are pooled and concentrated ~6 times in a model 50 Diaflo ultrafiltration assembly equipped with a UM-2 membrane (fraction VI, 105 mg of protein, specific activity 170).

*Hydroxyapatite Chromatography.* Portions are then further fractionated as follows: 10 ml of concentrated fraction VI (938 mg of protein) is applied to the column (1 × 15 cm) of hydroxyapatite equilibrated with 10 m$M$ sodium phosphate buffer (pH 7.2). The column is developed using a linear gradient consisting of 250 ml of 0.1 $M$ sodium phosphate buffer (pH 7.2) in the reservoir vessel and 250 ml of equilibrating buffer in the mixing chamber. With a flow rate of 10 ml/hr, 1.5-ml fractions are collected, and protein and enzyme are estimated. Fractions from 31 to 65 ml are pooled to give fraction VII (7.5 mg of protein, specific activity 425).

*DEAE-Sephadex Chromatography.* The specific activity of the eluted enzyme (fraction VII) with different batches of hydroxyapatite varies between 300 and 1000 units/mg. To obtain a higher specific activity fraction VII is adjusted to pH 8 and 44 ml of fraction VII (88 mg of protein, specific activity 384) are applied to a DEAE-Sephadex A-50 column (2 × 40 cm) equilibrated with sodium-potassium phosphate buffer (pH 8; $I$ = 0.05). The activity is eluted with a linear gradient consisting of

TABLE I

PURIFICATION OF WHEAT-GERM PHOSPHOGLYCERIC ACID MUTASE[a]

| Fraction | Volume (ml) | Total activity (units) | Total protein (mg) | Specific activity | Yield (%) |
|---|---|---|---|---|---|
| I | 1330 | 101,080 | 37,900 | 2.6 | 100 |
| II | 1000 | 96,000 | 11,550 | 8.3 | 94 |
| III | 2000 | 88,950 | 3,600 | 24 | 88 |
| IV | 1900 | 78,842 | 1,336 | 59 | 78 |
| V | 42 | 75,600 | 1;134 | 66 | 74 |
| VI | 133 | 43,100 | 319 | 135 | 42 |
| VII | 40 | 22,156 | 68.4 | 324 | 22 |
| VIII | 33 | 17,180 | 16.7 | 1027 | 17 |

[a] From 400 g of wheat germ.

250 ml of phosphate buffer–0.45 $M$ NaCl (pH 8) in the reservoir vessel and 250 ml of equilibrating buffer in the mixing chamber. The flow rate is 15 ml/hr and 5-ml fractions are collected and assayed for protein and mutase activity. Fractions from 345 to 375 ml are pooled to give fraction VIII (25 mg of protein, specific activity 1025). Tubes with mutase activity were pooled and concentrated to 5000 units of mutase per milliliter ($\sim$10 times) as described above (fraction VI). The enzyme is reasonably stable. Fractions III–VI can be kept frozen for several weeks. Fractions VII and VIII are not stable to freezing in phosphate buffer, but can be kept at 0–4° for 2 or 3 weeks without loss of activity. They become stable to freezing if 0.7% bovine serum albumin is added. Fractions VII and VIII are free of enolase ($<$0.1%). Table I summarizes the purification.

### PGA Mutase from Rice Germ

Extensive purification of the PGA mutase has also been accomplished from rice germ.[6] The procedure involves water extraction, acetone fractionation, ammonium sulfate precipitation, bentonite adsorption, kaolin treatment, and a second bentonite adsorption or, alternatively, hydroxyapatite treatment. After the second bentonite treatment, a stable preparation with 400-fold purification (specific activity of 500 units per milligram of protein), and 22% recovery is obtained. With the hydroxyapatite treatment, the purification is greater (specific activity of 750 units/mg of protein), but the recovery is lower (7%), and the enzyme is not stable.

### Properties

*Other Enzymic Activities.* The partially purified PGA mutases from wheat germ and from rice germ readily cleave phosphate from 3-PGA, 2-PGA and 2,3-DPGA.[5,6] The wheat germ preparations with highest specific activity retain 3-PGA phosphatase activity.[7,13] The optimum pH for this activity is 5.8.[7] The ratio mutase:phosphatase activity is 2500 at pH 5.8 and 100,000 at pH 9.6.[7] For the rice enzyme the ratio mutase:phosphatase is $\sim$16,000 at pH 9.0.[6] The enzyme preparations are free of enolase.[5,6]

*Stability and Storage.* The DPGA-independent PGA mutases are unstable above 50° and are not protected from heat denaturation by ammonium sulfate or 2,3-DPGA, as are the DPGA-dependent PGA mutases.[5,6,9] 3-PGA has little effect on the thermal stability of the wheat germ and rice germ mutases. The PGA mutase from wheat germ is fairly stable at alkaline pH so that no inactivation occurs up to about pH 9.5.

## TABLE II
### PROPERTIES OF THE DPGA-INDEPENDENT PGA MUTASES

| Enzyme source | Wheat germ | Rice germ |
|---|---|---|
| Molecular weight | $30,000^{e,f}$ | $30,000^{e,g}$ |
| | $54,000^{d,h}$ | |
| $s_{20,w}$ | $2.56^{f}$ | |
| Specific activity | $1200^{h}$ | |
| pH optimum | $8.9^{a,b,f,h}$ | $8.9^{a,b,g}$ |
| $K_{m(2\text{-PGA})}$ | $5\text{–}1 \times 10^{-4} M^{c,b,f}$ | $25 \times 10^{-4} M^{b,c,g}$ |
| $K_{m(3\text{-PGA})}$ | $3.3 \times 10^{-4} M^{a,h}$ | $1.15 \times 10^{-3} M^{a,g}$ |
| $K_{eq.}$ | $6^{b,f}$ | |
| Turnover number | $1.72 \times 10^{7f}$ | $2 \times 10^{7g}$ |
| Mutase/Pase | $2,500 \ (\text{pH} = 5.8)^{h}$ | $16,000 \ (\text{pH} = 9.0)^{g}$ |
| | $100,000 \ (\text{pH} = 9.6)^{h}$ | |

[a] Enolase-coupled assay (3-PGA as substrate).
[b] Direct method (2-PGA as substrate).
[c] Based on the Haldane relationship.
[d] Measured by gel filtration.
[e] Measured by sedimentation.
[f] N. Ito and S. Grisolia, J. Biol. Chem. **234**, 242 (1959).
[g] M. Fernandez and S. Grisolia, J. Biol. Chem. **235**, 2188 (1960).
[h] H. G. Britton, J. Carreras, and S. Grisolia, Biochemistry **10**, 4522 (1971).

At pH 5 and below, the enzyme is denatured in a few minutes at room temperature.[5]

Table II compares several molecular and kinetic properties of the mutase from wheat and rice germ.

# [66] Phosphoglycerate Mutase from Yeast, Chicken Breast Muscle, and Kidney (2,3-PGA-Dependent)

## By S. GRISOLIA and J. CARRERAS

Phosphoglyceric acid mutases (PGA mutases[1]) are of two types,[2-4] those that require 2,3-diphosphoglycerate (DPGA-dependent PGA

[1] Abbreviations used are: 2,3-DPGA, D-2,3-diphosphoglyceric acid; PGA, phosphoglyceric acid; 3-PGA, D-3-phosphoglyceric acid; 2-PGA, D-2-phosphoglyceric acid; PEP, phosphoenolpyruvic acid.
[2] This series, Vol. 5 [26].
[3] S. Grisolia and B. K. Joyce, J. Biol. Chem. **234**, 1335 (1959).
[4] S. Grisolia, in "Homologous Enzymes and Biochemical Evolution" (N. Van Thoai and J. Roche, eds.), p. **167**. Gordon & Breach, New York, 1968.

mutases) and those that do not (DPGA-independent PGA mutases).[5-8] Each type has a different mechanism of action.[5-8] The DPGA-independent mutases catalyze an intramolecular transfer of phosphate, whereas the DPGA-dependent enzymes possess a Ping-Pong mechanism involving a phosphoenzyme as intermediate (1):

$$3\text{-PGA} + \text{E-P}$$
$$\searrow$$
$$\text{E-DPGA} \rightleftharpoons \text{E} + \text{DPGA} \qquad (10)$$
$$\nearrow$$
$$2\text{-PGA} + \text{E-P}$$

Thus, the Enzyme Commission name for the enzyme, 2,3-biphospho-D-glycerate:2-phospho-D-glycerate phosphotransferase, is probably incorrect.

## Assay Method

Several methods have been described.[9] The more common is based on the formation of PEP from 3-PGA by rate-limiting amounts of mutase and an excess of enolase.[2,10-12] The appearance of PEP may be followed by the increase in optical density at 240 nm. Alternatively, the disappearance of PEP on addition of PGA mutase to an equilibrium mixture of PEP and 2-PGA also can be measured spectrophotometrically.[10-13] In some cases, using either 3-PGA or 2-PGA as substrate, enolase is added after inactivation of the mutase.[13,14] To increase sensitivity, PEP can be converted to pyruvate and this to lactate, and the DPNH change can be followed spectrophotometrically at 340 nm.[10] Also, hypoiodite[13,15] and mercuric chloride[13,16] have been used to decompose PEP to pyruvic acid and inorganic phosphate, which is subsequently measured.

A direct type of assay is based upon the difference in optical rotation

[5] H. G. Britton, J. Carreras, and S. Grisolia, *Biochemistry* **10**, 4522 (1971).
[6] H. G. Britton, J. Carreras, and S. Grisolia, *Biochemistry* **11**, 3008 (1972).
[7] H. G. Britton and J. B. Clarke, *Biochem. J.* **130**, 397 (1972).
[8] H. G. Britton, J. Carreras, and S. Grisolia, *Biochim. Biophys. Acta* **289**, 311 (1972).
[9] W. J. Ray, Jr., and E. J. Peck, Jr., in "The Enzymes" (P. D. Boyer, ed.), 3rd., ed., Vol. 6. p. 408 Academic Press, New York, 1972.
[10] E. Sutherland, T. Posternak, and C. F. Cori, *J. Biol. Chem.* **181**, 153 (1949).
[11] V. W. Rodwell, J. C. Towne, and S. Grisolia, *Biochim. Biophys. Acta* **20**, 394 (1956).
[12] R. W. Cowgill and L. I. Pizer, *J. Biol. Chem.* **223**, 885 (1956).
[13] V. W. Rodwell, PhD. Thesis, University of Kansas, Lawrence, 1956.
[14] V. W. Rodwell, J. C. Towne, and S. Grisolia, *J. Biol. Chem.* **228**, 875 (1957).
[15] H. Chiba, E. Sugimoto, R. Sasaki, and M. Hirose, *Agr. Biol. Chem.* **34**, 498 (1970).
[16] J. C. Lee, Master's Thesis, Purdue Univ., Lafayette, Indiana, 1964.

at neutral pH between the molybdate complexes of 3-PGA and 2-PGA[10,17-20]; it requires high substrate concentrations and is most useful to follow the reaction 2-PGA → 3-PGA, because the opposite reaction is not linear for greater than 2% conversion to product.[14] Finally, the possibility of differentiating 2-PGA and 3-PGA by molybdate-catalyzed hydrolysis[21] has been suggested as a basis for PGA mutase assay.[9]

The rapid enolase-coupled assay is the most convenient method applicable to crude tissue extracts.[2]

*Reagents*

Tris·HCl buffer, 1.0 $M$, pH 7.0
3-PGA, 1.0 $M$
Enolase (mutase-free), ∼100 units/ml
MgSO$_4$, 0.1 $M$

*Procedure.* The following components (in a quartz cell having a 1-cm light path) are mixed in 3 ml: potassium 3-PGA, 50 $\mu$moles[22]; MgSO$_4$, 10 $\mu$moles[23]; Tris, pH 7.0, 100 $\mu$moles; and 10 enzyme units of enolase. After temperature equilibration (30°), up to 1 enzyme unit of mutase (in 2-100 $\mu$l) is added. The rate of increase of optical density at 240 nm is then followed. This rate is not always linear initially; therefore the slope is calculated for a linear portion of the curve.[24]

*Enzyme Unit.* One enzyme unit causes an increase in optical density of 0.1 per minute.[2,11] The amount of 2-PGA formed may be calculated from the measured value for PEP. Under these experimental conditions, 1.5 $\mu$moles of 2-PGA formed correspond to 1 $\mu$mole of PEP measured.[25] The molar extinction coefficient of PEP at 240 nm is a function of pH, salt, and Mg$^{2+}$ concentration and temperature.[26] Under the standard conditions of the assay it is $1.75 \times 10^3$.[13] Thus, the enzyme units may be

[17] O. Meyerhof and W. Schulz, *Biochem. Z.* **297**, 60 (1938).
[18] E. W. Sutherland, T. Z. Posternak, and C. F. Cori, *J. Biol. Chem.* **179**, 501 (1949).
[19] C. E. Ballou and H. O. L. Fisher, *J. Amer. Chem. Soc.* **76**, 3188 (1954).
[20] H. Chiba and E. Sugimoto, *Bull. Agr. Chem. Soc. Jap.* **23**, 207 (1959).
[21] Z. B. Rose, *J. Biol. Chem.* **243**, 4806 (1968).
[22] Most commercial or isolated samples contain DPGA (0.2%); therefore in the routine assay there is no need to add extra DPGA.
[23] In kinetic studies, to maintain the free Mg$^{2+}$ constant, additional MgSO$_4$ is added to allow for Mg$^{2+}$ binding for 3-PGA and DPGA, assuming affinitey constants for 3-PGA at pH 7.4 of 255 $M^{-1}$ and for 2,3-DPGA of 1111 $M^{-1}$.
[24] The initial induction period may be eliminated by increasing the concentration of enolase to 25 units.
[25] R. J. Jacobs and S. Grisolia, *J. Biol. Chem.* **241**, 5926 (1966).
[26] F. Wold and C. E. Ballou, *J. Biol. Chem.* **227**, 301 (1957).

converted to micromoles of 2-PGA or 3-PGA turned over by dividing by 3.9.[27]

## Enzyme Preparations

DPGA-dependent PGA mutase has been detected in some 50 different tissues[3,4] and has been crystallized or extensively purified from *Escherichia coli*,[28] yeast,[2,11,20,29] chicken breast muscle,[30,31] rabbit muscle,[2,12,14,32] sheep muscle,[33] and pig kidney.[34] Partially purified preparations have been obtained from pig[35] and human heart[36] and from caprine[37] and human erythrocytes.[38]

### From Yeast

A simple procedure for the crystallization of PGA mutase from baker's yeast with excellent yield was first described by Rodwell *et al.*[2,11] The enzyme, by most criteria pure, showed five distinct peaks on electrophoresis.[39] Chiba and Sugimoto[40] confirmed these findings and separated the five components by zone electrophoresis.[41] While the crystalline form and several of their kinetic properties were almost identical, their specific activities ranged from 5300 (component I) to 200 (component V).[41,42] Short autolysis times favored a greater proportion of components with higher specific activities, and it was concluded that component I is the native PGA mutase in yeast cells and that it is modified into the other

[27] S. Grisolia and W. W. Cleland, *Biochemistry* **7**, 1115 (1968).

[28] G. D'Alessio and J. Josse, *J. Biol. Chem.* **246**, 4319 (1971).

[29] E. de la Morena, I. Santos, and S. Grisolia, *Biochim. Biophys. Acta* **151**, 526 (1968).

[30] A. Torralba and S. Grisolia, *J. Biol. Chem.* **241**, 1713 (1966).

[31] E. James, T. G. Flynn, and R. O. Hurst, *Fed. Proc., Fed. Amer. Soc. Exp. Biol.* **29**, 891 (Abstract) (1970).

[32] L. I. Pizer, *J. Biol. Chem.* **235**, 895 (1960).

[33] E. James, R. O. Hurst, and T. G. Flynn, *Can. J. Biochem.* **49**, 1183 (1971).

[34] D. Diederich, A. Khan, I. Santos, and S. Grisolia, *Biochim. Biophys. Acta* **212**, 441 (1970).

[35] J. Carreras, A. Torralba, and S. Grisolia, unpublished.

[36] C. H. Kirkpatrick and S. Grisolia, *Int. Congr. Biochem. 6th, Abstracts,* VI-60, (1964).

[37] D. R. Harkness and J. Ponce, *Arch. Biochem. Biophys.* **134**, 113 (1969).

[38] D. R. Harkness, W. Thompson, S. Roth, and V. Grayson, *Arch. Biochem. Biophys.* **138**, 208 (1970).

[39] H. Edelhoch, V. W. Rodwell, and S. Grisolia, *J. Biol. Chem.* **228**, 891 (1957).

[40] H. Chiba and E. Sugimoto, *Bull. Agr. Chem. Soc. Jap.* **23**, 213 (1959).

[41] H. Chiba, E. Sugimoto, and M. Kito, *Bull. Agr. Chem. Soc. Jap.* **24**, 428 (1960).

[42] H. Chiba, E. Sugimoto, and M. Kito, *Bull. Agr. Chem. Soc. Jap.* **24**, 555 (1960).

components by a PGA mutase-modifying enzyme during autolysis.[43] The modification process would be a limited proteolysis, and the differences in electrophoresis mobility of the individual components would derive from differences in lysine content.[44]

The procedure of Chiba and Sugimoto[20] results in low yields and is not reproducible. In contrast, our method[2] is very reproducible, but the specific activity of the crystals changed, presumably reflecting the activity of the contaminant proteolytic enzyme(s). Thus, a new method has been developed which consistently yields preparations of the highest specific activity. The following simplified procedure[29] is recommended. For yeast, chicken muscle, and pig kidney, unless otherwise indicated, all operations carried out at 0–3°. Solutions of $(NH_4)_2SO_4$ saturated at 25° and neutralized to pH 7 with concentrated $NH_4OH$ are used. The pH is determined with the glass electrode in a 1:20 dilution. All volumes refer to the initial volume for a particular step.

Centrifugations are for 15 min at approximately 8000 $g$. All reagents are measured and added at 0–3° except the acetone, which is measured at −20° and mixed in at about −60° (dry ice bath).

*Extraction.* Mix 200 g of yeast[45] with 400 ml of 1 $M$ $NH_4OH$. Keep the mixture for 16–18 hr at room temperature. Then add 24 ml of 0.5 $M$ $Na_4EDTA$, 890 ml distilled water, and 280 g $(NH_4)_2SO_4$. Heat at 70° ± 1° in a water bath for 5 min. Cool, centrifuge, and discard the precipitate.

*Ammonium Sulfate Fractionation.* To each 100 ml of the supernatant fluid (fraction 1) add 30 g of $(NH_4)_2SO_4$. After the $(NH_4)_2SO_4$ is dissolved, centrifuge. Discard the supernatant fluid and take the precipitate in water to about one-half the starting volume of fraction 1. This is fraction 2.

*Acetone Step.* Mix fraction 2 rapidly into 1 volume of acetone; allow to stand for a few minutes; the bulk of the clear supernatant fluid may then be poured off and the rest centrifuged at about 3000 $g$ for 5 min. Discard the supernatant fluid and make a paste (with the aid of a hand homogenizer) with a volume of water equal to the volume of the precipitate. The well-suspended mixture is centrifuged for 5 min. The supernatant is kept, and the precipitate is extracted again with a volume of water equal to the amount used in the first extraction. Centrifuge and discard the precipitate. Mix both supernatant fluids and bring to one-half the volume of fraction 2. This is fraction 3.

*Ammonium Sulfate Fractionation.* Mix fraction 3 with 1.5 volumes

[43] H. Chiba, E. Sugimoto, and M. Kito, *Bull. Agr. Chem. Soc. Jap.* **24**, 558 (1960).
[44] R. Sasaki, E. Sugimoto, and H. Chiba, *Arch. Biochem. Biophys.* **115**, 53 (1966).
[45] See this volume, [65].

TABLE I
PURIFICATION OF YEAST PHOSPHOGLYCERATE MUTASE

| Fraction | Volume (ml) | Total units | Total protein (mg) | Specific activity (units/mg) | Yield (%) |
|---|---|---|---|---|---|
| Extraction | 1060 | 1,908,000 | 17,808 | 108 | 100 |
| Ammonium sulfate | 500 | 1,860,000 | 7,000 | 266 | 98 |
| Acetone | 250 | 1,597,500 | 1,250 | 1278 | 84 |
| Ammonium sulfate[a] | 2 | 295,000 | 70 | 4200 | 15 |

[a] Crystalline.

of saturated (0°) $(NH_4)_2SO_4$; centrifuge, discard the precipitate, add 1.5 volumes of $(NH_4)_2SO_4$ to the supernatant fluid, and centrifuge. The precipitate containing the enzyme is extracted successively with 30 ml of $(NH_4)_2SO_4$ solutions of decreasing salt concentration as follows: 20 ml of the saturated $(NH_4)_2SO_4$ and 10 ml of water; 19 ml of $(NH_4)_2SO_4$ and 11 ml of water; 18 ml of $(NH_4)_2SO_4$ and 12 ml of water; 17 ml of $(NH_4)_2SO_4$ and 13 ml of water. The first, second, third, fourth (fraction 4 in Table I), and fifth extractions yielded 800, 11,000, 140,000, 475,000, and 120,000 units, respectively. Fraction 4 becomes turbid in about 1–2 hr and is centrifuged. The precipitate is discarded; the supernatant fluid is left to evaporate slowly in a beaker (covered with filter paper) in the refrigerator. Crystals appear overnight. The next day it is stirred, left standing in the cold for a couple of hours, and then centrifuged. A summary of the purification is shown in Table I.

Similar results have been otained with dry yeast preparations from a week to 5 years old. There was less than 20% variation in some 20 preparations. In agreement with the results obtained by others, the enzyme with a specific activity of over 4000 has the same crystalline appearance, i.e., rhombic plates, as previously described for the enzyme[2] with specific activity of about 2000. Also, disc electrophoresis and electrophoresis in cellulose acetate strips show essentially one band with either kind of preparation. Indeed, it is of interest that the gel electrophoresis does not differentiate between preparations having a 5-fold difference in specific activities.

## From Chicken Breast Muscle.[30]

Unless otherwise specified, centrifugations are at approximately 2000 g. All reagents are measured and added at 0–3°.

*Extraction.* The muscle is ground twice with a meat grinder. The ground muscle is extracted for 10 min with gentle stirring with 1.5 volume of cold water and then strained through gauze. The pulp is reextracted for 10 min with 1 volume of cold water and strained again. The combined extracts are centrifuged for 10 min, and the precipitate is discarded. The supernatant fluid is the crude fraction.

*Acetone Step.* To each volume of crude fraction is added, with constant stirring, 0.5 volume of acetone. The mixture is centrifuged for 10 min, and the precipitate is discarded. To the supernatant fluid is added 0.15 volume of acetone per volume of crude fraction, and the mixture is centrifuged for 15 min. The precipitate is taken up in water to half the volume of the crude fraction.

*Heat Step.* To the acetone fraction is added 0.1 volume of 60 m$M$ 2,3-PGA at pH 7.0. The mixture is heated with continuous gentle stirring for 10 min at 62° (marked turbidity and protein precipitation begin when the mixture reaches 50°); then it is cooled rapidly and centrifuged for 15 min, and the supernatant fluid is filtered through coarse filter paper. The filtrate is called the heated fraction.

A smaller amount of 2,3-PGA does not protect as well. A temperature of 62° gave a higher specific activity than 60°, and at 65° the enzyme losses were very high. More than 15 min at 62° gave losses of about one-half of the enzyme activity.

*First Ammonium Sulfate Fraction.* To each milliliter of the heated fraction is added 0.4 g of ammonium sulfate. The mixture is centrifuged for 10 min at 18,000 $g$, the supernatant fluid is discarded, and the precipitate is taken up in water to 0.045 the volume of the heated fraction. This is ammonium sulfate fraction I. This fraction can be kept at 2° for at least 1 week without loss of actvity and is stable at −20°; however, repeated freezing and thawing resulted in loss of activity.

Up to this stage the time required for the purification, starting with 200 g of ground muscle, is about 4 hr.

*Second Ammonium Sulfate Fraction.* To each milliliter of the ammonium sulfate fraction I, 0.7 ml of saturated ammonium sulfate is added. A small precipitate appears, and the mixture is centrifuged at 18,000 $g$ for 10 min. To the supernatant fluid saturated ammonium sulfate is added very slowly until a crystalline precipitate appears; this increases as the mixture stands for 4 hr at 2°. The suspension is centrifuged for 20 min at 18,000 $g$, and the precipitate is taken up to 1.5 ml of water. This is ammonium sulfate fraction II. Table II presents a resumé of the purification procedure. This procedure is extremely rapid and convenient. A conservative estimate is that it takes one-fifth to one-tenth the effort and expense necessary for the preparation of the rabbit enzyme. Either frozen or fresh chicken breasts purchased from any commercial source thus far tried have yielded satisfactory preparations.

The enzyme was shown to be homogeneous by sedimentation and by

## TABLE II
PURIFICATION OF CHICKEN BREAST MUSCLE PHOSPHOGLYCERATE MUTASE

| Step | Fraction | Volume (ml) | Activity (units) | Protein (mg) | Specific activity (units/mg) | Recovery (%) |
|---|---|---|---|---|---|---|
| 1 | Crude[a] | 400 | 650,000 | 6488 | 100 | 100 |
| 2 | Acetone | 200 | 500,000 | 3600 | 138 | 77 |
| 3 | Heated | 185 | 416,250 | 463 | 900 | 64 |
| 4 | Ammonium sulfate I[b] | 8 | 360,000 | 189 | 1900 | 55 |
| 5 | Ammonium sulfate II | 1.5 | 187,500 | 75 | 2500 | 28 |

[a] From 200 g of ground muscle.

[b] If the specific activity of the ammonium sulfate fraction I is less than 1900, another heating step can be carried out after adding 0.1 volume of 60 m$M$ 2,3-P-glycerate, pH 7.0, heating for 10 min at 62°, centrifuging for 15 min, and filtering the supernatant fluid.

electrophoresis at pH 5.4. However, James et al.[31] reported that it can be separated into six multiple forms, with the same specific activities, by DEAE-Sephadex chromatography and by polyacrylamide gel electrophoresis at pH 9.3.

### From Pig Kidney

The procedure of Diederich et al.[34] for purifying the enzyme from pig kidney some 2000-fold is very reproducible, but not as simple and more time consuming than the procedures described for yeast or muscle. The mutase of pork kidney proved to be very difficult to purify until acetone powders were prepared. For example, the activity of homogenates decreased at 4°, reaching approximately 25% of initial activity after 8 hr. Similarly, all attempts to fractionate crude homogenates with organic solvents resulted in inactivation. The enzyme was rapidly inactivated below pH 5.5, and it was also extremely heat sensitive, showing extensive inactivation at 45–50°. Water extracts of the acetone powder yields 75–80% of the original activity and the specific activity is 5–6 compared to <1 for fresh homogenates. Equally important for further purification, the activity of the acetone powder extracts is stable at 4° for at least 48 hr to ethanol and acetone at 4°, and sufficiently stable to heat to allow a heat step in purification.

*Acetone Powder Preparation.* Fresh pork kidneys are obtained from a local slaughterhouse and cooled to 4° while being transported to the laboratory. Fat is removed, and the tissue is ground with a large meat grinder; 2-kg portions of ground tissue are homogenized in 800 ml of water in a large Waring Blendor for 30 sec. Usually 7–8 kg of ground tissue are processed. The homogenized tissue is poured with vigorous stirring into 10 volumes of cold acetone. After the tissue has settled, the bulk of acetone is decanted, and the precipitate is dried under suction on two large Büchner funnels covered with rubber dams. After 2 hr, the nearly dried cakes are pulverized with the aid of a vegetable grater and sifted (under nitrogen) through a set of mechanically shaken screens of decreasing size. The resultant fine powder is finally dried in desiccators over alumina.

*Crude Extract.* Acetone powder, 200 g, is suspended in 10 volumes of water, stirred for 20 min, and then centrifuged at 7000 $g$ for 10 min. The supernatant is the crude fraction.

*Ethanol Fractionation.* Ethanol (95%), 1.6 volumes, is added to the crude fraction, and the mixture is centrifuged at 5000 $g$ for 15 min. Another 1.6 volumes of ethanol are added to the supernatant. After centrifugation, the precipitate is transferred, with thorough mixing, into 50 ml of water, and centrifuged for 30 min at 15,000 $g$; any insoluble material is discarded.

*Heat Step.* The supernatant (ethanol fraction) is diluted to a protein concentration of 12 mg/ml after the addition of 0.05 volume of 0.1 $M$ 2,3-DPGA. The mixture is brought to 54° in a water bath and held at that temperature for 4 min. After cooling in an icebath, it is centrifuged at 10,000 $g$ for 15 min. In practice the supernatants (heated fraction) from three 200-g acetone powder batches are pooled at this stage.

*Ammonium Sulfate Precipitation.* Saturated $(NH_4)_2SO_4$, (0°) 1.5 volumes is added to the heated fraction, and the mixture is centrifuged at 6000 $g$ for 20 min. Five volumes of $(NH_4)_2SO_4$ are added to the supernatant and then centrifuged at 6000 $g$ for 30 min.

*Sephadex G-100 Gel Filtration.* The loosely packed precipitate is transferred to a 40-ml centrifuge tube and centrifuged at 15,000 $g$ for 30 min. The pellet is taken up in a minimum of water (approximately 10 ml; protein concentration approximately 35 mg/ml). The fraction is centrifuged at 15,000 $g$, and the supernatant [$(NH_4)_2SO_4$ fraction] is layered on a 2.8 × 120 cm column of Sephadex G-100, equilibrated with 20 m$M$ sodium phosphate buffer (pH 7). The column is eluted with equilibrating buffer at a flow rate of approximately 10 ml/hr; 3-ml aliquots are collected. Two major protein peaks are eluted, the first eluting at the void volume. The mutase activity appears at 1.1–1.2 volumes. The tubes containing mutase with a specific activity above 250 are pooled (Sephadex fraction).

*DEAE-Cellulose Chromatography.* This fraction is percolated through a 2.5 × 50 cm column of DEAE-cellulose, equilibrated with 20

TABLE III
PURIFICATION OF PIG KIDNEY PHOSPHOGLYCERATE MUTASE

| Fraction | Volume (ml) | Activity (units) | Protein (mg) | Specific activity (units/mg) | Recovery[b] (%) |
|---|---|---|---|---|---|
| Crude | 1550 | 127,100 | 22,630 | 5.6 | 100 |
| Ethanol | 68 | 72,080 | 1,360 | 53 | 56 |
| Heated | 98 | 59,000 | 588 | 122 | 46 |
| $(NH_4)_2SO_4$[a] | 12 | 120,000 | 636 | 190 | 31 |
| Sephadex pool[a] | 43 | 110,000 | 210 | 522 | 29 |
| DEAE-cellulose pool[a] | 17 | 35,000 | 20.6 | 1730 | 9 |

[a] Data given for three 200-g batches of acetone powder.
[b] Recovery calculated from crude acetone powder extract.

m$M$ sodium phosphate (pH 7). The column was washed with 50 ml of equilibrating buffer and then developed with a linear gradient (1000 ml of 0.2 $M$ $Na_2HPO_4$ in the reservoir vessel and 1000 ml of 20 m$M$ equilibrating buffer in the mixing chamber). The flow rate is 20 ml/hr. Three major protein peaks are eluted in a reproducible fashion; the third protein peak contains the mutase activity. Tubes in the latter peak containing mutase with a specific activity above 1200 (DEAE-cellulose fraction) are concentrated to approximately 5 ml in a Model 50 Diaflo ultrafiltration assembly (Amicon Corporation) equipped with a PM-10 membrane. Ten volumes of 10 m$M$ sodium phosphate (pH 7) are added to the fraction in the chamber, and the fraction is then concentrated to approximately 10 mg/ml. This fraction may be stored frozen. A summary of the purification procedure is given in Table III.

The best fraction has a specific activity approximating that from chicken breast, rabbit muscle, and yeast. However, acrylamide gel disc electrophoresis demonstrated four separate bands of protein all containing mutase activity. This suggests subunit aggregation (dimers, tetramers, etc.).

## PGA Mutase from Other Sources

A relatively simple procedure has been described[28] for the simultaneous purification of glyceraldehyde phosphate dehydrogenase, PGA mutase, and PGA kinase from extracts of E. coli. The method, which involves streptomycin precipitation, ammonium sulfate fractionation, Sephadex G-150 gel filtration[1] and DEAE-cellulose chromatography, yields a homogeneous PGA mutase.

Rabbit muscle PGA mutase was purified in this laboratory[14] to an estimated 85% purity by removal of impurities at pH 4.2 from water extracts of acetone powders, acetone precipitaton, heating in the presence of ammonium sulfate, and ammonium sulfate fractionation. This method was modified[25] by using 2,3-DPGA (0.35 $\mu$mole per milligram of protein) to protect the enzyme from heat denaturation. This step resulted in over 80% recovery of activity from the preceding fraction (34% recovery in original procedure). A procedure described by Pizer to crystallize the enzyme,[32] is analogous to that used by Rodwell et al. for yeast.[11]

James et al.[33] have reported purification by acetone precipitation, heat denaturation in the presence of 2,3-DPGA, DEAE-Sephadex chromatography, and gel filtration on Sephadex G-100 (specific activity of 825 units/mg). The enzyme is reportedly homogeneous by disc gel (pH 5.8) and cellulose acetate (pH 5.4 and 7.0) electrophoresis and by sedimentation techniques. However, polyacrylamide gel electrophoresis at pH 9.3 revealed the presence of three isomers. Some separation of the three components was achieved by isoelectric focusing with much loss in activity.

The enzyme from liver has been crystallized[46] by ammonium sulfate, acetone precipitation, heat treatment, DEAE-cellulose, gel filtration, and DEAE-Sephadex chromatography. The crystalline enzyme (fine needles) has a molecular weight of about 65,000 and 2 (probably identical) subunits. The amino acid composition is markedly different from muscle's. It has 18 SH-groups.

The mutase has also been partially purified from human heart acetone powders,[4] and from pig heart,[35] by acetone fractionations, heating in the presence of 2,3-DPGA, ethanol precipitation in the presence of Hg$^{2+}$ and 2,3-DPGA, Sephadex G-100 gel filtration, and DEAE-Sephadex chromatography.[35]

The partial purification of PGA mutase from caprine and human erythrocytes has been described by Harkness et al.[37,38]

## Properties

*Stability and Storage.* The mutase is not stable at low protein concentrations[14]; therefore, in all studies requiring high enzyme dilutions, 1% albumin solution is used as a diluent. Salt protects remarkably the enzyme against heat denaturation.[27] At neutral pH, ammonium sulfate or 2,3-DPGA stabilizes PGA mutase against heating.[25,30,33,34] It has been reported[31] that 3-PGA also protects PGA mutase against heat inactiva-

[46] K. D. Kulbe and Ch. Ahrendt, *Int. Congr. Biochem.* Abstract No. 2e23, Stockholm, 1973.

TABLE IV

PHYSICAL AND CHEMICAL PROPERTIES OF THE PGA MUTASES

| Enzyme source: | Escherichia coli | Yeast | Chicken muscle | Rabbit muscle | Sheep muscle | Pig kidney |
|---|---|---|---|---|---|---|
| Molecular weight | 56,300[a,d] | 112,000[b,e] 110,000[a,i] | 65,690[b,f] | 57,000 ± 2,000[a,a] 64,000[b,e] | 52,200[c,h] 50,540[b,h] | 65,000[c,i] |
| Polymeric structure | — | tetramer[i,k] | — | dimer[l] | — | — |
| $E_{280\ nm}$ | — | 14.5[e] 14.2–13.8[m] | — | 10.5[e] | 7.1[h] | — |
| $s_{20,w}$ | 4.8[d] | 6.3[e] 6.4[i,n] | 3.9[f] | 4.47[e] 4.08[g] | 4.1[h] | — |
| $D_{20,w}$ | — | $5.28 \times 10^{-7}$[e] $5.6 \times 10^{-7}$[i] | — | $6.6 \times 10^{-7}$[e] | $7.21 \times 10^{-7}$[h] | — |
| $f/f_0$ | — | 1.27[e] 1.2[i] | — | 1.22[e] | — | — |
| Isoelectric point | — | 5.0–5.5[e] | — | 5.3[e] | 9.25[h] (8.7–8.9) | — |
| SH content | — | 0[o] 1[p] | 4[f] | 4[g] | 4[h] | — |

[a] Determined by sedimentation equilibrium.
[b] Determined by sedimentation and diffusion data.
[c] Determined by gel filtration.
[d] G. D'Alessio and J. Josse, J. Biol. Chem. 246, 4319 (1971).
[e] H. Edelhoch, V. W. Rodwell, and S. Grisolia, J. Biol. Chem. 228, 891 (1957).
[f] A. Torralba and S. Grisolia, J. Biol. Chem. 241, 1713 (1966).
[g] L. I. Pizer, J. Biol. Chem. 235, 895 (1960).
[h] E. James, R. O. Hurst, and T. G. Flynn, Can. J. Biochem. 49, 1183 (1971).
[i] D. Diederich, A. Khan, I. Santos, and S. Grisolia, Biochim. Biophys. Acta 212, 441 (1970).
[j] R. Sasaki, E. Sugimoto, and H. Chiba, Agr. Biol. Chem. 34, 135 (1970).
[k] Z. B. Rose, Arch. Biochem. Biophys. 146, 359 (1971).
[l] Z. B. Rose, Arch. Biochem. Biophys. 140, 508 (1970).
[m] H. Chiba, E. Sugimoto, and M. Kito, Bull. Agr. Chem. Soc. Jap. 24, 555 (1960).
[n] E. Sugimoto, R. Sasaki, and H. Chiba, Arch. Biochem. Biophys. 113, 444 (1966).
[o] S. Grisolia, in "Homologous Enzymes and Biochemical Evolution" (N. Van. Thoai and J. Roche, eds.), p. 167. Gordon & Breach, New York, 1968.
[p] E. Sugimoto, R. Sasaki, and H. Chiba, Agr. Biol. Chem. 27, 222 (1963).

tion; probably the protective effect was due to the 2,3-DPGA present as a contaminant in the 3-PGA preparation.

PGA mutases from yeast and from rabbit muscle are denatured (unfolded and dissociated into subunits) and inactivated by alkali (pH > 11.5)[32,47] and by high concentrations of urea (>4 $M$).[32,48] Reactivation and reconstitution of the urea-denatured enzyme is accomplished by dilution or by urea removal.[48] The alkali-treated enzyme partially regains its activity by neutralization, the ability to reactivate being lowered with the increase in pH and with time of exposure. Inactivation at pH 13 is almost irreversible. However, the reversibility at this pH is enhanced by the presence of substrates or the coenzyme.[47]

All the purified preparations are usually very stable at −20°.[2] However, repeated freeze-thaw produces inactivation.[2,12] It has been reported that the yeast enzyme may be stored at 0° in ammonium sulfate.[20]

*Molecular and Kinetic Properties.* The molecular weights for PGA mutases from mammalian tissues vary from 54,000 to 65,000.[30,32-34,39] A similar value has been reported for the 3-PGA mutase from *E. coli*.[28] In contrast, the yeast enzyme has a molecular weight of 110,000—112,000.[39,47] A dimeric and a tetrameric structure has been demonstrated for the rabbit muscle and yeast PGA mutases, respectively; the molecular weight of the monomers is ∼27,000.[47,49,50] X-ray crystallographic studies with the yeast enzyme show that the four chemically identical subunits are arranged in the molecule with almost perfect **222** symmetry.[51,52] Some physical properties of the purified PGA mutases are summarized in Table IV.

Spectrophotometric titration of the PGA mutases from chicken breast,[30] rabbit muscle,[37] and sheep muscle[33] with *p*-mercuribenzoate, has shown the presence of sulfhydryl groups (4 per enzyme mole) and their involvement in the enzymic activity. By contrast, the yeast enzyme contains one[53] or no[4] sulfhydryl groups. With this enzyme, it has been shown[54] by trinitrophenylation that four amino groups (located in the active site) are indispensable for the enzyme activity. Inactivation of the PGA mutase from rabbit muscle by treatment with dinitrofluorobenzene or

[47] R. Sasaki, E. Sugimoto, and H. Chiba, *Agr. Biol. Chem.* **34,** 135 (1970).
[48] E. Sugimoto, R. Sasaki, and H. Chiba, *Arch. Biochem. Biophys.* **113,** 444 (1966).
[49] Z. B. Rose, *Arch. Biochem. Biophys.* **140,** 508 (1970).
[50] Z. B. Rose, *Arch. Biochem. Biophys.* **146,** 359 (1971).
[51] J. W. Campbell, G. I. Hodgson, H. C. Watson, and R. K. Scopes, *J. Mol. Biol.* **61,** 257 (1971).
[52] J. W. Campbell, G. I. Hodgson, and H. C. Watson, *Nature New Biol.* **240,** 137 (1972).
[53] E. Sugimoto, R. Sasaki, and H. Chiba, *Agr. Biol. Chem.* **27,** 222 (1963).
[54] R. Sasaki, E. Sugimoto, and H. Chiba, *Biochim. Biophys. Acta* **227,** 584 (1971).

TABLE V

KINETIC AND THERMODYNAMIC PROPERTIES OF THE PGA MUTASES

| Enzyme source: | Yeast | Chicken breast muscle | Rabbit muscle | Sheep muscle | Pig kidney |
|---|---|---|---|---|---|
| Specific activity | $4200^{a,t}$ | $2500^{a,u}$ | $2100^{a,v}$ | — | $1730^{a,w}$ |
| pH optimum | $7.0^{a,v}$<br>$5.85^{d,v}$<br>$5.9^{e,cc}$ | $7.3^{a,u}$ | $7.0^{b,x}$<br>$5.9^{d,v}$<br>$7.0^{f,x}$ | $7.0^{c,y}$ | $6.5^{a,z}$ |
| $V'_{max}/E_T$ | — | $8,000 \pm 300^{g,aa}$<br>$6,600 \pm 300^{h,aa}$ | — | — | — |
| $K_{m(3\text{-}PGA)}$ | $<1 \times 10^{-4}\ M^{m,v}$<br>$2 \times 10^{-4}\ M^{g,bb}$<br>$6 \times 10^{-4}\ M^{k,r}$<br>$(1-3) \times 10^{-4}\ M^{l,dd}$ | $(6 \pm 0.6) \times 10^{-4}\ M^{g,aa}$<br>$(4.6 \pm 0.3) \times 10^{-3}\ M^{h,aa}$ | $<2 \times 10^{-3}\ M^{m,v}$<br>$5 \times 10^{-3}\ M^{i,z}$ | $9 \times 10^{-3}\ M^{i,y}$ | $2.7 \times 10^{-4}\ M^{a,z}$ |
| $K_{m(DPGA)}$ | $1.13 \times 10^{-4}\ M^{m,v}$<br>$3.31 \times 10^{-4}\ M^{n,cc}$<br>$8.0 \times 10^{-7}\ M^{k,r}$<br>$(5-8) \times 10^{-7}\ M^{l,dd}$ | $(1.4 \pm 0.3) \times 10^{-6}\ M^{g,aa}$<br>$(4.2 \pm 0.8) \times 10^{-5}\ M^{h,aa}$ | $1.25 \times 10^{-4}\ M^{h,v}$ | $3 \times 10^{-6}\ M^{i,y}$ | — |
| $K_{i(3\text{-}PGA)}$ | $3.2 \times 10^{-3}\ M^{k,r}$<br>$(2-5) \times 10^{-4}\ M^{l,dd}$ | $(2.9 \pm 0.8) \times 10^{-3}\ M^{g,aa}$<br>$(6.7 \pm 1.7) \times 10^{-3}\ M^{h,aa}$<br>$(2.5 \pm 0.4) \times 10^{-3}\ M^{2,aa}$<br>$(10.0 \pm 2) \times 10^{-3}\ M^{h,aa}$ | — | — | — |
| $K_{i(DPGA)}$ | $2.5 \times 10^{-4}\ M^{l,dd}$ | — | — | — | — |
| $K_{eq}$ | $6.3 \pm 0.3^{p,v}$<br>$4.9-5.25^{o,cc}$<br>$(5.8)^s$ | — | $5.0^{q,x}$ | — | — |
| Energy of activation | $10,500$ cal.$^{ee}$ | — | $10,500$ cal.$^{ee}$ | $11,000$ cal.$^v$ | — |
| Pasa activity | $1/100,000^v$ | $1/40,000^u$ | $1/50,000^v$ | $1/50,000^v$ | — |

[a] Enolase-coupled assay (standard conditions).

[b] Enolase-coupled assay (37°).

[c] Enolase-coupled assay (25°).

[d] Direct method (2-PGA as substrate, 30°).

[e] Polarimetric assay (2-PGA as substrate, 25°).

[f] Polarimetric assay (3-PGA as substrate, 37°).

[g] Enolase-coupled assay (30°, pH 7.4).

[h] Enolase-coupled assay (30°, pH 7.4, 400 mM KCl).

[i] Enolase-coupled assay (37°, pH 7.0).

[j] Enolase-coupled assay (25°, pH 7.0).

[k] Enolase-coupled assay (25°, pH 5.9).

[l] Enolase-coupled assay (25°, pH 7.5).

[m] Direct method (2-PGA as substrate, 30°, pH 5.9).

[n] Polarimetric assay (2-PGA as substrate, 25°, pH 5.9).

[o] Polarimetric assay (3-PGA as substrate, 25°, pH from 5.0 to 7.2).

[p] Direct method (2-PGA as substrate, 30°, pH from 4.6 to 6.65).

[q] Polarimetric assay (2-PGA and 3-PGA as substrates, 37°, pH 6.8).

[r] H. Chiba, E. Sugimoto, R. Sasaki, and M. Hirose, *Agr. Biol. Chem.* **34**, 498 (1970).

[s] Recalculated by L. Pizer, *in* "The Enzymes" (P. D. Boyer, H. Lardy, and K. Myrback, eds.) 2nd ed. Vol. VI, p. 179. Academic Press, New York, 1962.

[t] E. de la Morena, I. Santos, and S. Grisolia, *Biochim. Biophys. Acta* **151**, 526 (1968).

[u] A. Torralba and S. Grisolia, *J. Biol. Chem.* **241**, 1713 (1966).

[v] V. W. Rodwell, J. C. Towne, and S. Grisolia, *J. Biol. Chem.* **228**, 875 (1957).

[w] D. Diederich, A. Khan, I. Santos, and S. Grisolia, *Biochim. Biophys. Acta* **212**, 441 (1970).

[x] R. W. Cowgill and L. I. Pizer, *J. Biol. Chem.* **223**, 885 (1956).

[y] E. James, R. O. Hurst, and T. G. Flynn, *Can. J. Biochem.* **49**, 1183 (1971).

[z] H. G. Britton, J. Carreras, and S. Grisolia, *Biochim. Biophys. Acta* **289**, 311 (1972).

[aa] S. Grisolia and W. W. Cleland, *Biochemistry* **7**, 1115 (1968).

[bb] H. G. Britton, J. Carreras, and S. Grisolia, *Biochemistry* **11**, 3008 (1972).

[cc] H. Chiba and E. Sugimoto, *Bull. Agr. Chem. Soc. Jap.* **23**, 207 (1959).

[dd] R. Sasaki, E. Sugimoto, and H. Chiba, *Biochim. Biophys. Acta* **227**, 584 (1971).

[ee] This series, Vol 5 [26].

methylaminonaphthalenesulfonyl chloride has been reported.[32] The reagents reacted with both the imidazole ring of histidine and the $\epsilon$-amino group of lysine. The exposure of the enzyme to diazotized sulfanilic acid also caused its inactivation.[32] Similarly, rabbit muscle PGA mutase was partially inactivated by ethoxyformylation of histidine residues.[55] Photooxidation of this mutase with both methylene blue and Rose Bengal caused a complete loss of the enzymic activity.[55] Some kinetic and thermodynamic properties of the PGA mutases are presented in Table V.

[55] J. Carreras and S. Grisolia, unpublished observations.

# [67] 2,3-Diphosphoglycerate Mutase from Human Erythrocytes[1,2]

## By ZELDA B. ROSE

1,3-Diphosphoglycerate + 3-phosphoglycerate → 2,3-diphosphoglycerate
$$+ \text{ 3-phosphoglycerate}$$

## Assay Method

*Principle.* The assay utilizes substrate with $^{32}$P in the acid-labile acyl phosphoryl group. Incubations are terminated by the addition of acid which hydrolyzes the acyl phosphate liberating $^{32}$P$_i$ from unreacted substrate while having no effect on the product. Ammonium molybdate is added to form with inorganic phosphate a complex extractable with a mixture of isobutanol and benzene. Organic phosphate compounds remain in the aqueous phase. Portions of each phase are counted in a liquid scintillation counter to determine the extent of reaction. An alternative spectrophotometric assay can be used to follow the purification,[3] but it is not suitable for kinetic studies.[1]

*Reagents*

Sodium glycylglycine buffer, 0.1 $M$, pH 7.5
2-Mercaptoethanol, 0.1 $M$
3-Phosphoglycerate, 0.2 m$M$

[1] Z. B. Rose, *J. Biol. Chem.* **243**, 4810 (1968).
[2] Z. B. Rose and R. G. Whalen, *J. Biol. Chem.* **248**, 1513 (1973).
[3] S. Rapoport and J. Leubering, *J. Biol. Chem.* **196**, 583 (1952).

1,3-[1-³²P]Diphosphoglycerate,[4] 50 $\mu M$
Trichloroacetic acid, 15%
$H_2SO_4$, 4 $N$
Ammonium molybdate, 5%
Sodium phosphate, 10 m$M$
Isobutanol-benzene (1:1 by volume)
Scintillation counting mixture: $H_2O$–toluene–ethanol (1:20:10)

The toluene contains 2,3-diphenyloxazole (0.4%) and 1,4-bis[2-(4-methyl-5-phenyloxazolyl)]benzene (0.005%) from Packard Instrument Co.

*Procedure.* The following are incubated at 25° for 10 min in a 0.2 ml volume: 0.02 ml of glycylglycine buffer; 0.01 ml of 2-mercaptoethanol; 0.02 ml of 3-phosphoglycerate, and sufficient diphosphoglycerate mutase to convert up to 15% of the 1,3-diphosphoglycerate to products. The reaction is started by the addition of 0.02 ml of 1,3-[1-³²P]diphosphoglycerate and stopped with 0.10 ml of 15% trichloroacetic acid. Water (0.2 ml) is added and the protein removed by centrifugation. To 0.40 ml of the trichloroacetic acid supernatant are added 0.10 ml of $H_2SO_4$, 0.10 ml of ammonium molybdate, 0.02 ml of phosphate as carrier, 0.38 ml water, and 2 ml of isobutanol-benzene.[5] The sample is mixed on a Vortex Jr. mixer and centrifuged briefly at room temperature. Inorganic phosphate is found in the upper, organic phase and organic phosphates in the lower, aqueous phase. Half of either or both phases is counted in a scintillation counter.

*Definition of Unit and Specific Activity.* One unit of enzyme is that amount required to form 1 $\mu$mole of product per minute under the condi-

---

[4] To prepare 1,3[1-³²P]diphosphoglycerate, the following were incubated at 25° for 20 min in a 2-ml volume: 50 $\mu$moles of triethanolamine-chloride buffer, pH 8.0, 4 $\mu$moles of DPN; DL-glyceraldehyde-3-P (8 $\mu$moles of D-form); 22 $\mu$moles of sodium pyruvate, 2 $\mu$moles of EDTA; 20 $\mu$moles of potassium phosphate, 2 mCi of ³²P$_i$ (from New England Nuclear, carrier free in 0.02 $N$ HCl, neutralized with NaOH); 1.5 units of lactate dehydrogenase, and 7 units of glyceraldehyde-3-phosphate dehydrogenase. All subsequent steps were at 0–4°. EDTA (6 $\mu$moles) and 0.5 ml of trichloroacetic acid (50%) were added. After 2 min the acid was removed by three successive extractions with cold ether. The solution was neutralized with 2 $M$ triethanolamine. Cold water (50 ml) was added, and the sample was applied to a DEAE-cellulose column (1 × 17 cm) equilibrated with 10 m$M$ sodium glycylglycine buffer, pH 7.5. The column was washed successively with 30 ml of starting buffer and 120 ml of the same buffer containing 50 m$M$ NaCl. The 1,3-diphosphoglycerate was eluted with starting buffer containing 0.12 $M$ NaCl. The yield was 6 $\mu$moles, specific activity 1.4 × 10⁸ cpm/$\mu$mole (counted in a Packard Tri-Carb spectrometer). The compound as eluted could be stored in small portions at −80°. For its properties, see this series, Vol. 3 [36].

[5] I. Berenblum and E. Chain, *Biochem. J.* **32**, 295 (1938).

tions of the standard assay. Specific activity is the number of units per milligram of protein. Protein was determined in hemolysates by the Folin procedure[6] after preliminary extraction of heme with methyl ethyl ketone.[7] The Warburg-Christian method[8] was used subsequently.

## Purification Procedure

This method is suitable for the preparation of enzyme from rabbit or human red cells. Since it may be desirable to remove monophosphoglycerate mutase, the behavior of that enzyme is mentioned for steps that achieve separation. All procedures were carried out at 0–4°.

*Hemolysate.* The cells from 1 pint of fresh human blood were washed three times with 0.9% NaCl, and packed by centrifugation at 4000 $g$ for 5 min. White cells were removed by suction. Washed cells were frozen and stored not longer than 2 weeks. The thawed cells were diluted to four times their packed volume in 5 m$M$ potassium phosphate buffer, pH 7.2 containing 2 m$M$ EDTA, and 1 m$M$ 2-mercaptoethanol. After stirring for 15 min, the stroma was sedimented by centrifugation at 27,000 $g$ for 10 min. The supernatant was removed carefully.

*DEAE-Cellulose Chromatography.* The hemolysate was applied to a column of DEAE-cellulose (Eastman) (3 × 23 cm) equilibrated with 5 m$M$ potassium phosphate buffer containing 5 m$M$ EDTA and 1 m$M$ 2-mercaptoethanol. The column was washed with this buffer to remove the hemoglobin. The enzyme was eluted with a linear gradient formed from 900 ml of starting buffer and 0.1 $M$ potassium phosphate buffer, pH 7.2, containing 5 m$M$ EDTA, 1 m$M$ 2-mercaptoethanol, and 0.1 $M$ KCl as mixing buffer. Fractions with the highest activity were combined.

*Ammonium Sulfate Fractionation.* Solid ammonium sulfate (51.6 g/100 ml) was added. The precipitate was extracted successively with 0.55 (10 ml), 0.50 (10 ml), and 0.40 (5 ml) saturated (25°) solutions of ammonium sulfate, pH 6.9, containing 2 m$M$ EDTA. Most of the monophosphoglycerate mutase was in the highest ammonium sulfate fraction and diphosphoglycerate mutase was largely in the two lower salt fractions. In order to achieve the best relative purification only the 0.40–0.50 saturated ammonium sulfate fraction was used subsequently.

*Hydroxyapatite Chromatography.* The ammonium sulfate fraction

[6] O. H. Lowry, N. J. Rosebrough, A. L. Farr, and R. J. Randall, *J. Biol. Chem.* **193**, 265 (1951) and this series. Vol. 3 [73].

[7] S. Udenfriend, "Fluorescence Assay in Biology and Medicine," p. 207. Academic Press, New York, 1962.

[8] O. Warburg and W. Christian, *Biochem. Z.* **310**, 384 (1941) and this series, Vol. 3 [73].

TABLE I
PURIFICATION OF ERYTHROCYTE DIPHOSPHOGLYCERATE MUTASE

| Fraction | Volume (ml) | Total units | Total protein (mg) | Specific activity (units/mg) |
|---|---|---|---|---|
| Hemolysate | 404 | 94 | 38,000 | 0.0025 |
| DEAE-cellulose | 430 | 93 | 430 | 0.212 |
| Ammonium sulfate (0.4–0.5 saturated) | 10 | 36 | 60 | 0.60 |
| Hydroxyapatite | 30 | 18 | 1.8 | 10 |

was dialyzed against 5 m$M$ potassium phosphate, pH 7.2, containing 2 m$M$ 2-mercaptoethanol and applied to a column of hydroxyapatite (Bio-Gel HT from Bio-Rad) (1 × 7 cm) equilibrated with the same buffer. The column was eluted at 20 ml/hr with a linear gradient formed with 100 ml each of starting buffer and 0.2 $M$ potassium phosphate, pH 7.2, containing 2 m$M$ 2-mercaptoethanol. Monophosphoglycerate mutase was eluted with 30–40 m$M$ P$_i$ and diphosphoglycerate mutase with 70–85 m$M$ P$_i$.

The results of the purification procedure are summarized in Table I.

## Properties

*Stability.* Concentrated solutions of the enzyme could be stored at −20° for several months. A dilution containing 5 m$M$ glycylglycine buffer, pH 7.5, 2.5 mg/ml of bovine serum albumin, and 0.5 m$M$ EDTA retained 70% of its activity after storage at 4° for 2 months.

*Physical Properties.* The molecular weight determined by gel filtration was 60,000, with a subunit weight of 32,000. The isoelectric point was pH 5.23.

TABLE II
KINETIC VALUES FOR DIPHOSPHOGLYCERATE MUTASE

| | pH 7.5, Low salt | pH 7.24, 0.1 $M$ KCl |
|---|---|---|
| 1,3-Diphosphoglycerate ($K_m$) | 0.5 $\mu M$ | 3 $\mu M$ |
| Phosphoglycerate ($K_m$) | 1 $\mu M$ | 1 $\mu M$ |
| 2,3-Diphosphoglycerate ($K_i$) | 0.8 $\mu M$ | 20 $\mu M$ |
| P$_i$($K_i$) | 0.3 m$M$ | 0.6 m$M$ |

*Kinetic Constants.* Values were determined at low ionic strength[1] and also in 0.1 $M$ KCl,[9] which is close to the ionic composition of the red cell (Table II).

*Specificity.* Under the conditions of the assay, the enzyme does not show phosphatase activity. The highly purified preparation contained about 2% monophosphoglycerate mutase activity. This appeared to be an inherent property of the enzyme.

[9] Z. B. Rose, *Arch. Biochem. Biophys.* **158**, 903 (1973).

# Section VI

# Carboxylases and Decarboxylases

## [68] Ribulose-1,5-diphosphatc Carboxylase from *Rhodospirillum rubrum*

*By* Louise E. Anderson

D-Ribulose 1,5-diphosphate + $CO_2$ + $H_2O$ → 2 D-3-phosphoglycerate

## Assay Method

*Principle.* Ribulose-1,5-diphosphate carboxylase [3-phospho-D-glycerate carboxy-lyase (dimerizing), EC 4.1.1.39] catalyzes the formation of 2 moles of 3-P-glycerate from 1 mole of $CO_2$ and 1 mole of ribulose-1,5-diP. The enzyme is ubiquitously distributed among autotrophic organisms.[1-4] During purification[5] the enzyme from *Rhodospirillum rubrum* is assayed most conveniently using the spectrophotometric method of Racker.[6] The formation of P-glycerate is coupled to the oxidation of NADH with the enzymes P-glycerate kinase and glyceraldehyde-3-P dehydrogenase.

*Reagents*

Tris, 0.5 $M$, HCl buffer, pH 8.1
Ribulose-1,5-diP, 2.5 m$M$
NADH, 1.2 m$M$
GSH, 50 m$M$, neutralized with dilute NaOH
ATP, 0.12 $M$
$MgCl_2$, 0.1 $M$
$NaHCO_3$, 0.75 $M$
3-P-glycerate kinase, ammonium sulfate suspension
Glyceraldehyde-3-P dehydrogenase, centrifuged down from ammonium sulfate suspension and reconstituted in double-distilled water

*Procedure.* 3-P-Glycerate kinase, 5 μg, glyceraldehyde-3-P dehydrogenase, 250 μg, 0.1 ml each of the other components listed above, and double-distilled water to make a final volume of 1 ml are pipetted into a quartz cuvette. Reaction is initiated with enzyme and is followed by

[1] R. C. Fuller and M. Gibbs, *Plant Physiol.* **34,** 324 (1959).
[2] P. A. Trudinger, *Biochem. J.* **64,** 274 (1956).
[3] M. Gibbs, E. Latzko, R. G. Everson, and W. Cockburn, *in* "Harvesting the Sun" (A. San Pietro, F. A. Greer, and T. J. Army, eds.), p. 111. Academic Press, New York, 1967.
[4] A. Peterkofsky and E. Racker, *Plant Physiol.* **36,** 409 (1961).
[5] L. E. Anderson and R. C. Fuller, *J. Biol. Chem.* **244,** 3105 (1969).
[6] E. Racker, this series, Vol. 5, p. 266.

the change in absorbance at 340 nm with a Cary 14 recording spectrophotometer equipped with a 0.1A slide-wire. Temperature is maintained at 25°. There is usually a lag (few seconds to 1 min) before activity can be detected. Activity is measured only after the reaction rate has become maximal.

*Definition of Unit and Specific Activity.* Micromoles of 3-P-glycerate formed per minute per milliliter is equal to change in $A_{340}$ per minute divided by 6.22. Specific activity is obtained by dividing this value by the milligram of enzyme protein per milliliter in the assay cuvette.

Kinetic properties are estimated using an assay based on the incorporation of $^{14}CO_2$ into 3-P-glycerate.[5]

*Procedure.* Enzyme is incubated for 5 min at 30° in 1.8 ml of mixture containing 0.2 ml of Tris, 0.1 ml of GSH, and 0.2 ml of $NaHCO_3$, 1.25 $\mu$Ci of $NaH^{14}CO_3$ and double-distilled water. Reaction is initiated by the addition of 0.2 ml of ribulose-1,5-diP. At 30-sec intervals, 350 $\mu$l of reaction mixture are removed into 2 ml of glacial acetic acid, taken to dryness in a forced-draft oven at 50°, dissolved in 0.2 ml of water and 5 ml of Bray's liquid scintillator solution,[6] and counted for 20 min in a liquid scintillation counter. The 5-min preincubation of the enzyme in the reaction mixture eliminates the lag noted in the spectrophotometric assay. Initial velocity is estimated from plots of cpm incorporated versus time of incubation.

## Purification Procedure

*Growth of Bacterium.* If an explosion-proof hood is available, *R. rubrum* can be grown on the basal medium of Ormerod, *et al.*,[7] sparged continually with 1% $CO_2$ in $H_2$.[8] The S-1 strain of *R. rubrum* obtained from Dr. Howard Gest was used in the isolation described here. The only organic compound added to the medium is biotin; the nitrogen source is 2.5 m$M$ $(NH_4)_2SO_4$. The inoculum is cultured and transferred as described by Ormerod *et al.*[7]; a 10% inoculum is used. Cells are grown in Roux bottles in a water bath at 30°. Lumiline lights are used as light source; light intensity should be about 1000 foot-candles.

Alternatively, it is possible to grow cells in Fernbach flasks under an atmosphere of 1% $CO_2$ in $H_2$.[9] The culture is stirred continuously on a magnetic stirrer. The flask is connected through a sterile filter to

[7] J. G. Ormerod, K. S. Ormerod, and H. Gest, *Arch. Biochem. Biophys.* **94,** 449 (1961).

[8] L. Anderson and R. C. Fuller, *Plant Physiol.* **42,** 487 (1967).

[9] L. Anderson and R. C. Fuller, *Plant Physiol.* **42,** 497 (1967).
   **42,** 497 (1967).

PURIFICATION OF RIBULOSE-1,5-DIPHOSPHATE CARBOXYLASE
FROM *Rhodospirillum rubrum*

| Step | Protein (mg) | Activity (units) | Specific activity (units/mg protein) | Purification (fold) | Recovery (%) |
|---|---|---|---|---|---|
| 225,000 *g* supernatant | 130 | 38 | 0.31 | — | — |
| MnCl$_2$; (NH$_4$)$_2$SO$_4$ fractionation | 57 | 51 | 0.9 | 2.9 | >100 |
| Sephadex G-200 | 14 | 17 | 1.2 | 3.9 | 44 |
| Hydroxyapatite | 0.47 | 2.3 | 4.8 | 15 | 5.8 |

a gas reservoir over a 1% $CO_2$ buffer prepared as described by Pardee.[10] The gas reservoir, in turn, is connected to a reservoir of dilute $H_2SO_4$, set up so that, as $H_2$ is taken up by the culture, dilute acid replaces the gas in the gas reservoir. Growth is conveniently followed by gas uptake.

*Preparation of Cell-Free Extract.* All operations after harvesting are carried out at 0–4°. Cells from the logarithmic phase of growth are harvested by centrifugation, washed twice with 50 m$M$ potassium phosphate buffer, pH 7, suspended in sufficient 10 m$M$ potassium phosphate buffer, pH 7.6, to give 20–40 mg of protein per milliliter, and released at 12,000 psi from a French pressure cell. The extract is centrifuged for 10 min at 10,000 *g*, and the supernatant solution is centrifuged an additional 25 min at 225,000 *g*.

*Ammonium Sulfate Fractionation.* To remove nucleic acids 0.1 volume of 0.5 *M* MnCl$_2$ is added to the 225,000 *g* supernatant, the solution is allowed to stand for 3 hr, and then centrifuged for 10 min at 10,000 *g*. Solid, metal-free (NH$_4$)$_2$SO$_4$, 243 g per liter of supernatant solution, is added, and the precipitate is removed by centrifugation (10,000 *g*, 10 min). Additional (NH$_4$)$_2$SO$_4$, 97 g per liter of this supernatant solution, is added, and the precipitate is collected by centrifugation.

*Gel Filtration.* The precipitate is made up to 2 ml with 10 m$M$ potassium TES [N-tris(hydroxymethyl)methyl-2-aminoethane sulfonate], pH 7.6, 0.5 m$M$ dithiothreitol and layered onto a G-200 Sephadex column (2.5 × 30 cm, 40–120 μm) which has been equilibrated with the same buffer. The enzyme emerges from the column at about 1.5 times the exclude volume. Tubes containing peak enzyme activity are combined.

*Hydroxyapatite Chromatography.* The enzyme from the gel filtration

[10] A. B. Pardee, *J. Biol. Chem.* **179**, 1085 (1949).

step is placed on a hydroxyapatite[11] column (2.2 × 1.9 cm) equilibrated with TES, dithiothreitol buffer, and eluted with 3 m$M$ potassium phosphate, in TES, dithiothreitol buffer.

This procedure results in about 15-fold purification of the enzyme. Analytical ultracentrifugation reveals the presence of a minor (not more than 10% of the total protein) contaminant of lower $s$ value than the carboxylase.[5]

A summary of the purification and recovery of the enzyme is given in the table.

## Properties

*Stability.* Activity is retained in frozen crude extracts (225,000 $g$ supernatant) for several months, and for several days at 0° following ammonium sulfate fractionation. The purified enzyme is denatured by freezing but is relatively stable at 0° if 2-mercaptoethanol (5 m$M$) is present.[5]

*Molecular Size.* The $s°_{20,w}$ of the carboxylase, estimated by density gradient analysis or by analytical ultracentrifugation, is 6.2. The Stokes radius estimated by gel filtration analysis is 47 Å. The molecular weight can then be estimated to be 120,000.[5] The *R. rubrum* enzyme is the smallest of the ribulose-1,5-diP carboxylases. The enzyme from the rhodopseudomonads has an MW of about 360,000.[12] Carboxylases from all other photosynthetic organisms examined to date have molecular weights around 550,000.[13]

*Metal Requirement.* Unlike the carboxylase from higher plants, the *R. rubrum* enzyme does not require high levels of a divalent cation for activity.[5] After treatment with 1 m$M$ EDTA and filtration through metal-free G-25 Sephadex, activity was the same in the presence of 10 m$M$ or 100 $\mu M$ $MgCl_2$. (The $Mg^{2+}$ concentration was 100 $\mu M$ in the reaction mixture, probably owing to traces of $Mg^{2+}$ present in $NaHCO_3$.) Akazawa and co-workers have reported a variable requirement for $Mg^{2+}$[14,15] as well as lower molecular weights (68,000,[14] 83,000[15]). The enzyme used in their experiments was from heterotrophically cultured cells. It seems possible that the active enzyme is composed of two 60,000 MW subunits, which associate rapidly in the presence of $Mg^{2+}$, and some other component of the assay mixture.

[11] Obtained as powder from Calbiochem.
[12] L. E. Anderson, G. B. Price, and R. C. Fuller, *Science* **161**, 482 (1968).
[13] N. Kawashima and S. G. Wildman, *Annu. Rev. Plant Physiol.* **21**, 325 (1970).
[14] T. Akazawa, K. Sato, and T. Sugiyama, *Arch. Biochem. Biophys.* **132**, 255 (1969).
[15] T. Akazawa, T. Sugiyama, and H. Kataoka, *Plant Cell Physiol.* **11**, 541 (1970).

*Kinetic Properties.* The optimal pH is 8.1. The $K_m$ for ribulose-1,5-diP is 0.083 ± 0.007 m$M$; and for $CO_2$, 65 ± 13 m$M$.[5]

*Inhibitors.* The activity of the enzyme is inhibited by citrate, $K_i$ (competitive) 0.047 ± 0.013 m$M$ when ribulose-1,5-diP is variable substrate. With $CO_2$ as variable substrate mixed type kinetics ($V_p > V_{max}$, $K_p > K_m$) are obtained. Likewise 3-P-glycerate is competitive with ribulose-1,5-diP, $K_i$ 0.019 ± 0.004 m$M$ and noncompetitive with $CO_2$, $K_i$ 0.28 ± 0.13 m$M$. Phosphate is a less effective inhibitor, being noncompetitive with either substrate, $K_i$, 23 ± 2.6 m$M$ against ribulose-1,5-diP, and 13.6 ± 3.5 m$M$ against $CO_2$.[5]

## Acknowledgment

The purification and characterization of this enzyme was done in Dr. R. C. Fuller's laboratory, Biology Division, Oak Ridge National Laboratory and was supported by Grant GM 15595 from the National Institute of General Medical Sciences, United States Public Health Services to Dr. Fuller, and by the United States Atomic Energy Commission under contract with Union Carbide Corporation.

## [69] Ribulose-diphosphate Carboxylase from the Hydrogen Bacteria and *Rhodospirillum rubrum*

*By* Bruce A. McFadden, F. Robert Tabita, and Glenn D. Kuehn

D-Ribulose 1,5-diphosphate$^{-4}$ + $CO_2$ + $H_2O$ → 2 3-phospho-D-glycerate$^{-3}$ + 2 $H^+$

D-Ribulose-1,5-diphosphate (RuDP) carboxylase [3-Phospho-D-glycerate carboxy-lyase (dimerizing) EC 4.1.1.39] catalyzes primary fixation of carbon dioxide in all autotrophic species.[1] The enzyme is of interest from an evolutionary point of view because three sizes are found in nature[1,2]: small (MW 1.2 × 10$^5$), intermediate (MW 3.6 × 10$^5$), and large (MW 5–6 × 10$^5$). The latter, which are found in some bacteria and in algae and higher plants, are inhibited by 6-phospho-D-gluconate,[3] a compound which probably depresses fixation of $CO_2$ under heterotrophic conditions.[3,4] RuDP carboxylases of small and intermediate size, which are found only in certain bacteria, are insensitive to this compound.[3] Of fur-

[1] B. A. McFadden, *Bacteriol. Rev.* **37**, 289 (1973).

[2] L. E. Anderson, G. B. Price, and R. C. Fuller, *Science* **161**, 482 (1968).

[3] F. R. Tabita and B. A. McFadden, *Biochem. Biophys. Res. Commun.* **48**, 1153 (1972).

[4] D. K. Chu and J. A. Bassham, *Plant Physiol.* **50**, 224 (1972).

ther evolutionary interest is the quaternary structure of the enzyme. From most sources examined to date, it is apparently composed of two types of subunits of MW $\simeq 55,000$ and 15,000, but from three bacterial species it is apparently composed of one type of subunit.[5,6] In the present article, the isolation of the latter type of enzyme from the hydrogen bacteria and *Rhodospirillum rubrum* is described.

## Assay Method

*Principle.* To assay RuDP carboxylase, it is essential to establish that $CO_2$ fixation is RuDP-dependent. The normal assay involves measurement of RuDP-dependent fixation of $^{14}CO_2$ (provided as $H^{14}CO_3^-$) after terminating the fixation with acid. The product, [$^{14}C$]3-phosphoglycerate, is acid stable. Alternatively, a spectrophotometric assay employing coupled enzymic reactions may be used.[7]

*Reagents*

Tris-Cl, 160 m$M$, containing 50 m$M$ $MgCl_2 \cdot 6 H_2O$. After preparation the pH is adjusted to 8.0 (25°).

Tetrasodium ribulose diphosphate, 8 m$M$ (Sigma Chemical Company, St. Louis, Missouri). The solution is stable for weeks when frozen.

$NaH^{14}CO_3$, 100 m$M$

Trichloroacetic acid, 60% (w/v)

Bray solution[8]

Tris-$SO_4$, 20 m$M$, containing 1 m$M$ disodium EDTA, 10 m$M$ $MgCl_2 \cdot H_2O$, and 50 m$M$ $NaHCO_3$. After preparation the pH is adjusted to 8.0 (25°) (abbreviated TEMB).

TEMB buffer containing 5 m$M$ 2-mercaptoethanol (abbreviated TEMMB).

Tris-$SO_4$, 10 m$M$, containing 10 m$M$ $MgCl_2 \cdot 6 H_2O$. After preparation the pH is adjusted to 8.0 (25°) (abbreviated TM).

TM buffer containing 5 m$M$ 2-mercaptoethanol (abbreviated TMM).

Streptomycin sulfate, 10% (w/v)

$MgCl_2 \cdot 6 H_2O$, 1 $M$

[5] G. D. Kuehn and B. A. McFadden, *Biochemistry* 8, 2403 (1969).
[6] F. R. Tabita and B. A. McFadden, *J. Biol. Chem.* 249, 3459 (1974).
[7] E. Racker, *in* "Methods of Enzymatic Analysis" (H. U. Bergmeyer, ed.), p. 188. Academic Press, New York, 1963.
[8] G. A. Bray, *Anal. Biochem.* 1, 279 (1960).

*Procedure.* For assay of the enzyme from hydrogen bacteria, incubation mixtures at pH 8.0 contain: 0.10 ml 160 m$M$ Tris·Cl containing 50 m$M$ MgCl$_2$·6 H$_2$O; 0.05 ml 8 m$M$ tetrasodium RuDP; 0.05 ml 100 m$M$ NaH$^{14}$CO$_3$; 0.05 ml enzyme preparation in appropriate buffer (see Purification sections). All components except RuDP are incubated for 5 min at 30°, and the solution of RuDP (or water in the case of controls) is added at zero time. In the case of enzyme from *R. rubrum*, 0.05 ml of 1 m$M$ tetrasodium RuDP (instead of 8 m$M$) is added at zero time. After 5 min the reaction is stopped by the addition of 0.1 ml of 60% (w/v) trichloroacetic acid. Incorporation of $^{14}$C is measured in an aliquot portion by the method of Bray[8] at least 4 hr after acidification to permit complete release of $^{14}$CO$_2$. Under these conditions, quenching by trichloroacetic acid is negligible. Just before all experiments, the radioactivity of the H$^{14}$CO$_3^-$ used is determined. The specific radioactivity may be readily calculated if a large, known excess of HCO$_3^-$ has been previously added in preparing the stock solution.

One enzyme unit catalyzes the fixation of 1 $\mu$mole of CO$_2$[9] per minute at 30° under the conditions described. Specific activity is units per milligram of protein.[10]

## Purification from the Hydrogen Bacteria

*Growth.* Proposed changes in classification of the hydrogen bacteria have been discussed.[1] Nomenclature which has prevailed for decades will be used here to retain historical clarity. The culture of *Hydrogenomonas facilis* is a subculture of the original isolate of Schatz and Bovell[11] and has been maintained autotrophically at Washington State since 1953 with transfers at least once per month. *Hydrogenomonas eutropha* is maintained similarly. Cultures of both organisms may be obtained by request to B. A. McFadden. Stocks of both are maintained on autotrophic medium[12] supplemented with trace minerals.[13] Fresh inocula are added to starter medium containing fructose, grown, and transferred to 5-liter batches of fructose-containing medium as described.[14] After growth at 30° with agitation at 120–140 cycles/min, cells are harvested near the midexponential phase of growth. Because of the marked instability of

[9] T. G. Cooper, D. Filmer, M. Wishnick, and M. D. Lane, *J. Biol. Chem.* **244,** 1081 (1969).
[10] O. H. Lowry, N. J. Rosebrough, A. L. Farr, and R. J. Randall, *J. Biol. Chem.* **193,** 265 (1951).
[11] A. Schatz and C. Bovell, *J. Bacteriol.* **63,** 87 (1952).
[12] B. A. McFadden and H. R. Homann, *J. Bacteriol.* **89,** 839 (1965).
[13] R. Repaske, *J. Bacteriol.* **83,** 418 (1962).

RuDP carboxylase in the absence of substrate *in vivo* and *in vitro*,[14] the presence of bicarbonate is essential during all cell washings and subsequent treatments of cell-free extracts. Thus the collected cell paste is washed once at 2° in 5 volumes of TEMB buffer, followed by centrifugation at 9000 $g$. The resulting cell paste can be stored indefinitely at −20° with no loss in ribulose-1,5-diphosphate carboxylase activity.

### Hydrogenomonas facilis

All succeeding purification procedures are carried out near 2°.

*Crude Extract.* Frozen cells of *H. facilis* are suspended in an equal mass of TEMMB buffer. The cell suspension is disrupted by a French press at 15,000 psi, the extruded mixture is clarified by centrifugation at 41,190 $g$ for 15 min, and the supernatant is decanted and recentrifuged at 105,000 $g$ for 1 hr. The resulting red-brown supernatant is drawn off the gelatinous pellet with a pipette in order to ensure exclusion of particulate material. The protein concentration is generally 25–30 mg/ml. This 105,000 $g$ supernatant preparation, labeled fraction S-105, is dialyzed 12–15 hr against two changes of 60 volumes of TEMMB buffer. After dialysis the protein concentration is adjusted to 13 mg/ml with TEMMB to yield fraction S-105$_D$.

*Ethanol Fractionation.* Absolute ethanol, chilled to −77° in an acetone–dry-ice mixture, is added dropwise with gentle stirring to fraction S-105$_D$ to a final concentration of 15% (v/v). Stirring is discontinued, the mixture is incubated at 2° for 15 min, and a small amount of precipitated protein is collected by centrifugation at 30,900 $g$ for 10 min and discarded. The supernatant is then adjusted to 30% (v/v) with the cold absolute ethanol, the suspension incubated again at 2° for 15 min and the precipitated protein collected by centrifugation as before. The precipitate is immediately dissolved in cold TMM buffer. The buffer volume added is approximately one-sixth that of fraction S-105 and yields a protein concentration of 7–9 mg/ml. Most of the ethanol-precipitated material dissolves after 60–90 min with intermittent stirring at 2°. A small amount of insoluble residue remaining is pelleted by centrifugation at 105,000 $g$ for 1 hr, yielding a pale yellow supernatant fraction, E$_{15-30}$.

*Streptomycin Treatment.* Nucleic acid is partially removed by addition of 10% (w/v) streptomycin sulfate to fraction E$_{15-30}$. The streptomycin sulfate solution is prepared in cold TMM buffer and adjusted to pH 7.0 by addition of 0.5 $M$ NaOH. It is then slowly added to fraction E$_{15-30}$ with gentle stirring to a final ratio of 0.75 mg of streptomycin per milligram of protein. Stirring is discontinued, and the insoluble complex is allowed to precipitate over a 15-min interval. The complex is re-

[14] G. D. Kuehn and B. A. McFadden, *J. Bacteriol.* **95**, 937 (1968).

moved by centrifugation at 30,900 $g$ for 10 min, the supernatant immediately recentrifuged at 105,000 $g$ for 30 min, and the resulting supernatant, fraction SM, is stored overnight at 2°. This fraction can be stored for weeks at 2° with little loss in ribulose-1,5-diphosphate carboxylase activity.

*Chromatography on DEAE-Cellulose.* Fraction SM is adjusted to pH 8.0 by addition of 1.0 $M$ Tris-SO$_4$ (pH 8.0) and immediately applied to a 2.3 × 35 cm DEAE-cellulose column previously equilibrated with TM buffer. Gradient elution is accomplished by using two 600-ml beakers as a mixing chamber and reservoir. Both beakers are mounted at the same level and connected by tubing so as to maintain hydrostatic equilibrium; 400 ml of TM buffer is placed in the mixing chamber, and 400 ml of 0.3 $M$ NaCl buffered in TM is added to the reservoir. After elution at a rate of 30–40 ml/hr, all fractions are analyzed by measuring the absorbancy at 280 nm and the ribulose-1,5-diphosphate carboxylase activity.

After dialysis of pooled fractions against 10 m$M$ Tris-SO$_4$ containing 20 m$M$ Mg$^{2+}$ and 20 m$M$ HCO$_3^-$, pH 8.0 (25°), the enzyme is concentrated using an Amicon Diaflo Filtration Unit Model 202 equipped with PM-10 ultrafiltration membranes to a concentration of about 20 mg/ml. The resulting solution of ribulose-1,5-diphosphate carboxylase is stable for months at 2° with little decay in activity.

*Hydrogenomonas eutropha*

All operations are conducted at 2°.

*Streptomycin Precipitation.* To S-105$_D$, prepared as described for *H. facilis,* 10% w/v streptomycin sulfate solution (see previous section) is added in 0.025-ml increments to a final ratio of 0.75 mg of streptomycin sulfate per milligram of protein; 15 min after the final addition, the insoluble precipitate is removed by centrifugation at 29,000 $g$ for 15 min.

*Ammonium Sulfate Saturation.* The resulting supernatant, fraction SM, is slowly brought to 0.40 ammonium sulfate saturation by addition of 0.67 volume of alkaline ammonium sulfate solution that has been saturated at 0° and the pH of which has been adjusted to 7.8 with concentrated ammonium hydroxide. Precipitated protein is collected after 40-min centrifugation at 29,000 $g$ for 15 min, and the pellet is redissolved in 16 ml of TEMB buffer yielding fraction AAS.

*Sephadex G-50.* Each of four 4.0-ml aliquots of fraction AAS is applied to the top of four 2.3 × 35 cm. Sephadex G-50 gel columns equilibrated with TEMB buffer. Elution of each column is achieved with the same buffer at a flow rate of 30–40 ml/hr and 5-ml fractions are collected. The first four to five fractions after the void volume effluent containing

100% of the ribulose-1,5-diphosphate carboxylase activity are pooled and stored at 2° overnight.

*DEAE-Sephadex Chromatography.* This fraction is applied to a 2.3 × 35 cm DEAE-cellulose column. Elution is initially carried out with TEMB buffer containing 0.1 *M* NaCl until an ultraviolet-absorbing peak has been eluted. A 1 liter gradient from 0.1 to 0.3 *M* NaCl buffered in TEMB is then constructed as described earlier from 500-ml portions of salt-TEMB solutions. The flow rate is maintained at 30–40 ml/hr, and 10-ml fractions are collected. Peak fractions which display ribulose-1,5-diphosphate carboxylase activity, 280 nm/260 nm ratios greater than 1.8, and contain more than 1.0 mg of protein/ml are pooled.

*Ammonium Sulfate Fractionation.* The resulting solution is brought to 0.45 saturation with the saturated (0°) alkaline (pH 7.8) ammonium sulfate, and the precipitate is collected by centrifugation and dissolved in 3.3 ml of TEMB buffer.

*Sephadex G-200.* This protein solution is then subjected to descending Sephadex G-200 chromatography on a 2.3 × 50 cm column. Flow rates of 10–15 ml/hr are achieved by positioning the reservoir 15–20 cm above the effluent orifice. Elution is achieved with TEMB buffer and peak fractions containing ribulose-1,5-diphosphate carboxylase activity are again concentrated by alkaline ammonium sulfate precipitation. The protein is collected, dissolved in TEMB buffer to yield a concentration of 5–7 mg of protein per milliliter, and finally dialyzed against 1 liter (three changes) of the same buffer. This enzyme solution is stable for weeks when stored at 2°.

## Notes on Fractionation Procedures

Preliminary experiments revealed that RuDP carboxylase in extracts from *H. facilis* could not be subjected to standard salt fractionation with retention of activity. Hence the organic solvent precipitation method was developed. Result of purification from 30 g (wet weight) of *H. facilis* are summarized in Table I. The experimentally generated expression for enzyme concentration was: $OD_{500\ nm} = 0.894C + 0.022$, where $OD_{500\ nm}$ is the reading of a standard Lowry protein assay and $C$ is the concentration in milligrams per milliliter of RuDP carboxylase.

In contrast to the carboxylase from *H. facilis*, the enzyme in extracts of *H. eutropha* could be fractionated with alkaline ammonium sulfate precipitation after treatment of fractions S-105D with streptomycin sulfate. Table II summarizes data obtained for a typical isolation from 62 g (wet weight) of cells. Significant purification of the carboxylase was achieved during Sephadex G-50 chromatography by selecting only the first several fractions eluted after the column void volume. These frac-

TABLE I

ISOLATION SUMMARY OF RIBULOSE-1,5-DIPHOSPHATE CARBOXYLASE
FROM *Hydrogenomonas facilis* (30 G WET WEIGHT)

| Fractions | Total protein (mg) | $OD_{280 nm}$ : $OD_{260 nm}$ | Specific activity (units/mg) | Total units | Purification (fold) |
|---|---|---|---|---|---|
| S-105$_D$ | 800 | 0.67 | 0.069 | 55.0 | (1) |
| E$_{15-30}$ | 77 | 0.63 | 0.18 | 39.0 | 2.6 |
| SM | 59 | 0.79 | 0.22 | 36.8 | 3.2 |
| DEAE-cellulose chromatography | 25 | 1.71 | 1.36 | 40.3 | 20 |

tions contain approximately 50% of the protein applied to the column and 100% of the carboxylase activity. These early fractions are pooled and chromatographed on DEAE-cellulose. The peak fractions are pooled, concentrated with alkaline ammonium sulfate, and further purified by Sephadex G-200 chromatography to yield a homogeneous product with a specific activity of 1.9. The correlation between colorimetric Lowry assay and enzyme concentration ($C$ in milligrams per milliliter) is: $OD_{500 nm}$ = 0.830 $C$ + 0.030. At any stage of the purification procedure, the enzyme is safely stored at 2° in the presence of bicarbonate. Fraction E$_{15-30}$ and succeeding fractions prepared from *H. facilis* are also stable at 2° and pH 8.0 in the absence of bicarbonate. Similarly, fraction AAS and succeeding fractions from *H. eutropha* are stable for several weeks in the absence of bicarbonate. Although 2-mercaptoethanol, dithiothreitol, or glutathione may be occasionally included in some buffers, they are not essential for retention of carboxylase activity either during isolation or storage. Thiol reagents are not required for catalytic activity.

TABLE II

ISOLATION SUMMARY OF RIBULOSE-1,5-DIPHOSPHATE CARBOXYLASE
FROM *Hydrogenomonas eutropha* (62 G WET WEIGHT)

| Fractions | Total protein (mg) | $OD_{280 nm}$ : $OD_{260 nm}$ | Specific activity (units/mg) | Total units | Purification (fold) |
|---|---|---|---|---|---|
| S-105$_D$ | 2695 | 0.53 | 0.081 | 218 | (1) |
| SM | 2690 | 0.89 | 0.080 | 216 | 1.1 |
| AAS | 1078 | 1.00 | 0.14 | 148 | 1.8 |
| Sephadex G-50 | 564 | 1.21 | 0.28 | 155 | 3.7 |
| DEAE-cellulose chromatography | 88 | 1.87 | 1.62 | 143 | 22 |
| Sephadex G-200 | 47 | 1.90 | 1.94 | 91 | 26 |

The presence of sulfate ion is important for stability of the enzyme from *H. facilis* and sulfuric acid is therefore employed in the adjustment of the pH of Tris solutions used in isolation procedures.

In the presence of 1 m$M$ EDTA and 2–5 m$M$ thiol, dilute enzyme solutions (0.05 mg/ml) of both proteins lose 90% of their activity at pH 8.0 in 12 hr at 2°. Activity is only partially restored (to about 75%) by addition back of 20 m$M$ MgCl$_2$. Freezing or storage of enzyme in glycerol or sucrose at −20° results in loss of activity. RuDP carboxylase from both *H. eutropha* and *H. facilis* is free of ribose-5-phosphate isomerase, ribulose-5-phosphate kinase, 3-phosphoglycerate kinase, and 3-phosphoglyceraldehyde dehydrogenase as well as carbohydrate and phospholipid.

Electrophoretograms of purified preparations of both enzymes yields a single protein zone in both 7 and 10% polyacrylamide gels.

The absorption spectra of the carboxylases are similar and typical of proteins. The absorbancies at 280 nm and 25° for the enzymes from *H. eutropha* and *H. facilis* are 1.55 and 1.23, respectively, in 20 m$M$ Tris-sulfate containing 0.1 $M$ MgCl$_2$ (pH 8.0, 25°) at a protein concentration of 1.0 mg/ml and a path length of 1.0 cm.

## Purification from R. rubrum

*Growth.* A culture of *R. rubrum* standard available upon request from B. A. McFadden is maintained in the synthetic malate medium of Ormerod *et al.*[15] with 0.2% ammonium sulfate as the nitrogen source instead of L-glutamate and with the addition of 0.2% sodium bicarbonate. Inocula for 1-liter bottles containing 0.6% sodium butyrate instead of malate are cultured in completely filled 75-ml screw-cap tubes. One of these 75-ml precultures serves to inoculate a 1-liter prescription bottle of the same media. Cultures are placed in front of light banks of 100-W incandescent bulbs and grown at 30° at a light intensity of 800 footcandles. All cultures are harvested at the mid-exponential phase of growth and harvested by continuous-flow centrifugation. The resultant cell paste is then weighed and stored at −20° until needed. A typical cell yield is 3–4 g (wet packed cells) from 1 liter of culture.

*Purification Steps*

Upon thawing, the cell paste is suspended at 2° in TEMMB buffer. A 1 : 1 mass ratio of cell paste to buffer is used. All purification procedures

[15] J. G. Ormerod, K. D. Ormerod, and H. Gest, *Arch. Biochem. Biophys.* **94**, 449 (1961).

are carried out at 2°. Concentrations of ammonium sulfate are calculated from solubility values at 25°.

*Crude Extract.* The cell suspension (butyrate-grown cells) is passed twice through a French pressure cell at 15,000 psi. Unbroken cells and debris are removed by centrifugation of 30,000 $g$ for 15 min, and the supernatant is decanted. The pellet is resuspended in one-fifth the original volume of TEMMB and recentrifuged at 30,000 $g$ for 15 min. The supernatants are combined and subjected to ultracentrifugation at 200,000 $g$ for 60 min. The resulting high speed supernatant fraction, S-200, devoid of chromatophores and membrane fragments, is carefully removed with a pipette.

*Heat Treatment.* To the high speed supernatant fraction, 1 $M$ $MgCl_2 \cdot H_2O$ is added to a final concentration of 50 m$M$. Aliquots (10 ml) of this fraction typically containing 13-17 mg of protein per milliliter are transferred to test tubes to ensure thermal equilibration, incubated at 50° for 10 min, and then immediately chilled in an ice bath. Denatured protein is removed by centrifugation.

*Streptomycin Treatment.* The resulting supernatant fractions (50°, $Mg^{2+}$) are pooled and subjected to streptomycin treatment (0.75 mg of streptomycin per milligram protein) by addition of a 10% (w/v) solution of streptomycin sulfate with continuous stirring after which the cloudy brownish suspension is allowed to stand for 15 min in an ice bath. The solution is clarified by centrifugation and the supernatant, SM, is used for further purification.

*Ammonium Sulfate Fractionation.* To the streptomycin-supernatant fraction, solid ammonium sulfate is added and the solution slowly brought to 0.6 saturation with ammonium sulfate (0.39 g per milliliter of solution. The solution is allowed to stand for 60 min, and precipitated protein is collected by centrifugation at 30,000 $g$ for 20 min. The supernatant is discarded as it is devoid of enzyme activity. The precipitate is resuspended in a minimal amount (5–10 ml) of ice-cold TEMMB and recentrifuged at 100,000 $g$ to remove any insoluble matter.

*DEAE-Cellulose Chromatography.* Ammonium ion is then removed from the resulting supernatant on a 2.2 × 30 cm Sephadex G-25 column which has been equilibrated with TEMMB. The eluted protein fraction, AS, is then loaded onto a 4.4 × 60-cm DEAE-cellulose column previously equilibrated with TEMMB. Approximately 1 liter of buffer is passed through the column, resulting in the elution of nonadsorbed protein lacking in RuDP carboxylase activity. Further elution is carried out with TEMMB containing 0.1 $M$ NaCl until another peak absorbing at 280 nm has been eluted off the column. A linear gradient of 0.1 $M$–0.3 $M$ NaCl in TEMMB is then passed through the column, resulting in the elution of RuDP carboxylase; finally 0.5 $M$ NaCl in TEMMB elutes ultraviolet-absorbing material devoid of enzyme activity. The flow rate is maintained at 60 ml/hr, and 20-ml fractions are collected. Peak fractions which contain RuDP carboxylase (tubes 163–180) are pooled and brought once again to 0.60 saturation with solid ammonium sulfate. The precipitate is collected by centrifugation, dissolved in 15 ml of TEMB buffer, a portion is assayed, and $NH_4$ is removed on the Sephadex G-25 column equilibrated with the same buffer. At this point the enzyme is judged to be approximately 90% pure by polyacrylamide disc gel electrophoresis and the protein concentration is about 5 mg/ml.

*DEAE-Sephadex Chromatography.* The final step for the preparation of ribulose-1,5-diphosphate carboxylase involves adsorption of one-half of the product from the previous step on 50 g of DEAE-Sephadex A-25 which has been equilibrated with TEMB and adjusted to a total volume of 250 ml with TEMB in a beaker. The enzyme preparation is added to the DEAE-Sephadex and slowly stirred for 2–3 hr (or until 95% of the enzyme is adsorbed to the beads). The adsorbant, which contains bound enzyme, is then carefully filtered on a medium-porosity sintered-glass filter with gentle suction. The moist cake is then suspended in 120 ml of TEMB buffer containing 50 mM NaCl, stirred for 30 min, and the filtering process is repeated. Subsequent elutions are performed with 120-ml batches of buffer containing 80 mM, 0.1 $M$, and 0.13 $M$ NaCl. Finally, RuDP carboxylase is eluted from the DEAE-Sephadex

TABLE III

Isolation Summary of Ribulose-1,5-diphosphate Carboxylase
from *Rhodospirillum rubrum* (100 g Wet Weight)

| Fraction | Total protein (mg) | $OD_{280\ nm}$: $OD_{260\ nm}$ | Specific activity (units/mg) | Total units | Purification (fold) |
|---|---|---|---|---|---|
| S-200 | 3700 | 0.69 | 0.117 | 441 | (1) |
| 50°, $Mg^{2+}$ | 3670 | 0.74 | 0.110 | 404 | 0.9 |
| SM | 3265 | 0.85 | 0.106 | 346 | 0.9 |
| AS | 1073 | 1.05 | 0.337 | 362 | 2.9 |
| DEAE-cellulose chromatography | 220 | 1.60 | 1.28 | 282 | 11 |
| DEAE-Sephadex[a] | 85 | 1.71 | 1.76 (2.50)[b] | 143 | 15 |

[a] Data reflect two purifications of pooled peak fractions from DEAE-cellulose chromatography as described.

[b] Value in parentheses is the specific activity obtained in the presence of 4 mM EDTA and 10 mM DTT.

with three 100-ml washes of TEMB containing 0.15 $M$ NaCl. A small amount of enzyme can subsequently be eluted with 0.18 $M$ NaCl in TEMB, yielding, however, impure RuDP carboxylase. The 0.15 $M$ NaCl eluates are pooled and concentrated by ammonium sulfate precipitation as described earlier. The precipitate protein is then resuspended to a protein concentration of 2 mg/ml in either TEMMB or 50 m$M$ potassium phosphate buffer, pH 7.4, and dialyzed against 4 liters of the same buffer. This enzyme solution is stable for a short time (less than 2 weeks at 2°) but can be reactivated.[16]

Table III summarizes the results of purification of the enzyme from 100 g (wet weight) of *R. rubrum*. The product is electrophoretically homogeneous when examined in a series of gels polymerized from four concentrations of acrylamide.[16]

## General Properties of the Isolated Enzymes

*Composition and Structure.* The amino acid compositions of the enzyme from *R. rubrum*[6] and the hydrogen bacteria are similar to that for the large subunit of the spinach enzyme.[1] For the three bacterial enzymes less than 50% of the half-cystine analysis can be accounted for by cysteine residues[1,6] in contrast to the enzyme from higher plants in which most, if not all, half-cystines can be attributed to readily titratable cysteines.[1] Moreover, from the former sources some of the sulfhydryls are buried.[5,6]

Antiserum to RuDP carboxylase from *H. eutropha* does not inactivate the enzyme from *R. rubrum* and vice versa.[6]

In addition to the small enzyme found in *R. rubrum*, enzymes of intermediate (MW 350,000) and large size are found (MW > 500,000).[1]

*Kinetic.* Enzymes from *H. facilis*, *H. eutropha*, and *R. rubrum* have turnover numbers of 1.04, 1.38, and 1.69 moles of $CO_2$ utilized per second per mole of catalytic site, respectively, if it is assumed that there are 12 catalytic sites per molecule of enzyme from the hydrogen bacteria and 2 per molecule of *R. rubrum* enzyme.[5,6,16] The enzyme from spinach has a turnover number of about 1.5 moles $CO_2$ utilized per second per mole of catalytic site. RuDP carboxylase is one of the most ineffective catalysts in nature.[1]

Carbon dioxide fixation by the enzyme from *R. rubrum* is competitively inhibited by oxygen (McFadden, unpublished observation) suggesting that this enzyme has oxygenase activity.[17,18]

[16] F. R. Tabita and B. A. McFadden, *J. Biol. Chem.* in press.
[17] T. J. Andrews, G. H. Lorimer, and N. E. Tolbert, *Biochemistry* **12**, 11 (1973).
[18] G. H. Lorimer, T. J. Andrews, and N. E. Tolbert, *Biochemistry* **12**, 18 (1973).

The Michaelis constants for $CO_2$, and RuDP of 0.2–0.3 m$M$, and 0.05–0.24 m$M$, respectively, for the three bacterial enzymes[6,19] are similar to those for RuDP carboxylase from higher plants.[1] With enzyme from *R. rubrum*, however, there is pronounced positive homotropism for $CO_2$ in the presence of EDTA (but with saturating $Mg^{2+}$).[6] The $K_m$ for $Mg^{2+}$ is unusually low (0.2 m$M$) for the enzyme from *R. rubrum*,[1,6] but there is an absolute dependence upon $Mg^{2+}$.[16] RuDP carboxylase from *R. rubrum* shows a striking preference for $Mg^{2+}$ in studies of activation of metallic ion-depleted enzyme.[16] In contrast, similar preparations from *H. facilis* and *H. eutropha* are slightly reactivated (ca. 8% when compared to $Mg^{2+}$) by 20 m$M$ $Co^{2+}$; the enzyme from *H. facilis* is reactivated by 17% in the presence of 20 m$M$ $Mn^{2+}$ whereas that from *H. eutropha* remains inactive.[19]

The molecular weights of the enzyme from *H. eutropha* and *H. facilis* measured by sedimentation equilibrium are 515,000 and 551,000, respectively. The enzyme from *H. eutropha* consists of two types of subunits with molecular weights of 54,000 and 13,500 (Tabita and McFadden, unpublished observation). The molecular weight of the enzyme from *R. rubrum* measured by light scattering is 114,000. One component of a molecular weight of 56,000 is evident after dissociation with sodium dodecyl sulfate (SDS) with subsequent gel electrophoresis in polyacrylamide containing SDS.[6] In contrast, RuDP carboxylase from all other sources examined is apparently composed of large (MW $\cong$ 55,000) and small subunits (MW $\cong$ 15,000).[1]

[19] G. D. Kuehn and B. A. McFadden, *Biochemistry* **8**, 2394 (1969).

# [70] Ribulose-diphosphate Carboxylase from Spinach Leaves

*By* Marvin I. Siegel and M. Daniel Lane

$$\text{D-Ribulose 1,5-diphosphate} + CO_2 + H_2O \xrightarrow{Mg^{2+}} 2 \text{ D-3-phosphoglycerate} + 2 \text{ H}^+$$

## Assay Method

*Principle.* Ribulose-diphosphate carboxylase catalyzes the irreversible carboxylation of ribulose 1,5-diphosphate to form 3-phosphoglycerate with the stoichiometry shown in the reaction above. Carboxylase activity is conveniently determined[1] by following the rate of incorporation of $H^{14}CO_3^-$ into phosphoglycerate (acid-stable $^{14}C$-activity). Alternatively,

[1] J. M. Paulsen, and M. D. Lane, *Biochemistry* **5**, 2350 (1966).

activity can be determined in the presence of 3-phosphoglycerate kinase, ATP, glyceraldehyde-3-phosphate dehydrogenase and NADH by following the rate of NADH oxidation spectrophotometrically.[2] This enzyme is widely distributed in plant tissues and microorganisms.[3,4]

### Reagents

Tris (Cl⁻) buffer, 1.0 $M$, pH 7.8 at 25°C

$KH^{14}CO_3$, 0.25 $M$ (approximately $10^5$ cpm/μmole; the specific activity must be accurately determined).

D-Ribulose 1,5-diphosphate-Na₄, 7 m$M$, synthesized enzymically from D-ribose-5-P[5,6], and assayed by its quantitative conversion to 3-phosphoglycerate in the presence of the constituents of the $^{14}$C-labeled bicarbonate or spectrophotometric assays and excess ribulose diphosphate carboxylase. The quantity of 3-phosphoglycerate formed is determined either from acid-stable $^{14}$C-activity formed or NADH oxidized.[2] Ribulose diphosphate can be chromatographically purified as described by Wishnick and Lane.[7]

MgCl₂, 0.1 $M$, containing 0.6 m$M$ EDTA, pH 7.0

GSH, 0.06 $M$

Liquid scintillator: 0.25 g of 1,4-bis[2-(5-phenyloxazolyl)]benzene (POPOP), 10 g of 2,5-diphenyloxazole (PPO), and 100 g of recrystallized naphthalene per liter of dioxane.

*Procedure: $^{14}$C-Labeled Bicarbonate Fixation Assay.* The complete reaction mixture contains the following components (in micromoles): Tris (Cl⁻), 200; D-ribulose 1,5-diphosphate, 0.70; $KH^{14}CO_3$ (approximately 2 μCi), 50; MgCl₂, 10; EDTA, 0.06; GSH, 6; and ribulose-diphosphate carboxylase in a total volume of 1.0 ml. Prior to use, the carboxylase is dialyzed overnight at 4° against 10 m$M$ tris (Cl⁻), pH 7.8 at 4°, containing 0.1 m$M$ EDTA and 10 m$M$ 2-mercaptoethanol. The final pH of the reaction mixture is 7.9 at 30°. The carboxylation rate is determined by pipetting at 2-min intervals 0.1-ml aliquots of the reaction mixture into 15 × 45 mm glass vials (flat bottom shell vials, Demuth Glass Division, catalog No. 15045) containing 0.2 ml of 6 $N$ HCl, and drying in a forced-draft oven for 1 hr at 95°, water (0.3 ml) is added to each vial

[2] E. Racker, *in* "Methods of Enzymatic Analysis" (H. U. Bergmeyer, ed.), p. 188, Academic Press, New York, 1963.

[3] M. I. Siegel, M. Wishnick, and M. D. Lane, *in* "The Enzymes" (P. Boyer, ed.), 3rd ed., Vol. 6, p. 169. Academic Press, New York, 1972.

[4] N. Kawashima and S. G. Wildman, *Annu. Rev. Plant Physiol.* **21**, 325 (1970).

[5] B. L. Horecker, J. Hurwitz, and P. K. Stumpf, this series, Vol. 3, p. 193.

[6] B. L. Horecker, J. Hurwitz, and A. Weissbach, *Biochem. Prep.,* **6**, 83 (1958).

[7] M. Wishnick and M. D. Lane, *J. Biol. Chem.* **244**, 55 (1969).

followed by 3 ml of liquid scintillator. The vials are capped and acid-stable [14]C-activity (as 3-[[14]C] phosphoglycerate) determined with a liquid scintillation spectrometer.[8] The carboxylation rate is proportional to enzyme concentration up to a level of approximately 0.012 unit; proportionality to enzyme is maintained until at least 36% of the added ribulose diphosphate has been consumed. The carboxylation reaction follows zero-order kinetics under the conditions described.

*Definition of Unit and Specific Activity.* One unit of ribulose-diphosphate carboxylase is defined as that amount of enzyme which catalyzes the carboxylation of 1.0 μmole of ribulose diphosphate per minute under the assay conditions described. Protein is determined spectrophotometrically[1]: an absorbance (1-cm light path) of 1.00 at 280 nm for the pure enzyme is equivalent to 0.61 mg of protein. Specific activity is expressed as units per milligram of protein.

## Purification Procedure

The purification procedure described is based on that of Paulsen and Lane[1] with some modifications. All the operations are carried out at 4°. The results of the procedure are summarized in the table. Six hundred milligrams of homogeneous ribulose-diphosphate carboxylase are obtained by the following method.

*Extract.* Fresh spinach, purchased locally, is destemmed and washed with cold tap water. Six hundred grams of leaves are homogenized with 2 liters of 10 m$M$ potassium phosphate buffer, pH 7.6, containing 0.1 m$M$ EDTA, in a 4-liter capacity Waring Blendor for 1 min at medium speed. The resulting suspension is filtered through S. and S. No. 588 fluted filter paper. The filtrate is adjusted if necessary to pH 7 with 5 $N$ NH$_4$OH.

*Ammonium Sulfate Fractionation.* The extract is brought to 37% saturation with solid ammonium sulfate (226 g per liter of extract) and after standing for 30 min is centrifuged at 9000 $g$ for 30 min. The dark green precipitate is discarded, and the supernatant solution is brought to 50% saturation with solid ammonium sulfate (92.5 g/liter). After centrifugation as described above, the precipitate is dissolved in 0.1 $M$ potassium phosphate buffer, pH 7.6, containing 0.1 m$M$ EDTA and 10 m$M$ 2-mercaptoethanol (100 ml of buffer per 2 liters of initial extract). This solution (37–50% saturated ammonium sulfate fraction) referred to as Ammonium Sulfate I, is stored at 4° overnight. After dilution with an equal volume of glass-distilled water, saturated ammonium sulfate[9] (pH 7.0, 41 ml/100 ml of diluted Ammonium Sulfate I) is added. EDTA and 2-mercaptoethanol are added to bring their concentrations to 0.1 m$M$

---

[8] The inner vials may be discarded and the polyethylene bottles reused.

[9] Ammonium sulfate is saturated at room temperature and neutralized with NH$_4$OH so that, when diluted 5-fold, the pH is that indicated.

PURIFICATION OF RIBULOSE-DIPHOSPHATE CARBOXYLASE

| Step | Total protein (g) | Total activity (units) | Specific activity (units/mg of protein) | Yield (%) |
|---|---|---|---|---|
| 1. Extract from 600 g of leaves | 10.12[a] | 3542[c] | 0.35 | 100 |
| 2. Ammonium sulfate I | 5.17[a] | 2740[c] | 0.53 | 77 |
| 3. Ammonium sulfate II | 2.50[a] | 1975[c] | 0.79 | 56 |
| 4. DEAE-cellulose chromatography | 0.97[b] | 1067[c] | 1.10 | 30 |
| 5. Hydroxyapatite chromatography | 0.59[b] | 873[c] | 1.48 | 25 |

[a] Determined by the method of O. H. Lowry, N. J. Rosebrough, A. L. Farr, and R. J. Randall, *J. Biol. Chem.* **193**, 265 (1951).

[b] Determined from absorbance at 280 nm × 0.61.

[c] Determined by the $^{14}$C-labeled bicarbonate fixation assay.

and 10 m$M$, respectively. After centrifugation as described above, the precipitate is discarded and saturated ammonium sulfate[9] (pH 7.0, 8.6 ml/100 ml) is added to the supernatant solution. After centrifugation, the precipitate is discarded and additional saturated ammonium sulfate[9] (pH 7.0, 9.5 ml/100 ml) is added to the supernatant solution. The pellet recovered after centrifugation is dissolved in 0.1 $M$ potassium phosphate buffer, pH 7.6, containing 0.1 m$M$ EDTA and 10 m$M$ 2-mercaptoethanol (27 ml/100 ml of Ammonium Sulfate I before dilution with water). This solution referred to as Ammonium Sulfate II which constitutes a 38–45% saturated ammonium sulfate fraction, is divided into four equal fractions and can be stored at −20° for 6–12 months with little loss of activity.

Although the following steps are given for the purification of the total 38–45% saturated ammonium sulfate solution, the procedures can be readily "scaled down," permitting the preparation of smaller quantities of enzyme from each of the four equal fractions described above. The four fractions are thawed and centrifuged at 15,000 $g$ for 10 min, and the supernatant solution applied to a Sephadex G-25 column (4.5 × 42 cm), previously equilibrated with 5 m$M$ potassium phosphate buffer, pH 7.6 containing 0.1 m$M$ EDTA and 10 m$M$ 2-mercaptoethanol. The enzyme is eluted with the same buffer. The light brown fastest moving band, which is clearly resolved from a bright yellow, slower moving band, is collected visually and contains about 90% of the protein applied to the column. This fraction is immediately subjected to DEAE-cellulose chromatography.

*DEAE-Cellulose Chromatography.* A DEAE-cellulose column (4.5 ×

42 cm; exchange capacity 0.8 meq/g; Type 20, Carl Schliecher and Schuell Co.) is equilibrated with 0.5 $M$ potassium phosphate buffer, pH 7.6, then washed with 5 m$M$ potassium phosphate buffer, pH 7.6. Prior to use, the column is further washed with 5 m$M$ potassium phosphate buffer, pH 7.6, containing 0.1 m$M$ EDTA and 10 m$M$ 2-mercaptoethanol. The enzyme solution (approximately 2.5 g of protein) from the previous step is diluted with an equal volume of 5 m$M$ potassium phosphate buffer, pH 7.6, containing 0.1 m$M$ EDTA and 10 m$M$ 2-mercaptoethanol and applied to the DEAE-cellulose column. After the application of an additional 200 ml of the same 5 m$M$ phosphate buffer, elution is accomplished with a 4-liter linear potassium phosphate gradient (5 m$M$ to 250 m$M$, pH 7.6, containing 0.1 m$M$ EDTA and 10 m$M$ 2-mercaptoethanol). The column effluent is continuously monitored for protein and is collected fractionally. Fractions exhibiting peak ribulose-diphosphate carboxylase activity (using the [14]C-bicarbonate fixation assay) and containing approximately 50% of the total activity applied to the column are pooled. The pooled enzyme is referred to as the DEAE-cellulose fraction. The enzyme is precipitated by bringing the saturation to 55% with saturated ammonium sulfate,[9] (pH 7.0, 122 ml/100 ml pooled enzyme). Additional EDTA and 2-mercaptoethanol are added to maintain their concentrations at 0.1 m$M$ and 10 m$M$, respectively. In preparation for the next step, the enzyme suspension is centrifuged and the precipitate redissolved in 50 ml of 5 m$M$ potassium phosphate buffer, pH 7.6, containing 0.1 m$M$ EDTA and 10 m$M$ 2-mercaptoethanol. This solution is subjected to gel filtration on a Sephadex G-25 column (4.5 × 42 cm) previously equilibrated with the same 5 m$M$ phosphate buffer. The column effluent is monitored for protein and the protein peak collected as a single fraction.

*Hydroxyapatite Chromatography.* The gel-filtered enzyme is applied to a hydroxyapatite column (4.5 × 42 cm; Clarkson Hypatite C) previously equilibrated with the same 5 m$M$ potassium phosphate buffer. After application of an additional 500 ml of the 5 m$M$ potassium phosphate buffer, the carboxylase is eluted with 25 m$M$ potassium phosphate buffer, pH 7.6, containing 0.1 m$M$ EDTA and 10 m$M$ 2-mercaptoethanol. The column effluent is monitored for protein and ribulose-diphosphate carboxylase activity. The enzyme elutes as a distinct peak and those fractions containing maximum carboxylase activity are pooled (60–80% of the total activity applied). These fractions are brought to 55% saturation with saturated ammonium sulfate[9] (pH 7.0, 122 ml/100 ml of pooled volume); EDTA and 2-mercaptoethanol are added to produce final concentrations of 0.1 m$M$ and 10 m$M$, respectively. Ribulose-diphosphate carboxylase is stored as a suspension at 0–2° and retains nearly full activity for about 1 month. Freezing the enzyme in this form or in dilute phosphate buffer, pH 7.6, results in the complete loss of activity. The specific activity of the purified carboxylase is 1.4–1.6 units/mg, and these preparations are free of detectable quantities of 5-phosphoriboisomerase and 5-phosphoribulokinase activities.[1]

*Homogeneity and Molecular Characteristics.* Ribulose-diphosphate carboxylase prepared as described above is homogeneous in the analytical ultracentrifuge[1] and by polyacrylamide gel electrophoresis.[10] The carboxylase has a sedimentation coefficient ($s°_{20,w}$) of 21.0 S.[1] The relation between absorbancy at 280 nm and the refractometrically determined protein concentration is given by the equation $c = 0.61$ ($A_{280nm}^{1cm}$) where $c$ is protein concentration in milligrams per milliliter and $A$ is absorbancy at 280 nm (1.0 cm light path).[1] The absorbancy ratio $A_{280nm}/A_{260nm}$ of the pure enzyme is 1.9. To convert protein concentration determined by the method of Lowry *et al.*[11] using bovine serum albumin as standard to refractometrically determined protein concentration, the former should be multiplied by a factor of 0.79. The molecular weight of the carboxylase determined by the sedimentation equilibrium method[12] is 557,000.[1] The frictional coefficient ratio ($f/f_0$) of the enzyme calculated from the appropriate molecular parameters is 1.11.[1] This ratio is in the usual range for a protein molecule of spherical shape having typical hydration characteristics (i.e., about 0.2 g of water per gram of protein). The spherical shape of the molecule has been verified by electron microscopy.[3,13] The homogeneous carboxylase is composed of two different kinds of subunits, eight heavy chains each with a molecular weight of approximately 56,000 and eight light chains of molecular weight 14,000.[14,15]

*Copper Content.* Tightly bound copper has been detected in homogeneous preparations of ribulose-diphosphate carboxylase.[16] The enzyme has a typical Cu(II) EPR spectrum ($g\perp \sim g_{max} = 2.09$). Analysis by atomic absorption spectrophotometry, EPR, and neutron activation indicate the presence of 1 gram-atom of copper per mole (560,000 g) of carboxylase. The bound copper appears to be present as Cu(II). The precise functional role of the copper has not yet been elucidated, but it may be related to the oxygenase activity of the enzyme (see next section).

[10] A. C. Rutner and M. D. Lane, *Biochem. Biophys. Res. Commun.* **28**, 531 (1967).
[11] O. H. Lowry, N. J. Rosebrough, A. L. Farr, and R. J. Randall, *J. Biol. Chem.* **193**, 265 (1951).
[12] D. A. Yphantis, *Biochemistry* **3**, 297 (1964).
[13] A. K. Kleinschmidt, M. Wishnick, and M. D. Lane, unpublished observations.
[14] A. C. Rutner, *Biochem. Biophys. Res. Commun.* **39**, 923 (1970).
[15] T. Sugiyama, and T. Akazawa, *Biochemistry* **9**, 4499 (1970).
[16] M. Wishnick, M. D. Lane, M. Scrutton, and A. S. Mildvan, *J. Biol. Chem.* **244**, 5761 (1969).

*Kinetic Properties and Inhibitors.* The pH optimum of the ribulose-diphosphate carboxylase-catalyzed reaction is approximately 7.9. The $K_m$ values determined for ribulose 1,5-diphosphate, $Mg^{2+}$, total "$CO_2$" ($CO_2 + HCO_3^-$) at pH 7.9 are 0.12 m$M$, 1.1 m$M$, and 22 m$M$, respectively.[1] $Mn^{2+}$, which can replace $Mg^{2+}$ in the reaction, has a $K_m$ of 39 $\mu M$ and a $V_{max}$ 56% that with $Mg^{2+}$.

The active species in the carboxylation of ribulose diphosphate has been shown to be $CO_2$.[17] Therefore, correction of the Michaelis constant to the concentration of the active species of $CO_2$ results in a $K_m$ for $CO_2$ of approximately 0.45 m$M$.[17] Fructose 6-phosphate decreases the $K_m$'s for $CO_2$ and $Mg^{2+}$ and increases the apparent $K_m$ for ribulose diphosphate.[18] The enzyme catalyzes the carboxylation of 1340 moles of ribulose diphosphate per minute per mole of enzyme, or the formation of 2680 moles of 3-phosphoglycerate per minute per mole of enzyme under standard assay conditions.

Ribulose diphosphate becomes inhibitory at concentrations exceeding 0.7 m$M$. Orthophosphate and $(NH_4)_2SO_4$ are competitive inhibitors with respect to ribulose diphosphate and have $K_i$'s of 4.2 and 8.1 m$M$, respectively.[1] 3-Phosphoglycerate is a noncompetitive inhibitor ($K_i = 8.3$ m$M$) with respect to ribulose diphosphate and a competitive inhibitor ($K_i = 9.5$ m$M$) with respect to $HCO_3^-$.[1] Glyceraldehyde 3-phosphate shows a similar pattern of inhibition; it is a competitive inhibitor ($K_i = 22$ m$M$) with respect to $HCO_3^-$ and a noncompetitive inhibitor ($K_i = 19$ m$M$) with respect to ribulose diphosphate.[19] The carboxylase is inhibited 63% by 0.5 m$M$ arsenite in the presence of 0.4 m$M$ BAL. In the absence of BAL, the observed inhibition is 19% while BAL alone causes no inhibition.[19] At levels as high as 10 m$M$, azide hydroxylamine, and oxalate are without effect on the carboxylase reaction.[19] The enzyme is inhibited by $HgCl_2$ and $p$-chloromercuribenzoate,[20] as well as by iodoacetamide.[21]

Ribulose-diphosphate carboxylase is reversibly inhibited by cyanide at low concentrations; 10 $\mu M$ and 0.1 m$M$ cyanide inhibit the carboxylation reaction 51 and 91%, respectively.[7] Inhibition by cyanide ($K_i = 16$ $\mu M$) is uncompetitive with respect to ribulose diphosphate and is of mixed character with respect to $Mg^{2+}$ and $HCO_3^-$. This and the fact that cyanide binds stoichiometrically to the carboxylase, but only in the presence of ribulose diphosphate, suggest that cyanide combines readily with the enzyme–ribulose diphosphate complex, but not with enzyme.

[17] T. G. Cooper, D. Filmer, M. Wishnick, and M. D. Lane, *J. Biol. Chem.* **244**, 1081 (1969).
[18] B. B. Buchanan, and P. Schürmann, *J. Biol. Chem.* **248**, 4956 (1973).
[19] M. Wishnick, and M. D. Lane, unpublished observations.
[20] A. Weissbach, B. L. Horecker, and J. Hurwitz, *J. Biol. Chem.* **218**, 795 (1956).
[21] B. R. Rabin, and P. W. Trown, *Proc. Nat. Acad. Sci. U.S.* **51**, 501 (1964).

2-Carboxyribitol 1,5-diphosphate, which appears to be a transition state analog[22-25] and resembles the carboxylated intermediate proposed by Calvin,[26] is a potent and essentially irreversible inhibitor of the carboxylase-catalyzed reaction.[7,27] Maximal inhibition requires preliminary incubation of the enzyme with the inhibitor and divalent metal ion, e.g., $Mg^{2+}$[27]; nearly complete inhibition is obtained with a 10-fold stoichiometric excess of inhibitor over carboxylase in the preliminary incubation.[27] Although ribulose diphosphate added to the preliminary incubation mixture completely prevents inhibition by carboxyribitol diphosphate, once the inhibitor is bound at 30°, inhibition cannot be reversed by a $10^4$-fold excess of ribulose diphosphate over inhibitor in the assay mixture.[27] Tight binding of carboxyribitol diphosphate ($K_D \approx 10$ n$M$[27]) to the carboxylase occurs only in the presence of divalent metal ions, e.g., $Mg^{2+}$ or $Mn^{2+}$. The complex formed contains equimolar amounts of inhibitor and divalent metal ion when isolated by gel filtration.[23] There are eight tight binding sites for carboxyribitol diphosphate,[23,27] the same number of binding sites the carboxylase possesses for the substrate ribilose diphosphate.[23]

In addition to catalyzing the carboxylation of ribulose diphosphate to yield two molecules of phosphoglycerate, ribulose-diphosphate carboxylase is capable of adding oxygen to this same sugar diphosphate, thereby giving rise to phosphoglycolate and phosphoglycerate.[28-31] The pH optimum for this oxygenase activity is approximately 9.4, being much more alkaline than that of the carboxylase. No activity occurs in the absence of $Mg^{2+}$, and oxygenase activity is inhibited by cyanide as is carboxylase activity. The $K_m$ values for ribulose diphosphate for carboxylase and oxygenase activities are the same. As well as being the substrate ($K_m = 0.75$ m$M$, $V_{max} = 0.1$ $\mu$mole min$^{-1}$ mg$^{-1}$) for the oxygenation of ribulose diphosphate, oxygen is a competitive inhibitor ($K_i = 0.8$ m$M$) with respect to $CO_2$ and an uncompetitive inhibitor with respect to ribulose diphosphate in the carboxylation reaction.

*Mechanism.* The hypothesis that 2-carboxy-3-ketoribitol 1,5-diphosphate is an intermediate in the enzymic carboxylation of ribulose diphos-

[22] M. Calvin, *Science* **135**, 879 (1962).
[23] M. Wishnick, M. D. Lane, and M. C. Scrutton, *J. Biol .Chem.* **245**, 4939 (1970).
[24] E. Racker, *Nature (London)* **175**, 249 (1955).
[25] B. R. Rabin, D. F. Shaw, N. G. Pon, J. M. Anderson, and M. Calvin, *J. Amer. Chem. Soc.* **80**, 2528 (1958).
[26] M. Calvin, *Fed. Proc., Fed. Amer. Soc. Exp. Biol.* **13**, 697 (1954).
[27] M. I. Siegel, and M. D. Lane, *Biochem. Biophys. Res. Commun.* **48**, 508 (1972).
[28] G. Bowes and W. L. Ogren, *J. Biol. Chem.* **247**, 2171 (1972).
[29] W. L. Ogren, and G. Bowes, *Nature (London) New Biol.* **230**, 159 (1971).
[30] T. J. Andrews, G. H. Lorimer, and N. E. Tolbert, *Biochemistry* **12**, 11 (1973).
[31] G. H. Lorimer, T. J. Andrews, and N. E. Tolbert, *Biochemistry* **12**, 18 (1973).

phate[26] is supported by the finding that 2-carboxyribitol 1,5-diphosphate, a structural analog of the proposed intermediate, is a potent inhibitor of ribulose-diphosphate carboxylase (see above). 2-Carboxy-3-ketoribitol 1,5-diphosphate, prepared chemically by catalytic oxidation of carboxyribitol diphosphate, spontaneously decarboxylates to give rise to $CO_2$ and ketoribitol diphosphate, and more important, undergoes nonenzymic, as well as enzymic, cleavage to yield 2 molecules of 3-phosphoglycerate.[32,33] Chemically synthesized 2-[[14]C]carboxy-3-ketoribitol 1,5-diphosphate undergoes spontaneous hydrolytic cleavage at 25° and pH 9, giving rise to 1 molecule of D-3-phosphoglycerate and 1 molecule of L-3-phosphoglycerate, and unlabeled D-isomer arising from carbon atoms 3, 4, and 5 and the [14]C-labeled L-isomer from the carboxyl group and carbon atoms 1 and 2 of 2-[[14]C]carboxy-3-ketoribitol 1,5-diphosphate. The hypothesis that this six-carbon compound is an intermediate in the enzymic carboxylation of D-ribulose 1,5-diphosphate is substantiated by the finding that chemically synthesized 2-carboxy-3-ketoribitol 1,5-diphosphate is cleaved by ribulose-diphosphate carboxylase giving rise to 3-phosphoglycerate.[33]

Additional support for this proposed mechanism for the carboxylation of ribulose diphosphate derives from the recent discovery that ribulose-diphosphate carboxylase catalyzes the oxygenation (pH 9 and 25°) of ribulose diphosphate to yield 2-phosphoglycolate and 3-phosphoglycerate[30,31] Supported by mass spectral data, the proposed mechanism[32] for this reaction also involves the formation of a 3-ketoribitol diphosphate derivative (2-hydroperoxide-3-ketoribitol 1,5-diphosphate instead of the 2-carboxy-3-ketoribitol 1,5-diphosphate intermediate in the carboxylation reaction).

## Acknowledgments

The methods described in this report were developed with the support of a research grant (AM-14575) from the National Institutes of Health, U.S.P.H.S. M. I. Siegel was a predoctoral trainee of the U.S.P.H.S. (GM-00184) and is presently a postdoctoral fellow of the National Institute of Arthritis, Metabolism and Digestive Diseases of the National Institutes of Health in the Department of Pharmacology and Experimental Therapeutics, The Johns Hopkins University School of Medicine.

[32] M. I. Siegel, *Fed. Proc.* **32**, 627 (Abstract) (1973).
[33] M. I. Siegel, and M. D. Lane, *J. Biol. Chem.* **248**, 5486 (1973).

## [71] Ribulose-1,5-diphosphate Carboxylase from Leaf

*By* Jonathan Goldthwaite and Lawrence Bogorad

d-Ribulose 1,5-diphosphate + $CO_2 \rightarrow$ 2-d-3-phosphoglycerate

The enzyme ribulose-1,5-diphosphate carboxylase (RuDPCase, EC 4.1.1.39) is of considerable interest because of its abundance in plants, its key role in photosynthesis as well as photorespiration, and the effect of illumination of etiolated leaves on its activity level. Molecular biologists have studied its site of transcription and translation in the cell while agriculturalists have been interested in its role in the food chain and its indigestibility in cattle. These studies all require purification of RuDPCase to a greater or lesser extent.

The enzyme has traditionally been prepared from leaves by lengthy procedures involving several ammonium sulfate precipitations plus gel filtration and/or column chromatography on ion exchange media.[e.g.1,2] Purification by sucrose gradient centrifugation as described here is rapid and has the advantages of allowing up to 6 small samples to be easily carried through an extensive purification under conditions which should give optimal retention of the enzyme in its native state, since long preparations times, precipitation, and binding to ion-exchange materials are avoided.[3] It takes advantage of the high molecular weight of RuDPCase and its abundance in leaf tissue. Soluble polyvinylpyrrolidone (PVP) is included in the extraction buffer to prevent RuDPCase inactivation in leaf extracts rich in polyphenols. The main disadvantage of this procedure is that preparation of greater than 5–10 mg of RuDPCase would require preliminary concentration of the sample or multiple centrifuge runs.

A summary of the procedure as now used in the author's laboratory follows. Temperature should be 0–5° throughout.

### Homogenization and Extraction

Samples of leaf tissue (0.25–1.0 g) which are fresh or have been stored at 0° are completely ground with grinding buffer (2–9 times the tissue fresh weight) for 15–60 sec in a motor-driven glass homogenizer (Duall Tissue Grinder, Kontes Glass Co., Vineland, New Jersey, Size C or D). Grinding buffer usually contains 2% w/v soluble PVP (Plasdone C, GAF

[1] M. Wishnick and M. D. Lane, this series, Vol. 23, p. 570.
[2] E. Racker, this series, Vol. 5 p. 266.
[3] J. J. Goldthwaite and L. Bogorad, *Anal. Biochem.* **41**, 57 (1971).

Corp., Calvert City, Kentucky), 10 m$M$ MgSO$_4$, 20 m$M$ 2-mercaptoethanol, and a variable amount of Tris-sulfate, pH 8.0 (0.1–0.5 $M$), chosen to maintain the pH between 7.5 and 8.0. The homogenate is centrifuged in small tubes at 35,000–40,000 $g$ for 15 min. One aliquot of the supernatant is used directly for sucrose gradient purification. Another aliquot of the supernatant is used to assay RuDPCase and protein.

## RuDPCase Assay

RuDPCase activity is measured by incorporation of NaH$^{14}$CO$_3$ into acid-stable products in the presence of ribulose 1,5-diphosphate (RuDP). The following modification of the method of Rabin and Trown[4] is used.

RuDPCase activity should be estimated only when the extract is diluted into the linear range in a buffer similar to the grinding buffer but substituting Tris chloride, or after desalting because sulfate is a competitive inhibitor of the enzyme for RuDP.[1] A 25-$\mu$l aliquot of diluted enzyme is mixed on the surface of a planchet with 100 $\mu$l of reaction mixture containing the following (all are final concentrations): 100 m$M$ Tris chloride, pH 8.0; 50 m$M$ NaH$^{14}$CO$_3$ (specific activity 0.20 $\mu$Ci/$\mu$mole); 0.3 m$M$ RuDP; 10 m$M$ MgCl$_2$; 6 m$M$ reduced gluthathione; 0.1 m$M$ ethylenediaminetetraacetic acid. After 10 min at room temperature the reaction is stopped by addition of 5 drops 6 $N$ acetic acid. The planchets are dried and counted in a gas-flow counter. The reaction is linear with enzyme concentration until 30–50% of the RuDP is consumed. Incorporation in the absence of RuDP is less than 2–3% that when RuDP is added. This planchet procedure reduces pipetting and glassware contamination compared to the usual test-tube assay.

## Sucrose Gradient Centrifugation

Samples of up to 1.0 ml or leaf extracts containing 0.015 to 0.75 $\mu$g/ml RuDPCase can be subjected to sucrose gradient purification as follows and the resulting enzyme peak quantitated by UV absorption and by activity measurements. More concentrated samples can be handled the same way, and larger volumes can be applied to higher capacity rotors (e.g., the SW 27[5]).

The leaf supernatants are layered on sucrose gradients in SW 40 cellulose nitrate tubes (Beckman Instruments) made up in a pH 8.0 buffer consisting of 10 m$M$ MgSO$_4$, 1 m$M$ 2-mercaptoethanol, and 10 m$M$ Tris-

[4] B. R. Rabin and P. W. Trown, *Proc. Nat. Acad. Sci. U.S.* **51**, 497 (1964).
[5] J. E. Boynton, N. W. Gillham, and J. F. Chabot, *J. Cell Sci.* **10**, 267 (1972).

sulfate or 10–50 m$M$ NaHCO$_3$ (see below). (The bicarbonate buffer may be preferable for maintenance of RuDPCase activity. It is possible to reduce the mercaptoethanol concentration or substitute a lower concentration of dithiothreitol to reduce the baseline UV absorption even further.)

Linear gradients of sucrose concentration from 0.2 to 0.7 $M$ and logarithmic gradients[6] from 0.2 to 0.8 $M$ have been useful for purification of extracts of *Rumex,* spinach, bean, and maize. The logarithmic gradients can be prepared using an inexpensive device made from two plastic syringes or from commercial gradient makers. These gradients gave a distinctly narrower and more isolated RuDPCase zone than the linear gradients and are recommended for cases where improved resolution is desired.[3]

After centrifugation for 14–16 hr (linear gradients) or 21–22 hr (logarithmic gradients) the samples are fractionated by hand or preferably with a commercial fractionator (e.g., ISCO, Inc., Lincoln, Nebraska). Suitably small fractions (0.25–0.5 ml) are taken in the RuDPCase peak region to obtain the amount and purity of enzyme desired. RuDPCase from the center 0.5 ml at the peak of *Rumex* preparations was found to be homogeneous by gel electrophoresis and free of nucleic acid by absorption 280/260.[3] The enzyme was of excellent specific activity, up to 4 units/mg protein at 23°. If rigorous purification is contemplated, this method can serve as an expedient and successful first step.

*Other Adaptations Possible.* If desired, insoluble PVP can often be substituted to obtain improved clarity of the first supernatant. In polyphenol-rich species, the extract obtained with soluble PVP is a rich yellow or even tan, owing to retention of phenolics and quinones in the supernatant. RuDPCase activity of such preparations, however, is high and rather stable (see Goldthwaite and Bogorad,[3] p. 60, for discussion). In many species PVP may be unnecessary, but investigators should be wary of electrophoretic artifacts which phenolics can introduce[7] (see also Goldthwaite and Bogorad,[3] p. 65).

A variety of other extraction and gradient buffer compositions are possible. In general monovalent anions (e.g., Cl) give good enzyme activity and poor stability while polyvalent anions (SO$_4$, PO$_4$) inhibit but stabilize enzyme activity.[8] For example, phosphate buffer with 0.1 m$M$ EDTA and no Mg$^{2+}$ has been used by Wishnick and Lane.[1] Use of phosphate or sulfate (competitive inhibitors with RuDP[1]) together with

[6] H. Noll, *Nature (London)* **215,** 360 (1967).

[7] W. T. Jones and J. W. Lyttleton, *Phytochemistry* **11,** 1595 (1972).

[8] P. W. Trown, *Biochemistry* **4,** 908 (1965).

$HCO_3^-$ or 3-phosphoglycerate (competitive inhibitor with $CO_2$[1]) might give even better preservation of enzyme activity during preparation than either alone.

## Summary

The extraction, purification, and assay procedures described herein allow the preparation of mg amounts of highly pure RuDPCase from up to 6 samples within 24 hr. The purification occurs under mild conditions, where minimal alteration or contamination of the native enzyme should be expected. The procedure is suited for a variety of applications where large amounts of purified protein are not required, especially where estimates of the specific activity of the enzyme in several tissue samples are required.

# [72] D-Ribulose-1,5-diphosphate Oxygenase

*By* G. H. LORIMER, T. J. ANDREWS, and N. E. TOLBERT

$$\text{D-Ribulose 1,5-diphosphate} + O_2 \xrightarrow{Mg^{2+}} \text{2-phosphoglycolate}^- + \text{3-phosphoglycerate}^- + 2H^+$$

## Assay Method

*Principle.* Ribulose-diphosphate oxygenase catalyzes the irreversible oxidation of D-ribulose 1,5-diphosphate to 2-phosphoglycolate and 3-phosphoglycerate with the stoichiometry indicated by the above equation.[1-3] The activity of ribulose-diphosphate oxygenase is associated with that of ribulose-diphosphate carboxylase, and the oxygenase activity has been detected in the leaves of a variety of higher plants. The separation of ribulose-diphosphate oxygenase and ribulose-diphosphate carboxylase has not been achieved, and circumstantial evidence indicates that the same protein catalyzes both reactions.[2] Assay methods based upon the disappearance of ribulose diphosphate or the appearance of 3-phosphoglycerate are considered inadvisable, since carboxylation of the substrate would seriously interfere if traces of $CO_2$ were present. The assay methods employed therefore follow oxygen consumption either

[1] G. Bowes, W. L. Ogren, and R. H. Hageman, *Biochem. Biophys. Res. Commun.* **45**, 716 (1971).

[2] T. J. Andrews, G. H. Lorimer, and N. E. Tolbert, *Biochemistry* **12**, 11 (1973).

[3] G. H. Lorimer, T. J. Andrews, and N. E. Tolbert, *Biochemistry* **12**, 18 (1973).

manometrically[2] or by the use of an oxygen electrode. Parallel assays may be run for the ribulose diphosphate carboxylase activity.[4]

### Reagents

Ammediol (2-amino-2-methyl-1,3-propanediol)-HCl, 0.2 $M$, pH 9.3
MgCl$_2$, 0.1 $M$
EDTA, 0.1 $M$
Dithiothreitol, 8 m$M$
Ribulose diphosphate, Na$_4$ (Sigma), 25 m$M$

*Procedure.* MANOMETRIC METHOD. Sensitivity is maximized by using small flasks (total volume about 3 ml). The reaction mixture contains 100 $\mu$moles 2-amino-2-methyl-1,3-propanediol (ammediol)-HCl at pH 9.3; 10 $\mu$moles MgCl$_2$; 1 $\mu$mole EDTA; 0.4 $\mu$moles dithiothreitol, and 0.5–2 mg of enzyme protein in a volume of 0.92 ml. The side arm contains 2 $\mu$moles of ribulose diphosphate in 0.08 ml. Prior to use, the enzyme is passed through a small column of Sephadex G-25 equilibrated with a solution of 25 m$M$ glycylglycine-KOH at pH 7.7, 1 m$M$ EDTA, and 1 m$M$ dithiothreitol. The flasks are attached to the manometer (bath temperature, 25°) and shaken (150 oscillations min$^{-1}$) for 3 min with pure oxygen passing through and are equilibrated for a further 9 min while closed. The reaction is started by tipping in the ribulose diphosphate. Readings are taken every 90 sec. A short lag is usually observed, followed by a linear reaction which continues until the ribulose diphosphate becomes limiting. Data are expressed as micromoles O$_2$ uptake per milligram of protein per unit of time. A control in which ribulose diphosphate is omitted is run with each batch of assays. A control with boiled enzyme has a negligible rate. An all-glass submarine-type Warburg respirometer is greatly preferred, for other models lose O$_2$ from tubing connectors.

OXYGEN ELECTRODE METHOD. An electrode having a minimal "dead volume" behind the membrane is required. Before use, the enzyme is passed through Sephadex G-25 as described above. The dithiothreitol concentration is adjusted at this stage so that the amount carried over into the reaction mixture with the enzyme solution will result in a final dithiothreitol concentration in the range 0.1–0.4 m$M$. The assay is carried out at 25°. Two milliliters of air-saturated buffer solution, containing 0.125 $M$ ammediol-HCl at pH 9.3 and 12.5 m$M$ MgCl$_2$, is pipetted into the electrode chamber, followed by enzyme solution and water to give a total volume of 2.4 ml. A blank rate of O$_2$ consumption, due largely to the oxidation of dithiothreitol, is observed at this stage. At dithiothreitol concentrations greater than 0.4 m$M$, this rate becomes unacceptably

[4] M. Wishnick and M. D. Lane, this series, Vol. 23, p. 570.

large. The oxygenase reaction is initiated by the addition of 2 $\mu$moles of ribulose diphosphate in 0.1 ml. At $O_2$ concentrations in the region of air saturation (approximately 0.23 m$M$), the enzymic rate shows very nearly first-order dependence on $O_2$ concentration. This is to be expected since the $K_m$ of ribulose-diphosphate oxygenase for $O_2$ is 0.75 m$M$ (equivalent to an atmosphere containing about 65% $O_2$).[2] A knowledge of the $O_2$ concentration prevailing at the time of the rate measurement is therefore of critical importance. Oxygenase activity is conveniently expressed as the first-order rate constant with respect to $O_2$, calculated thus:

$$(\text{Enzymic rate} - \text{blank rate})/O_2 \text{ concentration min}^{-1}$$

If the rates are expressed in chart divisions min$^{-1}$ and the $O_2$ concentration in chart divisions, the need to calibrate the electrode in absolute terms is avoided.

### Purification Procedure

Since the ribulose-diphosphate oxygenase reaction appears to be catalyzed by the same protein as ribulose-diphosphate carboxylase, the reader is referred to previously published procedures for the purification of ribulose diphosphate carboxylase.[4]

A preparation greatly enriched with ribulose-diphosphate carboxylase and oxygenase activities, although not completely pure, may be obtained by sucrose density gradient centrifugation of leaf extracts from which the membranous components have been removed. This may be done on a small scale in swinging buckets[5] or on a larger scale by the use of a zonal rotor.[2] The following procedure is similar to that of Goldthwaite and Bogorad[5] but adapted to a zonal rotor.

A 600-ml linear (by volume) density gradient from 15 to 30% (w/w) sucrose in 25 m$M$ glycylglycine-KOH buffer at pH 7.7 is loaded into a B-30 (International Equipment Co.) zonal rotor from the rim, followed by sufficient 35% (w/w) sucrose to fill the rotor. Thirty milliliters of the enzyme solution is then pumped into the rotor core followed by a 50-ml overlay of the buffer. Centrifugation is then conducted for 4.5 hr at 50,000 rpm, after which the gradient is displaced by pumping 35% (w/w) sucrose into the rim of the rotor. The protein concentration is monitored by passage of the eluate through a recording UV monitor. A large peak of fraction 1 protein, which contains both the oxygenase and carboxylase activities, is well separated from less rapidly sedimenting contaminants. The apparent asymmetry of the fraction 1 protein peak is due to dilution incurred by the radial nature of the migration.

[5] J. J. Goldthwaite and L. Bogorad, *Anal. Biochem.* **41**, 57 (1971).

## Properties

The oxygenase reaction is specific for ribulose diphosphate. No reaction can be detected with fructose 1,6-diphosphate, fructose 6-phosphate, 3-phosphoglycerate, ribulose 5-phosphate, or ribulose.[2] The $K_m$ (ribulose diphosphate), measured under a gas phase of 100% oxygen, which is not saturating, is approximately 0.2 m$M$.[2] The $K_m$ ($O_2$) is approximately 0.75 m$M$.[2] The pH optimum of the ribulose-diphosphate oxygenase-catalyzed reaction is approximately 9.3.[2] The presence of $MgCl_2$ and a sulfhydryl-reducing reagent, such as dithiothreitol, is required for full activity. The enzyme should be preincubated with the $MgCl_2$ for an hour before assaying. Upon storage in the cold at 4° or at −18°, the enzyme from some sources (i.e., spinach and tobacco leaves) may lose 25 to 75% of its activity in a few days, but full activity can be restored by heating to 55° for 10 min in the presence of 50 m$M$ $MgCl_2$ and a sulfhydryl reagent.

Section VII

Glycosidases

## [73] 6-Phospho-β-D-galactosidase

*By* Wolfgang Hengstenberg and M. L. Morse

R-β-D-Galactoside 6-phosphate ⇌ ROH + galactose 6-phosphate

### Assay Method

The assay of this enzyme is based[1] on the hydrolysis of o-nitrophenyl-β-D-galactopyronoside 6-phosphate (ONPG-6P) to yield o-nitrophenol, which at pH 10, absorbs at 405 nm. The ONPG-6P can either be prepared[2] or obtained commercially from several sources (Calbiochem, NK Laboratories, Sigma).

### Reagents

Tris-Cl buffer, 0.1 $M$ pH 7.0, containing 50 m$M$ NaCl
Dicyclohexylammonium salt of ONPG-6P

*Procedure.* A measured amount of the enzyme is brought to a volume of 0.9 ml with 0.1 $M$ Tris-Cl buffer, pH 7.0 containing 50 m$M$ NaCl. The mixture is incubated at 30° in a 1.0-ml cuvette. At zero time 3 μmoles of the dicyclohexylammonium salt of o-nitrophenyl-β-D-galactoside 6-phosphate in 0.1 ml is mixed with the enzyme solution. The increasing absorbance is recorded at 405 nm in a Beckman DB spectrophotometer at 30° attached to a model 43 Photovolt recorder. $E = 4.8 \times 10^6$ cm²/M in 0.25 $M$ Na₂CO₃.

*Unit of Activity.* One unit of 6-phosphogalactosidase is equal to 1 μmole of o-nitrophenol released in 1 min at 30°.

### Bacterial Strains

Staphylococcal strains S305A or S306A, derivatives of NCTC 8511, which produce 6-phospho-β-D-galactosidase constitutively[3] were used as enzyme sources.

### Purification

*Cell Cultivation and Disruption.* Large amounts of cells of *Staphylococcus aureus* were grown at 37° in a model F-14 New Brunswick fermen-

---

[1] W. Hengstenberg, W. K. Penberthy, and M. L. Morse, *Eur. J. Biochem.* **14,** 27 (1970).
[2] See this series, Volume 41 [28].
[3] W. Hengstenberg, W. K. Penberthy, K. L. Hill, and M. L. Morse, *J. Bacteriol.* **99,** 383 (1969).

tor in 2% peptone medium. Maximal cell densities of $10^{10}$ cells per milliliter were obtained after overnight growth. Stirring was at 400 rpm, and aeration was achieved by applying suction to the fermentor vessel to avoid aerosol buildup. The cells were harvested by using a Lourdes centrifuge with a continuous-flow head. To avoid aerosols, the collecting vessel for the effluent of the centrifuge was kept under reduced pressure.

Thirty grams of staphylococcal cells (strain S305A, constitutive for lactose utilization), washed in 1 m$M$ Tris-EDTA, were mixed with 40 ml of standard buffer pH 7.5 (Tris·HCl 0.1 $M$, 0.1 $M$ NaCl, 10 m$M$ $MgCl_2$ 0.1 m$M$ EDTA, 0.1% mercaptoethanol, 0.1 m$M$ dithiothreitol), 1 mg DNase, and 90 ml of wet glass beads (0.3 mm in diameter obtained from B. Braun Apparatbau, Melsunglen, W. Germany). This suspension was shaken in a Buehler[4] cell homogenizer for 45 min at 13°. The turbidity measurement of the cell suspension after the disruption procedure indicated 80% reduction of turbidity. Cell debris and whole cells were sedimented at 10,000 $g$ for 30 min. Cell membranes were removed by centrifugation at 150,000 $g$ for 90 min; 2 batches of 30 g of bacteria were ground to obtain the cell-free extract. Although dithiothreitol is added routinely, no sensitivity of the enzyme to oxygen was noted.

This procedure was developed in the study of galactoside metabolism in *S. aureus*, and a strain which synthesizes all the components needed for galactoside metabolism constitutively was isolated. An inducible strain could be employed if a suitable inducer, i.e., lactose, galactose, galactose 6-phosphate, is used to induce the cells.

*Ammonium Sulfate Precipitation.* The $(NH_4)_2SO_4$ fractionation is performed with neutralized saturated solutions at 20°.

*DEAE-Cellulose Chromatography.* The precipitate of the 55–75% $(NH_4)_2SO_4$ fractionation is dissolved in 10 m$M$ sodium phosphate buffer, pH 7.5, containing 1 m$M$ EDTA and 1 m$M$ dithiothreitol, and the sample is desalted in a Canalco EFC apparatus for 3 hr. The buffer composition in the Canalco concentration fluid (concentrex) is nearly identical to the equilibration buffer of the column. Therefore, equilibration of the desalted sample with the column buffer is omitted. The desalted sample is applied to a DEAE-cellulose (Cellex D, 0.41 meq/g, Bio-Rad) 3 × 35 cm column which has been equilibrated with 10 m$M$ sodium phosphate buffer, pH 7.5, containing 0.1 $M$ NaCl, 1 m$M$ EDTA, and 1 m$M$ dithiothreitol. The DEAE cellulose is converted to the chloride form by repeated treatment with NaOH followed by HCl. The column is eluted with a linear gradient of 1200 ml varying from 0.1 $M$ to 0.5 $M$ NaCl. At a flow rate of 0.65 ml/min, the enzymic activity is found around tubes 58–64 (13 ml volumes) at 0.27–0.29 $M$ NaCl. Tubes containing activity (80-100 ml) are pooled and concentrated in a Canalco EFC apparatus (total pro-

[4] E. Bühler, 14 Tübingen, W. Germany.

tein about 100 mg) with a yield during concentration of greater than 90%.

*Sephadex G-150.* The concentrated solution from the DEAE-cellulose chromatography (6–8 ml) is applied to a G-150 column, 3.5 × 88 cm, the flow rate kept at 0.6 ml/min (elution buffer, 50 m$M$ Tris-Cl, pH 7.6, 0.1 $M$ NaCl, 1 m$M$ EDTA, and 1 m$M$ dithiothreitol). Enzyme activity is found in tubes 28–40 (12 ml volumes). Tubes with constant specific activity are pooled and concentrated with the EFC apparatus (total protein about 30 mg).

*Phosphocellulose.* The concentration solution of the DEAE-cellulose chromatography is applied to a P-cellulose column, 2.5 × 29 cm. The column is eluted with a linear gradient of 1 1, 0.1–0.8 $M$ NaCl, in 10 m$M$ sodium phosphate buffer pH 7.5, containing 1 m$M$ EDTA and 1 m$M$ dithiothreitol. The flow rate is kept at 0.87 ml/min. The enzyme is found in tubes 43–55. Tubes with constant specific activity are pooled and concentrated (total protein about 40 mg). All chromatographic procedures are carried out at 4°. Fractions are collected at 20-min intervals.

The yields and specific activities obtained during the purification steps as summarized in Table I.

## Distribution

The phospho-β-galactosidase was discovered in *Staphylococcus aureus* and is a characteristic of this microorganism. A similar enzyme has been discovered in *Streptococcus lactis*.[5] An enzyme attacking phosphorylated β-glucosides has also been reported.[6]

We have employed this enzyme as a tool in assaying the components of the phosphoenolpyruvate-dependent phosphotransferase system.[3]

TABLE I
PURIFICATION OF STAPHYLOCOCCAL 6-PHOSPHO-β-D-GALACTOSIDASE

| Purification step | Specific activity (units/mg protein) | Yield from previous step (%) | Purification (fold) | Overall purification (fold) | Overall yield (%) |
|---|---|---|---|---|---|
| Extract | 1.3–1.6 | — | — | — | — |
| 55–75% (NH₄)₂SO₄ | 2.5–3.2 | 80–90 | 2 | — | — |
| DEAE-cellulose | 25–32 | 60–70 | 10 | — | — |
| Phosphocellulose | 38–47 | 70 | 1.5 | 30 | 40 |
| Sephadex G-150 | 38–47 | 50 | 1.5 | 30 | 30 |

[5] L. L. MacKay, A. L. Walter, W. E. Sandine, and P. R. Elliker, *J. Bacteriol.* **99**, 603 (1969).

[6] C. F. Fox and G. Wilson, *Proc. Nat. Acad. Sci. U.S.* **59**, 988 (1968).

TABLE II
SOME KINETIC PROPERTIES OF THE 6-PHOSPHO-$\beta$-D-GALACTOSIDASE

| Property | Measurement |
| --- | --- |
| $K_m$ | 3 mM |
| $V_{max}$ | 81 $\mu$moles $\times$ mg protein$^{-1}$ $\times$ min$^{-1}$ at 30° |
| Half-life at 55° in the standard Tris buffer | About 5 min |
| pH optimum | 7.0 |

## Properties

The molecular weights determined were as follows: Sephadex chromatography, 46,000–47,000 ± 10%; in sodium dodecyl sulfate, 53,000 ± 10%; in 9 M urea, >40,000–50,000. The $s_{20,w}$ = 3.74 S and 4.5 S. The latter value probably indicates the presence of an aggregate.

The *amino-terminal amino acid* was threonine. The enzyme is most likely a single polypeptide chain of about 50,000 MW. The purified enzyme has an absorbance ratio at 250–260 nm of 1.44. A 0.1% solution has an absorbance of 1.5 at 280 nm in a 1.0-cm cuvette.

*Kinetic Characteristics.* Some kinetic properties are given in Table II.

The enzyme is sensitive to mercury ($Hg^{2+}$) and copper ($Cu^{2+}$), approximately 50% inactivated by millimolar concentrations of these ions. $Mn^{2+}$ produces about 50% inactivation to 10 mM, whereas $Mg^{2+}$ and $K^+$ show little effect to 50 mM.

# [74] Phosphocellobiase

By RICHARD L. ANDERSON and RICHARD E. PALMER

Cellobiose monophosphate[1] + $H_2O$ → D-glucose 6-phosphate + D-glucose

This phospho-$\beta$-glucosidase also catalyzes the cleavage of other phospho-$\beta$-glucosides. It functions in the metabolism of cellobiose,[2,3] gentiobiose,[4] and possibly other $\beta$-glucosides in *Aerobacter aerogenes*.

[1] 6-*O*-Phosphoryl-$\beta$-D-glucopyranosyl-(1 → 4)-D-glucose.
[2] R. E. Palmer and R. L. Anderson, *Biochem. Biophys. Res. Commun.* **45**, 125 (1971).
[3] R. E. Palmer and R. L. Anderson, *J. Biol. Chem.* **247**, 3420 (1972).
[4] R. E. Palmer and R. L. Anderson, *J. Bacteriol.* **112**, 1316 (1972).

## Assay Method[3]

*Principle.* The continuous spectrophotometric assay measures the rate of D-glucose 6-phosphate formation by coupling the reaction to glucose-6-phosphate dehydrogenase. With the coupling enzyme in excess, the rate of cleavage of cellobiose monophosphate or other phospho-$\beta$-glucoside is equivalent to the rate of $NADP^+$ reduction, which is measured by the absorbance increase at 340 nm.

*Reagents*

Glycylglycine-NaOH buffer, 0.2 $M$, pH 7.5
$NADP^+$, 10 m$M$
Cellobiose monophosphate,[5] 10 m$M$
Crystalline glucose-6-phosphate dehydrogenase

*Procedure.* The following are added to a microcuvette with a 1.0-cm-light path: 0.05 ml of buffer, 0.01 ml of $NADP^+$, 0.04 ml of cellobiose monophosphate, a nonlimiting amount of glucose-6-phosphate dehydrogenase, a rate-limiting amount of phosphocellobiase, and water to a volume of 0.15 ml. The reaction is initiated by the addition of phosphocellobiase. The rates are conveniently measured with a Gilford multiple-sample absorbance recording spectrophotometer. The cuvette compartment should be thermostated at 25°.

*Definition of Unit and Specific Activity.* One unit of enzyme is defined as the amount that cleaves 1 $\mu$mole of cellobiose monophosphate per minute. Specific activity is in terms of units per milligram of protein. Protein is determined by the ratio of absorbancies[6] at 280 and 260 nm or by the method of Lowry *et al.*[7]

## Purification Procedure[3]

*Growth of Organism.* A uracil auxotroph of *Aerobacter aerogenes* PRL-R3 was grown at 30° in a medium consisting of: 0.71% $Na_2HPO_4$, 0.15% $KH_2PO_4$, 0.3% $(NH_4)_2SO_4$, 0.009% $MgSO_4$, 0.0005% $FeSO_4 \cdot 7\ H_2O$, 0.0005% uracil, and 0.5% cellobiose. The sugar was autoclaved separately and added aseptically to the basal medium. The cultures were grown overnight in 500-ml volumes in Fernbach flasks on a rotary shaker.

[5] R. E. Palmer and R. L. Anderson, *J. Biol. Chem.* **247**, 3415 (1972).
[6] O. Warburg and W. Christian, *Biochem. Z.* **310**, 384 (1941).
[7] O. H. Lowry, N. J. Rosebrough, A. L. Farr, and R. J. Randall, *J. Biol. Chem.* **193**, 265 (1951).

*Preparation of Cell Extracts.* Cells were harvested by centrifugation, suspended in water, and broken by treatment for 10 min in a Raytheon 250-W 10 kHz sonic oscillator cooled with circulating ice water. The cell extract was the supernatant obtained after centrifugation of the broken-cell suspension for 10 min at 27,000 *g*.

*General.* The following procedures were performed at 0° to 4°. A summary of the purification is given in the table.

*Protamine Sulfate Treatment.* To a cell extract, a 5% (w/w) solution of protamine sulfate was added slowly with stirring to give a final concentration of 0.37% (w/v). After 30 min, the suspension was centrifuged and the resulting pellet was discarded.

*Ammonium Sulfate Fractionation.* The protein in the supernatant from the protamine sulfate step was fractionated by the addition of crystalline ammonium sulfate. The protein precipitating between 40 and 60% saturation was collected by centrifugation and dissolved in water.

*Chromatography on Sephadex G-75.* The above fraction was placed on a column (2.5 × 35 cm) of Sephadex G-75 equilibrated with 20 m$M$ glycylglycine buffer (pH 7.5) and was eluted with the same buffer. Five-milliliter fractions were collected, and those that contained the highest specific activity were combined. The enzyme was 14-fold purified with a 33% overall recovery of activity.

## Properties[3]

*Substrate Specificity.* Several phospho-$\beta$-glucosides are cleaved, including cellobiose monophosphate, gentiobiose monophosphate, cellobiitol monophosphate, phenyl $\beta$-D-glucoside monophosphate, salicin monophosphate, arbutin monophosphate, and methyl $\beta$-D-glucoside monophosphate. $K_m$ values range from 0.23 to 0.50 m$M$. Nonphosphorylated forms of the preceding $\beta$-glucosides are not cleaved.

*pH Optimum.* Activity as a function of pH is maximal in the pH

PARTIAL PURIFICATION OF PHOSPHOCELLOBIASE

| Fraction | Volume (ml) | Total activity (units) | Specific activity (units/mg protein) |
|---|---|---|---|
| Cell extract | 154 | 98 | 0.038 |
| Protamine sulfate supernatant | 160 | 77 | 0.051 |
| Ammonium sulfate precipitate | 25 | 53 | 0.12 |
| Sephadex G-75 | 30 | 27 | 0.54 |

range of 7 to 8, the actual measured optimum varying with the buffer used.

*Estimated Molecular Weight.* The sedimentation coefficient as determined by sucrose density gradient centrifugation is about 4.0 S, suggesting a molecular weight of about 52,000.

*Stability.* Partially purified enzyme kept at 4° retained 40% of its activity after 2 months.

*Note on Occurrence and Specificity.* Phospho-β-glucosidase activity occurs in several members of the Enterobacteriaceae.[8] Some organisms (*A. aerogenes* and *Escherichia coli*) are reported to possess two phospho-β-glucosidases which have different ratios of activity on various phospho-β-glucosides,[8,9] but the enzymes in those investigations were not tested for activity on cellobiose monophosphate. Although the phosphocellobiase reported here has not been purified highly, several lines of evidence are consistent with the view that the activity of the partially purified preparation on various phospho-β-glucosides is the result of a single enzyme.[3]

[8] S. Schaefler and I. Schenkein, *Proc. Nat. Acad. Sci. U.S.* **59**, 285 (1968).
[9] C. F. Fox and G. Wilson, *Proc. Nat. Acad. Sci. U.S.* **59**, 988 (1968).

## [75] β-Galactosidases from *Neurospora crassa*

*By* RICHARD STEPHENS and A. GIB DEBUSK

$$(\beta)\text{-R-O-D-galactopyranoside} + H_2O \rightleftarrows ROH + \text{D-galactose}$$

The occurrence of multiple forms of β-galactosidase in *Neurospora* was first reported by Bates and Woodward,[1] who described an alkaline β-galactosidase physically separable from the pH 4 optimum enzymes previously described by Landman and Bonner.[2] The acid β-galactosidase has been further characterized by Johnson and De Busk,[3] who have described two stable forms of the enzyme with, in addition to other differences in physical properties, pH optima of 4.2 and 4.5. The β-galactosidase system of *Neurospora* thus consists of the pH 7.5 enzyme and two activities at about pH 4, the 4.2 β-galactosidase and the 4.5 β-galactosidase. The purification of the 4.2 enzyme is essentially that of Johnson and DeBusk.[4] The pH 7.5 β-galactosidase has yet to be completely purified, and the properties given are those of partially purified extracts.

[1] W. K. Bates and D. O. Woodward, *Science* **146**, 777.
[2] O. E. Landman and D. M. Bonner, *Arch. Biochem. Biophys.* **41**, 283 (1952).
[3] H. N. Johnson and A. G. DeBusk, *Arch. Biochem. Biophys.* **138**, 412 (1970).
[4] H. N. Johnson and A. G. DeBusk, *Arch. Biochem. Biophys.* **138**, 408 (1970).

## I. pH 4.2 and 4.5 β-Galactosidase

**Assay Method**

*Principle.* β-Galactosidase activity is assayed spectrophotometerically by measuring the release of *o*-nitrophenol from *o*-nitrophenyl-β-D-galactopyranose (ONPG). Absorbance is measured at 400 nm in alkaline solution.

*Reagents*

ONPG (Sigma), 10 m*M* in sodium phosphate-citrate buffer, 80 m*M* pH 4.3

Potassium carbonate, 0.5 *M*

Enzyme. Clear supernatant of crude extract or product of subsequent steps. Dilute in 8 m*M* phosphate-citrate buffer, pH 4.3, so that 0.1 ml contains approximately 20 units as defined below.

*Procedure.* All assays are at 37°. The reaction is initiated by the addition of 0.4 ml of ONPG to 0.1 ml of enzyme. At the end of 20 or 30 min, the reaction is terminated by the addition of 2.5 ml of 0.5 *M* potassium carbonate. The entire reaction mixture (3 ml) is placed in a silica cell, and absorbance is measured at 400 nm against a blank in which water replaces ONPG. Change in absorbance due to *o*-nitrophenol released is obtained by subtracting the absorbance of a control in which water replaced enzyme.

Both the 4.2 and 4.5 β-galactosidase occur in the culture medium of induced cells. The extracellular enzymes may be assayed by using 1.0 ml of the culture filtrate as enzyme, initiating the reaction by the addition of 1.0 ml of ONPG and terminating the reaction with 1.0 ml of 0.5 *M* potassium carbonate.

*Definition of Enzyme Unit.* One unit of activity is defined as that amount of enzyme required to release 1 nmole of *o*-nitrophenol per minute under the assay conditions. Specific activity is defined as units of activity per milligram of protein as measured by the method of Lowry *et al.*[5] with Armour Serum Albumin fraction V used as the reference standard.

**Purification Procedure**

*Growth of Cultures.* The strain of *Neurospora crassa* used is the Oak Ridge wild-type 74-OR23-1A obtained from the Fungal Genetic Stock Center, California State University, Humboldt, Arcata, California.

[5] O. H. Lowry, N. J. Rosebrough, A. L. Farr, and R. J. Randall, *J. Biol. Chem.* **193**, 265 (1951).

Stocks are maintained in continuous vegetative culture on Vogel's Medium N⁶ supplemented with 2% commercial grade sucrose as a carbon source and 2% Difco Bacto-agar. Mycelia to be used for enzyme isolations are grown in 2800-ml Fernbach flasks containing 1500 ml of the same medium but without agar. These cultures are started from a heavy (approximately 50 mg) dry conidial inoculum and are grown for 60–72 hr at 25° on a gyrorotary shaker at approximately 100 rpm. At the end of this period the medium is removed by aspiration, 1500 ml of fresh Vogel's Medium N supplemented with β-lactose to 30 m$M$ is added, and the cultures are left on the shaker for an additional 60 hr in order to induce the enzymes. Induced mycelia are harvested by vacuum filtration on a Büchner funnel lined with Whatman No. 1 filter paper, washed with 1 liter of cold water and lyophilized. All subsequent steps are carried out at 0–4° unless otherwise indicated.

*Step 1. Preparation of Crude Extracts.* Lyophilized mycelia are ground by hand under liquid nitrogen in a precooled porcelain mortar. The resulting fine powder is suspended in 10 m$M$ phosphate buffer, pH 7.5, at a ratio of 15 ml of buffer per gram of mycelial powder. The suspension is homogenized in a VirTis homogenizer and sonicated in 100-ml batches in a Bronwill Biosonik II for 30 sec at full strength. This preparation is centrifuged at 15,000 $g$ for 30 min. The clear supernatant contains virtually 100% of both the pH 4.2 and the pH 4.5 β-galactosidase.

*Step 2. First Ammonium Sulfate Fractionation.* Solid ammonium sulfate (Mann, special enzyme grade) is added to the supernatant to 33% of saturation (196 g/100 ml) with continuous stirring. After 15 min the precipitate is removed by centrifugation at 27,000 $g$ for 15 min, and the supernatant is taken to 75% of saturation (29.2 g/100 ml) with ammonium sulfate and stirred slowly for 4 hr. The precipitate, containing approximately 75% of the original 4.2 β-galactosidase activity, is again collected by centrifugation and redissolved in one-tenth the original volume of 10 m$M$ phosphate buffer, pH 7.5. The supernatant may be discarded.

*Step 3. Acid Precipitation.* The redissolved ammonium sulfate precipitate is taken to pH 4.0 by the slow addition of 1 $M$ citric acid. This is stirred 8–10 hr, at which time the copious precipitate is removed by centrifugation and discarded.

*Step 4. Dialysis.* The acid supernatant is dialyzed 12 hr against at least 10 volumes of distilled water. The dialyzate should be changed once after about 4 hr.

*Step 5. Second Ammonium Sulfate Fractionation.* The dialyzed preparation is clarified by centrifugation at 15,000 $g$ for 30 min. The superna-

⁶ H. J. Vogel, *Microbial Genet. Bull.* **13**, 42 (1956).

tant is brought to 75% of saturation with solid ammonium sulfate and stirred for 8 hr. The precipitate is collected by centrifugation and redissolved in 1/100 the original volume of 8 m$M$ sodium phosphate-citrate buffer, pH 4.3, made 10 m$M$ with 2-mercaptoethanol.

*Step 6. CM-Sephadex Chromatography.* The redissolved precipitate is chromatographed on CM-Sephadex, C-50, medium mesh column (Pharmacia of Uppsala, Sweden). The column is equilibrated with 8 m$M$ phosphate-citrate buffer, pH 4.3, made 10 m$M$ with 2-mercaptoethanol and eluted with a 500-ml linear 0.1 $M$ to 1.0 $M$ sodium chloride gradient. The 4.2 $\beta$-galactosidase elutes at approximately 0.4 $M$ NaCl. Peak fractions are homogeneous on polyacrylamide gel electrophoresis in both cathodic (pH 5.0) and anodic (pH 9.5) systems.

A summary of the purification procedure is given in Table I.

Since dialysis against distilled water causes a rapid conversion of the pH 4.5 $\beta$-galactosidase to the 4.2 enzyme, the 4.5 enzyme may be isolated using steps 1 through 3 as above followed by steps 4 and 5 below.

*Step 4. Chromatography on Sephadex G-25.* The supernatant from the acid precipitate is placed in dialysis tubing and packed in dry commercial sucrose. When the volume is reduced to approximately $\frac{1}{10}$ the starting volume, the preparation is layered onto a Sephadex G-25 column equilibrated with 8 m$M$ phosphate-citrate buffer, pH 4.3, made 10 m$M$ with 2-mercaptoethanol.

*Step 5. CM-Sephadex Chromatography.* The active fractions from the Sephadex G-25 column are chromatographed on a CM-Sephadex column as described in step 6 above. The 4.5 $\beta$-galactosidase elutes at approximately 0.6 $M$ sodium chloride. Polyacrylamide gel electrophoresis of the

TABLE I
SUMMARY OF PURIFICATION OF pH 4.2 $\beta$-GALACTOSIDASE[a]

| Step | Volume (ml) | Protein (mg) | Total activity[b] | Specific activity | Recovery (%) |
|---|---|---|---|---|---|
| 1. Crude extract | 200 | 3745 | 20,255 | 5.4 | 100 |
| 2. 75% (NH$_4$)$_2$SO$_4$ ppt. | 100 | 1075 | 15,333 | 14.1 | 76 |
| 3. pH 4.0 soluble | 100 | 178 | 15,333 | 86.1 | 76 |
| 4. After dialysis | 110 | 38 | 15,000 | 392 | 74 |
| 5. 75% (NH$_4$)$_2$SO$_4$ ppt. after dialysis | 20 | 6.6 | 13,775 | 2078 | 68 |
| 6. CM eluent | 5 | 0.298 | 9,925 | 33,250 | 49 |

[a] From H. N. Johnson and A. G. DeBusk, *Arch. Biochem. Biophys.* **138**, 408 (1970).
[b] Enzyme units.

peak fractions results in two protein bands, both with β-galactosidase activity. The major band corresponds to the 4.5 β-galactosidase and a minor, faster moving, band having the properties of the 4.2 enzyme. It is felt that contamination by the 4.2 β-galactosidase is due to conversion of the 4.5 enzyme to the 4.2 enzyme under conditions required for the electrophoresis rather than contamination of the CM-Sephadex eluate.

Properties of the 4.5 β-galactosidase were determined using the fractions of highest activity which were eluted from the CM-Sephadex column. That fraction of the initial activity of a crude extract attributable to the 4.5 β-galactosidase varies between 30 and 60% of the total depending on the age and condition of the culture.

## Properties

*Stability.* The purified 4.2 β-galactosidase is stable at 4° with no significant loss of activity over a 30-day period. The 4.5 enzyme is slowly converted to the 4.2 form in solution. Both enzymes are considerably less stable in crude extracts. The 4.2 enzyme has a half-life of 28 min at 52°. The 4.5 enzyme has a half-life of 45 min under the same conditions.

*pH Optimum.* The enzymes are designated according to their pH optima. The pH dependence curves are very similar, both enzymes retaining 18–20% activity at pH 7.5.

*Isoelectric Point.* Using an LKB Ampholine column with pH 3–6 range ampholytes, the isoelectric point of the 4.2 β-galactosidase is pH 4.82. The pI for the 4.5 enzyme is 5.83.

*Molecular Weight.* Gel filtration on calibrated Sephadex G-200 columns gives a molecular weight estimate of 96,000. The enzymes are not separable by either gel filtration or centrifugation in sucrose gradients.

*Amino Acid Composition.* An amino acid analysis of the 4.2 β-galactosidase is shown in Table II. The number of residues is based on a molecular weight of 96,000. It is interesting that the enzyme contains no arginine.

*Carbohydrate Content.* Assuming a molecular weight of 96,000, the purified 4.2 β-galactosidase is about 5% by weight carbohydrate. The enzyme contains 28 moles of covalently bound carbohydrate per mole protein. Carbohydrate was determined by the phenol-sulfuric acid method and compared to a glucose standard. A qualitative analysis of the carbohydrate portion of the enzyme has not been done.

*Substrate Specificity.* Both forms of the enzyme are specific for β-galactosides. Thiogalactosides, α-galactosides, and β-glucosides are not substrates. Galactosides substituted on the galactose ring have not been tested.

TABLE II

AMINO ACID COMPOSITION OF THE pH 4.2 $\beta$-GALACTOSIDASE[a]

| | Residues/MW of 96,000 | |
| Amino acid | Found | Nearest integer |
|---|---|---|
| Lysine | 80.77 | 81 |
| Histidine | 14.48 | 15 |
| Arginine | 0.0 | 0 |
| Aspartic acid | 84.58 | 85 |
| Threonine | 43.18 | 43 |
| Serine | 202.06 | 202 |
| Glutamic acid | 140.84 | 141 |
| Proline | 26.04 | 26 |
| Glycine | 258.57 | 259 |
| Alanine | 92.33 | 92 |
| Cystine | — | — |
| Valine | 22.22 | 22 |
| Methionine | 1.27 | 1 |
| Isoleucine | 14.48 | 15 |
| Leucine | 26.04 | 26 |
| Tyrosine | 13.21 | 13 |
| Phenylalanine | 7.49 | 8 |
| | | 1029 |

[a] From H. N. Johnson and A. G. DeBusk, *Arch. Biochem. Biophys.* **138**, 408 (1970).

*Kinetics.* The $K_m$ of the 4.2 $\beta$-galactosidase for ONPG under the assay conditions is 0.45 m$M$. The $K_m$ of the 4.5 enzyme under the same conditions is 1.25 m$M$.

## II. pH 7.5 $\beta$-Galactosidase

### Assay Method

*Principle and Procedure.* The assay for the 7.5 $\beta$-galactosidase is identical to the assay for the 4.2 and 4.5 enzymes.

*Reagents*

ONPG, 5.0 m$M$ in potassium phosphate buffer, 10 m$M$, pH 7.5
Other reagents are as described for the pH 4.2 and 4.5 enzymes.

### Purification Procedure

The procedure that follows, although not a complete purification, yields a stable preparation of the 7.5 $\beta$-galactosidase which has been used to determine the properties of the enzyme. The growth of cultures and preparation of crude extracts are the same as described for the 4.2 and 4.5 enzymes with the exception that the inducing medium should be supplemented with 30 m$M$ D-galactose rather than β-lactose.

Solid ammonium sulfate is added to a crude extract to 40% of saturation and stirred for 5–10 min. This is immediately centrifuged at 27,000 $g$ for 15 min. The precipitate is redissolved in one-half the original volume of 10 m$M$ potassium phosphate buffer, pH 7.5, and stirred for 2 hr. This preparation is rapidly heated to 50–54° and maintained there with continuous stirring for 15 min. This is cooled in an ice bath to 0–5°, and the precipitate is removed by centrifugation. The clear supernatant contains approximately 20% of the 7.5 β-galactosidase activity of the crude extract. This preparation is stable at 4° for several weeks and contains no β-galactosidase activity assayable at pH 4.3. Because of a tendency for the 7.5 enzyme to form enzymatically active but insoluble aggregates, additional fractionations with ammonium sulfate and further concentration of this preparation should be avoided. Further purification by a variety of ion exchange and affinity matrices has not been successful.

## Properties

Properties of the 7.5 β-galactosidase were determined using the supernatant from the heat-treated preparation. The enzyme has a sharp pH optimum at pH 7.5. The isoelectric point is at pH 7.5. Gel filtration on calibrated Bio-Gel P-300 columns. (Bio-Rad Laboratories, Richmond, California) give a molecular weight estimate of 156,000. The enzyme is sensitive to disulfide reducing reagents, indicating the presence of at least one essential disulfide bond.

*Specificity.* The preparation contains no detectable carbohydrase activity other than β-galactosidase. The enzyme forms small amounts of an unidentified product different than the expected hydrolysis products with β-lactose as the substrate. This "transferase" activity is a function of the 7.5 β-galactosidase, not a contaminating protein because an electrophoretically altered 7.5 β-galactosidase which is isolated from a mutant strain of *Neurospora* forms only the expected hydrolysis products.

# [76] β-D-Fructofuranoside Fructohydrolase from Yeast

*By* Aida Goldstein and J. Oliver Lampen

R-β-Fructofuronoside + H₂O → ROH + fructose

β-D-Fructofuranoside fructohydrolase (EC 3.2.1.26) of yeast, more commonly known as invertase, was shown to occur as two distinct forms. External invertase, the larger and predominant form of the enzyme, is found primarily outside of the cell membrane and contains up to 50% carbohydrate depending on the species of yeast and purification method.[1] Internal invertase, the smaller enzyme, is found entirely within the cell. It constitutes only a small percentage of the total invertase activity of the cell and contains little or no carbohydrate.[2-4] Although two separate purification schemes are presented, a way is suggested for purifying both enzymes from one batch of cells.

## I. External Invertase

### Assay Method

*Principle.* The most common colorimetric methods use glucose oxidase in combination with a chromogen to estimate the amount of glucose released from the enzymic hydrolysis of sucrose. In the two-step method[4] described below, the first step is the enzymic hydrolysis of sucrose to glucose and fructose followed by heat inactivation of the enzyme. The second step is the determination of the liberated glucose using glucose oxidase, peroxidase, and *o*-dianisidine. The reaction is stopped with acid, and the developed color is read at 540 nm.

*Reagents*

Sodium acetate buffer, 0.2 $M$, pH 4.9
Sucrose 0.5 $M$
Potassium phosphate buffer, 0.5 $M$, pH 7.0
Solution A consists of 10 ml of glucose oxidase (84 mg),[5] 10 mg of

---

[1] N. P. Neumann and J. O. Lampen, *Biochemistry* 6, 468 (1967).
[2] J. Friis and P. Ottolenghi, *C. R. Trav. Lab. Carlsberg* 31, 259 (1959).
[3] S. Gascon and P. Ottolenghi, *C. R. Trav. Lab. Carlsberg* 36, 85 (1967).
[4] S. Gascon and J. O. Lampen, *J. Biol. Chem.* 243, 1567 (1968).
[5] Miles Laboratories, Inc., Kankakee, Illinois. When glucose oxidase was obtained from Nutritional Biochemical Corp., Cleveland, Ohio, Solution A contained 100 mg of glucose oxidase, 10 mg of peroxidase, and 100 ml of 0.1 $M$ phosphate buffer, pH 7. Specific activity of Miles glucose oxidase is 160 $\mu$mole units/mg, and NBC enzyme is 134 $\mu$mole units/mg.

peroxidase, and 90 ml of potassium phosphate buffer, 0.1 $M$, pH 7.0

Solution B consists of 300 mg o-dianisidine in 50 ml of distilled water. Store refrigerated in dark bottle.

Solution C consists of 1.0 ml of solution A, 0.5 ml of solution B, and 8.5 ml of 45% glycerol (v/v).[6] This solution can be stored frozen up to a month.

Hydrochloric acid, 6 $N$

Glucose, 2 m$M$ used as standard

Enzyme solution, which is diluted to give 0.2–0.5 unit invertase per ml (0.05–0.1 μg/ml)

*Procedure.* The reaction is carried out in two steps as follows: (a) Combine in a 12 × 100 mm test tube 50 μl of acetate buffer and 5–20 μl of enzyme solution. Place tube in a 30° bath and start the reaction by the addition of 25 μl of sucrose solution. After incubating for 5–10 min at 30°, stop the reaction by adding 0.1 ml of phosphate buffer, pH 7, and immediately heat for 3 min in a boiling water bath. The addition of the phosphate buffer before heating slows down the reaction and also renders the enzyme more sensitive to the heat treatment.

(b) After the tube cools down to 30°, add 1.0 ml of solution C to the total reaction mixture and incubate 20 min at 30°. Add 1.5 ml 6 $M$ HCl to stop the reaction and read the developed red color at 540 nm. This step measures the glucose released during the enzymic hydrolysis of sucrose in the first step.

With each set of measurements there should be included a sucrose blank, an enzyme blank, and three glucose standards (20–80 μmoles) which were carried through both steps of the reaction.

*Definition of Unit and Specific Activity.* One unit of invertase activity is defined as the amount of enzyme at pH 4.9 which hydrolyzes sucrose to produce 1 μmole of glucose per minute at 30°. Specific activity is expressed as units per milligram protein. In impure preparations, protein was determined by the method of Lowry et al.[7] using bovine serum albumin as standard. For purified enzyme preparations, the protein content was estimated spectrophotometrically at 280 nm using either lysozyme as a standard or a specific extinction coefficient ($E_{1 cm}^{1 \%}$) of 23.0.[1]

*Separation of External and Internal Invertases for Assay Purposes.* Crude extracts from yeast cells usually contain both external and internal invertase so that assaying the extract gives total invertase activity. To

---

[6] M. E. Wasko and E. W. Rice, *Clin. Chem.* **7**, 542 (1961).

[7] O. H. Lowry, N. J. Rosebrough, A. L. Farr, and R. J. Randall, *J. Biol. Chem.* **193**, 265 (1951). See also this series, Vol. 3 [73].

determine the contribution of each form of the enzyme, it is necessary to separate the activities by ion-exchange chromatography. This is accomplished by placing a 50-$\mu$l sample, diluted to ca. 2 ml with 0.05 $M$ Tris·HCl buffer, pH 7.3, on a 1 $\times$ 10 cm DEAE-Sephadex A-50 column equilibrated with 50 m$M$ Tris·HCl buffer, pH 7.3. First wash the column with 10 ml of equilibrating buffer and then elute the external enzyme with ca. 25 ml, 0.2 $M$ NaCl, and the internal enzyme with ca. 25 ml, 0.4 $M$ NaCl in 50 m$M$ Tris·HCl buffer, pH 7.3, collecting 2-ml fractions. Pool the active tubes of each form of invertase, determine volumes and reassay for activity.

## Purification Procedure[8]

Unless otherwise specified, procedures are carried out at 0–5°, centrifugations are performed on a Sorvall RC2-B centrifuge at maximum operating speeds for the rotors used (14,000 $g$ for large volumes and 40,000 $g$ for small volumes), and Buffer refers to 10 m$M$ potassium phosphate buffer, pH 6.5.

*Step 1. Preparation of Crude Extract.* The yeast paste, 500 g of mutant strain FH4C is suspended in 500 ml of Buffer containing 2 m$M$ PMSF,[9] and the cells are broken by passage two times through a prechilled laboratory Sub-Micron Disperser[10] at a continuous operating pressure of 9000 psi. The homogenate is centrifuged for 60 min, and the pellet is discarded. Both the external and internal invertases are contained in the supernatant liquid or crude extract.

*Step 2. Ammonium Sulfate Fractionation.* To the crude extract, add slowly with constant stirring solid ammonium sulfate to 85% saturation [560 g (NH$_4$)$_2$SO$_4$ per liter extract]. The solution is stirred an additional 4 hr and then centrifuged for 40 min. This step separates the external and internal invertases with the external invertase remaining in the 85% (NH$_4$)$_2$SO$_4$ soluble fraction. The 0–85% (NH$_4$)$_2$SO$_4$ fraction can be stored frozen and used later for the purification of internal invertase (see Section II).

*Step 3. Batch Adsorption on DEAE-Sephadex.* Dialyze the 85% (NH$_4$)$_2$SO$_4$ soluble fraction overnight in the cold room against running tap water. The dialysis bags are only half filled to allow for a 2-fold

[8] J. S. Tkacz, Ph.D. Dissertation, 1971. The procedure described is modified to shorten the purification time.

[9] PMSF or phenylmethylsulfonylfluoride is added as a protease inhibitor. It is prepared immediately before use as a 40 m$M$ solution in 95% ethanol and added to the buffer in a 1:20 ratio to give a final concentration of 2 m$M$ PMSF.

[10] Model 15 M-8TA, Manton-Gaulin Manufacturing Co., Inc., Everett, Massachusetts.

increase in volume during dialysis. To the dialyzed solution, add slowly with stirring 10 g of DEAE-Sephadex A-50 previously equilibrated with Buffer, and continue stirring for 2 hr. Allow the resin to settle and assay the liquid above the resin for activity. If more than 10% remains in the liquid, add more resin. When most of the enzyme is adsorbed, decant the liquid and wash the resin three times with 1-liter volumes each of Buffer containing 50 m$M$ NaCl. Pack the resin into a column 4 cm in diameter and wash the resin with 1.5 column volumes of Buffer containing 0.10 $M$ NaCl. The enzyme is then eluted with buffer containing 0.20 $M$ NaCl, collecting 10 ml per tube. All the tubes containing substantial activity are pooled.

*Step 4. DEAE-Sephadex Column.* Dialyze the pooled tubes from the batch step overnight against running tap water. Place the dialyzed solution on a 4 × 27 cm (350 ml) DEAE-Sephadex A-50 column previously equilibrated with Buffer. Wash the column successively with 300 ml each of 50 m$M$ NaCl and 80 m$M$ NaCl in Buffer. The enzyme is then eluted from the column with 0.16 $M$ NaCl in Buffer collecting 5 ml per tube. Pool the tubes containing activity.

*Step 5. SE-Sephadex Column.* The pooled fraction from above is dialyzed 24 hr against 2 changes, 6 liters each of 10 m$M$ sodium citrate buffer, pH 3.7. If a precipitate forms, centrifuge the dialyzed solution before placing on a 4 × 27 cm (350 ml) SE-Sephadex A-50 column which was equilibrated with 10 m$M$ sodium citrate buffer, pH 3.7. Wash the column with one column volume of the equilibrating buffer before eluting

TABLE I

PURIFICATION OF EXTERNAL INVERTASE FROM YEAST STRAIN FH4C

| Fraction | Volume (ml) | Total protein (mg) | Total activity (units) | Specific activity (units/mg) | Recovery (%) |
|---|---|---|---|---|---|
| Homogenate[a] | 1220 | 66,500 | 1,250,000 | 19 | — |
| Crude extract | 1290 | 28,000 | 865,000 | 31 | 100 |
| 85% Ammonium sulfate supernatant | 1520 | 3,420 | 685,000 | 200 | 79 |
| Batch DEAE-Sephadex | 276 | 323 | 587,000 | 1820 | 68 |
| DEAE-Sephadex chromatography | 220 | 152[b] | 524,000 | 3450 | 60 |
| SE-Sephadex and water dialysis | 174 | 96[b] | 466,000 | 4800 | 54 |

[a] Homogenate from 500 g of yeast paste (FH4C).

[b] Protein was determined at 280 nm using $E_{280}^{1\%}$ of 23. Other protein values were determined by the method of Lowry using lysozyme as standard.

the enzyme with a 900-ml linear gradient from 0 to 0.3 $M$ NaCl in 10 m$M$ citrate buffer, pH 3.7. Collect 5-ml fractions and pool the tubes with activity. At this stage the enzyme should be pure.

*Step 6. Dialysis and Lyophilization.* The pooled tubes are dialyzed 36 hr against 4 changes, 6 liters each, of distilled water and then lyophilized. Store the enzyme desiccated below 0°. From Table I it can be seen that the purified enzyme has a specific activity of 4800 with about 54% recovery.

## Properties

*Substrate Specificity.* Invertase will hydrolyze any compound with an unsubstituted β-D-fructofuranosyl residue, such as sucrose, raffinose, and methyl-β-D-fructofuranoside. Maltose is not a substrate for the enzyme.

At high substrate concentrations (1 $M$), invertase exhibits transferase activity, transferring the β-D-fructofuranosyl residue to primary alcohols, such as methanol, ethanol, and n-propanol. Isopropanol was the only secondary alcohol showing acceptor activity.[11]

Some amines, such as 2-amino-2-hydroxymethyl propane-1,3-diol (Tris)[12] and aniline, will reversibly inhibit invertase. Iodine will also inhibit the enzyme, and in some instances the inhibition can be reversed with mercaptoethanol.[13]

*Stability.* External invertase can be stored in the lyophilized state below 0° in a desiccator for at least a year without appreciable loss of activity. It also can be stored frozen in solution for a few months as long as the pH is around 4.5–5.0.

*pH Optimum.* The enzyme has a broad pH optimum, between pH 3.5 and 5.0, which drops off rapidly on the acid side and more slowly on the alkaline side of the optimum range. At pH 7, only a few percent of the maximum activity remains.

*pH and Heat Stability.* When incubated at 30°, the external enzyme is stable for at least 2 hr between pH 3 and 7.[14] At 56° the enzyme is stable for 15 min at pH 4.9, but becomes very unstable above pH 6.[15]

*Kinetic Properties.* The $K_m$ value for sucrose is 25–26 m$M$, and for raffinose 150 m$M$.[14] These values may vary sharply depending on the source of invertase and the pH.

[11] A. Baseer and S. Shall, *Biochim. Biophys. Acta* **250**, 192 (1971).
[12] K. Myrbäck, *Ark. Kemi* **25**, 315 (1965).
[13] A. Waheed and S. Shall, *Biochim. Biophys. Acta* **242**, 172 (1971).
[14] S. Gascon, N. P. Neumann, and J. O. Lampen, *J. Biol. Chem.* **243**, 1573 (1968).
[15] Unpublished data.

## II. Internal Invertase

### Assay Method

The assay procedure is the same as that described for external invertase except that the 0.2 $M$ sodium acetate buffer, pH 4.9, contains 0.5 mg bovine serum albumin per milliliter.

### Purification Procedure[16]

Conditions used are the same as those described for external invertase except that buffer A refers to 50 m$M$ Tris·HCl, pH 7.3 and buffer B to 50 m$M$ sodium acetate, pH 4.9. To use the 0–85% $(NH_4)_2SO_4$ fraction from step 2 under external invertase for purification of the internal invertase, it is dissolved in a minimum volume of buffer A, centrifuged to remove insoluble material and dialyzed overnight against buffer A. The purification is continued with step 3 below.

*Step 1. Preparation of Crude Extract.* The yeast paste, 2 kg from mutant strain FH4C, is suspended in 2 liters of buffer B containing 2 m$M$ PMSF. The cells are broken and centrifuged as described under step 1 for external invertase. The supernatant solution or crude extract contains both the internal and external invertases.

*Step 2. pH 4.9 Step.* With rapid stirring, the crude extract is brought to pH 4.9 by the addition of 2 $M$ acetic acid. Stir for an additional 2 hr and then allow the solution to stand overnight in the cold. Centrifuge and discard the inactive precipitate.

*Step 3. Batch Adsorption on DEAE-Sephadex.* The pH of the supernatant solution is adjusted with 1 $M$ Tris to pH 7.3 and the conductivity is adjusted with 2 $M$ NaCl until it is equivalent to buffer A containing 0.16 $M$ NaCl. Now add with stirring, 25 g DEAE-Sephadex previously equilibrated with buffer A containing 0.16 $M$ NaCl. Stir for an additional 2 hr and allow the resin to settle. This step separates the internal and external invertases, the internal binding to the resin and the external remaining in solution.[17] In order to check the liquid for loss of internal invertase, it is necessary to separate the two enzyme activities by ion exchange chromatography (see under external invertase assay for the method). If more than 25% remains in solution, add more resin. When

---

[16] Modified from that of S. Gascon and J. O. Lampen, *J. Biol. Chem.* **243**, 1567 (1968).

[17] The external invertase can be purified from the solution by first dialyzing the solution overnight against running tap water and then following the purification scheme for external invertase, starting at step 3.

most of the internal invertase is bound, decant the liquid and wash the resin 2 times with 1 liter each of 0.16 $M$ NaCl in buffer A. Pack the resin in a 4-cm diameter column and wash with two column volumes of 0.17 $M$ NaCl in buffer A. The enzyme is eluted with 0.3 $M$ NaCl in buffer A collecting 10-ml fractions. Pool all tubes with enzyme activity.

*Step 4. DEAE-Sephadex Column, pH 7.3.* The pooled solution is diluted with distilled water until the conductivity is equivalent to 0.16 $M$ NaCl in buffer A and is then placed on a $2 \times 33$ cm (100 ml) DEAE-Sephadex A-50 column equilibrated with 0.16 $M$ NaCl in buffer A. The column is washed with one column volume of equilibrating buffer and the enzyme is eluted with 1 liter linear gradient from 0.16–0.50 $M$ NaCl in buffer A, collecting 5-ml fractions. Pool the fractions with enzyme activity.

*Step 5. First DEAE-Sephadex Column, pH 4.9.* Dialyze the pooled fractions overnight against 4 liters of buffer B containing 50 m$M$ NaCl and remove the inactive precipitate by centrifugation. Place the supernatant solution on a $2 \times 18$ cm (55 ml) DEAE-Sephadex A-50 column equilibrated with 50 m$M$ NaCl in buffer B, wash with one column volume of equilibrating buffer and elute with a 600 ml of linear NaCl gradient from 0.05–0.35 $M$ in buffer B. Collect 4-ml fractions and pool the tubes with activity.

*Step 6. Second DEAE-Sephadex Column, pH 4.9.* Add distilled water to the pooled tubes until the conductivity is equivalent to buffer B containing 0.1 $M$ NaCl. Place the solution on a $2 \times 13$ cm column (40 ml)

TABLE II
PURIFICATION OF INTERNAL INVERTASE FROM YEAST STRAIN FH4C

| Fraction | Volume (ml) | Total protein[b] (mg) | Total activity (units) | Specific activity (units/mg) | Recovery (%) |
|---|---|---|---|---|---|
| Crude extract[a] | 3630 | 54,500 | 181,000 | 3 | 100 |
| pH 4.9 step | 4075 | 61,100 | 178,000 | 3 | 100 |
| Batch DEAE-Sephadex | 112 | 677 | 91,500 | 135 | 50 |
| DEAE-Sephadex column, pH 7.3 | 113 | 148 | 87,100 | 590 | 48 |
| 1st DEAE-Sephadex column, pH 4.9 | 158 | 36 | 80,000 | 2200 | 44 |
| 2nd DEAE-Sephadex column, pH 4.9 | 32 | 12 | 56,600 | 4700 | 32 |

[a] From 2 kg yeast paste.
[b] Protein determined by method of O. H. Lowry, N. J. Rosebrough, A. L. Farr, and R. J. Randall, *J. Biol. Chem.* **193**, 265 (1951), using bovine serum albumin as a standard.

equilibrated with 50 mM NaCl in buffer B, wash successively with 70 ml of 0.1 M NaCl, 70 ml of 0.13 M NaCl, and 40 ml of 0.15 M NaCl, all salts in buffer B. Elute the enzyme with 0.18 M NaCl in buffer B collecting 2-ml fractions, and pool the tubes with activity. Dialyze the pooled fractions overnight against 4 liters 20 mM potassium phosphate buffer, pH 7.0, and store frozen. The overall recovery is 32% and the specific activity of the purified enzyme is 4700. Table II summarizes the purification of internal invertase.

## Properties

*Substrate Specificity.* Same as for the external invertase.

*Inhibitors and Activators.* The internal invertase, in addition to being inhibited by the same amines as the external invertase, is also reversibly inhibited by some salts at high concentrations. Ammonium sulfate at 0.1 M concentration inhibits the enzyme 100%, while sodium sulfate and sodium chloride partially inhibit the enzyme.[11]

Bovine serum albumin when added to the assay mixture stimulates the activity of the purified internal invertase.

*Stability.* Internal invertase can be stored frozen in solution at pH 7–7.5 for a few months without loss of activity.

*pH Optimum.* Same as for the external invertase.

*pH and Heat Stability.* The purified internal invertase is stable for at least 2 hr at 30° between pH 6.5 and 9.[14] In the presence of bovine serum albumin, the enzyme becomes stable down to pH 4.9.[15]

At 56°, the purified enzyme is unstable at all pH's. But in the presence of BSA, mannan, or external invertase, it exhibits similar heat stability properties as external invertase, being stable at pH 4.9, but not above pH 6 when heated 10 min at 56°.[15]

*Kinetic Properties.* Same as for external invertase.

# Author Index

Numbers in parentheses are reference numbers and indicate that an author's work is referred to, although his name is not cited in the text.

## A

Abbate, M. J., 409
Abeles, R. H., 308, 312, 313(9), 321
Abou-Issa, H., 375, 378(2, 3), 379(2), 380(2, 3), 395(2)
Adam, H., 139, 140(2)
Adams, E., 280, 286
Adler, J., 44, 48
Ahrendt, C., 445
Akazawa, T., 460, 477
Allen, M. B., 358
Aloyo, V. J., 368
Alpers, J. B., 120
Altekar, W. W., 262
Ames, B. N., 232, 369
Anderson, D. E., 408
Anderson, J. M., 479
Anderson, L. E., 457, 458, 460, 461
Anderson, P. J., 223, 256
Anderson, R. L., 3, 4, 5(2, 6), 6(2, 3), 39, 40(1), 42(1), 63, 64(6, 7), 65, 66(6, 7), 269, 270, 271(2), 272(1, 2), 305, 306(1), 494, 495, 496(3)
Anderson, T. E., 40
Andreesen, J. R., 304
Andrews, P., 185, 313
Andrews, T. J., 214, 216(4), 217(4), 218(4), 219(4), 471, 479, 480(30, 31), 484, 485(2), 486(2), 487(2)
Aoe, H., 374
Arneson, R. A., 365
Arnold, H., 185, 186(7)
Arnon, D. I., 230, 398, 403, 404
Aronson, L. D., 286
Assensio, C., 19
Atkinson, D. E., 83, 96
Atzpodien, W., 79, 81(7), 83, 84(7)
Aust, A., 176, 181(6), 182(6)

## B

Baccari, V., 409, 412(2), 414, 421(2), 422(2), 423, 426
Bachman, B. J., 141
Bailey, K., 6
Balestrero, F., 358, 372
Ballard, F. J., 43, 44, 45(4), 47(4)
Ballou, C. E., 324, 330, 429, 437
Bamburg, J. R., 166, 167, 169(10), 170, 171(10), 174(6, 10), 175(10)
Banerjee, S., 54
Bank, W. J., 111, 114(2), 115(2)
Baptist, J. N., 314
Baranowski, T., 329, 335, 337(5), 338(5)
Bardawill, C. J., 127, 270, 306
Barnard, E. A., 6, 7(5, 7), 10, 15(5), 16, 17(7, 14), 18, 19
Barnes, L. D., 329
Barran, L. R., 262, 263, 264(11)
Bartoli, S., 409
Baseer, A., 508, 511(11)
Baselice, R. A., 281
Basford, R. E., 24
Bassham, J. A., 398, 399, 461
Bates, W. K., 497
Bauer, A. C., 223
Baxter, J. N., 291
Baxter, R. C., 358, 369, 372, 373(3), 374(3, 10)
Bender, R., 304
Benson, P. F., 48
Benziman, M., 187, 192, 195, 196, 198, 199, 200
Berenblum, I., 406, 451
Berger, L., 6
Bergmeyer, H. U., 162
Bergren, W. R., 47, 50(1)
Bernardi, G., 95

# Subject Index

## A